# STUDENT SOLUTIONS MANUAL
## for Stewart's
### MULTIVARIABLE CALCULUS: CONCEPTS AND CONTEXTS
## FOURTH EDITION

DAN CLEGG

Palomar College

BROOKS/COLE

CENGAGE Learning™

Australia · Brazil · Japan · Korea · Mexico · Singapore · Spain · United Kingdom · United States

**BROOKS/COLE**
CENGAGE Learning™

For product information and technology assistance, contact us at
**Cengage Learning Customer & Sales Support,**
**1-800-354-9706**

For permission to use material from this text or product, submit all requests online at **www.cengage.com/permissions.** Further permissions questions can be emailed to **permissionrequest@cengage.com.**

ISBN-13:  978-0-495-56055-5
ISBN-10:  0-495-56055-3

**Brooks/Cole**
10 Davis Drive
Belmont, CA 94002-3098
USA

Cengage Learning is a leading provider of customized learning solutions with office locations around the globe, including Singapore, the United Kingdom, Australia, Mexico, Brazil, and Japan. Locate your local office at **www.cengage.com/international.**

Cengage Learning products are represented in Canada by Nelson Education, Ltd.

To learn more about Brooks/Cole, visit **www.cengage.com/brookscole.**

Purchase any of our products at your local college store or at our preferred online store **www.ichapters.com.**

Printed in Canada
1 2 3 4 5 6 7 13 12 11 10 09

# ☐ PREFACE

This *Student Solutions Manual* contains detailed solutions to selected exercises in the text *Multivariable Calculus: Concepts and Contexts,* Fourth Edition (Chapters 8–13 of *Calculus: Concepts and Contexts,* Fourth Edition) by James Stewart. Specifically, it contains solutions to the odd-numbered exercises in each chapter section, review section, True-False Quiz, and Focus on Problem Solving section as well as all solutions to the Concept Check questions.

Each solution is presented in the context of the corresponding section of the text. In general, solutions to the initial exercises involving a new concept illustrate that concept in more detail; this knowledge is then utilized in subsequent solutions. So, while the intermediate steps of a solution are given, you may need to refer back to earlier exercises in the section or prior sections for additional explanation of the concepts involved. Note that, in many cases, different routes to an answer may exist which are equally valid; also, answers can be expressed in different but equivalent forms. Thus, the goal of this manual is not to give the definitive solution to each exercise, but rather to assist you as a student in understanding the concepts of the text and learning how to apply them to the challenge of solving a problem.

I would like to thank James Stewart for entrusting me with the writing of this manual and offering suggestions, Kathi Townes, Stephanie Kuhns, and Rebekah Steele of TECH-arts for typesetting and producing this manual, and Brian Betsill of TECH-arts for creating the illustrations. Brian Karasek prepared solutions for comparison of accuracy and style in addition to proofreading manuscript; his assistance and suggestions were very helpful and much appreciated. Finally, I would like to thank Richard Stratton and Elizabeth Neustaetter of Brooks/Cole, Cengage Learning for their trust, assistance, and patience.

DAN CLEGG
Palomar College

# ☐ CONTENTS

# 8 □ INFINITE SEQUENCES AND SERIES

## 8.1 Sequences

1. (a) A sequence is an ordered list of numbers. It can also be defined as a function whose domain is the set of positive integers.

   (b) The terms $a_n$ approach 8 as $n$ becomes large. In fact, we can make $a_n$ as close to 8 as we like by taking $n$ sufficiently large.

   (c) The terms $a_n$ become large as $n$ becomes large. In fact, we can make $a_n$ as large as we like by taking $n$ sufficiently large.

3. The first six terms of $a_n = \dfrac{n}{2n+1}$ are $\dfrac{1}{3}, \dfrac{2}{5}, \dfrac{3}{7}, \dfrac{4}{9}, \dfrac{5}{11}, \dfrac{6}{13}$. It appears that the sequence is approaching $\dfrac{1}{2}$.

$$\lim_{n \to \infty} \frac{n}{2n+1} = \lim_{n \to \infty} \frac{1}{2 + 1/n} = \frac{1}{2}$$

5. $\left\{1, \frac{1}{3}, \frac{1}{5}, \frac{1}{7}, \frac{1}{9}, \ldots\right\}$. The denominator of the $n$th term is the $n$th positive odd integer, so $a_n = \dfrac{1}{2n-1}$.

7. $\{2, 7, 12, 17, \ldots\}$. Each term is larger than the preceding one by 5, so $a_n = a_1 + d(n-1) = 2 + 5(n-1) = 5n - 3$.

9. $\left\{1, -\frac{2}{3}, \frac{4}{9}, -\frac{8}{27}, \ldots\right\}$. Each term is $-\frac{2}{3}$ times the preceding one, so $a_n = \left(-\frac{2}{3}\right)^{n-1}$.

11. $a_n = \dfrac{3+5n^2}{n+n^2} = \dfrac{(3+5n^2)/n^2}{(n+n^2)/n^2} = \dfrac{5+3/n^2}{1+1/n}$, so $a_n \to \dfrac{5+0}{1+0} = 5$ as $n \to \infty$.   Converges

13. $a_n = 1 - (0.2)^n$, so $\lim\limits_{n \to \infty} a_n = 1 - 0 = 1$ by (7).   Converges

15. Because the natural exponential function is continuous at 0, Theorem 5 enables us to write

$$\lim_{n \to \infty} a_n = \lim_{n \to \infty} e^{1/n} = e^{\lim_{n \to \infty}(1/n)} = e^0 = 1. \quad \text{Converges}$$

17. If $b_n = \dfrac{2n\pi}{1+8n}$, then $\lim\limits_{n \to \infty} b_n = \lim\limits_{n \to \infty} \dfrac{(2n\pi)/n}{(1+8n)/n} = \lim\limits_{n \to \infty} \dfrac{2\pi}{1/n+8} = \dfrac{2\pi}{8} = \dfrac{\pi}{4}$. Since $\tan$ is continuous at $\frac{\pi}{4}$, by

Theorem 5, $\lim\limits_{n \to \infty} \tan\left(\dfrac{2n\pi}{1+8n}\right) = \tan\left(\lim\limits_{n \to \infty} \dfrac{2n\pi}{1+8n}\right) = \tan\dfrac{\pi}{4} = 1$.   Converges

19. $a_n = \dfrac{(-1)^{n-1}\, n}{n^2+1} = \dfrac{(-1)^{n-1}}{n+1/n}$, so $0 \le |a_n| = \dfrac{1}{n+1/n} \le \dfrac{1}{n} \to 0$ as $n \to \infty$, so $a_n \to 0$ by the Squeeze Theorem and

Theorem 4.   Converges

21. $a_n = \dfrac{e^n + e^{-n}}{e^{2n} - 1} \cdot \dfrac{e^{-n}}{e^{-n}} = \dfrac{1 + e^{-2n}}{e^n - e^{-n}} \to 0$ as $n \to \infty$ because $1 + e^{-2n} \to 1$ and $e^n - e^{-n} \to \infty$.   Converges

23. $a_n = n^2 e^{-n} = \dfrac{n^2}{e^n}$. Since $\lim\limits_{x \to \infty} \dfrac{x^2}{e^x} \overset{\text{H}}{=} \lim\limits_{x \to \infty} \dfrac{2x}{e^x} \overset{\text{H}}{=} \lim\limits_{x \to \infty} \dfrac{2}{e^x} = 0$, it follows from Theorem 2 that $\lim\limits_{n \to \infty} a_n = 0$.   Converges

25. $0 \le \dfrac{\cos^2 n}{2^n} \le \dfrac{1}{2^n}$   [since $0 \le \cos^2 n \le 1$],   so since $\lim\limits_{n \to \infty} \dfrac{1}{2^n} = 0$, $\left\{\dfrac{\cos^2 n}{2^n}\right\}$ converges to 0 by the Squeeze Theorem.

1

**27.** $y = \left(1 + \dfrac{2}{x}\right)^{x}$ $\Rightarrow$ $\ln y = x \ln\left(1 + \dfrac{2}{x}\right)$, so

$$\lim_{x \to \infty} \ln y = \lim_{x \to \infty} \frac{\ln(1 + 2/x)}{1/x} \overset{\text{H}}{=} \lim_{x \to \infty} \frac{\left(\dfrac{1}{1 + 2/x}\right)\left(-\dfrac{2}{x^2}\right)}{-1/x^2} = \lim_{x \to \infty} \frac{2}{1 + 2/x} = 2 \Rightarrow$$

$$\lim_{x \to \infty}\left(1 + \frac{2}{x}\right)^{x} = \lim_{x \to \infty} e^{\ln y} = e^2, \text{ so by Theorem 2, } \lim_{n \to \infty}\left(1 + \frac{2}{n}\right)^{n} = e^2. \quad \text{Convergent}$$

**29.** $a_n = \dfrac{(2n-1)!}{(2n+1)!} = \dfrac{(2n-1)!}{(2n+1)(2n)(2n-1)!} = \dfrac{1}{(2n+1)(2n)} \to 0$ as $n \to \infty$. Converges

**31.** $\{0, 1, 0, 0, 1, 0, 0, 0, 1, \ldots\}$ diverges since the sequence takes on only two values, 0 and 1, and never stays arbitrarily close to either one (or any other value) for $n$ sufficiently large.

**33.** $a_n = \ln(2n^2 + 1) - \ln(n^2 + 1) = \ln\left(\dfrac{2n^2 + 1}{n^2 + 1}\right) = \ln\left(\dfrac{2 + 1/n^2}{1 + 1/n^2}\right) \to \ln 2$ as $n \to \infty$. Convergent

**35.**

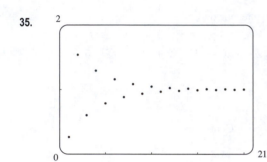

From the graph, it appears that the sequence converges to 1.

$\{(-2/e)^n\}$ converges to 0 by (7), and hence $\{1 + (-2/e)^n\}$

converges to $1 + 0 = 1$.

**37.**

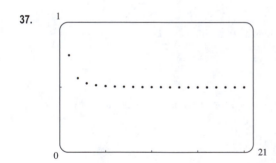

From the graph, it appears that the sequence converges to $\frac{1}{2}$.

As $n \to \infty$,

$$a_n = \sqrt{\frac{3 + 2n^2}{8n^2 + n}} = \sqrt{\frac{3/n^2 + 2}{8 + 1/n}} \Rightarrow \sqrt{\frac{0 + 2}{8 + 0}} = \sqrt{\frac{1}{4}} = \frac{1}{2},$$

so $\lim\limits_{n \to \infty} a_n = \frac{1}{2}$.

**39.**

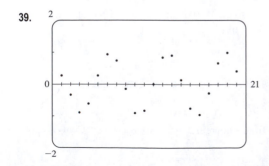

From the graph, it appears that the sequence $\{a_n\} = \left\{\dfrac{n^2 \cos n}{1 + n^2}\right\}$ is

divergent, since it oscillates between 1 and $-1$ (approximately). To

prove this, suppose that $\{a_n\}$ converges to $L$. If $b_n = \dfrac{n^2}{1 + n^2}$, then

$\{b_n\}$ converges to 1, and $\lim\limits_{n \to \infty} \dfrac{a_n}{b_n} = \dfrac{L}{1} = L$. But $\dfrac{a_n}{b_n} = \cos n$, so

$\lim\limits_{n \to \infty} \dfrac{a_n}{b_n}$ does not exist. This contradiction shows that $\{a_n\}$ diverges.

**41.** (a) $a_n = 1000(1.06)^n$ $\Rightarrow$ $a_1 = 1060$, $a_2 = 1123.60$, $a_3 = 1191.02$, $a_4 = 1262.48$, and $a_5 = 1338.23$.

(b) $\lim\limits_{n \to \infty} a_n = 1000 \lim\limits_{n \to \infty} (1.06)^n$, so the sequence diverges by (7) with $r = 1.06 > 1$.

**43.** (a) We are given that the initial population is 5000, so $P_0 = 5000$. The number of catfish increases by $8\%$ per month and is

decreased by 300 per month, so $P_1 = P_0 + 8\% P_0 - 300 = 1.08 P_0 - 300$, $P_2 = 1.08 P_1 - 300$, and so on. Thus,

$P_n = 1.08 P_{n-1} - 300$.

(b) Using the recursive formula with $P_0 = 5000$, we get $P_1 = 5100$, $P_2 = 5208$, $P_3 = 5325$ (rounding any portion of a

catfish), $P_4 = 5451$, $P_5 = 5587$, and $P_6 = 5734$, which is the number of catfish in the pond after six months.

**45.** (a) $a_1 = 1$, $a_{n+1} = 4 - a_n$ for $n \geq 1$. $a_1 = 1$, $a_2 = 4 - a_1 = 4 - 1 = 3$, $a_3 = 4 - a_2 = 4 - 3 = 1$,

$a_4 = 4 - a_3 = 4 - 1 = 3$, $a_5 = 4 - a_4 = 4 - 3 = 1$. Since the terms of the sequence alternate between 1 and 3,

the sequence is divergent.

(b) $a_1 = 2$, $a_2 = 4 - a_1 = 4 - 2 = 2$, $a_3 = 4 - a_2 = 4 - 2 = 2$. Since all of the terms are 2, $\lim\limits_{n \to \infty} a_n = 2$ and hence, the

sequence is convergent.

**47.** (a) Let $a_n$ be the number of rabbit pairs in the $n$th month. Clearly $a_1 = 1 = a_2$. In the $n$th month, each pair that is

2 or more months old (that is, $a_{n-2}$ pairs) will produce a new pair to add to the $a_{n-1}$ pairs already present. Thus,

$a_n = a_{n-1} + a_{n-2}$, so that $\{a_n\} = \{f_n\}$, the Fibonacci sequence.

(b) $a_n = \dfrac{f_{n+1}}{f_n}$ $\Rightarrow$ $a_{n-1} = \dfrac{f_n}{f_{n-1}} = \dfrac{f_{n-1} + f_{n-2}}{f_{n-1}} = 1 + \dfrac{f_{n-2}}{f_{n-1}} = 1 + \dfrac{1}{f_{n-1}/f_{n-2}} = 1 + \dfrac{1}{a_{n-2}}$. If $L = \lim\limits_{n \to \infty} a_n$,

then $L = \lim\limits_{n \to \infty} a_{n-1}$ and $L = \lim\limits_{n \to \infty} a_{n-2}$, so $L$ must satisfy $L = 1 + \dfrac{1}{L}$ $\Rightarrow$ $L^2 - L - 1 = 0$ $\Rightarrow$ $L = \frac{1+\sqrt{5}}{2}$

[since $L$ must be positive].

**49.** $a_n = \dfrac{1}{2n+3}$ is decreasing since $a_{n+1} = \dfrac{1}{2(n+1)+3} = \dfrac{1}{2n+5} < \dfrac{1}{2n+3} = a_n$ for each $n \geq 1$. The sequence is

bounded since $0 < a_n \leq \frac{1}{5}$ for all $n \geq 1$. Note that $a_1 = \frac{1}{5}$.

**51.** The terms of $a_n = n(-1)^n$ alternate in sign, so the sequence is not monotonic. The first five terms are $-1$, 2, $-3$, 4, and $-5$.

Since $\lim\limits_{n \to \infty} |a_n| = \lim\limits_{n \to \infty} n = \infty$, the sequence is not bounded.

**53.** Since $\{a_n\}$ is a decreasing sequence, $a_n > a_{n+1}$ for all $n \geq 1$. Because all of its terms lie between 5 and 8, $\{a_n\}$ is a

bounded sequence. By the Monotonic Sequence Theorem, $\{a_n\}$ is convergent; that is, $\{a_n\}$ has a limit $L$. $L$ must be less than

8 since $\{a_n\}$ is decreasing, so $5 \leq L < 8$.

**55.** $a_1 = 1$, $a_{n+1} = 3 - \dfrac{1}{a_n}$. We show by induction that $\{a_n\}$ is increasing and bounded above by 3. Let $P_n$ be the proposition

that $a_{n+1} > a_n$ and $0 < a_n < 3$. Clearly $P_1$ is true. Assume that $P_n$ is true. Then $a_{n+1} > a_n$ $\Rightarrow$ $\dfrac{1}{a_{n+1}} < \dfrac{1}{a_n}$ $\Rightarrow$

$-\dfrac{1}{a_{n+1}} > -\dfrac{1}{a_n}$. Now $a_{n+2} = 3 - \dfrac{1}{a_{n+1}} > 3 - \dfrac{1}{a_n} = a_{n+1}$   $\Leftrightarrow$   $P_{n+1}$. This proves that $\{a_n\}$ is increasing and bounded

above by 3, so $1 = a_1 < a_n < 3$, that is, $\{a_n\}$ is bounded, and hence convergent by the Monotonic Sequence Theorem.

If $L = \lim\limits_{n\to\infty} a_n$, then $\lim\limits_{n\to\infty} a_{n+1} = L$ also, so $L$ must satisfy $L = 3 - 1/L$   $\Rightarrow$   $L^2 - 3L + 1 = 0$   $\Rightarrow$   $L = \frac{3\pm\sqrt{5}}{2}$.

But $L > 1$, so $L = \frac{3+\sqrt{5}}{2}$.

**57.** $(0.8)^n < 0.000001$   $\Rightarrow$   $\ln(0.8)^n < \ln(0.000001)$   $\Rightarrow$   $n\ln(0.8) < \ln(0.000001)$   $\Rightarrow$   $n > \dfrac{\ln(0.000001)}{\ln(0.8)}$   $\Rightarrow$

$n > 61.9$, so $n$ must be at least 62 to satisfy the given inequality.

**59.** (a) Suppose $\{p_n\}$ converges to $p$. Then $p_{n+1} = \dfrac{bp_n}{a+p_n}$   $\Rightarrow$   $\lim\limits_{n\to\infty} p_{n+1} = \dfrac{b\,\lim\limits_{n\to\infty} p_n}{a + \lim\limits_{n\to\infty} p_n}$   $\Rightarrow$   $p = \dfrac{bp}{a+p}$   $\Rightarrow$

$p^2 + ap = bp$   $\Rightarrow$   $p(p + a - b) = 0$   $\Rightarrow$   $p = 0$ or $p = b - a$.

(b) $p_{n+1} = \dfrac{bp_n}{a+p_n} = \dfrac{\left(\dfrac{b}{a}\right)p_n}{1 + \dfrac{p_n}{a}} < \left(\dfrac{b}{a}\right)p_n$ since $1 + \dfrac{p_n}{a} > 1$.

(c) By part (b), $p_1 < \left(\dfrac{b}{a}\right)p_0$, $p_2 < \left(\dfrac{b}{a}\right)p_1 < \left(\dfrac{b}{a}\right)^2 p_0$, $p_3 < \left(\dfrac{b}{a}\right)p_2 < \left(\dfrac{b}{a}\right)^3 p_0$, etc. In general, $p_n < \left(\dfrac{b}{a}\right)^n p_0$,

so $\lim\limits_{n\to\infty} p_n \le \lim\limits_{n\to\infty} \left(\dfrac{b}{a}\right)^n \cdot p_0 = 0$ since $b < a$. $\left[\text{By (7), } \lim\limits_{n\to\infty} r^n = 0 \text{ if } -1 < r < 1. \text{ Here } r = \dfrac{b}{a} \in (0,1).\right]$

(d) Let $a < b$. We first show, by induction, that if $p_0 < b - a$, then $p_n < b - a$ and $p_{n+1} > p_n$.

For $n = 0$, we have $p_1 - p_0 = \dfrac{bp_0}{a+p_0} - p_0 = \dfrac{p_0(b - a - p_0)}{a+p_0} > 0$ since $p_0 < b - a$. So $p_1 > p_0$.

Now we suppose the assertion is true for $n = k$, that is, $p_k < b - a$ and $p_{k+1} > p_k$. Then

$b - a - p_{k+1} = b - a - \dfrac{bp_k}{a+p_k} = \dfrac{a(b-a) + bp_k - ap_k - bp_k}{a+p_k} = \dfrac{a(b - a - p_k)}{a+p_k} > 0$ because $p_k < b - a$. So

$p_{k+1} < b - a$. And $p_{k+2} - p_{k+1} = \dfrac{bp_{k+1}}{a+p_{k+1}} - p_{k+1} = \dfrac{p_{k+1}(b - a - p_{k+1})}{a+p_{k+1}} > 0$ since $p_{k+1} < b - a$. Therefore,

$p_{k+2} > p_{k+1}$. Thus, the assertion is true for $n = k + 1$. It is therefore true for all $n$ by mathematical induction.

A similar proof by induction shows that if $p_0 > b - a$, then $p_n > b - a$ and $\{p_n\}$ is decreasing.

In either case the sequence $\{p_n\}$ is bounded and monotonic, so it is convergent by the Monotonic Sequence Theorem.

It then follows from part (a) that $\lim\limits_{n\to\infty} p_n = b - a$.

## 8.2   Series

1. (a) A sequence is an ordered list of numbers whereas a series is the *sum* of a list of numbers.

   (b) A series is convergent if the sequence of partial sums is a convergent sequence. A series is divergent if it is not convergent.

3.

| $n$ | $s_n$ |
|---|---|
| 1 | $-2.40000$ |
| 2 | $-1.92000$ |
| 3 | $-2.01600$ |
| 4 | $-1.99680$ |
| 5 | $-2.00064$ |
| 6 | $-1.99987$ |
| 7 | $-2.00003$ |
| 8 | $-1.99999$ |
| 9 | $-2.00000$ |
| 10 | $-2.00000$ |

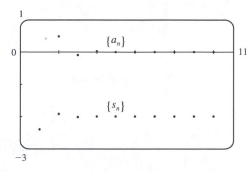

From the graph and the table, it seems that the series converges to $-2$. In fact, it is a geometric

series with $a = -2.4$ and $r = -\frac{1}{5}$, so its sum is $\displaystyle\sum_{n=1}^{\infty} \frac{12}{(-5)^n} = \frac{-2.4}{1 - \left(-\frac{1}{5}\right)} = \frac{-2.4}{1.2} = -2$.

Note that the dot corresponding to $n = 1$ is part of both $\{a_n\}$ and $\{s_n\}$.

**TI-86 Note:** To graph $\{a_n\}$ and $\{s_n\}$, set your calculator to Param mode and DrawDot mode. (DrawDot is under

GRAPH, MORE, FORMT (F3).) Now under E(t) = make the assignments: `xt1=t, yt1=12/(-5)^t, xt2=t,`

`yt2=sum seq(yt1,t,1,t,1).` (sum and seq are under LIST, OPS (F5), MORE.) Under WIND use

`1,10,1,0,10,1,-3,1,1` to obtain a graph similar to the one above. Then use TRACE (F4) to see the values.

5.

| $n$ | $s_n$ |
|---|---|
| 1 | 0.44721 |
| 2 | 1.15432 |
| 3 | 1.98637 |
| 4 | 2.88080 |
| 5 | 3.80927 |
| 6 | 4.75796 |
| 7 | 5.71948 |
| 8 | 6.68962 |
| 9 | 7.66581 |
| 10 | 8.64639 |

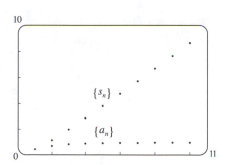

The series $\displaystyle\sum_{n=1}^{\infty} \frac{n}{\sqrt{n^2 + 4}}$ diverges, since its terms do not approach 0.

**7.**

| $n$ | $s_n$ |
|---|---|
| 1 | 0.29289 |
| 2 | 0.42265 |
| 3 | 0.50000 |
| 4 | 0.55279 |
| 5 | 0.59175 |
| 6 | 0.62204 |
| 7 | 0.64645 |
| 8 | 0.66667 |
| 9 | 0.68377 |
| 10 | 0.69849 |

From the graph and the table, it seems that the series converges.

$$\sum_{n=1}^{k}\left(\frac{1}{\sqrt{n}}-\frac{1}{\sqrt{n+1}}\right)=\left(\frac{1}{\sqrt{1}}-\frac{1}{\sqrt{2}}\right)+\left(\frac{1}{\sqrt{2}}-\frac{1}{\sqrt{3}}\right)+\cdots+\left(\frac{1}{\sqrt{k}}-\frac{1}{\sqrt{k+1}}\right)$$

$$=1-\frac{1}{\sqrt{k+1}},$$

so $\displaystyle\sum_{n=1}^{\infty}\left(\frac{1}{\sqrt{n}}-\frac{1}{\sqrt{n+1}}\right)=\lim_{k\to\infty}\left(1-\frac{1}{\sqrt{k+1}}\right)=1.$

**9.** (a) $\displaystyle\lim_{n\to\infty}a_n=\lim_{n\to\infty}\frac{2n}{3n+1}=\frac{2}{3}$, so the *sequence* $\{a_n\}$ is convergent by (8.1.1).

(b) Since $\displaystyle\lim_{n\to\infty}a_n=\frac{2}{3}\neq 0$, the *series* $\displaystyle\sum_{n=1}^{\infty}a_n$ is divergent by the Test for Divergence.

**11.** $3-4+\frac{16}{3}-\frac{64}{9}+\cdots$ is a geometric series with ratio $r=-\frac{4}{3}$. Since $|r|=\frac{4}{3}>1$, the series diverges.

**13.** $10-2+0.4-0.08+\cdots$ is a geometric series with ratio $-\frac{2}{10}=-\frac{1}{5}$. Since $|r|=\frac{1}{5}<1$, the series converges to

$$\frac{a}{1-r}=\frac{10}{1-(-1/5)}=\frac{10}{6/5}=\frac{50}{6}=\frac{25}{3}.$$

**15.** $\displaystyle\sum_{n=1}^{\infty}5\left(\frac{2}{3}\right)^{n-1}$ is a geometric series with $a=5$ and $r=\frac{2}{3}$. Since $|r|=\frac{2}{3}<1$, the series converges to

$$\frac{a}{1-r}=\frac{5}{1-2/3}=\frac{5}{1/3}=15.$$

**17.** $\displaystyle\sum_{n=0}^{\infty}\frac{\pi^n}{3^{n+1}}=\frac{1}{3}\sum_{n=0}^{\infty}\left(\frac{\pi}{3}\right)^n$ is a geometric series with ratio $r=\frac{\pi}{3}$. Since $|r|>1$, the series diverges.

**19.** $\displaystyle\sum_{n=1}^{\infty}\frac{n-1}{3n-1}$ diverges by the Test for Divergence since $\displaystyle\lim_{n\to\infty}a_n=\lim_{n\to\infty}\frac{n-1}{3n-1}=\frac{1}{3}\neq 0$.

**21.** $\displaystyle\sum_{k=2}^{\infty}\frac{k^2}{k^2-1}$ diverges by the Test for Divergence since $\displaystyle\lim_{k\to\infty}a_k=\lim_{k\to\infty}\frac{k^2}{k^2-1}=1\neq 0$.

**23.** Converges.

$$\sum_{n=1}^{\infty}\frac{1+2^n}{3^n}=\sum_{n=1}^{\infty}\left(\frac{1}{3^n}+\frac{2^n}{3^n}\right)=\sum_{n=1}^{\infty}\left[\left(\frac{1}{3}\right)^n+\left(\frac{2}{3}\right)^n\right]\quad\text{[sum of two convergent geometric series]}$$

$$=\frac{1/3}{1-1/3}+\frac{2/3}{1-2/3}=\frac{1}{2}+2=\frac{5}{2}$$

**25.** $\displaystyle\sum_{n=1}^{\infty}\sqrt[n]{2}=2+\sqrt{2}+\sqrt[3]{2}+\sqrt[4]{2}+\cdots$ diverges by the Test for Divergence since

$$\lim_{n\to\infty}a_n=\lim_{n\to\infty}\sqrt[n]{2}=\lim_{n\to\infty}2^{1/n}=2^0=1\neq 0.$$

**27.** $\sum\limits_{n=1}^{\infty} \arctan n$ diverges by the Test for Divergence since $\lim\limits_{n\to\infty} a_n = \lim\limits_{n\to\infty} \arctan n = \frac{\pi}{2} \neq 0$.

**29.** $\sum\limits_{n=1}^{\infty} \dfrac{1}{e^n} = \sum\limits_{n=1}^{\infty} \left(\dfrac{1}{e}\right)^n$ is a geometric series with first term $a = \dfrac{1}{e}$ and ratio $r = \dfrac{1}{e}$. Since $|r| = \dfrac{1}{e} < 1$, the series converges

to $\dfrac{1/e}{1 - 1/e} = \dfrac{1/e}{1 - 1/e} \cdot \dfrac{e}{e} = \dfrac{1}{e - 1}$. By Example 6, $\sum\limits_{n=1}^{\infty} \dfrac{1}{n(n+1)} = 1$. Thus, by Theorem 8(ii),

$$\sum\limits_{n=1}^{\infty} \left(\dfrac{1}{e^n} + \dfrac{1}{n(n+1)}\right) = \sum\limits_{n=1}^{\infty} \dfrac{1}{e^n} + \sum\limits_{n=1}^{\infty} \dfrac{1}{n(n+1)} = \dfrac{1}{e - 1} + 1 = \dfrac{1}{e - 1} + \dfrac{e - 1}{e - 1} = \dfrac{e}{e - 1}.$$

**31.** Using partial fractions, the partial sums of the series $\sum\limits_{n=2}^{\infty} \dfrac{2}{n^2 - 1}$ are

$$s_n = \sum\limits_{i=2}^{n} \dfrac{2}{(i-1)(i+1)} = \sum\limits_{i=2}^{n} \left(\dfrac{1}{i-1} - \dfrac{1}{i+1}\right)$$

$$= \left(1 - \dfrac{1}{3}\right) + \left(\dfrac{1}{2} - \dfrac{1}{4}\right) + \left(\dfrac{1}{3} - \dfrac{1}{5}\right) + \cdots + \left(\dfrac{1}{n-3} - \dfrac{1}{n-1}\right) + \left(\dfrac{1}{n-2} - \dfrac{1}{n}\right)$$

This sum is a telescoping series and $s_n = 1 + \dfrac{1}{2} - \dfrac{1}{n-1} - \dfrac{1}{n}$.

Thus, $\sum\limits_{n=2}^{\infty} \dfrac{2}{n^2 - 1} = \lim\limits_{n\to\infty} s_n = \lim\limits_{n\to\infty} \left(1 + \dfrac{1}{2} - \dfrac{1}{n-1} - \dfrac{1}{n}\right) = \dfrac{3}{2}$.

**33.** For the series $\sum\limits_{n=1}^{\infty} \dfrac{3}{n(n+3)}$, $s_n = \sum\limits_{i=1}^{n} \dfrac{3}{i(i+3)} = \sum\limits_{i=1}^{n} \left(\dfrac{1}{i} - \dfrac{1}{i+3}\right)$ [using partial fractions]. The latter sum is

$$\left(1 - \tfrac{1}{4}\right) + \left(\tfrac{1}{2} - \tfrac{1}{5}\right) + \left(\tfrac{1}{3} - \tfrac{1}{6}\right) + \left(\tfrac{1}{4} - \tfrac{1}{7}\right) + \cdots + \left(\tfrac{1}{n-3} - \tfrac{1}{n}\right) + \left(\tfrac{1}{n-2} - \tfrac{1}{n+1}\right) + \left(\tfrac{1}{n-1} - \tfrac{1}{n+2}\right) + \left(\tfrac{1}{n} - \tfrac{1}{n+3}\right)$$

$$= 1 + \tfrac{1}{2} + \tfrac{1}{3} - \tfrac{1}{n+1} - \tfrac{1}{n+2} - \tfrac{1}{n+3} \quad \text{[telescoping series]}$$

Thus, $\sum\limits_{n=1}^{\infty} \dfrac{3}{n(n+3)} = \lim\limits_{n\to\infty} s_n = \lim\limits_{n\to\infty} \left(1 + \tfrac{1}{2} + \tfrac{1}{3} - \tfrac{1}{n+1} - \tfrac{1}{n+2} - \tfrac{1}{n+3}\right) = 1 + \tfrac{1}{2} + \tfrac{1}{3} = \tfrac{11}{6}$. Converges

**35.** (a) Many people would guess that $x < 1$, but note that $x$ consists of an infinite number of 9s.

(b) $x = 0.99999\ldots = \dfrac{9}{10} + \dfrac{9}{100} + \dfrac{9}{1000} + \dfrac{9}{10,000} + \cdots = \sum\limits_{n=1}^{\infty} \dfrac{9}{10^n}$, which is a geometric series with $a_1 = 0.9$ and

$r = 0.1$. Its sum is $\dfrac{0.9}{1 - 0.1} = \dfrac{0.9}{0.9} = 1$, that is, $x = 1$.

(c) The number 1 has two decimal representations, $1.00000\ldots$ and $0.99999\ldots$.

(d) Except for 0, all rational numbers that have a terminating decimal representation can be written in more than one way. For

example, 0.5 can be written as $0.49999\ldots$ as well as $0.50000\ldots$.

**37.** $0.\overline{2} = \dfrac{2}{10} + \dfrac{2}{10^2} + \cdots$ is a geometric series with $a = \dfrac{2}{10}$ and $r = \dfrac{1}{10}$. It converges to $\dfrac{a}{1 - r} = \dfrac{2/10}{1 - 1/10} = \dfrac{2}{9}$.

**39.** $1.5\overline{342} = 1.53 + \dfrac{42}{10^4} + \dfrac{42}{10^6} + \cdots$. Now $\dfrac{42}{10^4} + \dfrac{42}{10^6} + \cdots$ is a geometric series with $a = \dfrac{42}{10^4}$ and $r = \dfrac{1}{10^2}$.

It converges to $\dfrac{a}{1 - r} = \dfrac{42/10^4}{1 - 1/10^2} = \dfrac{42/10^4}{99/10^2} = \dfrac{42}{9900}$.

Thus, $1.5\overline{342} = 1.53 + \dfrac{42}{9900} = \dfrac{153}{100} + \dfrac{42}{9900} = \dfrac{15,147}{9900} + \dfrac{42}{9900} = \dfrac{15,189}{9900}$ or $\dfrac{5063}{3300}$.

**41.** $\displaystyle\sum_{n=1}^{\infty} \frac{x^n}{3^n} = \sum_{n=1}^{\infty} \left(\frac{x}{3}\right)^n$ is a geometric series with $r = \frac{x}{3}$, so the series converges $\Leftrightarrow$ $|r| < 1$ $\Leftrightarrow$ $\dfrac{|x|}{3} < 1$ $\Leftrightarrow$ $|x| < 3$;

that is, $-3 < x < 3$. In that case, the sum of the series is $\dfrac{a}{1-r} = \dfrac{x/3}{1 - x/3} = \dfrac{x/3}{1 - x/3} \cdot \dfrac{3}{3} = \dfrac{x}{3 - x}$.

**43.** $\displaystyle\sum_{n=0}^{\infty} \frac{\cos^n x}{2^n}$ is a geometric series with first term 1 and ratio $r = \dfrac{\cos x}{2}$, so it converges $\Leftrightarrow$ $|r| < 1$. But $|r| = \dfrac{|\cos x|}{2} \le \dfrac{1}{2}$

for all $x$. Thus, the series converges for all real values of $x$ and the sum of the series is $\dfrac{1}{1 - (\cos x)/2} = \dfrac{2}{2 - \cos x}$.

**45.** After defining $f$, We use `convert(f,parfrac);` in Maple, `Apart` in Mathematica, or `Expand Rational` and

`Simplify` in Derive to find that the general term is $\dfrac{3n^2 + 3n + 1}{(n^2 + n)^3} = \dfrac{1}{n^3} - \dfrac{1}{(n+1)^3}$. So the $n$th partial sum is

$$s_n = \sum_{k=1}^{n} \left(\frac{1}{k^3} - \frac{1}{(k+1)^3}\right) = \left(1 - \frac{1}{2^3}\right) + \left(\frac{1}{2^3} - \frac{1}{3^3}\right) + \cdots + \left(\frac{1}{n^3} - \frac{1}{(n+1)^3}\right) = 1 - \frac{1}{(n+1)^3}$$

The series converges to $\lim\limits_{n\to\infty} s_n = 1$. This can be confirmed by directly computing the sum using `sum(f,1..infinity);`

(in Maple), `Sum[f,{n,1,Infinity}]` (in Mathematica), or `Calculus Sum` (from 1 to $\infty$) and `Simplify` (in Derive).

**47.** For $n = 1$, $a_1 = 0$ since $s_1 = 0$. For $n > 1$,

$$a_n = s_n - s_{n-1} = \frac{n-1}{n+1} - \frac{(n-1)-1}{(n-1)+1} = \frac{(n-1)n - (n+1)(n-2)}{(n+1)n} = \frac{2}{n(n+1)}$$

Also, $\displaystyle\sum_{n=1}^{\infty} a_n = \lim_{n\to\infty} s_n = \lim_{n\to\infty} \frac{1 - 1/n}{1 + 1/n} = 1$.

**49.** (a) After the first pill is taken, 100 mg of the drug is in the body. After the second pill is taken, 100 mg plus 100(5%) mg

remains in the body. After the third pill is taken 100 mg plus 100(5%) mg plus [100(5%)](5%) mg remains in the body.

This gives us $100 + 100(0.05) + 100(0.05)^2 = 105.25$ mg of the drug remaining in the body after the patient takes three

pills.

(b) Continuing the pattern established in part (a), we get $100 + 100(0.05) + 100(0.05)^2 + \cdots + 100(0.05)^{n-1}$ mg after $n$

pills are taken. By (3), this sum is $\dfrac{100(1 - 0.05^n)}{1 - 0.05}$.

(c) The amount of the drug remaining in the body in the long run can be approximated by summing the infinite series with

$a_1 = 100$ and $r = 0.05$. This sum is $\dfrac{100}{1 - 0.05} = \dfrac{100}{0.95} \approx 105.26$ mg.

**51.** (a) The first step in the chain occurs when the local government spends $D$ dollars. The people who receive it spend a fraction $c$

of those $D$ dollars, that is, $Dc$ dollars. Those who receive the $Dc$ dollars spend a fraction $c$ of it, that is, $Dc^2$ dollars.

Continuing in this way, we see that the total spending after $n$ transactions is

$$S_n = D + Dc + Dc^2 + \cdots + Dc^{n-1} = \frac{D(1 - c^n)}{1 - c} \quad \text{by (3)}.$$

(b) $\displaystyle\lim_{n\to\infty} S_n = \lim_{n\to\infty} \frac{D(1 - c^n)}{1 - c} = \frac{D}{1 - c} \lim_{n\to\infty} (1 - c^n) = \frac{D}{1 - c}$ $\left[\text{since } 0 < c < 1 \quad \Rightarrow \quad \lim_{n\to\infty} c^n = 0\right]$

$= \dfrac{D}{s}$ $\;[\text{since } c + s = 1]\; = kD$ $\;[\text{since } k = 1/s]$

If $c = 0.8$, then $s = 1 - c = 0.2$ and the multiplier is $k = 1/s = 5$.

**53.** $\sum_{n=2}^{\infty} (1+c)^{-n}$ is a geometric series with $a = (1+c)^{-2}$ and $r = (1+c)^{-1}$, so the series converges when

$\left| (1+c)^{-1} \right| < 1 \quad \Leftrightarrow \quad |1+c| > 1 \quad \Leftrightarrow \quad 1+c > 1 \text{ or } 1+c < -1 \quad \Leftrightarrow \quad c > 0 \text{ or } c < -2$. We calculate the sum of the

series and set it equal to 2: $\dfrac{(1+c)^{-2}}{1-(1+c)^{-1}} = 2 \quad \Leftrightarrow \quad \left(\dfrac{1}{1+c}\right)^2 = 2 - 2\left(\dfrac{1}{1+c}\right) \quad \Leftrightarrow \quad 1 = 2(1+c)^2 - 2(1+c) \quad \Leftrightarrow$

$2c^2 + 2c - 1 = 0 \quad \Leftrightarrow \quad c = \dfrac{-2 \pm \sqrt{12}}{4} = \dfrac{\pm\sqrt{3}-1}{2}$. However, the negative root is inadmissible because $-2 < \dfrac{-\sqrt{3}-1}{2} < 0$.

So $c = \dfrac{\sqrt{3}-1}{2}$.

**55.** $e^{s_n} = e^{1+\frac{1}{2}+\frac{1}{3}+\cdots+\frac{1}{n}} = e^1 e^{1/2} e^{1/3} \cdots e^{1/n} > (1+1)\left(1+\tfrac{1}{2}\right)\left(1+\tfrac{1}{3}\right) \cdots \left(1+\tfrac{1}{n}\right) \qquad [e^x > 1+x]$

$\qquad = \dfrac{2}{1} \dfrac{3}{2} \dfrac{4}{3} \cdots \dfrac{n+1}{n} = n+1$

Thus, $e^{s_n} > n+1$ and $\lim\limits_{n\to\infty} e^{s_n} = \infty$. Since $\{s_n\}$ is increasing, $\lim\limits_{n\to\infty} s_n = \infty$, implying that the harmonic series is

divergent.

**57.** Let $d_n$ be the diameter of $C_n$. We draw lines from the centers of the $C_i$ to

the center of $D$ (or $C$), and using the Pythagorean Theorem, we can write

$1^2 + \left(1 - \tfrac{1}{2}d_1\right)^2 = \left(1 + \tfrac{1}{2}d_1\right)^2 \quad \Leftrightarrow$

$1 = \left(1 + \tfrac{1}{2}d_1\right)^2 - \left(1 - \tfrac{1}{2}d_1\right)^2 = 2d_1$  [difference of squares]  $\Rightarrow \quad d_1 = \tfrac{1}{2}$.

Similarly,

$1 = \left(1 + \tfrac{1}{2}d_2\right)^2 - \left(1 - d_1 - \tfrac{1}{2}d_2\right)^2 = 2d_2 + 2d_1 - d_1^2 - d_1 d_2$

$\quad = (2 - d_1)(d_1 + d_2) \quad \Leftrightarrow$

$d_2 = \dfrac{1}{2 - d_1} - d_1 = \dfrac{(1-d_1)^2}{2-d_1}$, $1 = \left(1 + \tfrac{1}{2}d_3\right)^2 - \left(1 - d_1 - d_2 - \tfrac{1}{2}d_3\right)^2 \quad \Leftrightarrow \quad d_3 = \dfrac{[1-(d_1+d_2)]^2}{2-(d_1+d_2)}$, and in general,

$d_{n+1} = \dfrac{\left(1 - \sum_{i=1}^{n} d_i\right)^2}{2 - \sum_{i=1}^{n} d_i}$. If we actually calculate $d_2$ and $d_3$ from the formulas above, we find that they are $\dfrac{1}{6} = \dfrac{1}{2 \cdot 3}$ and

$\dfrac{1}{12} = \dfrac{1}{3 \cdot 4}$ respectively, so we suspect that in general, $d_n = \dfrac{1}{n(n+1)}$. To prove this, we use induction: Assume that for all

$k \le n,\ d_k = \dfrac{1}{k(k+1)} = \dfrac{1}{k} - \dfrac{1}{k+1}$. Then $\sum_{i=1}^{n} d_i = 1 - \dfrac{1}{n+1} = \dfrac{n}{n+1}$  [telescoping sum]. Substituting this into our

formula for $d_{n+1}$, we get $d_{n+1} = \dfrac{\left[1 - \dfrac{n}{n+1}\right]^2}{2 - \left(\dfrac{n}{n+1}\right)} = \dfrac{\dfrac{1}{(n+1)^2}}{\dfrac{n+2}{n+1}} = \dfrac{1}{(n+1)(n+2)}$, and the induction is complete.

Now, we observe that the partial sums $\sum_{i=1}^{n} d_i$ of the diameters of the circles approach 1 as $n \to \infty$; that is,

$\sum_{n=1}^{\infty} a_n = \sum_{n=1}^{\infty} \dfrac{1}{n(n+1)} = 1$, which is what we wanted to prove.

**59.** The series $1 - 1 + 1 - 1 + 1 - 1 + \cdots$ diverges (geometric series with $r = -1$) so we cannot say that

$0 = 1 - 1 + 1 - 1 + 1 - 1 + \cdots$.

**61.** Suppose on the contrary that $\sum (a_n + b_n)$ converges. Then $\sum (a_n + b_n)$ and $\sum a_n$ are convergent series. So by Theorem 8(iii), $\sum [(a_n + b_n) - a_n]$ would also be convergent. But $\sum [(a_n + b_n) - a_n] = \sum b_n$, a contradiction, since $\sum b_n$ is given to be divergent.

**63.** The partial sums $\{s_n\}$ form an increasing sequence, since $s_n - s_{n-1} = a_n > 0$ for all $n$. Also, the sequence $\{s_n\}$ is bounded since $s_n \le 1000$ for all $n$. So by the Monotonic Sequence Theorem, the sequence of partial sums converges, that is, the series $\sum a_n$ is convergent.

**65. (a)** At the first step, only the interval $\left(\frac{1}{3}, \frac{2}{3}\right)$ (length $\frac{1}{3}$) is removed. At the second step, we remove the intervals $\left(\frac{1}{9}, \frac{2}{9}\right)$ and $\left(\frac{7}{9}, \frac{8}{9}\right)$, which have a total length of $2 \cdot \left(\frac{1}{3}\right)^2$. At the third step, we remove $2^2$ intervals, each of length $\left(\frac{1}{3}\right)^3$. In general, at the $n$th step we remove $2^{n-1}$ intervals, each of length $\left(\frac{1}{3}\right)^n$, for a length of $2^{n-1} \cdot \left(\frac{1}{3}\right)^n = \frac{1}{3}\left(\frac{2}{3}\right)^{n-1}$. Thus, the total length of all removed intervals is $\sum_{n=1}^{\infty} \frac{1}{3}\left(\frac{2}{3}\right)^{n-1} = \frac{1/3}{1 - 2/3} = 1$ [geometric series with $a = \frac{1}{3}$ and $r = \frac{2}{3}$]. Notice that at the $n$th step, the leftmost interval that is removed is $\left(\left(\frac{1}{3}\right)^n, \left(\frac{2}{3}\right)^n\right)$, so we never remove 0, and 0 is in the Cantor set. Also, the rightmost interval removed is $\left(1 - \left(\frac{2}{3}\right)^n, 1 - \left(\frac{1}{3}\right)^n\right)$, so 1 is never removed. Some other numbers in the Cantor set are $\frac{1}{3}, \frac{2}{3}, \frac{1}{9}, \frac{2}{9}, \frac{7}{9}$, and $\frac{8}{9}$.

**(b)** The area removed at the first step is $\frac{1}{9}$; at the second step, $8 \cdot \left(\frac{1}{9}\right)^2$; at the third step, $(8)^2 \cdot \left(\frac{1}{9}\right)^3$. In general, the area removed at the $n$th step is $(8)^{n-1}\left(\frac{1}{9}\right)^n = \frac{1}{9}\left(\frac{8}{9}\right)^{n-1}$, so the total area of all removed squares is
$$\sum_{n=1}^{\infty} \frac{1}{9}\left(\frac{8}{9}\right)^{n-1} = \frac{1/9}{1 - 8/9} = 1.$$

**67. (a)** For $\sum_{n=1}^{\infty} \frac{n}{(n+1)!}$, $s_1 = \frac{1}{1 \cdot 2} = \frac{1}{2}$, $s_2 = \frac{1}{2} + \frac{2}{1 \cdot 2 \cdot 3} = \frac{5}{6}$, $s_3 = \frac{5}{6} + \frac{3}{1 \cdot 2 \cdot 3 \cdot 4} = \frac{23}{24}$,

$s_4 = \frac{23}{24} + \frac{4}{1 \cdot 2 \cdot 3 \cdot 4 \cdot 5} = \frac{119}{120}$. The denominators are $(n+1)!$, so a guess would be $s_n = \frac{(n+1)! - 1}{(n+1)!}$.

**(b)** For $n = 1$, $s_1 = \frac{1}{2} = \frac{2! - 1}{2!}$, so the formula holds for $n = 1$. Assume $s_k = \frac{(k+1)! - 1}{(k+1)!}$. Then
$$s_{k+1} = \frac{(k+1)! - 1}{(k+1)!} + \frac{k+1}{(k+2)!} = \frac{(k+1)! - 1}{(k+1)!} + \frac{k+1}{(k+1)!(k+2)} = \frac{(k+2)! - (k+2) + k + 1}{(k+2)!}$$
$$= \frac{(k+2)! - 1}{(k+2)!}$$

Thus, the formula is true for $n = k + 1$. So by induction, the guess is correct.

**(c)** $\lim_{n \to \infty} s_n = \lim_{n \to \infty} \frac{(n+1)! - 1}{(n+1)!} = \lim_{n \to \infty} \left[1 - \frac{1}{(n+1)!}\right] = 1$ and so $\sum_{n=1}^{\infty} \frac{n}{(n+1)!} = 1$.

## 8.3   The Integral and Comparison Tests; Estimating Sums

**1.** The picture shows that $a_2 = \dfrac{1}{2^{1.3}} < \displaystyle\int_1^2 \dfrac{1}{x^{1.3}}\, dx$,

$a_3 = \dfrac{1}{3^{1.3}} < \displaystyle\int_2^3 \dfrac{1}{x^{1.3}}\, dx$, and so on, so $\displaystyle\sum_{n=2}^{\infty} \dfrac{1}{n^{1.3}} < \int_1^{\infty} \dfrac{1}{x^{1.3}}\, dx$. The

integral converges by (5.10.2) with $p = 1.3 > 1$, so the series converges.

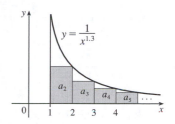

**3.** (a) We cannot say anything about $\sum a_n$. If $a_n > b_n$ for all $n$ and $\sum b_n$ is convergent, then $\sum a_n$ could be convergent or

divergent. (See the note after Example 2.)

(b) If $a_n < b_n$ for all $n$, then $\sum a_n$ is convergent. [This is part (i) of the Comparison Test.]

**5.** $\displaystyle\sum_{n=1}^{\infty} n^b$ is a $p$-series with $p = -b$. $\displaystyle\sum_{n=1}^{\infty} b^n$ is a geometric series. By (1), the $p$-series is convergent if $p > 1$. In this case,

$\displaystyle\sum_{n=1}^{\infty} n^b = \sum_{n=1}^{\infty} \left(1/n^{-b}\right)$, so $-b > 1 \iff b < -1$ are the values for which the series converge. A geometric series

$\displaystyle\sum_{n=1}^{\infty} ar^{n-1}$ converges if $|r| < 1$, so $\displaystyle\sum_{n=1}^{\infty} b^n$ converges if $|b| < 1 \iff -1 < b < 1$.

**7.** The function $f(x) = 1/\sqrt[5]{x} = x^{-1/5}$ is continuous, positive, and decreasing on $[1, \infty)$, so the Integral Test applies.

$\int_1^{\infty} x^{-1/5}\, dx = \lim\limits_{t \to \infty} \int_1^t x^{-1/5}\, dx = \lim\limits_{t \to \infty} \left[\frac{5}{4}x^{4/5}\right]_1^t = \lim\limits_{t \to \infty} \left(\frac{5}{4}t^{4/5} - \frac{5}{4}\right) = \infty$, so $\displaystyle\sum_{n=1}^{\infty} 1/\sqrt[5]{n}$ diverges.

**9.** $\dfrac{n}{2n^3 + 1} < \dfrac{n}{2n^3} = \dfrac{1}{2n^2} < \dfrac{1}{n^2}$ for all $n \geq 1$, so $\displaystyle\sum_{n=1}^{\infty} \dfrac{n}{2n^3 + 1}$ converges by comparison with $\displaystyle\sum_{n=1}^{\infty} \dfrac{1}{n^2}$, which converges

because it is a $p$-series with $p = 2 > 1$.

**11.** The series $\displaystyle\sum_{n=1}^{\infty} \dfrac{1}{n^{0.85}}$ is a $p$-series with $p = 0.85 \leq 1$, so it diverges by (1). Therefore, the series $\displaystyle\sum_{n=1}^{\infty} \dfrac{2}{n^{0.85}}$ must also diverge,

for if it converged, then $\displaystyle\sum_{n=1}^{\infty} \dfrac{1}{n^{0.85}}$ would have to converge [by Theorem 8(i) in Section 8.2].

**13.** $1 + \dfrac{1}{8} + \dfrac{1}{27} + \dfrac{1}{64} + \dfrac{1}{125} + \cdots = \displaystyle\sum_{n=1}^{\infty} \dfrac{1}{n^3}$. This is a $p$-series with $p = 3 > 1$, so it converges by (1).

**15.** $f(x) = xe^{-x}$ is continuous and positive on $[1, \infty)$. $f'(x) = -xe^{-x} + e^{-x} = e^{-x}(1 - x) < 0$ for $x > 1$, so $f$ is decreasing

on $[1, \infty)$. Thus, the Integral Test applies.

$\int_1^{\infty} xe^{-x}\, dx = \lim\limits_{b \to \infty} \int_1^b xe^{-x}\, dx = \lim\limits_{b \to \infty} \left[-xe^{-x} - e^{-x}\right]_1^b$  [by parts]  $= \lim\limits_{b \to \infty} [-be^{-b} - e^{-b} + e^{-1} + e^{-1}] = 2/e$

since $\lim\limits_{b \to \infty} be^{-b} = \lim\limits_{b \to \infty} (b/e^b) \overset{\text{H}}{=} \lim\limits_{b \to \infty} (1/e^b) = 0$ and $\lim\limits_{b \to \infty} e^{-b} = 0$. Thus, $\sum_{n=1}^{\infty} ne^{-n}$ converges.

**17.** $f(x) = \dfrac{1}{x \ln x}$ is continuous and positive on $[2, \infty)$, and also decreasing since $f'(x) = -\dfrac{1 + \ln x}{x^2 (\ln x)^2} < 0$ for $x > 2$, so we can

use the Integral Test. $\displaystyle\int_2^\infty \dfrac{1}{x \ln x}\, dx = \lim_{t \to \infty} [\ln(\ln x)]_2^t = \lim_{t \to \infty} [\ln(\ln t) - \ln(\ln 2)] = \infty$, so the series $\displaystyle\sum_{n=2}^\infty \dfrac{1}{n \ln n}$ diverges.

**19.** $\dfrac{\cos^2 n}{n^2 + 1} \le \dfrac{1}{n^2 + 1} < \dfrac{1}{n^2}$, so the series $\displaystyle\sum_{n=1}^\infty \dfrac{\cos^2 n}{n^2 + 1}$ converges by comparison with the $p$-series $\displaystyle\sum_{n=1}^\infty \dfrac{1}{n^2}$ $[p = 2 > 1]$.

**21.** $\dfrac{n - 1}{n\, 4^n}$ is positive for $n > 1$ and $\dfrac{n - 1}{n\, 4^n} < \dfrac{n}{n\, 4^n} = \dfrac{1}{4^n} = \left(\dfrac{1}{4}\right)^n$, so $\displaystyle\sum_{n=1}^\infty \dfrac{n - 1}{n\, 4^n}$ converges by comparison with the convergent

geometric series $\displaystyle\sum_{n=1}^\infty \left(\dfrac{1}{4}\right)^n$.

**23.** $1 + \dfrac{1}{3} + \dfrac{1}{5} + \dfrac{1}{7} + \dfrac{1}{9} + \cdots = \displaystyle\sum_{n=1}^\infty \dfrac{1}{2n - 1}$. The function $f(x) = \dfrac{1}{2x - 1}$ is

continuous, positive, and decreasing on $[1, \infty)$, so the Integral Test applies.

$\displaystyle\int_1^\infty \dfrac{1}{2x - 1}\, dx = \lim_{t \to \infty} \int_1^t \dfrac{1}{2x - 1}\, dx = \lim_{t \to \infty} \left[\tfrac{1}{2} \ln |2x - 1|\right]_1^t = \tfrac{1}{2} \lim_{t \to \infty} (\ln(2t - 1) - 0) = \infty$, so the series $\displaystyle\sum_{n=1}^\infty \dfrac{1}{2n - 1}$

diverges.

**25.** Use the Limit Comparison Test with $a_n = \dfrac{1 + 4^n}{1 + 3^n}$ and $b_n = \dfrac{4^n}{3^n}$:

$$\lim_{n \to \infty} \dfrac{a_n}{b_n} = \lim_{n \to \infty} \dfrac{\dfrac{1 + 4^n}{1 + 3^n}}{\dfrac{4^n}{3^n}} = \lim_{n \to \infty} \dfrac{1 + 4^n}{1 + 3^n} \cdot \dfrac{3^n}{4^n} = \lim_{n \to \infty} \dfrac{1 + 4^n}{4^n} \cdot \dfrac{3^n}{1 + 3^n} = \lim_{n \to \infty} \left(\dfrac{1}{4^n} + 1\right) \cdot \dfrac{1}{\dfrac{1}{3^n} + 1} = 1 > 0$$

Since the geometric series $\sum b_n = \sum \left(\tfrac{4}{3}\right)^n$ diverges, so does $\displaystyle\sum_{n=1}^\infty \dfrac{1 + 4^n}{1 + 3^n}$. Alternatively, use the Comparison Test with

$\dfrac{1 + 4^n}{1 + 3^n} > \dfrac{1 + 4^n}{3^n + 3^n} > \dfrac{4^n}{2(3^n)} = \dfrac{1}{2}\left(\dfrac{4}{3}\right)^n$ or use the Test for Divergence.

**27.** $\dfrac{2 + (-1)^n}{n \sqrt{n}} \le \dfrac{3}{n \sqrt{n}}$, and $\displaystyle\sum_{n=1}^\infty \dfrac{3}{n \sqrt{n}}$ converges because it is a constant multiple of the convergent $p$-series $\displaystyle\sum_{n=1}^\infty \dfrac{1}{n \sqrt{n}}$

$\left[p = \tfrac{3}{2} > 1\right]$, so the given series converges by the Comparison Test.

**29.** Use the Limit Comparison Test with $a_n = \sin\left(\dfrac{1}{n}\right)$ and $b_n = \dfrac{1}{n}$. Then $\sum a_n$ and $\sum b_n$ are series with positive terms and

$\displaystyle\lim_{n \to \infty} \dfrac{a_n}{b_n} = \lim_{n \to \infty} \dfrac{\sin(1/n)}{1/n} = \lim_{\theta \to 0} \dfrac{\sin \theta}{\theta} = 1 > 0$. Since $\displaystyle\sum_{n=1}^\infty b_n$ is the divergent harmonic series,

$\displaystyle\sum_{n=1}^\infty \sin(1/n)$ also diverges. [Note that we could also use l'Hospital's Rule to evaluate the limit:

$\displaystyle\lim_{x \to \infty} \dfrac{\sin(1/x)}{1/x} \overset{\text{H}}{=} \lim_{x \to \infty} \dfrac{\cos(1/x) \cdot (-1/x^2)}{-1/x^2} = \lim_{x \to \infty} \cos \dfrac{1}{x} = \cos 0 = 1.]$

**31.** We have already shown (in Exercise 17) that when $p = 1$ the series $\sum_{n=2}^{\infty} \dfrac{1}{n(\ln n)^p}$ diverges, so assume that $p \neq 1$.

$f(x) = \dfrac{1}{x(\ln x)^p}$ is continuous and positive on $[2, \infty)$, and $f'(x) = -\dfrac{p + \ln x}{x^2(\ln x)^{p+1}} < 0$ if $x > e^{-p}$, so that $f$ is eventually

decreasing and we can use the Integral Test.

$$\int_2^{\infty} \frac{1}{x(\ln x)^p}\, dx = \lim_{t \to \infty} \left[ \frac{(\ln x)^{1-p}}{1 - p} \right]_2^t \quad [\text{for } p \neq 1] = \lim_{t \to \infty} \left[ \frac{(\ln t)^{1-p}}{1-p} - \frac{(\ln 2)^{1-p}}{1-p} \right]$$

This limit exists whenever $1 - p < 0 \quad \Leftrightarrow \quad p > 1$, so the series converges for $p > 1$.

**33.** (a) $f(x) = \dfrac{1}{x^2}$ is positive and continuous and $f'(x) = -\dfrac{2}{x^3}$ is negative for $x > 0$, and so the Integral Test applies.

$$\sum_{n=1}^{\infty} \frac{1}{n^2} \approx s_{10} = \frac{1}{1^2} + \frac{1}{2^2} + \frac{1}{3^2} + \cdots + \frac{1}{10^2} \approx 1.549768.$$

$$R_{10} \leq \int_{10}^{\infty} \frac{1}{x^2}\, dx = \lim_{t \to \infty} \left[ \frac{-1}{x} \right]_{10}^t = \lim_{t \to \infty} \left( -\frac{1}{t} + \frac{1}{10} \right) = \frac{1}{10}, \text{ so the error is at most } 0.1.$$

(b) $s_{10} + \displaystyle\int_{11}^{\infty} \frac{1}{x^2}\, dx \leq s \leq s_{10} + \int_{10}^{\infty} \frac{1}{x^2}\, dx \quad \Rightarrow \quad s_{10} + \frac{1}{11} \leq s \leq s_{10} + \frac{1}{10} \quad \Rightarrow$

$\quad 1.549768 + 0.090909 = 1.640677 \leq s \leq 1.549768 + 0.1 = 1.649768$, so we get $s \approx 1.64522$ (the average of $1.640677$

$\quad$ and $1.649768$) with error $\leq 0.005$ (the maximum of $1.649768 - 1.64522$ and $1.64522 - 1.640677$, rounded up).

(c) $R_n \leq \displaystyle\int_n^{\infty} \frac{1}{x^2}\, dx = \frac{1}{n}$. So $R_n < 0.001$ if $\dfrac{1}{n} < \dfrac{1}{1000} \quad \Leftrightarrow \quad n > 1000$.

**35.** $f(x) = 1/(2x + 1)^6$ is continuous, positive, and decreasing on $[1, \infty)$, so the Integral Test applies. Using (3),

$$R_n \leq \int_n^{\infty} (2x + 1)^{-6}\, dx = \lim_{t \to \infty} \left[ \frac{-1}{10(2x+1)^5} \right]_n^t = \frac{1}{10(2n+1)^5}. \text{ To be correct to five decimal places, we want}$$

$$\frac{1}{10(2n+1)^5} \leq \frac{5}{10^6} \quad \Leftrightarrow \quad (2n+1)^5 \geq 20{,}000 \quad \Leftrightarrow \quad n \geq \tfrac{1}{2}\left(\sqrt[5]{20{,}000} - 1\right) \approx 3.12, \text{ so use } n = 4.$$

$$s_4 = \sum_{n=1}^{4} \frac{1}{(2n+1)^6} = \frac{1}{3^6} + \frac{1}{5^6} + \frac{1}{7^6} + \frac{1}{9^6} \approx 0.001\,446 \approx 0.00145.$$

**37.** $\displaystyle\sum_{n=1}^{10} \frac{1}{\sqrt{n^4 + 1}} = \frac{1}{\sqrt{2}} + \frac{1}{\sqrt{17}} + \frac{1}{\sqrt{82}} + \cdots + \frac{1}{\sqrt{10{,}001}} \approx 1.24856.$ Now $\dfrac{1}{\sqrt{n^4 + 1}} < \dfrac{1}{\sqrt{n^4}} = \dfrac{1}{n^2}$, so the error is

$$R_{10} \leq T_{10} \leq \int_{10}^{\infty} \frac{1}{x^2}\, dx = \lim_{t \to \infty} \left[ -\frac{1}{x} \right]_{10}^t = \lim_{t \to \infty} \left( -\frac{1}{t} + \frac{1}{10} \right) = \frac{1}{10} = 0.1.$$

**39.** (a) From the figure, $a_2 + a_3 + \cdots + a_n \leq \int_1^n f(x)\, dx$, so with

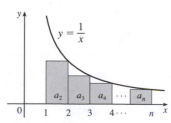

$\quad f(x) = \dfrac{1}{x}, \dfrac{1}{2} + \dfrac{1}{3} + \dfrac{1}{4} + \cdots + \dfrac{1}{n} \leq \displaystyle\int_1^n \frac{1}{x}\, dx = \ln n.$

$\quad$ Thus, $s_n = 1 + \dfrac{1}{2} + \dfrac{1}{3} + \dfrac{1}{4} + \cdots + \dfrac{1}{n} \leq 1 + \ln n.$

(b) By part (a), $s_{10^6} \leq 1 + \ln 10^6 \approx 14.82 < 15$ and

$\quad s_{10^9} \leq 1 + \ln 10^9 \approx 21.72 < 22.$

41. Since $\dfrac{d_n}{10^n} \le \dfrac{9}{10^n}$ for each $n$, and since $\displaystyle\sum_{n=1}^{\infty} \dfrac{9}{10^n}$ is a convergent geometric series $\left(|r| = \frac{1}{10} < 1\right)$, $0.d_1 d_2 d_3 \ldots = \displaystyle\sum_{n=1}^{\infty} \dfrac{d_n}{10^n}$

will always converge by the Comparison Test.

43. Yes. Since $\sum a_n$ is a convergent series with positive terms, $\displaystyle\lim_{n \to \infty} a_n = 0$ by Theorem 8.2.6, and $\sum b_n = \sum \sin(a_n)$ is a

series with positive terms (for large enough $n$). We have $\displaystyle\lim_{n \to \infty} \dfrac{b_n}{a_n} = \lim_{n \to \infty} \dfrac{\sin(a_n)}{a_n} = 1 > 0$ by Theorem 3.3.2. Thus, $\sum b_n$

is also convergent by the Limit Comparison Test.

45. $\displaystyle\lim_{n \to \infty} n a_n = \lim_{n \to \infty} \dfrac{a_n}{1/n}$, so we apply the Limit Comparison Test with $b_n = \dfrac{1}{n}$. Since $\displaystyle\lim_{n \to \infty} n a_n > 0$ we know that either both

series converge or both series diverge, and we also know that $\displaystyle\sum_{n=1}^{\infty} \dfrac{1}{n}$ diverges $[p$-series with $p = 1]$. Therefore, $\sum a_n$ must be

divergent.

## 8.4 Other Convergence Tests

1. (a) An alternating series is a series whose terms are alternately positive and negative.

   (b) An alternating series $\displaystyle\sum_{n=1}^{\infty} a_n = \sum_{n=1}^{\infty} (-1)^{n-1} b_n$, where $b_n = |a_n|$, converges if $0 < b_{n+1} \le b_n$ for all $n$ and $\displaystyle\lim_{n \to \infty} b_n = 0$.

   (This is the Alternating Series Test.)

   (c) The error involved in using the partial sum $s_n$ as an approximation to the total sum $s$ is the remainder $R_n = s - s_n$ and the

   size of the error is smaller than $b_{n+1}$; that is, $|R_n| \le b_{n+1}$. (This is the Alternating Series Estimation Theorem.)

3. $\dfrac{4}{7} - \dfrac{4}{8} + \dfrac{4}{9} - \dfrac{4}{10} + \dfrac{4}{11} - \cdots = \displaystyle\sum_{n=1}^{\infty} (-1)^{n-1} \dfrac{4}{n+6}$. Now $b_n = \dfrac{4}{n+6} > 0$, $\{b_n\}$ is decreasing, and $\displaystyle\lim_{n \to \infty} b_n = 0$, so the

   series converges by the Alternating Series Test.

5. $\displaystyle\sum_{n=1}^{\infty} a_n = \sum_{n=1}^{\infty} (-1)^{n-1} \dfrac{1}{2n+1} = \sum_{n=1}^{\infty} (-1)^{n-1} b_n$. Now $b_n = \dfrac{1}{2n+1} > 0$, $\{b_n\}$ is decreasing, and $\displaystyle\lim_{n \to \infty} b_n = 0$, so the

   series converges by the Alternating Series Test.

7. $\displaystyle\sum_{n=1}^{\infty} a_n = \sum_{n=1}^{\infty} (-1)^n \dfrac{3n-1}{2n+1} = \sum_{n=1}^{\infty} (-1)^n b_n$. Now $\displaystyle\lim_{n \to \infty} b_n = \lim_{n \to \infty} \dfrac{3 - 1/n}{2 + 1/n} = \dfrac{3}{2} \ne 0$. Since $\displaystyle\lim_{n \to \infty} a_n \ne 0$

   (in fact the limit does not exist), the series diverges by the Test for Divergence.

9. $b_n = \dfrac{n}{n^2 + 9} > 0$, and if $f(x) = \dfrac{x}{x^2 + 9}$, then $f'(x) = \dfrac{(x^2 + 9)(1) - x(2x)}{(x^2 + 9)^2} = \dfrac{9 - x^2}{(x^2 + 9)^2} < 0$ for $x > 3$, so $\{b_n\}$ is

   eventually decreasing. Also, $\displaystyle\lim_{n \to \infty} b_n = \lim_{n \to \infty} \dfrac{n}{n^2 + 9} = \lim_{n \to \infty} \dfrac{1}{n + 9/n} = 0$, so the series $\displaystyle\sum_{n=1}^{\infty} (-1)^{n+1} \dfrac{n}{n^2 + 9}$ converges by

   the Alternating Series Test.

**11.** $\displaystyle\sum_{n=1}^{\infty} \frac{(-1)^{n-1}}{n} = 1 - \frac{1}{2} + \frac{1}{3} - \frac{1}{4} + \cdots + \frac{1}{49} - \frac{1}{50} + \frac{1}{51} - \frac{1}{52} + \cdots$. The 50th partial sum of this series is an

underestimate, since $\displaystyle\sum_{n=1}^{\infty} \frac{(-1)^{n-1}}{n} = s_{50} + \left(\frac{1}{51} - \frac{1}{52}\right) + \left(\frac{1}{53} - \frac{1}{54}\right) + \cdots$, and the terms in parentheses are all positive.

The result can be seen geometrically in Figure 1.

**13.** If $p > 0$, $\dfrac{1}{(n+1)^p} \le \dfrac{1}{n^p}$ ($\{1/n^p\}$ is decreasing) and $\displaystyle\lim_{n\to\infty} \frac{1}{n^p} = 0$, so the series converges by the Alternating Series Test.

If $p \le 0$, $\displaystyle\lim_{n\to\infty} \frac{(-1)^{n-1}}{n^p}$ does not exist, so the series diverges by the Test for Divergence. Thus, $\displaystyle\sum_{n=1}^{\infty} \frac{(-1)^{n-1}}{n^p}$

converges $\Leftrightarrow$ $p > 0$.

**15.** The series $\displaystyle\sum_{n=1}^{\infty} \frac{(-1)^{n+1}}{n^6}$ satisfies (i) of the Alternating Series Test because $\dfrac{1}{(n+1)^6} < \dfrac{1}{n^6}$ and (ii) $\displaystyle\lim_{n\to\infty} \frac{1}{n^6} = 0$, so the

series is convergent. Now $b_5 = \dfrac{1}{5^6} = 0.000064 > 0.00005$ and $b_6 = \dfrac{1}{6^6} \approx 0.00002 < 0.00005$, so by the Alternating Series

Estimation Theorem, $n = 5$. (That is, since the 6th term is less than the desired error, we need to add the first 5 terms to get the

sum to the desired accuracy.)

**17.**

The graph gives us an estimate for the sum of the series

$$\sum_{n=1}^{\infty} \frac{(-0.8)^n}{n!} \text{ of } -0.55.$$

$$b_8 = \frac{(0.8)^n}{8!} \approx 0.000\,004, \text{ so}$$

$$\sum_{n=1}^{\infty} \frac{(-0.8)^n}{n!} \approx s_7 = \sum_{n=1}^{7} \frac{(-0.8)^n}{n!}$$

$$\approx -0.8 + 0.32 - 0.085\overline{3} + 0.0170\overline{6} - 0.002\,731 + 0.000\,364 - 0.000\,042 \approx -0.5507$$

Adding $b_8$ to $s_7$ does not change the fourth decimal place of $s_7$, so the sum of the series, correct to four decimal places,

is $-0.5507$.

**19.** $b_7 = \dfrac{7^2}{10^7} = 0.000\,004\,9$, so

$$\sum_{n=1}^{\infty} \frac{(-1)^{n-1} n^2}{10^n} \approx s_6 = \sum_{n=1}^{6} \frac{(-1)^{n-1} n^2}{10^n} = \frac{1}{10} - \frac{4}{100} + \frac{9}{1000} - \frac{16}{10,000} + \frac{25}{100,000} - \frac{36}{1,000,000} = 0.067\,614$$

Adding $b_7$ to $s_6$ does not change the fourth decimal place of $s_6$, so the sum of the series, correct to four decimal places, is

$0.0676$.

**21.** Using the Ratio Test, $\lim\limits_{n\to\infty}\left|\dfrac{a_{n+1}}{a_n}\right| = \lim\limits_{n\to\infty}\left|\dfrac{(-3)^{n+1}/(n+1)^3}{(-3)^n/n^3}\right| = \lim\limits_{n\to\infty}\left|\dfrac{(-3)n^3}{(n+1)^3}\right| = 3\lim\limits_{n\to\infty}\left(\dfrac{n}{n+1}\right)^3 = 3 > 1$,

so the series $\sum\limits_{n=1}^{\infty}\dfrac{(-3)^n}{n^3}$ diverges.

**23.** Using the Ratio Test, $\lim\limits_{n\to\infty}\left|\dfrac{a_{n+1}}{a_n}\right| = \lim\limits_{n\to\infty}\left|\dfrac{(-10)^{n+1}}{(n+1)!}\cdot\dfrac{n!}{(-10)^n}\right| = \lim\limits_{n\to\infty}\left|\dfrac{-10}{n+1}\right| = 0 < 1$, so the series $\sum\limits_{n=0}^{\infty}\dfrac{(-10)^n}{n!}$ is

absolutely convergent.

**25.** $\lim\limits_{k\to\infty}\left|\dfrac{a_{k+1}}{a_k}\right| = \lim\limits_{k\to\infty}\left[\dfrac{(k+1)\left(\frac{2}{3}\right)^{k+1}}{k\left(\frac{2}{3}\right)^k}\right] = \lim\limits_{k\to\infty}\dfrac{k+1}{k}\left(\dfrac{2}{3}\right)^1 = \dfrac{2}{3}\lim\limits_{k\to\infty}\left(1+\dfrac{1}{k}\right) = \dfrac{2}{3}(1) = \dfrac{2}{3} < 1$, so the series

$\sum\limits_{n=1}^{\infty} k\left(\dfrac{2}{3}\right)^k$ is absolutely convergent by the Ratio Test. Since the terms of this series are positive, absolute convergence is the

same as convergence.

**27.** Consider the series whose terms are the absolute values of the terms of the given series. $\sum\limits_{n=1}^{\infty}\left|\dfrac{(-1)^{n-1}}{\sqrt{n}}\right| = \sum\limits_{n=1}^{\infty}\dfrac{1}{n^{1/2}}$, which is

a divergent $p$-series $(p = \frac{1}{2} \leq 1)$. Thus, $\sum\limits_{n=1}^{\infty}\dfrac{(-1)^{n-1}}{\sqrt{n}}$ is *not* absolutely convergent.

**29.** $\lim\limits_{n\to\infty}\left|\dfrac{a_{n+1}}{a_n}\right| = \lim\limits_{n\to\infty}\left[\dfrac{10^{n+1}}{(n+2)4^{2(n+1)+1}}\cdot\dfrac{(n+1)4^{2n+1}}{10^n}\right] = \lim\limits_{n\to\infty}\left[\dfrac{10^{n+1}}{(n+2)4^{2n+3}}\cdot\dfrac{(n+1)4^{2n+1}}{10^n}\right]$

$= \lim\limits_{n\to\infty}\left(\dfrac{10}{4^2}\cdot\dfrac{n+1}{n+2}\right) = \dfrac{5}{8} < 1$,

so the series $\sum\limits_{n=1}^{\infty}\dfrac{10^n}{(n+1)4^{2n+1}}$ is absolutely convergent by the Ratio Test. Since the terms of this series are positive, absolute

convergence is the same as convergence.

**31.** $\left|\dfrac{(-1)^n \arctan n}{n^2}\right| < \dfrac{\pi/2}{n^2}$, so since $\sum\limits_{n=1}^{\infty}\dfrac{\pi/2}{n^2} = \dfrac{\pi}{2}\sum\limits_{n=1}^{\infty}\dfrac{1}{n^2}$ converges $(p = 2 > 1)$, the given series $\sum\limits_{n=1}^{\infty}\dfrac{(-1)^n \arctan n}{n^2}$

converges absolutely by the Comparison Test.

**33.** Use the Ratio Test with the series

$1 - \dfrac{1\cdot 3}{3!} + \dfrac{1\cdot 3\cdot 5}{5!} - \dfrac{1\cdot 3\cdot 5\cdot 7}{7!} + \cdots + (-1)^{n-1}\dfrac{1\cdot 3\cdot 5\cdot\cdots\cdot(2n-1)}{(2n-1)!} + \cdots = \sum\limits_{n=1}^{\infty}(-1)^{n-1}\dfrac{1\cdot 3\cdot 5\cdot\cdots\cdot(2n-1)}{(2n-1)!}$.

$\lim\limits_{n\to\infty}\left|\dfrac{a_{n+1}}{a_n}\right| = \lim\limits_{n\to\infty}\left|\dfrac{(-1)^n\cdot 1\cdot 3\cdot 5\cdot\cdots\cdot(2n-1)[2(n+1)-1]}{[2(n+1)-1]!}\cdot\dfrac{(2n-1)!}{(-1)^{n-1}\cdot 1\cdot 3\cdot 5\cdot\cdots\cdot(2n-1)}\right|$

$= \lim\limits_{n\to\infty}\left|\dfrac{(-1)(2n+1)(2n-1)!}{(2n+1)(2n)(2n-1)!}\right| = \lim\limits_{n\to\infty}\dfrac{1}{2n} = 0 < 1$,

so the given series is absolutely convergent and therefore convergent.

**35.** By the recursive definition, $\lim\limits_{n\to\infty}\left|\dfrac{a_{n+1}}{a_n}\right| = \lim\limits_{n\to\infty}\left|\dfrac{5n+1}{4n+3}\right| = \dfrac{5}{4} > 1$, so the series diverges by the Ratio Test.

**37. (a)** $\lim\limits_{n\to\infty}\left|\dfrac{1/(n+1)^3}{1/n^3}\right| = \lim\limits_{n\to\infty}\dfrac{n^3}{(n+1)^3} = \lim\limits_{n\to\infty}\dfrac{1}{(1+1/n)^3} = 1.$  Inconclusive

**(b)** $\lim\limits_{n\to\infty}\left|\dfrac{(n+1)}{2^{n+1}}\cdot\dfrac{2^n}{n}\right| = \lim\limits_{n\to\infty}\dfrac{n+1}{2n} = \lim\limits_{n\to\infty}\left(\dfrac{1}{2}+\dfrac{1}{2n}\right) = \dfrac{1}{2}.$  Conclusive (convergent)

**(c)** $\lim\limits_{n\to\infty}\left|\dfrac{(-3)^n}{\sqrt{n+1}}\cdot\dfrac{\sqrt{n}}{(-3)^{n-1}}\right| = 3\lim\limits_{n\to\infty}\sqrt{\dfrac{n}{n+1}} = 3\lim\limits_{n\to\infty}\sqrt{\dfrac{1}{1+1/n}} = 3.$  Conclusive (divergent)

**(d)** $\lim\limits_{n\to\infty}\left|\dfrac{\sqrt{n+1}}{1+(n+1)^2}\cdot\dfrac{1+n^2}{\sqrt{n}}\right| = \lim\limits_{n\to\infty}\left[\sqrt{1+\dfrac{1}{n}}\cdot\dfrac{1/n^2+1}{1/n^2+(1+1/n)^2}\right] = 1.$  Inconclusive

**39.** $\lim\limits_{n\to\infty}\sqrt[n]{|a_n|} = \lim\limits_{n\to\infty}\dfrac{n^2+1}{2n^2+1} = \lim\limits_{n\to\infty}\dfrac{1+1/n^2}{2+1/n^2} = \dfrac{1}{2} < 1$, so the series $\sum\limits_{n=1}^{\infty}\left(\dfrac{n^2+1}{2n^2+1}\right)^n$ is absolutely convergent by the

Root Test.

**41. (a)** $\lim\limits_{n\to\infty}\left|\dfrac{a_{n+1}}{a_n}\right| = \lim\limits_{n\to\infty}\left|\dfrac{x^{n+1}}{(n+1)!}\cdot\dfrac{n!}{x^n}\right| = \lim\limits_{n\to\infty}\left|\dfrac{x}{n+1}\right| = |x|\lim\limits_{n\to\infty}\dfrac{1}{n+1} = |x|\cdot 0 = 0 < 1$, so by the Ratio Test the

series $\sum\limits_{n=0}^{\infty}\dfrac{x^n}{n!}$ converges for all $x$.

**(b)** Since the series of part (a) always converges, we must have $\lim\limits_{n\to\infty}\dfrac{x^n}{n!} = 0$ by Theorem 8.2.6.

## 8.5  Power Series

**1.** A power series is a series of the form $\sum_{n=0}^{\infty} c_n x^n = c_0 + c_1 x + c_2 x^2 + c_3 x^3 + \cdots$, where $x$ is a variable and the $c_n$'s are

constants called the coefficients of the series.

More generally, a series of the form $\sum_{n=0}^{\infty} c_n(x-a)^n = c_0 + c_1(x-a) + c_2(x-a)^2 + \cdots$ is called a power series in

$(x-a)$ or a power series centered at $a$ or a power series about $a$, where $a$ is a constant.

**3.** If $a_n = \dfrac{x^n}{\sqrt{n}}$, then $\lim\limits_{n\to\infty}\left|\dfrac{a_{n+1}}{a_n}\right| = \lim\limits_{n\to\infty}\left|\dfrac{x^{n+1}}{\sqrt{n+1}}\cdot\dfrac{\sqrt{n}}{x^n}\right| = \lim\limits_{n\to\infty}\left|\dfrac{x}{\sqrt{n+1}/\sqrt{n}}\right| = \lim\limits_{n\to\infty}\dfrac{|x|}{\sqrt{1+1/n}} = |x|.$

By the Ratio Test, the series $\sum\limits_{n=1}^{\infty}\dfrac{x^n}{\sqrt{n}}$ converges when $|x| < 1$, so the radius of convergence $R = 1$. Now we'll check the

endpoints, that is, $x = \pm 1$. When $x = 1$, the series $\sum\limits_{n=1}^{\infty}\dfrac{1}{\sqrt{n}}$ diverges because it is a $p$-series with $p = \dfrac{1}{2} \le 1$. When $x = -1$,

the series $\sum\limits_{n=1}^{\infty}\dfrac{(-1)^n}{\sqrt{n}}$ converges by the Alternating Series Test. Thus, the interval of convergence is $I = [-1, 1)$.

**5.** If $a_n = \dfrac{(-1)^{n-1}x^n}{n^3}$, then

$$\lim_{n\to\infty}\left|\frac{a_{n+1}}{a_n}\right| = \lim_{n\to\infty}\left|\frac{(-1)^n x^{n+1}}{(n+1)^3}\cdot\frac{n^3}{(-1)^{n-1}x^n}\right| = \lim_{n\to\infty}\left|\frac{(-1)xn^3}{(n+1)^3}\right| = \lim_{n\to\infty}\left[\left(\frac{n}{n+1}\right)^3|x|\right] = 1^3\cdot|x| = |x|. \text{ By the}$$

Ratio Test, the series $\displaystyle\sum_{n=1}^{\infty}\frac{(-1)^{n-1}x^n}{n^3}$ converges when $|x| < 1$, so the radius of convergence $R = 1$. Now we'll check the

endpoints, that is, $x = \pm 1$. When $x = 1$, the series $\displaystyle\sum_{n=1}^{\infty}\frac{(-1)^{n-1}}{n^3}$ converges by the Alternating Series Test. When $x = -1$,

the series $\displaystyle\sum_{n=1}^{\infty}\frac{(-1)^{n-1}(-1)^n}{n^3} = -\sum_{n=1}^{\infty}\frac{1}{n^3}$ converges because it is a constant multiple of a convergent $p$-series $[p = 3 > 1]$.

Thus, the interval of convergence is $I = [-1, 1]$.

**7.** If $a_n = \dfrac{x^n}{n!}$, then $\displaystyle\lim_{n\to\infty}\left|\frac{a_{n+1}}{a_n}\right| = \lim_{n\to\infty}\left|\frac{x^{n+1}}{(n+1)!}\cdot\frac{n!}{x^n}\right| = \lim_{n\to\infty}\left|\frac{x}{n+1}\right| = |x|\lim_{n\to\infty}\frac{1}{n+1} = |x|\cdot 0 = 0 < 1$ for *all* real $x$.

So, by the Ratio Test, $R = \infty$ and $I = (-\infty, \infty)$.

**9.** If $a_n = (-1)^n\dfrac{n^2 x^n}{2^n}$, then

$$\lim_{n\to\infty}\left|\frac{a_{n+1}}{a_n}\right| = \lim_{n\to\infty}\left|\frac{(n+1)^2 x^{n+1}}{2^{n+1}}\cdot\frac{2^n}{n^2 x^n}\right| = \lim_{n\to\infty}\left|\frac{x(n+1)^2}{2n^2}\right| = \lim_{n\to\infty}\left[\frac{|x|}{2}\left(1+\frac{1}{n}\right)^2\right] = \frac{|x|}{2}(1)^2 = \tfrac{1}{2}|x|. \text{ By the}$$

Ratio Test, the series $\displaystyle\sum_{n=1}^{\infty}(-1)^n\frac{n^2 x^n}{2^n}$ converges when $\tfrac{1}{2}|x| < 1 \iff |x| < 2$, so the radius of convergence is $R = 2$.

When $x = \pm 2$, both series $\displaystyle\sum_{n=1}^{\infty}(-1)^n\frac{n^2(\pm 2)^n}{2^n} = \sum_{n=1}^{\infty}(\mp 1)^n n^2$ diverge by the Test for Divergence since

$\displaystyle\lim_{n\to\infty}\left|(\mp 1)^n n^2\right| = \infty$. Thus, the interval of convergence is $I = (-2, 2)$.

**11.** $a_n = \dfrac{(-2)^n x^n}{\sqrt[4]{n}}$, so $\displaystyle\lim_{n\to\infty}\left|\frac{a_{n+1}}{a_n}\right| = \lim_{n\to\infty}\frac{2^{n+1}|x|^{n+1}}{\sqrt[4]{n+1}}\cdot\frac{\sqrt[4]{n}}{2^n|x|^n} = \lim_{n\to\infty}2|x|\sqrt[4]{\frac{n}{n+1}} = 2|x|$, so by the Ratio Test, the

series converges when $2|x| < 1 \iff |x| < \tfrac{1}{2}$, so $R = \tfrac{1}{2}$. When $x = -\tfrac{1}{2}$, we get the divergent $p$-series $\displaystyle\sum_{n=1}^{\infty}\frac{1}{\sqrt[4]{n}}$

$\left[p = \tfrac{1}{4} \le 1\right]$. When $x = \tfrac{1}{2}$, we get the series $\displaystyle\sum_{n=1}^{\infty}\frac{(-1)^n}{\sqrt[4]{n}}$, which converges by the Alternating Series Test.

Thus, $I = \left(-\tfrac{1}{2}, \tfrac{1}{2}\right]$.

**13.** If $a_n = \dfrac{(x-2)^n}{n^2+1}$, then $\displaystyle\lim_{n\to\infty}\left|\frac{a_{n+1}}{a_n}\right| = \lim_{n\to\infty}\left|\frac{(x-2)^{n+1}}{(n+1)^2+1}\cdot\frac{n^2+1}{(x-2)^n}\right| = |x-2|\lim_{n\to\infty}\frac{n^2+1}{(n+1)^2+1} = |x-2|. \text{ By the}$

Ratio Test, the series $\displaystyle\sum_{n=0}^{\infty}\frac{(x-2)^n}{n^2+1}$ converges when $|x-2| < 1$ $[R = 1]$ $\iff -1 < x - 2 < 1 \iff 1 < x < 3$. When

$x = 1$, the series $\displaystyle\sum_{n=0}^{\infty}(-1)^n\frac{1}{n^2+1}$ converges by the Alternating Series Test; when $x = 3$, the series $\displaystyle\sum_{n=0}^{\infty}\frac{1}{n^2+1}$ converges by

comparison with the $p$-series $\displaystyle\sum_{n=1}^{\infty}\frac{1}{n^2}$ $[p = 2 > 1]$. Thus, the interval of convergence is $I = [1, 3]$.

**15.** If $a_n = \dfrac{3^n (x+4)^n}{\sqrt{n}}$, then $\lim\limits_{n \to \infty} \left| \dfrac{a_{n+1}}{a_n} \right| = \lim\limits_{n \to \infty} \left| \dfrac{3^{n+1}(x+4)^{n+1}}{\sqrt{n+1}} \cdot \dfrac{\sqrt{n}}{3^n (x+4)^n} \right| = 3\,|x+4|\, \lim\limits_{n \to \infty} \dfrac{\sqrt{n}}{\sqrt{n+1}} = 3\,|x+4|.$

By the Ratio Test, the series $\sum\limits_{n=1}^{\infty} \dfrac{3^n (x+4)^n}{\sqrt{n}}$ converges when $3\,|x+4| < 1 \;\Leftrightarrow\; |x+4| < \frac{1}{3} \;\; \left[R = \frac{1}{3}\right] \;\Leftrightarrow$

$-\frac{1}{3} < x + 4 < \frac{1}{3} \;\Leftrightarrow\; -\frac{13}{3} < x < -\frac{11}{3}$. When $x = -\frac{13}{3}$, the series $\sum\limits_{n=1}^{\infty} (-1)^n \dfrac{1}{\sqrt{n}}$ converges by the Alternating Series

Test; when $x = -\frac{11}{3}$, the series $\sum\limits_{n=1}^{\infty} \dfrac{1}{\sqrt{n}}$ diverges $\left[p = \frac{1}{2} \le 1\right]$. Thus, the interval of convergence is $I = \left[-\frac{13}{3}, -\frac{11}{3}\right)$.

**17.** $\lim\limits_{n \to \infty} \left| \dfrac{a_{n+1}}{a_n} \right| = \lim\limits_{n \to \infty} \left[ \dfrac{|4x+1|^{n+1}}{(n+1)^2} \cdot \dfrac{n^2}{|4x+1|^n} \right] = \lim\limits_{n \to \infty} \dfrac{|4x+1|}{(1+1/n)^2} = |4x+1|$, so by the Ratio Test, the series

converges when $|4x+1| < 1 \;\Leftrightarrow\; -1 < 4x+1 < 1 \;\Leftrightarrow\; -2 < 4x < 0 \;\Leftrightarrow\; -\frac{1}{2} < x < 0$, so $R = \frac{1}{4}$. When $x = -\frac{1}{2}$,

the series becomes $\sum\limits_{n=1}^{\infty} \dfrac{(-1)^n}{n^2}$, which converges by the Alternating Series Test. When $x = 0$, the series becomes $\sum\limits_{n=1}^{\infty} \dfrac{1}{n^2}$,

a convergent $p$-series $[p = 2 > 1]$. $\;\; I = \left[-\frac{1}{2}, 0\right]$.

**19.** If $a_n = n!\,(2x-1)^n$, then $\lim\limits_{n \to \infty} \left| \dfrac{a_{n+1}}{a_n} \right| = \lim\limits_{n \to \infty} \left| \dfrac{(n+1)!\,(2x-1)^{n+1}}{n!(2x-1)^n} \right| = \lim\limits_{n \to \infty} (n+1)\,|2x-1| \to \infty$ as $n \to \infty$

for all $x \neq \frac{1}{2}$. Since the series diverges for all $x \neq \frac{1}{2}$, $R = 0$ and $I = \left\{\frac{1}{2}\right\}$.

**21.** $a_n = \dfrac{n}{b^n} (x-a)^n$, where $b > 0$.

$\lim\limits_{n \to \infty} \left| \dfrac{a_{n+1}}{a_n} \right| = \lim\limits_{n \to \infty} \dfrac{(n+1)\,|x-a|^{n+1}}{b^{n+1}} \cdot \dfrac{b^n}{n\,|x-a|^n} = \lim\limits_{n \to \infty} \left(1 + \dfrac{1}{n}\right) \dfrac{|x-a|}{b} = \dfrac{|x-a|}{b}.$

By the Ratio Test, the series converges when $\dfrac{|x-a|}{b} < 1 \;\Leftrightarrow\; |x-a| < b \;\; [\text{so } R = b] \;\Leftrightarrow\; -b < x - a < b \;\Leftrightarrow$

$a - b < x < a + b$. When $|x-a| = b$, $\lim\limits_{n \to \infty} |a_n| = \lim\limits_{n \to \infty} n = \infty$, so the series diverges. Thus, $I = (a-b, a+b)$.

**23.** If $a_n = \dfrac{x^n}{1 \cdot 3 \cdot 5 \cdot \,\cdots\, \cdot (2n-1)}$, then

$\lim\limits_{n \to \infty} \left| \dfrac{a_{n+1}}{a_n} \right| = \lim\limits_{n \to \infty} \left| \dfrac{x^{n+1}}{1 \cdot 3 \cdot 5 \cdot \,\cdots\, \cdot (2n-1)(2n+1)} \cdot \dfrac{1 \cdot 3 \cdot 5 \cdot \,\cdots\, \cdot (2n-1)}{x^n} \right| = \lim\limits_{n \to \infty} \dfrac{|x|}{2n+1} = 0 < 1$. Thus, by

the Ratio Test, the series $\sum\limits_{n=1}^{\infty} \dfrac{x^n}{1 \cdot 3 \cdot 5 \cdot \,\cdots\, \cdot (2n-1)}$ converges for *all* real $x$ and we have $R = \infty$ and $I = (-\infty, \infty)$.

**25.** (a) We are given that the power series $\sum_{n=0}^{\infty} c_n x^n$ is convergent for $x = 4$. So by Theorem 3, it must converge for at least

$\quad\;\; -4 < x \le 4$. In particular, it converges when $x = -2$; that is, $\sum_{n=0}^{\infty} c_n (-2)^n$ is convergent.

(b) It does not follow that $\sum_{n=0}^{\infty} c_n (-4)^n$ is necessarily convergent. [See the comments after Theorem 3 about convergence at

$\quad\;\;$ the endpoint of an interval. An example is $c_n = (-1)^n / (n 4^n)$.]

**27.** If $a_n = \dfrac{(n!)^k}{(kn)!} x^n$, then

$$\lim_{n \to \infty} \left| \frac{a_{n+1}}{a_n} \right| = \lim_{n \to \infty} \frac{[(n+1)!]^k (kn)!}{(n!)^k [k(n+1)]!} |x| = \lim_{n \to \infty} \frac{(n+1)^k}{(kn+k)(kn+k-1) \cdots (kn+2)(kn+1)} |x|$$

$$= \lim_{n \to \infty} \left[ \frac{(n+1)}{(kn+1)} \frac{(n+1)}{(kn+2)} \cdots \frac{(n+1)}{(kn+k)} \right] |x|$$

$$= \lim_{n \to \infty} \left[ \frac{n+1}{kn+1} \right] \lim_{n \to \infty} \left[ \frac{n+1}{kn+2} \right] \cdots \lim_{n \to \infty} \left[ \frac{n+1}{kn+k} \right] |x|$$

$$= \left( \frac{1}{k} \right)^k |x| < 1 \quad \Leftrightarrow \quad |x| < k^k \text{ for convergence, and the radius of convergence is } R = k^k.$$

**29. (a)** If $a_n = \dfrac{(-1)^n x^{2n+1}}{n!(n+1)! \, 2^{2n+1}}$, then

$$\lim_{n \to \infty} \left| \frac{a_{n+1}}{a_n} \right| = \lim_{n \to \infty} \left| \frac{x^{2n+3}}{(n+1)!(n+2)! \, 2^{2n+3}} \cdot \frac{n!(n+1)! \, 2^{2n+1}}{x^{2n+1}} \right| = \left( \frac{x}{2} \right)^2 \lim_{n \to \infty} \frac{1}{(n+1)(n+2)} = 0 \text{ for all } x.$$

So $J_1(x)$ converges for all $x$ and its domain is $(-\infty, \infty)$.

**(b), (c)** The initial terms of $J_1(x)$ up to $n = 5$ are $a_0 = \dfrac{x}{2}$,

$a_1 = -\dfrac{x^3}{16}, a_2 = \dfrac{x^5}{384}, a_3 = -\dfrac{x^7}{18,432}, a_4 = \dfrac{x^9}{1,474,560}$,

and $a_5 = -\dfrac{x^{11}}{176,947,200}$. The partial sums seem to

approximate $J_1(x)$ well near the origin, but as $|x|$ increases,

we need to take a large number of terms to get a good

approximation.

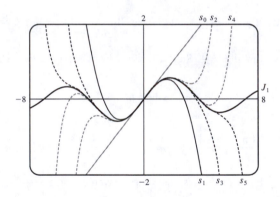

**31.** $s_{2n-1} = 1 + 2x + x^2 + 2x^3 + x^4 + 2x^5 + \cdots + x^{2n-2} + 2x^{2n-1}$

$$= 1(1 + 2x) + x^2(1 + 2x) + x^4(1 + 2x) + \cdots + x^{2n-2}(1 + 2x) = (1 + 2x)(1 + x^2 + x^4 + \cdots + x^{2n-2})$$

$$= (1 + 2x)\frac{1 - x^{2n}}{1 - x^2} \text{ [by (8.2.3) with } r = x^2] \quad \to \frac{1 + 2x}{1 - x^2} \text{ as } n \to \infty \text{ by (8.2.4)], when } |x| < 1.$$

Also $s_{2n} = s_{2n-1} + x^{2n} \to \dfrac{1 + 2x}{1 - x^2}$ since $x^{2n} \to 0$ for $|x| < 1$. Therefore, $s_n \to \dfrac{1 + 2x}{1 - x^2}$ since $s_{2n}$ and $s_{2n-1}$ both

approach $\dfrac{1 + 2x}{1 - x^2}$ as $n \to \infty$. Thus, the interval of convergence is $(-1, 1)$ and $f(x) = \dfrac{1 + 2x}{1 - x^2}$.

**33.** For $2 < x < 3$, $\sum c_n x^n$ diverges and $\sum d_n x^n$ converges. By Exercise 8.2.61, $\sum (c_n + d_n) x^n$ diverges. Since both series

converge for $|x| < 2$, the radius of convergence of $\sum (c_n + d_n) x^n$ is 2.

**35.** No. If a power series is centered at $a$, its interval of convergence is symmetric about $a$. If a power series has an infinite radius

of convergence, then its interval of convergence must be $(-\infty, \infty)$, not $[0, \infty)$.

## 8.6   Representations of Functions as Power Series

**1.** If $f(x) = \sum_{n=0}^{\infty} c_n x^n$ has radius of convergence 10, then $f'(x) = \sum_{n=1}^{\infty} n c_n x^{n-1}$ also has radius of convergence 10 by Theorem 2.

**3.** Our goal is to write the function in the form $\dfrac{1}{1-r}$, and then use Equation (1) to represent the function as a sum of a power

series. $f(x) = \dfrac{1}{1+x} = \dfrac{1}{1-(-x)} = \sum_{n=0}^{\infty} (-x)^n = \sum_{n=0}^{\infty} (-1)^n x^n$ with $|-x| < 1 \iff |x| < 1$, so $R = 1$ and $I = (-1, 1)$.

**5.** $f(x) = \dfrac{2}{3-x} = \dfrac{2}{3}\left(\dfrac{1}{1-x/3}\right) = \dfrac{2}{3} \sum_{n=0}^{\infty} \left(\dfrac{x}{3}\right)^n$ or, equivalently, $2 \sum_{n=0}^{\infty} \dfrac{1}{3^{n+1}} x^n$. The series converges when $\left|\dfrac{x}{3}\right| < 1$,

that is, when $|x| < 3$, so $R = 3$ and $I = (-3, 3)$.

**7.** $f(x) = \dfrac{x}{9+x^2} = \dfrac{x}{9}\left[\dfrac{1}{1+(x/3)^2}\right] = \dfrac{x}{9}\left[\dfrac{1}{1-\{-(x/3)^2\}}\right] = \dfrac{x}{9} \sum_{n=0}^{\infty} \left[-\left(\dfrac{x}{3}\right)^2\right]^n = \dfrac{x}{9} \sum_{n=0}^{\infty} (-1)^n \dfrac{x^{2n}}{9^n} = \sum_{n=0}^{\infty} (-1)^n \dfrac{x^{2n+1}}{9^{n+1}}$

The geometric series $\sum_{n=0}^{\infty} \left[-\left(\dfrac{x}{3}\right)^2\right]^n$ converges when $\left|-\left(\dfrac{x}{3}\right)^2\right| < 1 \iff \dfrac{|x^2|}{9} < 1 \iff |x|^2 < 9 \iff |x| < 3$, so

$R = 3$ and $I = (-3, 3)$.

**9.** $f(x) = \dfrac{1+x}{1-x} = (1+x)\left(\dfrac{1}{1-x}\right) = (1+x)\sum_{n=0}^{\infty} x^n = \sum_{n=0}^{\infty} x^n + \sum_{n=0}^{\infty} x^{n+1} = 1 + \sum_{n=1}^{\infty} x^n + \sum_{n=1}^{\infty} x^n = 1 + 2\sum_{n=1}^{\infty} x^n$.

The series converges when $|x| < 1$, so $R = 1$ and $I = (-1, 1)$.

*A second approach:* $f(x) = \dfrac{1+x}{1-x} = \dfrac{-(1-x)+2}{1-x} = -1 + 2\left(\dfrac{1}{1-x}\right) = -1 + 2\sum_{n=0}^{\infty} x^n = 1 + 2\sum_{n=1}^{\infty} x^n$.

*A third approach:*

$f(x) = \dfrac{1+x}{1-x} = (1+x)\left(\dfrac{1}{1-x}\right) = (1+x)(1 + x + x^2 + x^3 + \cdots)$

$= (1 + x + x^2 + x^3 + \cdots) + (x + x^2 + x^3 + x^4 + \cdots) = 1 + 2x + 2x^2 + 2x^3 + \cdots = 1 + 2\sum_{n=1}^{\infty} x^n$.

**11.** (a) $f(x) = \dfrac{1}{(1+x)^2} = \dfrac{d}{dx}\left(\dfrac{-1}{1+x}\right) = -\dfrac{d}{dx}\left[\sum_{n=0}^{\infty} (-1)^n x^n\right]$   [from Exercise 3]

$= \sum_{n=1}^{\infty} (-1)^{n+1} n x^{n-1}$   [from Theorem 2(i)] $= \sum_{n=0}^{\infty} (-1)^n (n+1) x^n$ with $R = 1$.

In the last step, note that we *decreased* the initial value of the summation variable $n$ by 1, and then *increased* each occurrence of $n$ in the term by 1 [also note that $(-1)^{n+2} = (-1)^n$].

(b) $f(x) = \dfrac{1}{(1+x)^3} = -\dfrac{1}{2}\dfrac{d}{dx}\left[\dfrac{1}{(1+x)^2}\right] = -\dfrac{1}{2}\dfrac{d}{dx}\left[\sum_{n=0}^{\infty} (-1)^n (n+1) x^n\right]$   [from part (a)]

$= -\dfrac{1}{2}\sum_{n=1}^{\infty} (-1)^n (n+1) n x^{n-1} = \dfrac{1}{2}\sum_{n=0}^{\infty} (-1)^n (n+2)(n+1) x^n$ with $R = 1$.

(c) $f(x) = \dfrac{x^2}{(1+x)^3} = x^2 \cdot \dfrac{1}{(1+x)^3} = x^2 \cdot \dfrac{1}{2}\sum_{n=0}^{\infty} (-1)^n (n+2)(n+1) x^n$   [from part (b)]

$= \dfrac{1}{2}\sum_{n=0}^{\infty} (-1)^n (n+2)(n+1) x^{n+2}$

To write the power series with $x^n$ rather than $x^{n+2}$, we will *decrease* each occurrence of $n$ in the term by 2 and *increase*

the initial value of the summation variable by 2. This gives us $\dfrac{1}{2}\sum_{n=2}^{\infty} (-1)^n (n)(n-1) x^n$ with $R = 1$.

**13.** $f(x) = \ln(5-x) = -\int \dfrac{dx}{5-x} = -\dfrac{1}{5}\int \dfrac{dx}{1-x/5} = -\dfrac{1}{5}\int \left[\sum_{n=0}^{\infty} \left(\dfrac{x}{5}\right)^n\right] dx = C - \dfrac{1}{5}\sum_{n=0}^{\infty} \dfrac{x^{n+1}}{5^n(n+1)} = C - \sum_{n=1}^{\infty} \dfrac{x^n}{n\,5^n}$

Putting $x = 0$, we get $C = \ln 5$. The series converges for $|x/5| < 1 \;\Leftrightarrow\; |x| < 5$, so $R = 5$.

**15.** We know that $\dfrac{1}{1+4x} = \dfrac{1}{1-(-4x)} = \sum_{n=0}^{\infty}(-4x)^n$. Differentiating, we get

$\dfrac{-4}{(1+4x)^2} = \sum_{n=1}^{\infty}(-4)^n n x^{n-1} = \sum_{n=0}^{\infty}(-4)^{n+1}(n+1)x^n$, so

$f(x) = \dfrac{x}{(1+4x)^2} = \dfrac{-x}{4}\cdot\dfrac{-4}{(1+4x)^2} = \dfrac{-x}{4}\sum_{n=0}^{\infty}(-4)^{n+1}(n+1)x^n = \sum_{n=0}^{\infty}(-1)^n 4^n(n+1)x^{n+1}$

for $|-4x| < 1 \;\Leftrightarrow\; |x| < \frac{1}{4}$, so $R = \frac{1}{4}$.

**17.** By Example 5, $\dfrac{1}{(1-x)^2} = \sum_{n=0}^{\infty}(n+1)x^n$. Thus,

$$f(x) = \dfrac{1+x}{(1-x)^2} = \dfrac{1}{(1-x)^2} + \dfrac{x}{(1-x)^2} = \sum_{n=0}^{\infty}(n+1)x^n + \sum_{n=0}^{\infty}(n+1)x^{n+1}$$

$$= \sum_{n=0}^{\infty}(n+1)x^n + \sum_{n=1}^{\infty} n x^n \qquad \text{[make the starting values equal]}$$

$$= 1 + \sum_{n=1}^{\infty}[(n+1)+n]x^n = 1 + \sum_{n=1}^{\infty}(2n+1)x^n = \sum_{n=0}^{\infty}(2n+1)x^n \;\text{ with } R = 1.$$

**19.** $f(x) = \dfrac{x}{x^2+16} = \dfrac{x}{16}\left(\dfrac{1}{1-(-x^2/16)}\right) = \dfrac{x}{16}\sum_{n=0}^{\infty}\left(-\dfrac{x^2}{16}\right)^n = \dfrac{x}{16}\sum_{n=0}^{\infty}(-1)^n \dfrac{1}{16^n}x^{2n} = \sum_{n=0}^{\infty}(-1)^n \dfrac{1}{16^{n+1}}x^{2n+1}$.

The series converges when $\left|-x^2/16\right| < 1 \;\Leftrightarrow\; x^2 < 16 \;\Leftrightarrow\; |x| < 4$, so $R = 4$. The partial sums are $s_1 = \dfrac{x}{16}$,

$s_2 = s_1 - \dfrac{x^3}{16^2}$, $s_3 = s_2 + \dfrac{x^5}{16^3}$, $s_4 = s_3 - \dfrac{x^7}{16^4}$, $s_5 = s_4 + \dfrac{x^9}{16^5}$, .... Note that $s_1$ corresponds to the first term of the infinite

sum, regardless of the value of the summation variable and the value of the exponent.

As $n$ increases, $s_n(x)$ approximates $f$ better on the interval of convergence, which is $(-4, 4)$.

**21.** $f(x) = \ln\left(\dfrac{1+x}{1-x}\right) = \ln(1+x) - \ln(1-x) = \int\dfrac{dx}{1+x} + \int\dfrac{dx}{1-x} = \int\dfrac{dx}{1-(-x)} + \int\dfrac{dx}{1-x}$

$= \int\left[\sum_{n=0}^{\infty}(-1)^n x^n + \sum_{n=0}^{\infty} x^n\right] dx = \int\left[(1-x+x^2-x^3+x^4-\cdots) + (1+x+x^2+x^3+x^4+\cdots)\right] dx$

$= \int(2 + 2x^2 + 2x^4 + \cdots)\,dx = \int\sum_{n=0}^{\infty}2x^{2n}\,dx = C + \sum_{n=0}^{\infty}\dfrac{2x^{2n+1}}{2n+1}$

But $f(0) = \ln \frac{1}{1} = 0$, so $C = 0$ and we have $f(x) = \sum\limits_{n=0}^{\infty} \frac{2x^{2n+1}}{2n+1}$ with $R = 1$. If $x = \pm 1$, then $f(x) = \pm 2 \sum\limits_{n=0}^{\infty} \frac{1}{2n+1}$,

which both diverge by the Limit Comparison Test with $b_n = \frac{1}{n}$. The partial sums are $s_1 = \frac{2x}{1}$, $s_2 = s_1 + \frac{2x^3}{3}$,

$s_3 = s_2 + \frac{2x^5}{5}, \dots$.

As $n$ increases, $s_n(x)$ approximates $f$ better on the interval of convergence, which is $(-1, 1)$.

**23.** $\dfrac{t}{1 - t^8} = t \cdot \dfrac{1}{1 - t^8} = t \sum\limits_{n=0}^{\infty} (t^8)^n = \sum\limits_{n=0}^{\infty} t^{8n+1} \quad \Rightarrow \quad \displaystyle\int \dfrac{t}{1 - t^8}\, dt = C + \sum\limits_{n=0}^{\infty} \dfrac{t^{8n+2}}{8n+2}$. The series for $\dfrac{1}{1 - t^8}$ converges

when $\left| t^8 \right| < 1 \quad \Leftrightarrow \quad |t| < 1$, so $R = 1$ for that series and also the series for $t/(1 - t^8)$. By Theorem 2, the series for

$\displaystyle\int \dfrac{t}{1 - t^8}\, dt$ also has $R = 1$.

**25.** By Example 7, $\tan^{-1} x = \sum\limits_{n=0}^{\infty} (-1)^n \dfrac{x^{2n+1}}{2n+1}$ with $R = 1$, so

$x - \tan^{-1} x = x - \left( x - \dfrac{x^3}{3} + \dfrac{x^5}{5} - \dfrac{x^7}{7} + \cdots \right) = \dfrac{x^3}{3} - \dfrac{x^5}{5} + \dfrac{x^7}{7} - \cdots = \sum\limits_{n=1}^{\infty} (-1)^{n+1} \dfrac{x^{2n+1}}{2n+1}$ and

$\dfrac{x - \tan^{-1} x}{x^3} = \sum\limits_{n=1}^{\infty} (-1)^{n+1} \dfrac{x^{2n-2}}{2n+1}$, so

$\displaystyle\int \dfrac{x - \tan^{-1} x}{x^3}\, dx = C + \sum\limits_{n=1}^{\infty} (-1)^{n+1} \dfrac{x^{2n-1}}{(2n+1)(2n-1)} = C + \sum\limits_{n=1}^{\infty} (-1)^{n+1} \dfrac{x^{2n-1}}{4n^2 - 1}$. By Theorem 2, $R = 1$.

**27.** $\dfrac{1}{1 + x^5} = \dfrac{1}{1 - (-x^5)} = \sum\limits_{n=0}^{\infty} (-x^5)^n = \sum\limits_{n=0}^{\infty} (-1)^n x^{5n} \quad \Rightarrow$

$\displaystyle\int \dfrac{1}{1 + x^5}\, dx = \int \sum\limits_{n=0}^{\infty} (-1)^n x^{5n}\, dx = C + \sum\limits_{n=0}^{\infty} (-1)^n \dfrac{x^{5n+1}}{5n+1}$. Thus,

$I = \displaystyle\int_0^{0.2} \dfrac{1}{1 + x^5}\, dx = \left[ x - \dfrac{x^6}{6} + \dfrac{x^{11}}{11} - \cdots \right]_0^{0.2} = 0.2 - \dfrac{(0.2)^6}{6} + \dfrac{(0.2)^{11}}{11} - \cdots$. The series is alternating, so if we use

the first two terms, the error is at most $(0.2)^{11}/11 \approx 1.9 \times 10^{-9}$. So $I \approx 0.2 - (0.2)^6/6 \approx 0.199\,989$ to six decimal places.

**29.** We substitute $3x$ for $x$ in Example 7, and find that

$$\int x \arctan(3x)\,dx = \int x \sum_{n=0}^{\infty}(-1)^n \frac{(3x)^{2n+1}}{2n+1}\,dx = \int \sum_{n=0}^{\infty}(-1)^n \frac{3^{2n+1}\,x^{2n+2}}{2n+1}\,dx = C + \sum_{n=0}^{\infty}(-1)^n \frac{3^{2n+1}\,x^{2n+3}}{(2n+1)(2n+3)}$$

So

$$\int_0^{0.1} x \arctan(3x)\,dx = \left[\frac{3x^3}{1\cdot 3} - \frac{3^3 x^5}{3\cdot 5} + \frac{3^5 x^7}{5\cdot 7} - \frac{3^7 x^9}{7\cdot 9} + \cdots\right]_0^{0.1}$$

$$= \frac{1}{10^3} - \frac{9}{5\times 10^5} + \frac{243}{35\times 10^7} - \frac{2187}{63\times 10^9} + \cdots.$$

The series is alternating, so if we use three terms, the error is at most $\dfrac{2187}{63\times 10^9} \approx 3.5\times 10^{-8}$. So

$$\int_0^{0.1} x \arctan(3x)\,dx \approx \frac{1}{10^3} - \frac{9}{5\times 10^5} + \frac{243}{35\times 10^7} \approx 0.000\,983 \text{ to six decimal places.}$$

**31.** By Example 7, $\arctan x = x - \dfrac{x^3}{3} + \dfrac{x^5}{5} - \dfrac{x^7}{7} + \cdots$, so $\arctan 0.2 = 0.2 - \dfrac{(0.2)^3}{3} + \dfrac{(0.2)^5}{5} - \dfrac{(0.2)^7}{7} + \cdots$.

The series is alternating, so if we use three terms, the error is at most $\dfrac{(0.2)^7}{7} \approx 0.000\,002$.

Thus, to five decimal places, $\arctan 0.2 \approx 0.2 - \dfrac{(0.2)^3}{3} + \dfrac{(0.2)^5}{5} \approx 0.197\,40$.

**33.** (a) $J_0(x) = \displaystyle\sum_{n=0}^{\infty}\frac{(-1)^n\,x^{2n}}{2^{2n}(n!)^2}$, $J_0'(x) = \displaystyle\sum_{n=1}^{\infty}\frac{(-1)^n\,2n x^{2n-1}}{2^{2n}(n!)^2}$, and $J_0''(x) = \displaystyle\sum_{n=1}^{\infty}\frac{(-1)^n\,2n(2n-1)x^{2n-2}}{2^{2n}(n!)^2}$, so

$$x^2 J_0''(x) + x J_0'(x) + x^2 J_0(x) = \sum_{n=1}^{\infty}\frac{(-1)^n\,2n(2n-1)x^{2n}}{2^{2n}(n!)^2} + \sum_{n=1}^{\infty}\frac{(-1)^n\,2n x^{2n}}{2^{2n}(n!)^2} + \sum_{n=0}^{\infty}\frac{(-1)^n\,x^{2n+2}}{2^{2n}(n!)^2}$$

$$= \sum_{n=1}^{\infty}\frac{(-1)^n\,2n(2n-1)x^{2n}}{2^{2n}(n!)^2} + \sum_{n=1}^{\infty}\frac{(-1)^n\,2n x^{2n}}{2^{2n}(n!)^2} + \sum_{n=1}^{\infty}\frac{(-1)^{n-1}\,x^{2n}}{2^{2n-2}\,[(n-1)!]^2}$$

$$= \sum_{n=1}^{\infty}\frac{(-1)^n\,2n(2n-1)x^{2n}}{2^{2n}(n!)^2} + \sum_{n=1}^{\infty}\frac{(-1)^n\,2n x^{2n}}{2^{2n}(n!)^2} + \sum_{n=1}^{\infty}\frac{(-1)^n(-1)^{-1}2^2 n^2 x^{2n}}{2^{2n}(n!)^2}$$

$$= \sum_{n=1}^{\infty}(-1)^n\left[\frac{2n(2n-1)+2n-2^2 n^2}{2^{2n}(n!)^2}\right]x^{2n}$$

$$= \sum_{n=1}^{\infty}(-1)^n\left[\frac{4n^2-2n+2n-4n^2}{2^{2n}(n!)^2}\right]x^{2n} = 0$$

(b) $\displaystyle\int_0^1 J_0(x)\,dx = \int_0^1\left[\sum_{n=0}^{\infty}\frac{(-1)^n\,x^{2n}}{2^{2n}(n!)^2}\right]dx = \int_0^1\left(1 - \frac{x^2}{4} + \frac{x^4}{64} - \frac{x^6}{2304} + \cdots\right)dx$

$$= \left[x - \frac{x^3}{3\cdot 4} + \frac{x^5}{5\cdot 64} - \frac{x^7}{7\cdot 2304} + \cdots\right]_0^1 = 1 - \frac{1}{12} + \frac{1}{320} - \frac{1}{16{,}128} + \cdots$$

Since $\frac{1}{16{,}128} \approx 0.000062$, it follows from The Alternating Series Estimation Theorem that, correct to three decimal places,

$\int_0^1 J_0(x)\,dx \approx 1 - \frac{1}{12} + \frac{1}{320} \approx 0.920$.

**35.** (a) $f(x) = \sum\limits_{n=0}^{\infty} \dfrac{x^n}{n!} \Rightarrow f'(x) = \sum\limits_{n=1}^{\infty} \dfrac{nx^{n-1}}{n!} = \sum\limits_{n=1}^{\infty} \dfrac{x^{n-1}}{(n-1)!} = \sum\limits_{n=0}^{\infty} \dfrac{x^n}{n!} = f(x)$

    (b) By Theorem 7.4.2, the only solution to the differential equation $df(x)/dx = f(x)$ is $f(x) = Ke^x$, but $f(0) = 1$, so

    $K = 1$ and $f(x) = e^x$.

    *Or:* We could solve the equation $df(x)/dx = f(x)$ as a separable differential equation.

**37.** If $a_n = \dfrac{x^n}{n^2}$, then by the Ratio Test, $\lim\limits_{n\to\infty} \left| \dfrac{a_{n+1}}{a_n} \right| = \lim\limits_{n\to\infty} \left| \dfrac{x^{n+1}}{(n+1)^2} \cdot \dfrac{n^2}{x^n} \right| = |x| \lim\limits_{n\to\infty} \left( \dfrac{n}{n+1} \right)^2 = |x| < 1$ for

convergence, so $R = 1$. When $x = \pm 1$, $\sum\limits_{n=1}^{\infty} \left| \dfrac{x^n}{n^2} \right| = \sum\limits_{n=1}^{\infty} \dfrac{1}{n^2}$ which is a convergent $p$-series ($p = 2 > 1$), so the interval of

convergence for $f$ is $[-1, 1]$. By Theorem 2, the radii of convergence of $f'$ and $f''$ are both 1, so we need only check the

endpoints. $f(x) = \sum\limits_{n=1}^{\infty} \dfrac{x^n}{n^2} \Rightarrow f'(x) = \sum\limits_{n=1}^{\infty} \dfrac{nx^{n-1}}{n^2} = \sum\limits_{n=0}^{\infty} \dfrac{x^n}{n+1}$, and this series diverges for $x = 1$ (harmonic series)

and converges for $x = -1$ (Alternating Series Test), so the interval of convergence is $[-1, 1)$.   $f''(x) = \sum\limits_{n=1}^{\infty} \dfrac{nx^{n-1}}{n+1}$ diverges

at both 1 and $-1$ (Test for Divergence) since $\lim\limits_{n\to\infty} \dfrac{n}{n+1} = 1 \neq 0$, so its interval of convergence is $(-1, 1)$.

**39.** By Example 7, $\tan^{-1} x = \sum\limits_{n=0}^{\infty} (-1)^n \dfrac{x^{2n+1}}{2n+1}$ for $|x| < 1$. In particular, for $x = \dfrac{1}{\sqrt{3}}$, we

    have $\dfrac{\pi}{6} = \tan^{-1}\left( \dfrac{1}{\sqrt{3}} \right) = \sum\limits_{n=0}^{\infty} (-1)^n \dfrac{(1/\sqrt{3})^{2n+1}}{2n+1} = \sum\limits_{n=0}^{\infty} (-1)^n \left( \dfrac{1}{3} \right)^n \dfrac{1}{\sqrt{3}} \dfrac{1}{2n+1}$, so

    $\pi = \dfrac{6}{\sqrt{3}} \sum\limits_{n=0}^{\infty} \dfrac{(-1)^n}{(2n+1)3^n} = 2\sqrt{3} \sum\limits_{n=0}^{\infty} \dfrac{(-1)^n}{(2n+1)3^n}$.

## 8.7   Taylor and Maclaurin Series

**1.** Using Theorem 5 with $\sum\limits_{n=0}^{\infty} b_n(x-5)^n$, $b_n = \dfrac{f^{(n)}(a)}{n!}$, so $b_8 = \dfrac{f^{(8)}(5)}{8!}$.

**3.** Since $f^{(n)}(0) = (n+1)!$, Equation 7 gives the Maclaurin series

    $\sum\limits_{n=0}^{\infty} \dfrac{f^{(n)}(0)}{n!} x^n = \sum\limits_{n=0}^{\infty} \dfrac{(n+1)!}{n!} x^n = \sum\limits_{n=0}^{\infty} (n+1)x^n$. Applying the Ratio Test with $a_n = (n+1)x^n$ gives us

    $\lim\limits_{n\to\infty} \left| \dfrac{a_{n+1}}{a_n} \right| = \lim\limits_{n\to\infty} \left| \dfrac{(n+2)x^{n+1}}{(n+1)x^n} \right| = |x| \lim\limits_{n\to\infty} \dfrac{n+2}{n+1} = |x| \cdot 1 = |x|$. For convergence, we must have $|x| < 1$, so the

    radius of convergence $R = 1$.

**5.**

| $n$ | $f^{(n)}(x)$ | $f^{(n)}(0)$ |
|---|---|---|
| 0 | $(1-x)^{-2}$ | 1 |
| 1 | $2(1-x)^{-3}$ | 2 |
| 2 | $6(1-x)^{-4}$ | 6 |
| 3 | $24(1-x)^{-5}$ | 24 |
| 4 | $120(1-x)^{-6}$ | 120 |
| ⋮ | ⋮ | ⋮ |

$$(1-x)^{-2} = f(0) + f'(0)x + \frac{f''(0)}{2!}x^2 + \frac{f'''(0)}{3!}x^3 + \frac{f^{(4)}(0)}{4!}x^4 + \cdots$$

$$= 1 + 2x + \frac{6}{2}x^2 + \frac{24}{6}x^3 + \frac{120}{24}x^4 + \cdots$$

$$= 1 + 2x + 3x^2 + 4x^3 + 5x^4 + \cdots = \sum_{n=0}^{\infty} (n+1)x^n$$

$$\lim_{n\to\infty} \left| \frac{a_{n+1}}{a_n} \right| = \lim_{n\to\infty} \left| \frac{(n+2)x^{n+1}}{(n+1)x^n} \right| = |x| \lim_{n\to\infty} \frac{n+2}{n+1} = |x|\,(1) = |x| < 1$$

for convergence, so $R = 1$.

**7.**

| $n$ | $f^{(n)}(x)$ | $f^{(n)}(0)$ |
|---|---|---|
| 0 | $\sin \pi x$ | 0 |
| 1 | $\pi \cos \pi x$ | $\pi$ |
| 2 | $-\pi^2 \sin \pi x$ | 0 |
| 3 | $-\pi^3 \cos \pi x$ | $-\pi^3$ |
| 4 | $\pi^4 \sin \pi x$ | 0 |
| 5 | $\pi^5 \cos \pi x$ | $\pi^5$ |
| ⋮ | ⋮ | ⋮ |

$$\sin \pi x = f(0) + f'(0)x + \frac{f''(0)}{2!}x^2 + \frac{f'''(0)}{3!}x^3$$

$$+ \frac{f^{(4)}(0)}{4!}x^4 + \frac{f^{(5)}(0)}{5!}x^5 + \cdots$$

$$= 0 + \pi x + 0 - \frac{\pi^3}{3!}x^3 + 0 + \frac{\pi^5}{5!}x^5 + \cdots$$

$$= \pi x - \frac{\pi^3}{3!}x^3 + \frac{\pi^5}{5!}x^5 - \frac{\pi^7}{7!}x^7 + \cdots$$

$$= \sum_{n=0}^{\infty} (-1)^n \frac{\pi^{2n+1}}{(2n+1)!} x^{2n+1}$$

$$\lim_{n\to\infty} \left| \frac{a_{n+1}}{a_n} \right| = \lim_{n\to\infty} \left| \frac{\pi^{2n+3} x^{2n+3}}{(2n+3)!} \cdot \frac{(2n+1)!}{\pi^{2n+1} x^{2n+1}} \right| = \lim_{n\to\infty} \frac{\pi^2 x^2}{(2n+3)(2n+2)}$$

$$= 0 < 1 \quad \text{for all } x, \text{ so } R = \infty.$$

**9.**

| $n$ | $f^{(n)}(x)$ | $f^{(n)}(0)$ |
|---|---|---|
| 0 | $e^{5x}$ | 1 |
| 1 | $5e^{5x}$ | 5 |
| 2 | $5^2 e^{5x}$ | 25 |
| 3 | $5^3 e^{5x}$ | 125 |
| 4 | $5^4 e^{5x}$ | 625 |
| ⋮ | ⋮ | ⋮ |

$$e^{5x} = \sum_{n=0}^{\infty} \frac{f^{(n)}(0)}{n!} x^n = \sum_{n=0}^{\infty} \frac{5^n}{n!} x^n.$$

$$\lim_{n\to\infty} \left| \frac{a_{n+1}}{a_n} \right| = \lim_{n\to\infty} \left[ \frac{5^{n+1}|x|^{n+1}}{(n+1)!} \cdot \frac{n!}{5^n |x|^n} \right]$$

$$= \lim_{n\to\infty} \frac{5|x|}{n+1} = 0 < 1 \quad \text{for all } x, \text{ so } R = \infty.$$

**11.**

| $n$ | $f^{(n)}(x)$ | $f^{(n)}(1)$ |
|---|---|---|
| 0 | $x^4 - 3x^2 + 1$ | $-1$ |
| 1 | $4x^3 - 6x$ | $-2$ |
| 2 | $12x^2 - 6$ | 6 |
| 3 | $24x$ | 24 |
| 4 | 24 | 24 |
| 5 | 0 | 0 |
| 6 | 0 | 0 |
| ⋮ | ⋮ | ⋮ |

$f^{(n)}(x) = 0$ for $n \geq 5$, so $f$ has a finite series expansion about $a = 1$.

$$f(x) = x^4 - 3x^2 + 1 = \sum_{n=0}^{4} \frac{f^{(n)}(1)}{n!} (x-1)^n$$

$$= \frac{-1}{0!} (x-1)^0 + \frac{-2}{1!} (x-1)^1 + \frac{6}{2!} (x-1)^2 + \frac{24}{3!} (x-1)^3 + \frac{24}{4!} (x-1)^4$$

$$= -1 - 2(x-1) + 3(x-1)^2 + 4(x-1)^3 + (x-1)^4$$

A finite series converges for all $x$, so $R = \infty$.

**13.** $f(x) = e^x \implies f^{(n)}(x) = e^x$, so $f^{(n)}(3) = e^3$ and $e^x = \sum\limits_{n=0}^{\infty} \dfrac{e^3}{n!}(x-3)^n$. If $a_n = \dfrac{e^3}{n!}(x-3)^n$, then

$$\lim_{n\to\infty}\left|\frac{a_{n+1}}{a_n}\right| = \lim_{n\to\infty}\left|\frac{e^3(x-3)^{n+1}}{(n+1)!}\cdot\frac{n!}{e^3(x-3)^n}\right| = \lim_{n\to\infty}\frac{|x-3|}{n+1} = 0 < 1 \text{ for all } x, \text{ so } R=\infty.$$

**15.**

| $n$ | $f^{(n)}(x)$ | $f^{(n)}(\pi)$ |
|---|---|---|
| 0 | $\cos x$ | $-1$ |
| 1 | $-\sin x$ | $0$ |
| 2 | $-\cos x$ | $1$ |
| 3 | $\sin x$ | $0$ |
| 4 | $\cos x$ | $-1$ |
| ⋮ | ⋮ | ⋮ |

$$\cos x = \sum_{k=0}^{\infty}\frac{f^{(k)}(\pi)}{k!}(x-\pi)^k = -1 + \frac{(x-\pi)^2}{2!} - \frac{(x-\pi)^4}{4!} + \frac{(x-\pi)^6}{6!} - \cdots$$

$$= \sum_{n=0}^{\infty}(-1)^{n+1}\frac{(x-\pi)^{2n}}{(2n)!}$$

$$\lim_{n\to\infty}\left|\frac{a_{n+1}}{a_n}\right| = \lim_{n\to\infty}\left[\frac{|x-\pi|^{2n+2}}{(2n+2)!}\cdot\frac{(2n)!}{|x-\pi|^{2n}}\right]$$

$$= \lim_{n\to\infty}\frac{|x-\pi|^2}{(2n+2)(2n+1)} = 0 < 1 \quad \text{for all } x, \text{ so } R=\infty.$$

**17.**

| $n$ | $f^{(n)}(x)$ | $f^{(n)}(9)$ |
|---|---|---|
| 0 | $x^{-1/2}$ | $\frac{1}{3}$ |
| 1 | $-\frac{1}{2}x^{-3/2}$ | $-\frac{1}{2}\cdot\frac{1}{3^3}$ |
| 2 | $\frac{3}{4}x^{-5/2}$ | $-\frac{1}{2}\cdot\left(-\frac{3}{2}\right)\cdot\frac{1}{3^5}$ |
| 3 | $-\frac{15}{8}x^{-7/2}$ | $-\frac{1}{2}\cdot\left(-\frac{3}{2}\right)\cdot\left(-\frac{5}{2}\right)\cdot\frac{1}{3^7}$ |
| ⋮ | ⋮ | ⋮ |

$$\frac{1}{\sqrt{x}} = \frac{1}{3} - \frac{1}{2\cdot3^3}(x-9) + \frac{3}{2^2\cdot3^5}\frac{(x-9)^2}{2!}$$

$$- \frac{3\cdot5}{2^3\cdot3^7}\frac{(x-9)^3}{3!} + \cdots$$

$$= \frac{1}{3} + \sum_{n=1}^{\infty}(-1)^n\frac{1\cdot3\cdot5\cdot\cdots\cdot(2n-1)}{2^n\cdot3^{2n+1}\cdot n!}(x-9)^n.$$

$$\lim_{n\to\infty}\left|\frac{a_{n+1}}{a_n}\right| = \lim_{n\to\infty}\left[\frac{1\cdot3\cdot5\cdot\cdots\cdot(2n-1)[2(n+1)-1]\,|x-9|^{n+1}}{2^{n+1}\cdot3^{[2(n+1)+1]}\cdot(n+1)!}\cdot\frac{2^n\cdot3^{2n+1}\cdot n!}{1\cdot3\cdot5\cdot\cdots\cdot(2n-1)\,|x-9|^n}\right]$$

$$= \lim_{n\to\infty}\left[\frac{(2n+1)\,|x-9|}{2\cdot3^2(n+1)}\right] = \frac{1}{9}|x-9| < 1 \quad \text{for convergence, so } |x-9| < 9 \text{ and } R=9.$$

**19.** If $f(x) = \sin \pi x$, then $f^{(n+1)}(x) = \pm\pi^{n+1}\sin\pi x$ or $\pm\pi^{n+1}\cos\pi x$. In each case, $\left|f^{(n+1)}(x)\right| \le \pi^{n+1}$, so by Formula 9

with $a=0$ and $M=\pi^{n+1}$, $|R_n(x)| \le \dfrac{\pi^{n+1}}{(n+1)!}|x|^{n+1} = \dfrac{|\pi x|^{n+1}}{(n+1)!}$. Thus, $|R_n(x)| \to 0$ as $n\to\infty$ by Equation 10.

So $\lim\limits_{n\to\infty} R_n(x) = 0$ and, by Theorem 8, the series in Exercise 7 represents $\sin\pi x$ for all $x$.

**21.** The general binomial series in (17) is

$$(1+x)^k = \sum_{n=0}^{\infty}\binom{k}{n}x^n = 1 + kx + \frac{k(k-1)}{2!}x^2 + \frac{k(k-1)(k-2)}{3!}x^3 + \cdots.$$

$$(1+x)^{1/2} = \sum_{n=0}^{\infty}\binom{\frac{1}{2}}{n}x^n = 1 + \left(\tfrac{1}{2}\right)x + \frac{\left(\tfrac{1}{2}\right)\left(-\tfrac{1}{2}\right)}{2!}x^2 + \frac{\left(\tfrac{1}{2}\right)\left(-\tfrac{1}{2}\right)\left(-\tfrac{3}{2}\right)}{3!}x^3 + \cdots$$

$$= 1 + \frac{x}{2} - \frac{x^2}{2^2\cdot2!} + \frac{1\cdot3\cdot x^3}{2^3\cdot3!} - \frac{1\cdot3\cdot5\cdot x^4}{2^4\cdot4!} + \cdots$$

$$= 1 + \frac{x}{2} + \sum_{n=2}^{\infty}\frac{(-1)^{n-1}1\cdot3\cdot5\cdot\cdots\cdot(2n-3)x^n}{2^n\cdot n!} \quad \text{for } |x| < 1, \quad \text{so } R=1.$$

**23.** $\dfrac{1}{(2+x)^3} = \dfrac{1}{[2(1+x/2)]^3} = \dfrac{1}{8}\left(1 + \dfrac{x}{2}\right)^{-3} = \dfrac{1}{8}\sum_{n=0}^{\infty}\binom{-3}{n}\left(\dfrac{x}{2}\right)^n$. The binomial coefficient is

$$\binom{-3}{n} = \dfrac{(-3)(-4)(-5)\cdots\cdots(-3-n+1)}{n!} = \dfrac{(-3)(-4)(-5)\cdots\cdots[-(n+2)]}{n!}$$

$$= \dfrac{(-1)^n \cdot 2 \cdot 3 \cdot 4 \cdot 5 \cdots\cdots (n+1)(n+2)}{2 \cdot n!} = \dfrac{(-1)^n(n+1)(n+2)}{2}$$

Thus, $\dfrac{1}{(2+x)^3} = \dfrac{1}{8}\sum_{n=0}^{\infty}\dfrac{(-1)^n(n+1)(n+2)}{2}\dfrac{x^n}{2^n} = \sum_{n=0}^{\infty}\dfrac{(-1)^n(n+1)(n+2)x^n}{2^{n+4}}$ for $\left|\dfrac{x}{2}\right| < 1 \;\Leftrightarrow\; |x| < 2$, so $R = 2$.

**25.** $\sin x = \sum_{n=0}^{\infty}(-1)^n\dfrac{x^{2n+1}}{(2n+1)!} \;\Rightarrow\; f(x) = \sin(\pi x) = \sum_{n=0}^{\infty}(-1)^n\dfrac{(\pi x)^{2n+1}}{(2n+1)!} = \sum_{n=0}^{\infty}(-1)^n\dfrac{\pi^{2n+1}}{(2n+1)!}x^{2n+1}, R = \infty$.

**27.** $e^x = \sum_{n=0}^{\infty}\dfrac{x^n}{n!} \;\Rightarrow\; e^{2x} = \sum_{n=0}^{\infty}\dfrac{(2x)^n}{n!} = \sum_{n=0}^{\infty}\dfrac{2^n x^n}{n!}$, so $f(x) = e^x + e^{2x} = \sum_{n=0}^{\infty}\dfrac{1}{n!}x^n + \sum_{n=0}^{\infty}\dfrac{2^n}{n!}x^n = \sum_{n=0}^{\infty}\dfrac{2^n+1}{n!}x^n$,
$R = \infty$.

**29.** $\cos x = \sum_{n=0}^{\infty}(-1)^n\dfrac{x^{2n}}{(2n)!} \;\Rightarrow\; \cos(\tfrac{1}{2}x^2) = \sum_{n=0}^{\infty}(-1)^n\dfrac{\left(\tfrac{1}{2}x^2\right)^{2n}}{(2n)!} = \sum_{n=0}^{\infty}(-1)^n\dfrac{x^{4n}}{2^{2n}(2n)!}$, so

$f(x) = x\cos(\tfrac{1}{2}x^2) = \sum_{n=0}^{\infty}(-1)^n\dfrac{1}{2^{2n}(2n)!}x^{4n+1}, R = \infty$.

**31.** We must write the binomial in the form $(1+$ expression$)$, so we'll factor out a 4.

$$\dfrac{x}{\sqrt{4+x^2}} = \dfrac{x}{\sqrt{4(1+x^2/4)}} = \dfrac{x}{2\sqrt{1+x^2/4}} = \dfrac{x}{2}\left(1 + \dfrac{x^2}{4}\right)^{-1/2} = \dfrac{x}{2}\sum_{n=0}^{\infty}\binom{-\tfrac{1}{2}}{n}\left(\dfrac{x^2}{4}\right)^n$$

$$= \dfrac{x}{2}\left[1 + \left(-\tfrac{1}{2}\right)\dfrac{x^2}{4} + \dfrac{\left(-\tfrac{1}{2}\right)\left(-\tfrac{3}{2}\right)}{2!}\left(\dfrac{x^2}{4}\right)^2 + \dfrac{\left(-\tfrac{1}{2}\right)\left(-\tfrac{3}{2}\right)\left(-\tfrac{5}{2}\right)}{3!}\left(\dfrac{x^2}{4}\right)^3 + \cdots\right]$$

$$= \dfrac{x}{2} + \dfrac{x}{2}\sum_{n=1}^{\infty}(-1)^n\dfrac{1 \cdot 3 \cdot 5 \cdots\cdots (2n-1)}{2^n \cdot 4^n \cdot n!}x^{2n}$$

$$= \dfrac{x}{2} + \sum_{n=1}^{\infty}(-1)^n\dfrac{1 \cdot 3 \cdot 5 \cdots\cdots (2n-1)}{n! \, 2^{3n+1}}x^{2n+1} \text{ and } \dfrac{x^2}{4} < 1 \;\Leftrightarrow\; \dfrac{|x|}{2} < 1 \;\Leftrightarrow\; |x| < 2, \quad \text{so } R = 2.$$

**33.** $\sin^2 x = \dfrac{1}{2}(1 - \cos 2x) = \dfrac{1}{2}\left[1 - \sum_{n=0}^{\infty}\dfrac{(-1)^n(2x)^{2n}}{(2n)!}\right] = \dfrac{1}{2}\left[1 - 1 - \sum_{n=1}^{\infty}\dfrac{(-1)^n(2x)^{2n}}{(2n)!}\right] = \sum_{n=1}^{\infty}\dfrac{(-1)^{n+1}2^{2n-1}x^{2n}}{(2n)!}$,
$R = \infty$

**35.** $\cos x = \sum_{n=0}^{\infty}(-1)^n\dfrac{x^{2n}}{(2n)!} \;\Rightarrow\; f(x) = \cos(x^2) = \sum_{n=0}^{\infty}\dfrac{(-1)^n\left(x^2\right)^{2n}}{(2n)!} = \sum_{n=0}^{\infty}\dfrac{(-1)^n x^{4n}}{(2n)!}, R = \infty$

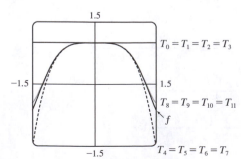

Notice that, as $n$ increases, $T_n(x)$

becomes a better approximation to $f(x)$.

**37.** $e^x \overset{(11)}{=} \sum_{n=0}^{\infty} \dfrac{x^n}{n!}$, so $e^{-x} = \sum_{n=0}^{\infty} \dfrac{(-x)^n}{n!} = \sum_{n=0}^{\infty} (-1)^n \dfrac{x^n}{n!}$, so

$$f(x) = xe^{-x} = \sum_{n=0}^{\infty} (-1)^n \dfrac{1}{n!} x^{n+1}$$

$$= x - x^2 + \tfrac{1}{2}x^3 - \tfrac{1}{6}x^4 + \tfrac{1}{24}x^5 - \tfrac{1}{120}x^6 + \cdots$$

$$= \sum_{n=1}^{\infty} (-1)^{n-1} \dfrac{x^n}{(n-1)!}$$

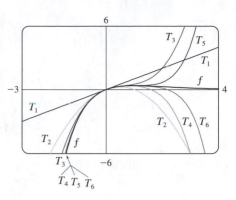

The series for $e^x$ converges for all $x$, so the same is true of the series

for $f(x)$; that is, $R = \infty$. From the graphs of $f$ and the first few Taylor

polynomials, we see that $T_n(x)$ provides a closer fit to $f(x)$ near 0 as $n$ increases.

**39.** $e^x = \sum_{n=0}^{\infty} \dfrac{x^n}{n!}$, so $e^{-0.2} = \sum_{n=0}^{\infty} \dfrac{(-0.2)^n}{n!} = 1 - 0.2 + \dfrac{1}{2!}(0.2)^2 - \dfrac{1}{3!}(0.2)^3 + \dfrac{1}{4!}(0.2)^4 - \dfrac{1}{5!}(0.2)^5 + \dfrac{1}{6!}(0.2)^6 - \cdots.$

But $\dfrac{1}{6!}(0.2)^6 = 8.\overline{8} \times 10^{-8}$, so by the Alternating Series Estimation Theorem, $e^{-0.2} \approx \sum_{n=0}^{5} \dfrac{(-0.2)^n}{n!} \approx 0.81873$, correct to

five decimal places.

**41. (a)** $1/\sqrt{1-x^2} = \left[1 + (-x^2)\right]^{-1/2} = 1 + \left(-\tfrac{1}{2}\right)(-x^2) + \dfrac{\left(-\tfrac{1}{2}\right)\left(-\tfrac{3}{2}\right)}{2!}(-x^2)^2 + \dfrac{\left(-\tfrac{1}{2}\right)\left(-\tfrac{3}{2}\right)\left(-\tfrac{5}{2}\right)}{3!}(-x^2)^3 + \cdots$

$$= 1 + \sum_{n=1}^{\infty} \dfrac{1 \cdot 3 \cdot 5 \cdot \cdots \cdot (2n-1)}{2^n \cdot n!} x^{2n}$$

**(b)** $\sin^{-1} x = \displaystyle\int \dfrac{1}{\sqrt{1-x^2}}\, dx = C + x + \sum_{n=1}^{\infty} \dfrac{1 \cdot 3 \cdot 5 \cdot \cdots \cdot (2n-1)}{(2n+1)2^n \cdot n!} x^{2n+1}$

$$= x + \sum_{n=1}^{\infty} \dfrac{1 \cdot 3 \cdot 5 \cdot \cdots \cdot (2n-1)}{(2n+1)2^n \cdot n!} x^{2n+1} \quad \text{since } 0 = \sin^{-1} 0 = C.$$

**43.** $\cos x \overset{(16)}{=} \sum_{n=0}^{\infty} (-1)^n \dfrac{x^{2n}}{(2n)!} \quad \Rightarrow \quad \cos(x^3) = \sum_{n=0}^{\infty} (-1)^n \dfrac{(x^3)^{2n}}{(2n)!} = \sum_{n=0}^{\infty} (-1)^n \dfrac{x^{6n}}{(2n)!} \quad \Rightarrow$

$x \cos(x^3) = \sum_{n=0}^{\infty} (-1)^n \dfrac{x^{6n+1}}{(2n)!} \quad \Rightarrow \quad \displaystyle\int x \cos(x^3)\, dx = C + \sum_{n=0}^{\infty} (-1)^n \dfrac{x^{6n+2}}{(6n+2)(2n)!}$, with $R = \infty$.

**45.** $\cos x \overset{(16)}{=} \sum_{n=0}^{\infty} (-1)^n \dfrac{x^{2n}}{(2n)!} \quad \Rightarrow \quad \cos x - 1 = \sum_{n=1}^{\infty} (-1)^n \dfrac{x^{2n}}{(2n)!} \quad \Rightarrow \quad \dfrac{\cos x - 1}{x} = \sum_{n=1}^{\infty} (-1)^n \dfrac{x^{2n-1}}{(2n)!} \quad \Rightarrow$

$\displaystyle\int \dfrac{\cos x - 1}{x}\, dx = C + \sum_{n=1}^{\infty} (-1)^n \dfrac{x^{2n}}{2n \cdot (2n)!}$, with $R = \infty$.

**47.** By Exercise 43, $\displaystyle\int x \cos(x^3)\, dx = C + \sum_{n=0}^{\infty} (-1)^n \dfrac{x^{6n+2}}{(6n+2)(2n)!}$, so

$$\int_0^1 x \cos(x^3)\, dx = \left[\sum_{n=0}^{\infty} (-1)^n \dfrac{x^{6n+2}}{(6n+2)(2n)!}\right]_0^1 = \sum_{n=0}^{\infty} \dfrac{(-1)^n}{(6n+2)(2n)!} = \dfrac{1}{2} - \dfrac{1}{8 \cdot 2!} + \dfrac{1}{14 \cdot 4!} - \dfrac{1}{20 \cdot 6!} + \cdots, \text{ but}$$

$\dfrac{1}{20 \cdot 6!} = \dfrac{1}{14{,}400} \approx 0.000\,069$, so $\displaystyle\int_0^1 x \cos(x^3)\, dx \approx \dfrac{1}{2} - \dfrac{1}{16} + \dfrac{1}{336} \approx 0.440$ (correct to three decimal places) by the

Alternating Series Estimation Theorem.

**49.** $\sqrt{1+x^4} = (1+x^4)^{1/2} = \sum\limits_{n=0}^{\infty} \binom{1/2}{n}(x^4)^n$, so $\int \sqrt{1+x^4}\,dx = C + \sum\limits_{n=0}^{\infty} \binom{1/2}{n}\dfrac{x^{4n+1}}{4n+1}$ and hence, since $0.4 < 1$,

we have

$$I = \int_0^{0.4} \sqrt{1+x^4}\,dx = \sum_{n=0}^{\infty} \binom{1/2}{n}\dfrac{(0.4)^{4n+1}}{4n+1}$$

$$= (1)\dfrac{(0.4)^1}{0!} + \dfrac{\frac{1}{2}}{1!}\dfrac{(0.4)^5}{5} + \dfrac{\frac{1}{2}\left(-\frac{1}{2}\right)}{2!}\dfrac{(0.4)^9}{9} + \dfrac{\frac{1}{2}\left(-\frac{1}{2}\right)\left(-\frac{3}{2}\right)}{3!}\dfrac{(0.4)^{13}}{13} + \dfrac{\frac{1}{2}\left(-\frac{1}{2}\right)\left(-\frac{3}{2}\right)\left(-\frac{5}{2}\right)}{4!}\dfrac{(0.4)^{17}}{17} + \cdots$$

$$= 0.4 + \dfrac{(0.4)^5}{10} - \dfrac{(0.4)^9}{72} + \dfrac{(0.4)^{13}}{208} - \dfrac{5(0.4)^{17}}{2176} + \cdots$$

Now $\dfrac{(0.4)^9}{72} \approx 3.6 \times 10^{-6} < 5 \times 10^{-6}$, so by the Alternating Series Estimation Theorem, $I \approx 0.4 + \dfrac{(0.4)^5}{10} \approx 0.40102$

(correct to five decimal places).

**51.** $\lim\limits_{x\to 0}\dfrac{x - \ln(1+x)}{x^2} = \lim\limits_{x\to 0}\dfrac{x - \left(x - \frac{1}{2}x^2 + \frac{1}{3}x^3 - \frac{1}{4}x^4 + \frac{1}{5}x^5 - \cdots\right)}{x^2} = \lim\limits_{x\to 0}\dfrac{\frac{1}{2}x^2 - \frac{1}{3}x^3 + \frac{1}{4}x^4 - \frac{1}{5}x^5 + \cdots}{x^2}$

$$= \lim\limits_{x\to 0}\left(\tfrac{1}{2} - \tfrac{1}{3}x + \tfrac{1}{4}x^2 - \tfrac{1}{5}x^3 + \cdots\right) = \tfrac{1}{2}$$

since power series are continuous functions.

**53.** $\lim\limits_{x\to 0}\dfrac{\sin x - x + \frac{1}{6}x^3}{x^5} = \lim\limits_{x\to 0}\dfrac{\left(x - \frac{1}{3!}x^3 + \frac{1}{5!}x^5 - \frac{1}{7!}x^7 + \cdots\right) - x + \frac{1}{6}x^3}{x^5}$

$$= \lim\limits_{x\to 0}\dfrac{\frac{1}{5!}x^5 - \frac{1}{7!}x^7 + \cdots}{x^5} = \lim\limits_{x\to 0}\left(\dfrac{1}{5!} - \dfrac{x^2}{7!} + \dfrac{x^4}{9!} - \cdots\right) = \dfrac{1}{5!} = \dfrac{1}{120}$$

since power series are continuous functions.

**55.** From Equation 11, we have $e^{-x^2} = 1 - \dfrac{x^2}{1!} + \dfrac{x^4}{2!} - \dfrac{x^6}{3!} + \cdots$ and we know that $\cos x = 1 - \dfrac{x^2}{2!} + \dfrac{x^4}{4!} - \cdots$ from

Equation 16. Therefore, $e^{-x^2}\cos x = \left(1 - x^2 + \frac{1}{2}x^4 - \cdots\right)\left(1 - \frac{1}{2}x^2 + \frac{1}{24}x^4 - \cdots\right)$. Writing only the terms with

degree $\leq 4$, we get $e^{-x^2}\cos x = 1 - \frac{1}{2}x^2 + \frac{1}{24}x^4 - x^2 + \frac{1}{2}x^4 + \frac{1}{2}x^4 + \cdots = 1 - \frac{3}{2}x^2 + \frac{25}{24}x^4 + \cdots$.

**57.** $\dfrac{x}{\sin x} \overset{(15)}{=} \dfrac{x}{x - \frac{1}{6}x^3 + \frac{1}{120}x^5 - \cdots}.$

$$
\begin{array}{r}
1 + \frac{1}{6}x^2 + \frac{7}{360}x^4 + \cdots \\
x - \frac{1}{6}x^3 + \frac{1}{120}x^5 - \cdots \overline{\big)\ x} \\
\underline{x - \frac{1}{6}x^3 + \frac{1}{120}x^5 - \cdots} \\
\frac{1}{6}x^3 - \frac{1}{120}x^5 + \cdots \\
\underline{\frac{1}{6}x^3 - \frac{1}{36}x^5 + \cdots} \\
\frac{7}{360}x^5 + \cdots \\
\underline{\frac{7}{360}x^5 + \cdots} \\
\cdots
\end{array}
$$

From the long division above, $\dfrac{x}{\sin x} = 1 + \frac{1}{6}x^2 + \frac{7}{360}x^4 + \cdots$.

**59.** $\sum\limits_{n=0}^{\infty}(-1)^n\dfrac{x^{4n}}{n!} = \sum\limits_{n=0}^{\infty}\dfrac{\left(-x^4\right)^n}{n!} = e^{-x^4}$, by (11).

**61.** $\displaystyle\sum_{n=1}^{\infty}(-1)^{n-1}\frac{3^n}{n5^n}=\sum_{n=1}^{\infty}(-1)^{n-1}\frac{(3/5)^n}{n}=\ln\left(1+\frac{3}{5}\right)$ [from Table 1] $=\ln\dfrac{8}{5}$

**63.** $\displaystyle\sum_{n=0}^{\infty}\frac{(-1)^n\,\pi^{2n+1}}{4^{2n+1}(2n+1)!}=\sum_{n=0}^{\infty}\frac{(-1)^n\left(\frac{\pi}{4}\right)^{2n+1}}{(2n+1)!}=\sin\frac{\pi}{4}=\frac{1}{\sqrt{2}}$, by (15).

**65.** $3+\dfrac{9}{2!}+\dfrac{27}{3!}+\dfrac{81}{4!}+\cdots=\dfrac{3^1}{1!}+\dfrac{3^2}{2!}+\dfrac{3^3}{3!}+\dfrac{3^4}{4!}+\cdots=\displaystyle\sum_{n=1}^{\infty}\frac{3^n}{n!}=\sum_{n=0}^{\infty}\frac{3^n}{n!}-1=e^3-1$, by (11).

**67.** Assume that $|f'''(x)|\le M$, so $f'''(x)\le M$ for $a\le x\le a+d$. Now $\int_a^x f'''(t)\,dt\le\int_a^x M\,dt\ \Rightarrow$

$f''(x)-f''(a)\le M(x-a)\ \Rightarrow\ f''(x)\le f''(a)+M(x-a)$. Thus, $\int_a^x f''(t)\,dt\le\int_a^x\left[f''(a)+M(t-a)\right]dt\ \Rightarrow$

$f'(x)-f'(a)\le f''(a)(x-a)+\frac{1}{2}M(x-a)^2\ \Rightarrow\ f'(x)\le f'(a)+f''(a)(x-a)+\frac{1}{2}M(x-a)^2\ \Rightarrow$

$\int_a^x f'(t)\,dt\le\int_a^x\left[f'(a)+f''(a)(t-a)+\frac{1}{2}M(t-a)^2\right]dt\ \Rightarrow$

$f(x)-f(a)\le f'(a)(x-a)+\frac{1}{2}f''(a)(x-a)^2+\frac{1}{6}M(x-a)^3$. So

$f(x)-f(a)-f'(a)(x-a)-\frac{1}{2}f''(a)(x-a)^2\le\frac{1}{6}M(x-a)^3$. But

$R_2(x)=f(x)-T_2(x)=f(x)-f(a)-f'(a)(x-a)-\frac{1}{2}f''(a)(x-a)^2$, so $R_2(x)\le\frac{1}{6}M(x-a)^3$.

A similar argument using $f'''(x)\ge -M$ shows that $R_2(x)\ge-\frac{1}{6}M(x-a)^3$. So $|R_2(x_2)|\le\frac{1}{6}M\,|x-a|^3$.

Although we have assumed that $x>a$, a similar calculation shows that this inequality is also true if $x<a$.

**69.** (a) $g(x)=\displaystyle\sum_{n=0}^{\infty}\binom{k}{n}x^n\ \Rightarrow\ g'(x)=\sum_{n=1}^{\infty}\binom{k}{n}nx^{n-1}$, so

$(1+x)g'(x)=(1+x)\displaystyle\sum_{n=1}^{\infty}\binom{k}{n}nx^{n-1}=\sum_{n=1}^{\infty}\binom{k}{n}nx^{n-1}+\sum_{n=1}^{\infty}\binom{k}{n}nx^{n}$

$\displaystyle=\sum_{n=0}^{\infty}\binom{k}{n+1}(n+1)x^n+\sum_{n=0}^{\infty}\binom{k}{n}nx^n\qquad\left[\begin{array}{l}\text{Replace }n\text{ with }n+1\\ \text{in the first series}\end{array}\right]$

$\displaystyle=\sum_{n=0}^{\infty}(n+1)\frac{k(k-1)(k-2)\cdots(k-n+1)(k-n)}{(n+1)!}x^n+\sum_{n=0}^{\infty}\left[(n)\frac{k(k-1)(k-2)\cdots(k-n+1)}{n!}\right]x^n$

$\displaystyle=\sum_{n=0}^{\infty}\frac{(n+1)k(k-1)(k-2)\cdots(k-n+1)}{(n+1)!}\left[(k-n)+n\right]x^n$

$\displaystyle=k\sum_{n=0}^{\infty}\frac{k(k-1)(k-2)\cdots(k-n+1)}{n!}x^n=k\sum_{n=0}^{\infty}\binom{k}{n}x^n=kg(x)$

Thus, $g'(x)=\dfrac{kg(x)}{1+x}$.

(b) $h(x)=(1+x)^{-k}\,g(x)\ \Rightarrow$

$h'(x)=-k(1+x)^{-k-1}g(x)+(1+x)^{-k}g'(x)$   [Product Rule]

$=-k(1+x)^{-k-1}g(x)+(1+x)^{-k}\dfrac{kg(x)}{1+x}$   [from part (a)]

$=-k(1+x)^{-k-1}g(x)+k(1+x)^{-k-1}g(x)=0$

(c) From part (b) we see that $h(x)$ must be constant for $x\in(-1,1)$, so $h(x)=h(0)=1$ for $x\in(-1,1)$.

Thus, $h(x)=1=(1+x)^{-k}g(x)\ \Leftrightarrow\ g(x)=(1+x)^k$ for $x\in(-1,1)$.

## 8.8 Applications of Taylor Polynomials

**1.** (a)

| $n$ | $f^{(n)}(x)$ | $f^{(n)}(0)$ | $T_n(x)$ |
|---|---|---|---|
| 0 | $\cos x$ | 1 | 1 |
| 1 | $-\sin x$ | 0 | 1 |
| 2 | $-\cos x$ | $-1$ | $1 - \frac{1}{2}x^2$ |
| 3 | $\sin x$ | 0 | $1 - \frac{1}{2}x^2$ |
| 4 | $\cos x$ | 1 | $1 - \frac{1}{2}x^2 + \frac{1}{24}x^4$ |
| 5 | $-\sin x$ | 0 | $1 - \frac{1}{2}x^2 + \frac{1}{24}x^4$ |
| 6 | $-\cos x$ | $-1$ | $1 - \frac{1}{2}x^2 + \frac{1}{24}x^4 - \frac{1}{720}x^6$ |

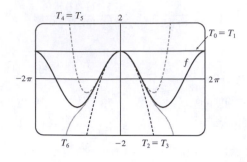

(b)

| $x$ | $f$ | $T_0 = T_1$ | $T_2 = T_3$ | $T_4 = T_5$ | $T_6$ |
|---|---|---|---|---|---|
| $\frac{\pi}{4}$ | 0.7071 | 1 | 0.6916 | 0.7074 | 0.7071 |
| $\frac{\pi}{2}$ | 0 | 1 | $-0.2337$ | 0.0200 | $-0.0009$ |
| $\pi$ | $-1$ | 1 | $-3.9348$ | 0.1239 | $-1.2114$ |

(c) As $n$ increases, $T_n(x)$ is a good approximation to $f(x)$ on a larger and larger interval.

**3.**

| $n$ | $f^{(n)}(x)$ | $f^{(n)}(2)$ |
|---|---|---|
| 0 | $1/x$ | $\frac{1}{2}$ |
| 1 | $-1/x^2$ | $-\frac{1}{4}$ |
| 2 | $2/x^3$ | $\frac{1}{4}$ |
| 3 | $-6/x^4$ | $-\frac{3}{8}$ |

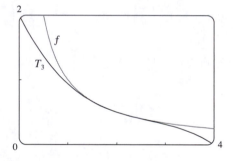

$$T_3(x) = \sum_{n=0}^{3} \frac{f^{(n)}(2)}{n!}(x-2)^n$$

$$= \frac{\frac{1}{2}}{0!} - \frac{\frac{1}{4}}{1!}(x-2) + \frac{\frac{1}{4}}{2!}(x-2)^2 - \frac{\frac{3}{8}}{3!}(x-2)^3$$

$$= \frac{1}{2} - \frac{1}{4}(x-2) + \frac{1}{8}(x-2)^2 - \frac{1}{16}(x-2)^3$$

**5.**

| $n$ | $f^{(n)}(x)$ | $f^{(n)}(\pi/2)$ |
|---|---|---|
| 0 | $\cos x$ | 0 |
| 1 | $-\sin x$ | $-1$ |
| 2 | $-\cos x$ | 0 |
| 3 | $\sin x$ | 1 |

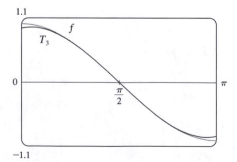

$$T_3(x) = \sum_{n=0}^{3} \frac{f^{(n)}(\pi/2)}{n!}\left(x - \frac{\pi}{2}\right)^n$$

$$= -\left(x - \frac{\pi}{2}\right) + \frac{1}{6}\left(x - \frac{\pi}{2}\right)^3$$

**7.**

| $n$ | $f^{(n)}(x)$ | $f^{(n)}(0)$ |
|---|---|---|
| 0 | $xe^{-2x}$ | 0 |
| 1 | $(1-2x)e^{-2x}$ | 1 |
| 2 | $4(x-1)e^{-2x}$ | $-4$ |
| 3 | $4(3-2x)e^{-2x}$ | 12 |

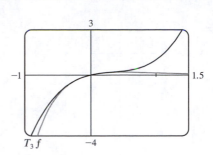

$$T_3(x) = \sum_{n=0}^{3} \frac{f^{(n)}(0)}{n!} x^n = \tfrac{0}{1}\cdot 1 + \tfrac{1}{1}x^1 + \tfrac{-4}{2}x^2 + \tfrac{12}{6}x^3 = x - 2x^2 + 2x^3$$

**9.** You may be able to simply find the Taylor polynomials for

$f(x) = \cot x$ using your CAS. We will list the values of $f^{(n)}(\pi/4)$

for $n = 0$ to $n = 5$.

| $n$ | 0 | 1 | 2 | 3 | 4 | 5 |
|---|---|---|---|---|---|---|
| $f^{(n)}(\pi/4)$ | 1 | $-2$ | 4 | $-16$ | 80 | $-512$ |

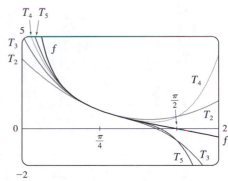

$$T_5(x) = \sum_{n=0}^{5} \frac{f^{(n)}(\pi/4)}{n!}\left(x - \tfrac{\pi}{4}\right)^n$$

$$= 1 - 2\left(x - \tfrac{\pi}{4}\right) + 2\left(x - \tfrac{\pi}{4}\right)^2 - \tfrac{8}{3}\left(x - \tfrac{\pi}{4}\right)^3 + \tfrac{10}{3}\left(x - \tfrac{\pi}{4}\right)^4 - \tfrac{64}{15}\left(x - \tfrac{\pi}{4}\right)^5$$

For $n = 2$ to $n = 5$, $T_n(x)$ is the polynomial consisting of all the terms up to and including the $\left(x - \tfrac{\pi}{4}\right)^n$ term.

**11.**

| $n$ | $f^{(n)}(x)$ | $f^{(n)}(4)$ |
|---|---|---|
| 0 | $\sqrt{x}$ | 2 |
| 1 | $\tfrac{1}{2}x^{-1/2}$ | $\tfrac{1}{4}$ |
| 2 | $-\tfrac{1}{4}x^{-3/2}$ | $-\tfrac{1}{32}$ |
| 3 | $\tfrac{3}{8}x^{-5/2}$ | |

(a) $f(x) = \sqrt{x} \approx T_2(x) = 2 + \dfrac{1}{4}(x-4) - \dfrac{1/32}{2!}(x-4)^2$

$$= 2 + \tfrac{1}{4}(x-4) - \tfrac{1}{64}(x-4)^2$$

(b) $|R_2(x)| \le \dfrac{M}{3!}|x-4|^3$, where $|f'''(x)| \le M$. Now $4 \le x \le 4.2$ ⇒

$|x-4| \le 0.2$ ⇒ $|x-4|^3 \le 0.008$. Since $f'''(x)$ is decreasing

on $[4, 4.2]$, we can take $M = |f'''(4)| = \tfrac{3}{8}4^{-5/2} = \tfrac{3}{256}$, so

$$|R_2(x)| \le \frac{3/256}{6}(0.008) = \frac{0.008}{512} = 0.000\,015\,625.$$

(c)

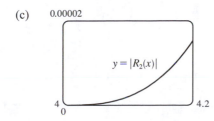

From the graph of $|R_2(x)| = |\sqrt{x} - T_2(x)|$, it seems that the

error is less than $1.52 \times 10^{-5}$ on $[4, 4.2]$.

**13.**

| $n$ | $f^{(n)}(x)$ | $f^{(n)}(1)$ |
|---|---|---|
| 0 | $x^{2/3}$ | 1 |
| 1 | $\frac{2}{3}x^{-1/3}$ | $\frac{2}{3}$ |
| 2 | $-\frac{2}{9}x^{-4/3}$ | $-\frac{2}{9}$ |
| 3 | $\frac{8}{27}x^{-7/3}$ | $\frac{8}{27}$ |
| 4 | $-\frac{56}{81}x^{-10/3}$ | |

(a) $f(x) = x^{2/3} \approx T_3(x) = 1 + \frac{2}{3}(x-1) - \frac{2/9}{2!}(x-1)^2 + \frac{8/27}{3!}(x-1)^3$

$$= 1 + \frac{2}{3}(x-1) - \frac{1}{9}(x-1)^2 + \frac{4}{81}(x-1)^3$$

(b) $|R_3(x)| \le \frac{M}{4!}|x-1|^4$, where $\left|f^{(4)}(x)\right| \le M$. Now $0.8 \le x \le 1.2 \Rightarrow$

$|x-1| \le 0.2 \Rightarrow |x-1|^4 \le 0.0016$. Since $\left|f^{(4)}(x)\right|$ is decreasing

on $[0.8, 1.2]$, we can take $M = \left|f^{(4)}(0.8)\right| = \frac{56}{81}(0.8)^{-10/3}$, so

$$|R_3(x)| \le \frac{\frac{56}{81}(0.8)^{-10/3}}{24}(0.0016) \approx 0.000\,096\,97.$$

(c)

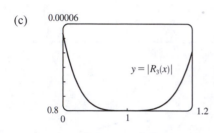

From the graph of $|R_3(x)| = \left|x^{2/3} - T_3(x)\right|$, it seems that the

error is less than $0.000\,053\,3$ on $[0.8, 1.2]$.

**15.**

| $n$ | $f^{(n)}(x)$ | $f^{(n)}(0)$ |
|---|---|---|
| 0 | $e^{x^2}$ | 1 |
| 1 | $e^{x^2}(2x)$ | 0 |
| 2 | $e^{x^2}(2+4x^2)$ | 2 |
| 3 | $e^{x^2}(12x+8x^3)$ | 0 |
| 4 | $e^{x^2}(12+48x^2+16x^4)$ | |

(a) $f(x) = e^{x^2} \approx T_3(x) = 1 + \frac{2}{2!}x^2 = 1 + x^2$

(b) $|R_3(x)| \le \frac{M}{4!}|x|^4$, where $\left|f^{(4)}(x)\right| \le M$. Now $0 \le x \le 0.1 \Rightarrow$

$x^4 \le (0.1)^4$, and letting $x = 0.1$ gives

$$|R_3(x)| \le \frac{e^{0.01}(12+0.48+0.0016)}{24}(0.1)^4 \approx 0.00006.$$

(c)

From the graph of $|R_3(x)| = \left|e^{x^2} - T_3(x)\right|$, it appears that the

error is less than $0.000\,051$ on $[0, 0.1]$.

**17.**

| $n$ | $f^{(n)}(x)$ | $f^{(n)}(0)$ |
|---|---|---|
| 0 | $x\sin x$ | 0 |
| 1 | $\sin x + x\cos x$ | 0 |
| 2 | $2\cos x - x\sin x$ | 2 |
| 3 | $-3\sin x - x\cos x$ | 0 |
| 4 | $-4\cos x + x\sin x$ | $-4$ |
| 5 | $5\sin x + x\cos x$ | |

(a) $f(x) = x\sin x \approx T_4(x) = \frac{2}{2!}(x-0)^2 + \frac{-4}{4!}(x-0)^4 = x^2 - \frac{1}{6}x^4$

(b) $|R_4(x)| \le \frac{M}{5!}|x|^5$, where $\left|f^{(5)}(x)\right| \le M$. Now $-1 \le x \le 1 \Rightarrow$

$|x| \le 1$, and a graph of $f^{(5)}(x)$ shows that $\left|f^{(5)}(x)\right| \le 5$ for $-1 \le x \le 1$.

Thus, we can take $M = 5$ and get $|R_4(x)| \le \frac{5}{5!} \cdot 1^5 = \frac{1}{24} = 0.041\overline{6}$.

(c)

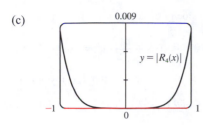

From the graph of $|R_4(x)| = |x \sin x - T_4(x)|$, it seems that the error is less than 0.0082 on $[-1, 1]$.

**19.** From Exercise 5, $\cos x = -\left(x - \frac{\pi}{2}\right) + \frac{1}{6}\left(x - \frac{\pi}{2}\right)^3 + R_3(x)$, where $|R_3(x)| \leq \frac{M}{4!}\left|x - \frac{\pi}{2}\right|^4$ with

$\left|f^{(4)}(x)\right| = |\cos x| \leq M = 1$. Now $x = 80° = (90° - 10°) = \left(\frac{\pi}{2} - \frac{\pi}{18}\right) = \frac{4\pi}{9}$ radians, so the error is

$\left|R_3\left(\frac{4\pi}{9}\right)\right| \leq \frac{1}{24}\left(\frac{\pi}{18}\right)^4 \approx 0.000\,039$, which means our estimate would *not* be accurate to five decimal places. However,

$T_3 = T_4$, so we can use $\left|R_4\left(\frac{4\pi}{9}\right)\right| \leq \frac{1}{120}\left(\frac{\pi}{18}\right)^5 \approx 0.000\,001$. Therefore, to five decimal places,

$\cos 80° \approx -\left(-\frac{\pi}{18}\right) + \frac{1}{6}\left(-\frac{\pi}{18}\right)^3 \approx 0.17365$.

**21.** All derivatives of $e^x$ are $e^x$, so $|R_n(x)| \leq \frac{e^x}{(n+1)!}|x|^{n+1}$, where $0 < x < 0.1$. Letting $x = 0.1$,

$R_n(0.1) \leq \frac{e^{0.1}}{(n+1)!}(0.1)^{n+1} < 0.00001$, and by trial and error we find that $n = 3$ satisfies this inequality since

$R_3(0.1) < 0.0000046$. Thus, by adding the four terms of the Maclaurin series for $e^x$ corresponding to $n = 0, 1, 2$, and 3,

we can estimate $e^{0.1}$ to within 0.00001. (In fact, this sum is $1.1051\overline{6}$ and $e^{0.1} \approx 1.10517$.)

**23.** $\sin x = x - \frac{1}{3!}x^3 + \frac{1}{5!}x^5 - \cdots$. By the Alternating Series

Estimation Theorem, the error in the approximation

$\sin x = x - \frac{1}{3!}x^3$ is less than $\left|\frac{1}{5!}x^5\right| < 0.01 \quad\Leftrightarrow$

$\left|x^5\right| < 120(0.01) \quad\Leftrightarrow\quad |x| < (1.2)^{1/5} \approx 1.037$. The curves

$y = x - \frac{1}{6}x^3$ and $y = \sin x - 0.01$ intersect at $x \approx 1.043$, so

the graph confirms our estimate. Since both the sine function

and the given approximation are odd functions, we need to check the estimate only for $x > 0$. Thus, the desired range of

values for $x$ is $-1.037 < x < 1.037$.

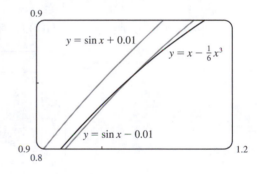

**25.** $\arctan x = x - \frac{x^3}{3} + \frac{x^5}{5} - \frac{x^7}{7} + \cdots$. By the Alternating Series

Estimation Theorem, the error is less than $\left|-\frac{1}{7}x^7\right| < 0.05 \quad\Leftrightarrow$

$\left|x^7\right| < 0.35 \quad\Leftrightarrow\quad |x| < (0.35)^{1/7} \approx 0.8607$. The curves

$y = x - \frac{1}{3}x^3 + \frac{1}{5}x^5$ and $y = \arctan x + 0.05$ intersect at

$x \approx 0.9245$, so the graph confirms our estimate. Since both the

arctangent function and the given approximation are odd functions,

we need to check the estimate only for $x > 0$. Thus, the desired

range of values for $x$ is $-0.86 < x < 0.86$.

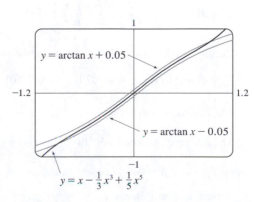

**27.** Let $s(t)$ be the position function of the car, and for convenience set $s(0) = 0$. The velocity of the car is $v(t) = s'(t)$ and the

acceleration is $a(t) = s''(t)$, so the second degree Taylor polynomial is $T_2(t) = s(0) + v(0)t + \dfrac{a(0)}{2}t^2 = 20t + t^2$. We

estimate the distance traveled during the next second to be $s(1) \approx T_2(1) = 20 + 1 = 21$ m. The function $T_2(t)$ would not be

accurate over a full minute, since the car could not possibly maintain an acceleration of $2$ m/s$^2$ for that long (if it did, its final

speed would be $140$ m/s $\approx 313$ mi/h!).

**29.** $E = \dfrac{q}{D^2} - \dfrac{q}{(D+d)^2} = \dfrac{q}{D^2} - \dfrac{q}{D^2(1 + d/D)^2} = \dfrac{q}{D^2}\left[1 - \left(1 + \dfrac{d}{D}\right)^{-2}\right]$.

We use the Binomial Series to expand $(1 + d/D)^{-2}$:

$$E = \frac{q}{D^2}\left[1 - \left(1 - 2\left(\frac{d}{D}\right) + \frac{2 \cdot 3}{2!}\left(\frac{d}{D}\right)^2 - \frac{2 \cdot 3 \cdot 4}{3!}\left(\frac{d}{D}\right)^3 + \cdots\right)\right] = \frac{q}{D^2}\left[2\left(\frac{d}{D}\right) - 3\left(\frac{d}{D}\right)^2 + 4\left(\frac{d}{D}\right)^3 - \cdots\right]$$

$$\approx \frac{q}{D^2} \cdot 2\left(\frac{d}{D}\right) = 2qd \cdot \frac{1}{D^3}$$

when $D$ is much larger than $d$; that is, when $P$ is far away from the dipole.

**31.** (a) $L$ is the length of the arc subtended by the angle $\theta$, so $L = R\theta \quad \Rightarrow$

$\theta = L/R$. Now $\sec\theta = (R+C)/R \quad \Rightarrow \quad R\sec\theta = R + C \quad \Rightarrow$

$C = R\sec\theta - R = R\sec(L/R) - R$.

(b) First we'll find a Taylor polynomial $T_4(x)$ for $f(x) = \sec x$ at $x = 0$.

| $n$ | $f^{(n)}(x)$ | $f^{(n)}(0)$ |
|---|---|---|
| 0 | $\sec x$ | 1 |
| 1 | $\sec x \tan x$ | 0 |
| 2 | $\sec x(2\tan^2 x + 1)$ | 1 |
| 3 | $\sec x \tan x(6\tan^2 x + 5)$ | 0 |
| 4 | $\sec x(24\tan^4 x + 28\tan^2 x + 5)$ | 5 |

Thus, $f(x) = \sec x \approx T_4(x) = 1 + \frac{1}{2!}(x - 0)^2 + \frac{5}{4!}(x - 0)^4 = 1 + \frac{1}{2}x^2 + \frac{5}{24}x^4$. By part (a),

$$C \approx R\left[1 + \frac{1}{2}\left(\frac{L}{R}\right)^2 + \frac{5}{24}\left(\frac{L}{R}\right)^4\right] - R = R + \frac{1}{2}R \cdot \frac{L^2}{R^2} + \frac{5}{24}R \cdot \frac{L^4}{R^4} - R = \frac{L^2}{2R} + \frac{5L^4}{24R^3}.$$

(c) Taking $L = 100$ km and $R = 6370$ km, the formula in part (a) says that

$C = R\sec(L/R) - R = 6370\sec(100/6370) - 6370 \approx 0.785\,009\,965\,44$ km.

The formula in part (b) says that $C \approx \dfrac{L^2}{2R} + \dfrac{5L^4}{24R^3} = \dfrac{100^2}{2 \cdot 6370} + \dfrac{5 \cdot 100^4}{24 \cdot 6370^3} \approx 0.785\,009\,957\,36$ km.

The difference between these two results is only $0.000\,000\,008\,08$ km, or $0.000\,008\,08$ m!

**33.** Using $f(x) = T_n(x) + R_n(x)$ with $n = 1$ and $x = r$, we have $f(r) = T_1(r) + R_1(r)$, where $T_1$ is the first-degree Taylor

polynomial of $f$ at $a$. Because $a = x_n$, $f(r) = f(x_n) + f'(x_n)(r - x_n) + R_1(r)$. But $r$ is a root of $f$, so $f(r) = 0$

and we have $0 = f(x_n) + f'(x_n)(r - x_n) + R_1(r)$. Taking the first two terms to the left side gives us

$f'(x_n)(x_n - r) - f(x_n) = R_1(r)$. Dividing by $f'(x_n)$, we get $x_n - r - \dfrac{f(x_n)}{f'(x_n)} = \dfrac{R_1(r)}{f'(x_n)}$. By the formula for Newton's

method, the left side of the preceding equation is $x_{n+1} - r$, so $|x_{n+1} - r| = \left| \dfrac{R_1(r)}{f'(x_n)} \right|$. Taylor's Inequality gives us

$|R_1(r)| \leq \dfrac{|f''(r)|}{2!} |r - x_n|^2$. Combining this inequality with the facts $|f''(x)| \leq M$ and $|f'(x)| \geq K$ gives us

$|x_{n+1} - r| \leq \dfrac{M}{2K} |x_n - r|^2$.

## 8   Review

### CONCEPT CHECK

**1.** (a) See Definition 8.1.1.

(b) See Definition 8.2.2.

(c) The terms of the sequence $\{a_n\}$ approach 3 as $n$ becomes large.

(d) By adding sufficiently many terms of the series, we can make the partial sums as close to 3 as we like.

**2.** (a) See the definition on page 561.

(b) A sequence is monotonic if it is either increasing or decreasing.

(c) By Theorem 8.1.8, every bounded, monotonic sequence is convergent.

**3.** (a) See (4) in Section 8.2.

(b) The $p$-series $\displaystyle\sum_{n=1}^{\infty} \frac{1}{n^p}$ is convergent if $p > 1$.

**4.** If $\sum a_n = 3$, then $\displaystyle\lim_{n \to \infty} a_n = 0$ and $\displaystyle\lim_{n \to \infty} s_n = 3$.

**5.** (a) *Test for Divergence:* If $\displaystyle\lim_{n \to \infty} a_n$ does not exist or if $\displaystyle\lim_{n \to \infty} a_n \neq 0$, then the series $\sum_{n=1}^{\infty} a_n$ is divergent.

(b) *Integral Test:* Suppose $f$ is a continuous, positive, decreasing function on $[1, \infty)$ and let $a_n = f(n)$. Then the series

$\sum_{n=1}^{\infty} a_n$ is convergent if and only if the improper integral $\int_1^{\infty} f(x)\, dx$ is convergent. In other words:

(i) If $\int_1^{\infty} f(x)\, dx$ is convergent, then $\sum_{n=1}^{\infty} a_n$ is convergent.

(ii) If $\int_1^{\infty} f(x)\, dx$ is divergent, then $\sum_{n=1}^{\infty} a_n$ is divergent.

(c) *Comparison Test:* Suppose that $\sum a_n$ and $\sum b_n$ are series with positive terms.

(i) If $\sum b_n$ is convergent and $a_n \leq b_n$ for all $n$, then $\sum a_n$ is also convergent.

(ii) If $\sum b_n$ is divergent and $a_n \geq b_n$ for all $n$, then $\sum a_n$ is also divergent.

(d) *Limit Comparison Test:* Suppose that $\sum a_n$ and $\sum b_n$ are series with positive terms. If $\displaystyle\lim_{n \to \infty} (a_n / b_n) = c$, where $c$ is a

finite number and $c > 0$, then either both series converge or both diverge.

(e) *Alternating Series Test:* If the alternating series $\sum_{n=1}^{\infty}(-1)^{n-1}b_n = b_1 - b_2 + b_3 - b_4 + b_5 - b_6 + \cdots$ $[b_n > 0]$

satisfies (i) $b_{n+1} \leq b_n$ for all $n$ and (ii) $\lim\limits_{n\to\infty} b_n = 0$, then the series is convergent.

(f) *Ratio Test:*

(i) If $\lim\limits_{n\to\infty}\left|\dfrac{a_{n+1}}{a_n}\right| = L < 1$, then the series $\sum\limits_{n=1}^{\infty} a_n$ is absolutely convergent (and therefore convergent).

(ii) If $\lim\limits_{n\to\infty}\left|\dfrac{a_{n+1}}{a_n}\right| = L > 1$ or $\lim\limits_{n\to\infty}\left|\dfrac{a_{n+1}}{a_n}\right| = \infty$, then the series $\sum\limits_{n=1}^{\infty} a_n$ is divergent.

(iii) If $\lim\limits_{n\to\infty}\left|\dfrac{a_{n+1}}{a_n}\right| = 1$, the Ratio Test is inconclusive; that is, no conclusion can be drawn about the convergence or

divergence of $\sum a_n$.

**6.** (a) A series $\sum a_n$ is called *absolutely convergent* if the series of absolute values $\sum |a_n|$ is convergent.

(b) If a series $\sum a_n$ is absolutely convergent, then it is convergent.

**7.** (a) Use (4) in Section 8.3.

(b) See Example 8 in Section 8.3.

(c) By adding terms until you reach the desired accuracy given by the Alternating Series Estimation Theorem.

**8.** (a) $\sum\limits_{n=0}^{\infty} c_n(x-a)^n$

(b) Given the power series $\sum\limits_{n=0}^{\infty} c_n(x-a)^n$, the radius of convergence is:

(i) 0 if the series converges only when $x = a$

(ii) $\infty$ if the series converges for all $x$, or

(iii) a positive number $R$ such that the series converges if $|x-a| < R$ and diverges if $|x-a| > R$.

(c) The interval of convergence of a power series is the interval that consists of all values of $x$ for which the series converges. Corresponding to the cases in part (b), the interval of convergence is: (i) the single point $\{a\}$, (ii) all real numbers, that is, the real number line $(-\infty, \infty)$, or (iii) an interval with endpoints $a - R$ and $a + R$ which can contain neither, either, or both of the endpoints. In this case, we must test the series for convergence at each endpoint to determine the interval of convergence.

**9.** (a), (b) See Theorem 8.6.2.

**10.** (a) $T_n(x) = \sum\limits_{i=0}^{n} \dfrac{f^{(i)}(a)}{i!}(x-a)^i$

(b) $\sum\limits_{n=0}^{\infty} \dfrac{f^{(n)}(a)}{n!}(x-a)^n$

(c) $\sum_{n=0}^{\infty} \dfrac{f^{(n)}(0)}{n!} x^n$   [$a = 0$ in part (b)]

(d) See Theorem 8.7.8.

(e) See Taylor's Inequality (8.7.9).

**11.** (a)–(f) See the table on page 613.

**12.** See the binomial series (8.7.17) for the expansion. The radius of convergence for the binomial series is 1.

## TRUE-FALSE QUIZ

**1.** False.    See Note 2 after Theorem 8.2.6.

**3.** True.    If $\lim\limits_{n\to\infty} a_n = L$, then as $n \to \infty$, $2n + 1 \to \infty$, so $a_{2n+1} \to L$.

**5.** False.    For example, take $c_n = (-1)^n/(n6^n)$.

**7.** False, since $\lim\limits_{n\to\infty}\left|\dfrac{a_{n+1}}{a_n}\right| = \lim\limits_{n\to\infty}\left|\dfrac{1}{(n+1)^3} \cdot \dfrac{n^3}{1}\right| = \lim\limits_{n\to\infty}\left|\dfrac{n^3}{(n+1)^3} \cdot \dfrac{1/n^3}{1/n^3}\right| = \lim\limits_{n\to\infty}\dfrac{1}{(1+1/n)^3} = 1.$

**9.** False.    See the note after Example 4 in Section 8.3.

**11.** True.    See (7) in Section 8.1.

**13.** True.    By Theorem 8.7.5 the coefficient of $x^3$ is $\dfrac{f'''(0)}{3!} = \dfrac{1}{3} \;\Rightarrow\; f'''(0) = 2.$

Or:  Use Theorem 8.6.2 to differentiate $f$ three times.

**15.** False.    For example, let $a_n = b_n = (-1)^n$. Then $\{a_n\}$ and $\{b_n\}$ are divergent, but $a_n b_n = 1$, so $\{a_n b_n\}$ is convergent.

**17.** True by Theorem 8.4.1.  $\left[\sum (-1)^n a_n \text{ is absolutely convergent and hence convergent.}\right]$

**19.** True.    $0.99999\ldots = 0.9 + 0.9(0.1)^1 + 0.9(0.1)^2 + 0.9(0.1)^3 + \cdots = \sum_{n=1}^{\infty}(0.9)(0.1)^{n-1} = \dfrac{0.9}{1-0.1} = 1$ by the formula

for the sum of a geometric series $[S = a_1/(1-r)]$ with ratio $r$ satisfying $|r| < 1$.

## EXERCISES

**1.** $\left\{\dfrac{2+n^3}{1+2n^3}\right\}$ converges since $\lim\limits_{n\to\infty}\dfrac{2+n^3}{1+2n^3} = \lim\limits_{n\to\infty}\dfrac{2/n^3+1}{1/n^3+2} = \dfrac{1}{2}.$

**3.** $\lim\limits_{n\to\infty} a_n = \lim\limits_{n\to\infty}\dfrac{n^3}{1+n^2} = \lim\limits_{n\to\infty}\dfrac{n}{1/n^2+1} = \infty$, so the sequence diverges.

**5.** $|a_n| = \left|\dfrac{n\sin n}{n^2+1}\right| \le \dfrac{n}{n^2+1} < \dfrac{1}{n}$, so $|a_n| \to 0$ as $n \to \infty$. Thus, $\lim\limits_{n\to\infty} a_n = 0$. The sequence $\{a_n\}$ is convergent.

**7.** $\left\{\left(1 + \dfrac{3}{n}\right)^{4n}\right\}$ is convergent. Let $y = \left(1 + \dfrac{3}{x}\right)^{4x}$. Then

$$\lim_{x \to \infty} \ln y = \lim_{x \to \infty} 4x \ln(1 + 3/x) = \lim_{x \to \infty} \frac{\ln(1 + 3/x)}{1/(4x)} \overset{\text{H}}{=} \lim_{x \to \infty} \frac{\dfrac{1}{1 + 3/x}\left(-\dfrac{3}{x^2}\right)}{-1/(4x^2)} = \lim_{x \to \infty} \frac{12}{1 + 3/x} = 12, \text{ so}$$

$$\lim_{x \to \infty} y = \lim_{n \to \infty} \left(1 + \frac{3}{n}\right)^{4n} = e^{12}.$$

**9.** $\dfrac{n}{n^3 + 1} < \dfrac{n}{n^3} = \dfrac{1}{n^2}$, so $\displaystyle\sum_{n=1}^{\infty} \dfrac{n}{n^3 + 1}$ converges by the Comparison Test with the convergent $p$-series $\displaystyle\sum_{n=1}^{\infty} \dfrac{1}{n^2}$ $[\,p = 2 > 1\,]$.

**11.** $\displaystyle\lim_{n \to \infty} \left|\dfrac{a_{n+1}}{a_n}\right| = \lim_{n \to \infty} \left[\dfrac{(n+1)^3}{5^{n+1}} \cdot \dfrac{5^n}{n^3}\right] = \lim_{n \to \infty} \left(1 + \dfrac{1}{n}\right)^3 \cdot \dfrac{1}{5} = \dfrac{1}{5} < 1$, so $\displaystyle\sum_{n=1}^{\infty} \dfrac{n^3}{5^n}$ converges by the Ratio Test.

**13.** Let $f(x) = \dfrac{1}{x \sqrt{\ln x}}$. Then $f$ is continuous, positive, and decreasing on $[2, \infty)$, so the Integral Test applies.

$$\int_2^{\infty} f(x)\,dx = \lim_{t \to \infty} \int_2^t \frac{1}{x \sqrt{\ln x}}\,dx \quad \left[u = \ln x,\, du = \frac{1}{x}\,dx\right] = \lim_{t \to \infty} \int_{\ln 2}^{\ln t} u^{-1/2}\,du = \lim_{t \to \infty} \left[2\sqrt{u}\right]_{\ln 2}^{\ln t}$$

$$= \lim_{t \to \infty} \left(2\sqrt{\ln t} - 2\sqrt{\ln 2}\right) = \infty,$$

so the series $\displaystyle\sum_{n=2}^{\infty} \dfrac{1}{n \sqrt{\ln n}}$ diverges.

**15.** $b_n = \dfrac{\sqrt{n}}{n + 1} > 0$, $\{b_n\}$ is decreasing, and $\displaystyle\lim_{n \to \infty} b_n = 0$, so the series $\displaystyle\sum_{n=1}^{\infty} (-1)^{n-1} \dfrac{\sqrt{n}}{n + 1}$ converges by the Alternating Series Test.

**17.** $\displaystyle\lim_{n \to \infty} \left|\dfrac{a_{n+1}}{a_n}\right| = \lim_{n \to \infty} \dfrac{1 \cdot 3 \cdot 5 \cdot \cdots \cdot (2n - 1)(2n + 1)}{5^{n+1}\,(n + 1)!} \cdot \dfrac{5^n\, n!}{1 \cdot 3 \cdot 5 \cdot \cdots \cdot (2n - 1)} = \lim_{n \to \infty} \dfrac{2n + 1}{5(n + 1)} = \dfrac{2}{5} < 1$, so the series converges by the Ratio Test.

**19.** $\displaystyle\sum_{n=1}^{\infty} \dfrac{(-3)^{n-1}}{2^{3n}} = \sum_{n=1}^{\infty} \dfrac{(-3)^{n-1}}{(2^3)^n} = \sum_{n=1}^{\infty} \dfrac{(-3)^{n-1}}{8^n} = \dfrac{1}{8} \sum_{n=1}^{\infty} \dfrac{(-3)^{n-1}}{8^{n-1}} = \dfrac{1}{8} \sum_{n=1}^{\infty} \left(-\dfrac{3}{8}\right)^{n-1} = \dfrac{1}{8}\left(\dfrac{1}{1 - (-3/8)}\right)$

$$= \dfrac{1}{8} \cdot \dfrac{8}{11} = \dfrac{1}{11}$$

**21.** $\displaystyle\sum_{n=1}^{\infty} [\tan^{-1}(n + 1) - \tan^{-1} n] = \lim_{n \to \infty} s_n$

$$= \lim_{n \to \infty} [(\tan^{-1} 2 - \tan^{-1} 1) + (\tan^{-1} 3 - \tan^{-1} 2) + \cdots + (\tan^{-1}(n + 1) - \tan^{-1} n)]$$

$$= \lim_{n \to \infty} [\tan^{-1}(n + 1) - \tan^{-1} 1] = \frac{\pi}{2} - \frac{\pi}{4} = \frac{\pi}{4}$$

**23.** $1.2345345345\ldots = 1.2 + 0.0\overline{345} = \dfrac{12}{10} + \dfrac{345/10{,}000}{1 - 1/1000} = \dfrac{12}{10} + \dfrac{345}{9990} = \dfrac{4111}{3330}$

**25.** $\displaystyle\sum_{n=1}^{\infty} \frac{(-1)^{n+1}}{n^5} = 1 - \frac{1}{32} + \frac{1}{243} - \frac{1}{1024} + \frac{1}{3125} - \frac{1}{7776} + \frac{1}{16{,}807} - \frac{1}{32{,}768} + \cdots.$

Since $b_8 = \dfrac{1}{8^5} = \dfrac{1}{32{,}768} < 0.000031$, $\displaystyle\sum_{n=1}^{\infty} \frac{(-1)^{n+1}}{n^5} \approx \sum_{n=1}^{7} \frac{(-1)^{n+1}}{n^5} \approx 0.9721.$

**27.** $\displaystyle\sum_{n=1}^{\infty} \frac{1}{2 + 5^n} \approx \sum_{n=1}^{8} \frac{1}{2 + 5^n} \approx 0.18976224.$ To estimate the error, note that $\dfrac{1}{2 + 5^n} < \dfrac{1}{5^n}$, so the remainder term is

$$R_8 = \sum_{n=9}^{\infty} \frac{1}{2 + 5^n} < \sum_{n=9}^{\infty} \frac{1}{5^n} = \frac{1/5^9}{1 - 1/5} = 6.4 \times 10^{-7} \quad \left[\text{geometric series with } a = \tfrac{1}{5^9} \text{ and } r = \tfrac{1}{5}\right].$$

**29.** Use the Limit Comparison Test. $\displaystyle\lim_{n\to\infty} \left| \frac{\left(\frac{n+1}{n}\right)a_n}{a_n} \right| = \lim_{n\to\infty} \frac{n+1}{n} = \lim_{n\to\infty} \left(1 + \frac{1}{n}\right) = 1 > 0.$

Since $\sum |a_n|$ is convergent, so is $\displaystyle\sum \left|\left(\frac{n+1}{n}\right)a_n\right|$, by the Limit Comparison Test.

**31.** $\displaystyle\lim_{n\to\infty} \left|\frac{a_{n+1}}{a_n}\right| = \lim_{n\to\infty} \left[\frac{|x+2|^{n+1}}{(n+1)\,4^{n+1}} \cdot \frac{n\,4^n}{|x+2|^n}\right] = \lim_{n\to\infty} \left[\frac{n}{n+1} \frac{|x+2|}{4}\right] = \frac{|x+2|}{4} < 1 \iff |x+2| < 4$, so $R = 4$.

$|x+2| < 4 \iff -4 < x+2 < 4 \iff -6 < x < 2$. If $x = -6$, then the series $\displaystyle\sum_{n=1}^{\infty} \frac{(x+2)^n}{n\,4^n}$ becomes

$\displaystyle\sum_{n=1}^{\infty} \frac{(-4)^n}{n4^n} = \sum_{n=1}^{\infty} \frac{(-1)^n}{n}$, the alternating harmonic series, which converges by the Alternating Series Test. When $x = 2$, the

series becomes the harmonic series $\displaystyle\sum_{n=1}^{\infty} \frac{1}{n}$, which diverges. Thus, $I = [-6, 2)$.

**33.** $\displaystyle\lim_{n\to\infty} \left|\frac{a_{n+1}}{a_n}\right| = \lim_{n\to\infty} \left|\frac{2^{n+1}(x-3)^{n+1}}{\sqrt{n+4}} \cdot \frac{\sqrt{n+3}}{2^n(x-3)^n}\right| = 2\,|x-3| \lim_{n\to\infty} \sqrt{\frac{n+3}{n+4}} = 2\,|x-3| < 1 \iff |x-3| < \tfrac{1}{2},$

so $R = \tfrac{1}{2}$. $|x-3| < \tfrac{1}{2} \iff -\tfrac{1}{2} < x - 3 < \tfrac{1}{2} \iff \tfrac{5}{2} < x < \tfrac{7}{2}$. For $x = \tfrac{7}{2}$, the series $\displaystyle\sum_{n=1}^{\infty} \frac{2^n(x-3)^n}{\sqrt{n+3}}$ becomes

$\displaystyle\sum_{n=0}^{\infty} \frac{1}{\sqrt{n+3}} = \sum_{n=3}^{\infty} \frac{1}{n^{1/2}}$, which diverges $\left[p = \tfrac{1}{2} \le 1\right]$, but for $x = \tfrac{5}{2}$, we get $\displaystyle\sum_{n=0}^{\infty} \frac{(-1)^n}{\sqrt{n+3}}$, which is a convergent

alternating series, so $I = \left[\tfrac{5}{2}, \tfrac{7}{2}\right)$.

**35.**

| $n$ | $f^{(n)}(x)$ | $f^{(n)}\left(\frac{\pi}{6}\right)$ |
|---|---|---|
| 0 | $\sin x$ | $\frac{1}{2}$ |
| 1 | $\cos x$ | $\frac{\sqrt{3}}{2}$ |
| 2 | $-\sin x$ | $-\frac{1}{2}$ |
| 3 | $-\cos x$ | $-\frac{\sqrt{3}}{2}$ |
| 4 | $\sin x$ | $\frac{1}{2}$ |
| $\vdots$ | $\vdots$ | $\vdots$ |

$$\sin x = f\left(\frac{\pi}{6}\right) + f'\left(\frac{\pi}{6}\right)\left(x - \frac{\pi}{6}\right) + \frac{f''\left(\frac{\pi}{6}\right)}{2!}\left(x - \frac{\pi}{6}\right)^2 + \frac{f^{(3)}\left(\frac{\pi}{6}\right)}{3!}\left(x - \frac{\pi}{6}\right)^3 + \frac{f^{(4)}\left(\frac{\pi}{6}\right)}{4!}\left(x - \frac{\pi}{6}\right)^4 + \cdots$$

$$= \frac{1}{2}\left[1 - \frac{1}{2!}\left(x - \frac{\pi}{6}\right)^2 + \frac{1}{4!}\left(x - \frac{\pi}{6}\right)^4 - \cdots\right] + \frac{\sqrt{3}}{2}\left[\left(x - \frac{\pi}{6}\right) - \frac{1}{3!}\left(x - \frac{\pi}{6}\right)^3 + \cdots\right]$$

$$= \frac{1}{2}\sum_{n=0}^{\infty}(-1)^n\frac{1}{(2n)!}\left(x - \frac{\pi}{6}\right)^{2n} + \frac{\sqrt{3}}{2}\sum_{n=0}^{\infty}(-1)^n\frac{1}{(2n+1)!}\left(x - \frac{\pi}{6}\right)^{2n+1}$$

**37.** $\dfrac{1}{1+x} = \dfrac{1}{1-(-x)} = \displaystyle\sum_{n=0}^{\infty}(-x)^n = \sum_{n=0}^{\infty}(-1)^n\,x^n$ for $|x| < 1$ $\Rightarrow$ $\dfrac{x^2}{1+x} = \displaystyle\sum_{n=0}^{\infty}(-1)^n\,x^{n+2}$ with $R = 1$.

**39.** $\displaystyle\int\frac{1}{4-x}\,dx = -\ln(4-x) + C$ and

$$\int\frac{1}{4-x}\,dx = \frac{1}{4}\int\frac{1}{1-x/4}\,dx = \frac{1}{4}\int\sum_{n=0}^{\infty}\left(\frac{x}{4}\right)^n dx = \frac{1}{4}\int\sum_{n=0}^{\infty}\frac{x^n}{4^n}\,dx = \frac{1}{4}\sum_{n=0}^{\infty}\frac{x^{n+1}}{4^n(n+1)} + C. \text{ So}$$

$$\ln(4-x) = -\frac{1}{4}\sum_{n=0}^{\infty}\frac{x^{n+1}}{4^n(n+1)} + C = -\sum_{n=0}^{\infty}\frac{x^{n+1}}{4^{n+1}(n+1)} + C = -\sum_{n=1}^{\infty}\frac{x^n}{n4^n} + C. \text{ Putting } x = 0, \text{ we get } C = \ln 4.$$

Thus, $f(x) = \ln(4-x) = \ln 4 - \displaystyle\sum_{n=1}^{\infty}\frac{x^n}{n4^n}$. The series converges for $|x/4| < 1$ $\Leftrightarrow$ $|x| < 4$, so $R = 4$.

*Another solution:*

$$\ln(4-x) = \ln[4(1-x/4)] = \ln 4 + \ln(1-x/4) = \ln 4 + \ln[1+(-x/4)]$$

$$= \ln 4 + \sum_{n=1}^{\infty}(-1)^{n+1}\frac{(-x/4)^n}{n} \text{ [from Table 1]} = \ln 4 + \sum_{n=1}^{\infty}(-1)^{2n+1}\frac{x^n}{n4^n} = \ln 4 - \sum_{n=1}^{\infty}\frac{x^n}{n4^n}.$$

**41.** $\sin x = \displaystyle\sum_{n=0}^{\infty}\frac{(-1)^n\,x^{2n+1}}{(2n+1)!}$ $\Rightarrow$ $\sin(x^4) = \displaystyle\sum_{n=0}^{\infty}\frac{(-1)^n\,(x^4)^{2n+1}}{(2n+1)!} = \sum_{n=0}^{\infty}\frac{(-1)^n\,x^{8n+4}}{(2n+1)!}$ for all $x$, so the radius of

convergence is $\infty$.

**43.** $f(x) = \dfrac{1}{\sqrt[4]{16 - x}} = \dfrac{1}{\sqrt[4]{16(1 - x/16)}} = \dfrac{1}{\sqrt[4]{16}\left(1 - \frac{1}{16}x\right)^{1/4}} = \frac{1}{2}\left(1 - \frac{1}{16}x\right)^{-1/4}$

$\qquad = \dfrac{1}{2}\left[1 + \left(-\frac{1}{4}\right)\left(-\dfrac{x}{16}\right) + \dfrac{\left(-\frac{1}{4}\right)\left(-\frac{5}{4}\right)}{2!}\left(-\dfrac{x}{16}\right)^2 + \dfrac{\left(-\frac{1}{4}\right)\left(-\frac{5}{4}\right)\left(-\frac{9}{4}\right)}{3!}\left(-\dfrac{x}{16}\right)^3 + \cdots\right]$

$\qquad = \dfrac{1}{2} + \displaystyle\sum_{n=1}^{\infty} \dfrac{1 \cdot 5 \cdot 9 \cdot \cdots \cdot (4n - 3)}{2 \cdot 4^n \cdot n! \cdot 16^n} x^n = \dfrac{1}{2} + \displaystyle\sum_{n=1}^{\infty} \dfrac{1 \cdot 5 \cdot 9 \cdot \cdots \cdot (4n - 3)}{2^{6n+1}\, n!} x^n$

for $\left|-\dfrac{x}{16}\right| < 1 \quad \Leftrightarrow \quad |x| < 16$, so $R = 16$.

**45.** $e^x = \displaystyle\sum_{n=0}^{\infty} \dfrac{x^n}{n!}$, so $\dfrac{e^x}{x} = \dfrac{1}{x}\displaystyle\sum_{n=0}^{\infty} \dfrac{x^n}{n!} = \displaystyle\sum_{n=0}^{\infty} \dfrac{x^{n-1}}{n!} = x^{-1} + \displaystyle\sum_{n=1}^{\infty} \dfrac{x^{n-1}}{n!} = \dfrac{1}{x} + \displaystyle\sum_{n=1}^{\infty} \dfrac{x^{n-1}}{n!}$ and

$\displaystyle\int \dfrac{e^x}{x}\, dx = C + \ln|x| + \sum_{n=1}^{\infty} \dfrac{x^n}{n \cdot n!}$.

**47. (a)**

| $n$ | $f^{(n)}(x)$ | $f^{(n)}(1)$ |
|-----|--------------|--------------|
| 0 | $x^{1/2}$ | 1 |
| 1 | $\frac{1}{2}x^{-1/2}$ | $\frac{1}{2}$ |
| 2 | $-\frac{1}{4}x^{-3/2}$ | $-\frac{1}{4}$ |
| 3 | $\frac{3}{8}x^{-5/2}$ | $\frac{3}{8}$ |
| 4 | $-\frac{15}{16}x^{-7/2}$ | $-\frac{15}{16}$ |
| $\vdots$ | $\vdots$ | $\vdots$ |

$\sqrt{x} \approx T_3(x) = 1 + \dfrac{1/2}{1!}(x - 1) - \dfrac{1/4}{2!}(x - 1)^2 + \dfrac{3/8}{3!}(x - 1)^3$

$\qquad = 1 + \frac{1}{2}(x - 1) - \frac{1}{8}(x - 1)^2 + \frac{1}{16}(x - 1)^3$

**(b)**

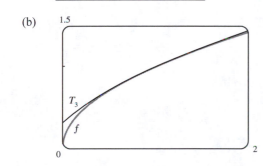

**(c)** $|R_3(x)| \le \dfrac{M}{4!}|x - 1|^4$, where $\left|f^{(4)}(x)\right| \le M$ with

$f^{(4)}(x) = -\dfrac{15}{16}x^{-7/2}$. Now $0.9 \le x \le 1.1 \quad \Rightarrow$

$-0.1 \le x - 1 \le 0.1 \quad \Rightarrow \quad (x - 1)^4 \le (0.1)^4$,

and letting $x = 0.9$ gives $M = \dfrac{15}{16(0.9)^{7/2}}$, so

$|R_3(x)| \le \dfrac{15}{16(0.9)^{7/2}\, 4!}(0.1)^4 \approx 0.000\,005\,648$

$\qquad\qquad\qquad\qquad \approx 0.000\,006 = 6 \times 10^{-6}$

**(d)**

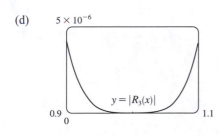

From the graph of $|R_3(x)| = |\sqrt{x} - T_3(x)|$, it appears that

the error is less than $5 \times 10^{-6}$ on $[0.9, 1.1]$.

**49.** $\sin x = \displaystyle\sum_{n=0}^{\infty} (-1)^n \dfrac{x^{2n+1}}{(2n + 1)!} = x - \dfrac{x^3}{3!} + \dfrac{x^5}{5!} - \dfrac{x^7}{7!} + \cdots$, so $\sin x - x = -\dfrac{x^3}{3!} + \dfrac{x^5}{5!} - \dfrac{x^7}{7!} + \cdots$ and

$\dfrac{\sin x - x}{x^3} = -\dfrac{1}{3!} + \dfrac{x^2}{5!} - \dfrac{x^4}{7!} + \cdots$. Thus, $\displaystyle\lim_{x \to 0} \dfrac{\sin x - x}{x^3} = \lim_{x \to 0}\left(-\dfrac{1}{6} + \dfrac{x^2}{120} - \dfrac{x^4}{5040} + \cdots\right) = -\dfrac{1}{6}$.

# ☐ FOCUS ON PROBLEM SOLVING

**1.** It would be far too much work to compute 15 derivatives of $f$. The key idea is to remember that $f^{(n)}(0)$ occurs in the

coefficient of $x^n$ in the Maclaurin series of $f$. We start with the Maclaurin series for sin: $\sin x = x - \dfrac{x^3}{3!} + \dfrac{x^5}{5!} - \cdots$.

Then $\sin(x^3) = x^3 - \dfrac{x^9}{3!} + \dfrac{x^{15}}{5!} - \cdots$, and so the coefficient of $x^{15}$ is $\dfrac{f^{(15)}(0)}{15!} = \dfrac{1}{5!}$. Therefore,

$$f^{(15)}(0) = \frac{15!}{5!} = 6 \cdot 7 \cdot 8 \cdot 9 \cdot 10 \cdot 11 \cdot 12 \cdot 13 \cdot 14 \cdot 15 = 10{,}897{,}286{,}400.$$

**3.** (a) At each stage, each side is replaced by four shorter sides, each of length
$\frac{1}{3}$ of the side length at the preceding stage. Writing $s_0$ and $\ell_0$ for the
number of sides and the length of the side of the initial triangle, we
generate the table at right. In general, we have $s_n = 3 \cdot 4^n$ and
$\ell_n = \left(\frac{1}{3}\right)^n$, so the length of the perimeter at the $n$th stage of construction
is $p_n = s_n \ell_n = 3 \cdot 4^n \cdot \left(\frac{1}{3}\right)^n = 3 \cdot \left(\frac{4}{3}\right)^n$.

| | |
|---|---|
| $s_0 = 3$ | $\ell_0 = 1$ |
| $s_1 = 3 \cdot 4$ | $\ell_1 = 1/3$ |
| $s_2 = 3 \cdot 4^2$ | $\ell_2 = 1/3^2$ |
| $s_3 = 3 \cdot 4^3$ | $\ell_3 = 1/3^3$ |
| $\vdots$ | $\vdots$ |

(b) $p_n = \dfrac{4^n}{3^{n-1}} = 4\left(\dfrac{4}{3}\right)^{n-1}$. Since $\frac{4}{3} > 1$, $p_n \to \infty$ as $n \to \infty$.

(c) The area of each of the small triangles added at a given stage is one-ninth of the area of the triangle added at the preceding
stage. Let $a$ be the area of the original triangle. Then the area $a_n$ of each of the small triangles added at stage $n$ is

$a_n = a \cdot \dfrac{1}{9^n} = \dfrac{a}{9^n}$. Since a small triangle is added to each side at every stage, it follows that the total area $A_n$ added to the

figure at the $n$th stage is $A_n = s_{n-1} \cdot a_n = 3 \cdot 4^{n-1} \cdot \dfrac{a}{9^n} = a \cdot \dfrac{4^{n-1}}{3^{2n-1}}$. Then the total area enclosed by the snowflake

curve is $A = a + A_1 + A_2 + A_3 + \cdots = a + a \cdot \dfrac{1}{3} + a \cdot \dfrac{4}{3^3} + a \cdot \dfrac{4^2}{3^5} + a \cdot \dfrac{4^3}{3^7} + \cdots$. After the first term, this is a

geometric series with common ratio $\dfrac{4}{9}$, so $A = a + \dfrac{a/3}{1 - \frac{4}{9}} = a + \dfrac{a}{3} \cdot \dfrac{9}{5} = \dfrac{8a}{5}$. But the area of the original equilateral

triangle with side 1 is $a = \dfrac{1}{2} \cdot 1 \cdot \sin\dfrac{\pi}{3} = \dfrac{\sqrt{3}}{4}$. So the area enclosed by the snowflake curve is $\dfrac{8}{5} \cdot \dfrac{\sqrt{3}}{4} = \dfrac{2\sqrt{3}}{5}$.

**5.** $\ln\left(1 - \dfrac{1}{n^2}\right) = \ln\left(\dfrac{n^2 - 1}{n^2}\right) = \ln\dfrac{(n+1)(n-1)}{n^2} = \ln[(n+1)(n-1)] - \ln n^2$

$\qquad\qquad = \ln(n+1) + \ln(n-1) - 2\ln n = \ln(n-1) - \ln n - \ln n + \ln(n+1)$

$\qquad\qquad = \ln\dfrac{n-1}{n} - [\ln n - \ln(n+1)] = \ln\dfrac{n-1}{n} - \ln\dfrac{n}{n+1}$.

Let $s_k = \displaystyle\sum_{n=2}^{k} \ln\left(1 - \dfrac{1}{n^2}\right) = \sum_{n=2}^{k}\left(\ln\dfrac{n-1}{n} - \ln\dfrac{n}{n+1}\right)$ for $k \geq 2$. Then

$s_k = \left(\ln\dfrac{1}{2} - \ln\dfrac{2}{3}\right) + \left(\ln\dfrac{2}{3} - \ln\dfrac{3}{4}\right) + \cdots + \left(\ln\dfrac{k-1}{k} - \ln\dfrac{k}{k+1}\right) = \ln\dfrac{1}{2} - \ln\dfrac{k}{k+1}$, so

$\displaystyle\sum_{n=2}^{\infty} \ln\left(1 - \dfrac{1}{n^2}\right) = \lim_{k \to \infty} s_k = \lim_{k \to \infty}\left(\ln\dfrac{1}{2} - \ln\dfrac{k}{k+1}\right) = \ln\dfrac{1}{2} - \ln 1 = \ln 1 - \ln 2 - \ln 1 = -\ln 2.$

**7.** $u = 1 + \dfrac{x^3}{3!} + \dfrac{x^6}{6!} + \dfrac{x^9}{9!} + \cdots , \quad v = x + \dfrac{x^4}{4!} + \dfrac{x^7}{7!} + \dfrac{x^{10}}{10!} + \cdots , \quad w = \dfrac{x^2}{2!} + \dfrac{x^5}{5!} + \dfrac{x^8}{8!} + \cdots.$

Use the Ratio Test to show that the series for $u$, $v$, and $w$ have positive radii of convergence ($\infty$ in each case), so

Theorem 8.6.2 applies, and hence, we may differentiate each of these series:

$$\frac{du}{dx} = \frac{3x^2}{3!} + \frac{6x^5}{6!} + \frac{9x^8}{9!} + \cdots = \frac{x^2}{2!} + \frac{x^5}{5!} + \frac{x^8}{8!} + \cdots = w$$

Similarly, $\dfrac{dv}{dx} = 1 + \dfrac{x^3}{3!} + \dfrac{x^6}{6!} + \dfrac{x^9}{9!} + \cdots = u$, and $\dfrac{dw}{dx} = x + \dfrac{x^4}{4!} + \dfrac{x^7}{7!} + \dfrac{x^{10}}{10!} + \cdots = v.$

So $u' = w$, $v' = u$, and $w' = v$. Now differentiate the left hand side of the desired equation:

$$\frac{d}{dx}(u^3 + v^3 + w^3 - 3uvw) = 3u^2u' + 3v^2v' + 3w^2w' - 3(u'vw + uv'w + uvw')$$

$$= 3u^2w + 3v^2u + 3w^2v - 3(vw^2 + u^2w + uv^2) = 0 \quad \Rightarrow$$

$u^3 + v^3 + w^3 - 3uvw = C$. To find the value of the constant $C$, we put $x = 0$ in the last equation and get

$1^3 + 0^3 + 0^3 - 3(1 \cdot 0 \cdot 0) = C \quad \Rightarrow \quad C = 1$, so $u^3 + v^3 + w^3 - 3uvw = 1$.

**9.** If $L$ is the length of a side of the equilateral triangle, then the area is $A = \frac{1}{2}L \cdot \frac{\sqrt{3}}{2}L = \frac{\sqrt{3}}{4}L^2$ and so $L^2 = \frac{4}{\sqrt{3}}A$.

Let $r$ be the radius of one of the circles. When there are $n$ rows of circles, the figure shows that

$$L = \sqrt{3}\,r + r + (n-2)(2r) + r + \sqrt{3}\,r = r\left(2n - 2 + 2\sqrt{3}\right), \text{ so } r = \frac{L}{2\left(n + \sqrt{3} - 1\right)}.$$

The number of circles is $1 + 2 + \cdots + n = \dfrac{n(n+1)}{2}$, and so the total area of the circles is

$$A_n = \frac{n(n+1)}{2}\pi r^2 = \frac{n(n+1)}{2}\pi \frac{L^2}{4\left(n + \sqrt{3} - 1\right)^2}$$

$$= \frac{n(n+1)}{2}\pi \frac{4A/\sqrt{3}}{4\left(n + \sqrt{3} - 1\right)^2} = \frac{n(n+1)}{\left(n + \sqrt{3} - 1\right)^2}\frac{\pi A}{2\sqrt{3}} \quad \Rightarrow$$

$$\frac{A_n}{A} = \frac{n(n+1)}{\left(n + \sqrt{3} - 1\right)^2}\frac{\pi}{2\sqrt{3}}$$

$$= \frac{1 + 1/n}{\left[1 + (\sqrt{3} - 1)/n\right]^2}\frac{\pi}{2\sqrt{3}} \to \frac{\pi}{2\sqrt{3}} \text{ as } n \to \infty$$

**11.** By Table 1 in Section 8.7, $\tan^{-1}x = \displaystyle\sum_{n=0}^{\infty}(-1)^n\frac{x^{2n+1}}{2n+1}$ for $|x| < 1$. In particular, for $x = \dfrac{1}{\sqrt{3}}$, we

have $\dfrac{\pi}{6} = \tan^{-1}\left(\dfrac{1}{\sqrt{3}}\right) = \displaystyle\sum_{n=0}^{\infty}(-1)^n\frac{(1/\sqrt{3})^{2n+1}}{2n+1} = \sum_{n=0}^{\infty}(-1)^n\left(\dfrac{1}{3}\right)^n\frac{1}{\sqrt{3}}\frac{1}{2n+1}$, so

$$\pi = \frac{6}{\sqrt{3}}\sum_{n=0}^{\infty}\frac{(-1)^n}{(2n+1)3^n} = 2\sqrt{3}\sum_{n=0}^{\infty}\frac{(-1)^n}{(2n+1)3^n} = 2\sqrt{3}\left(1 + \sum_{n=1}^{\infty}\frac{(-1)^n}{(2n+1)3^n}\right) \Rightarrow \sum_{n=1}^{\infty}\frac{(-1)^n}{(2n+1)3^n} = \frac{\pi}{2\sqrt{3}} - 1.$$

**13.** Let $f(x)$ denote the left-hand side of the equation $1 + \dfrac{x}{2!} + \dfrac{x^2}{4!} + \dfrac{x^3}{6!} + \dfrac{x^4}{8!} + \cdots = 0$. If $x \geq 0$, then $f(x) \geq 1$ and there are

no solutions of the equation. Note that $f(-x^2) = 1 - \dfrac{x^2}{2!} + \dfrac{x^4}{4!} - \dfrac{x^6}{6!} + \dfrac{x^8}{8!} - \cdots = \cos x$. The solutions of $\cos x = 0$ for

$x < 0$ are given by $x = \dfrac{\pi}{2} - \pi k$, where $k$ is a positive integer. Thus, the solutions of $f(x) = 0$ are $x = -\left(\dfrac{\pi}{2} - \pi k\right)^2$, where

$k$ is a positive integer.

**15.** Call the series $S$. We group the terms according to the number of digits in their denominators:

$$S = \underbrace{\left(\tfrac{1}{1} + \tfrac{1}{2} + \cdots + \tfrac{1}{8} + \tfrac{1}{9}\right)}_{g_1} + \underbrace{\left(\tfrac{1}{11} + \cdots + \tfrac{1}{99}\right)}_{g_2} + \underbrace{\left(\tfrac{1}{111} + \cdots + \tfrac{1}{999}\right)}_{g_3} + \cdots$$

Now in the group $g_n$, since we have 9 choices for each of the $n$ digits in the denominator, there are $9^n$ terms.

Furthermore, each term in $g_n$ is less than $\dfrac{1}{10^{n-1}}$ [except for the first term in $g_1$]. So $g_n < 9^n \cdot \dfrac{1}{10^{n-1}} = 9\left(\dfrac{9}{10}\right)^{n-1}$.

Now $\displaystyle\sum_{n=1}^{\infty} 9\left(\dfrac{9}{10}\right)^{n-1}$ is a geometric series with $a = 9$ and $r = \dfrac{9}{10} < 1$. Therefore, by the Comparison Test,

$$S = \sum_{n=1}^{\infty} g_n < \sum_{n=1}^{\infty} 9\left(\dfrac{9}{10}\right)^{n-1} = \dfrac{9}{1 - 9/10} = 90.$$

# 9 □ VECTORS AND THE GEOMETRY OF SPACE

## 9.1 Three-Dimensional Coordinate Systems

**1.** We start at the origin, which has coordinates $(0, 0, 0)$. First we move 4 units along the positive $x$-axis, affecting only the
$x$-coordinate, bringing us to the point $(4, 0, 0)$. We then move 3 units straight downward, in the negative $z$-direction. Thus
only the $z$-coordinate is affected, and we arrive at $(4, 0, -3)$.

**3.** The distance from a point to the $xz$-plane is the absolute value of the $y$-coordinate of the point. $Q(-5, -1, 4)$ has the
$y$-coordinate with the smallest absolute value, so $Q$ is the point closest to the $xz$-plane. $R(0, 3, 8)$ must lie in the $yz$-plane
since the distance from $R$ to the $yz$-plane, given by the $x$-coordinate of $R$, is 0.

**5.** The equation $x + y = 2$ represents the set of all points in
$\mathbb{R}^3$ whose $x$- and $y$-coordinates have a sum of 2, or
equivalently where $y = 2 - x$. This is the set
$\{(x, 2 - x, z) \mid x \in \mathbb{R}, z \in \mathbb{R}\}$ which is a vertical plane
that intersects the $xy$-plane in the line $y = 2 - x$, $z = 0$.

**7.** (a) We can find the lengths of the sides of the triangle by using the distance formula between pairs of vertices:

$$|PQ| = \sqrt{(7 - 3)^2 + [0 - (-2)]^2 + [1 - (-3)]^2} = \sqrt{16 + 4 + 16} = 6$$

$$|QR| = \sqrt{(1 - 7)^2 + (2 - 0)^2 + (1 - 1)^2} = \sqrt{36 + 4 + 0} = \sqrt{40} = 2\sqrt{10}$$

$$|RP| = \sqrt{(3 - 1)^2 + (-2 - 2)^2 + (-3 - 1)^2} = \sqrt{4 + 16 + 16} = 6$$

The longest side is $QR$, but the Pythagorean Theorem is not satisfied: $|PQ|^2 + |RP|^2 \neq |QR|^2$. Thus $PQR$ is not a right
triangle. $PQR$ is isosceles, as two sides have the same length.

(b) Compute the lengths of the sides of the triangle by using the distance formula between pairs of vertices:

$$|PQ| = \sqrt{(4 - 2)^2 + [1 - (-1)]^2 + (1 - 0)^2} = \sqrt{4 + 4 + 1} = 3$$

$$|QR| = \sqrt{(4 - 4)^2 + (-5 - 1)^2 + (4 - 1)^2} = \sqrt{0 + 36 + 9} = \sqrt{45} = 3\sqrt{5}$$

$$|RP| = \sqrt{(2 - 4)^2 + [-1 - (-5)]^2 + (0 - 4)^2} = \sqrt{4 + 16 + 16} = 6$$

Since the Pythagorean Theorem is satisfied by $|PQ|^2 + |RP|^2 = |QR|^2$, $PQR$ is a right triangle. $PQR$ is not isosceles,
as no two sides have the same length.

**9.** (a) First we find the distances between points:

$$|AB| = \sqrt{(3-2)^2 + (7-4)^2 + (-2-2)^2} = \sqrt{26}$$

$$|BC| = \sqrt{(1-3)^2 + (3-7)^2 + [3-(-2)]^2} = \sqrt{45} = 3\sqrt{5}$$

$$|AC| = \sqrt{(1-2)^2 + (3-4)^2 + (3-2)^2} = \sqrt{3}$$

In order for the points to lie on a straight line, the sum of the two shortest distances must be equal to the longest distance.

Since $\sqrt{26} + \sqrt{3} \neq 3\sqrt{5}$, the three points do not lie on a straight line.

(b) First we find the distances between points:

$$|DE| = \sqrt{(1-0)^2 + [-2-(-5)]^2 + (4-5)^2} = \sqrt{11}$$

$$|EF| = \sqrt{(3-1)^2 + [4-(-2)]^2 + (2-4)^2} = \sqrt{44} = 2\sqrt{11}$$

$$|DF| = \sqrt{(3-0)^2 + [4-(-5)]^2 + (2-5)^2} = \sqrt{99} = 3\sqrt{11}$$

Since $|DE| + |EF| = |DF|$, the three points lie on a straight line.

**11.** The radius of the sphere is the distance between $(4, 3, -1)$ and $(3, 8, 1)$: $r = \sqrt{(3-4)^2 + (8-3)^2 + [1-(-1)]^2} = \sqrt{30}$.

Thus, an equation of the sphere is $(x-3)^2 + (y-8)^2 + (z-1)^2 = 30$.

**13.** Completing squares in the equation $x^2 + y^2 + z^2 - 6x + 4y - 2z = 11$ gives

$(x^2 - 6x + 9) + (y^2 + 4y + 4) + (z^2 - 2z + 1) = 11 + 9 + 4 + 1 \quad \Rightarrow \quad (x-3)^2 + (y+2)^2 + (z-1)^2 = 25$, which we

recognize as an equation of a sphere with center $(3, -2, 1)$ and radius 5.

**15.** Completing squares in the equation $2x^2 - 8x + 2y^2 + 2z^2 + 24z = 1$ gives

$2(x^2 - 4x + 4) + 2y^2 + 2(z^2 + 12z + 36) = 1 + 8 + 72 \quad \Rightarrow \quad 2(x-2)^2 + 2y^2 + 2(z+6)^2 = 81 \quad \Rightarrow$

$(x-2)^2 + y^2 + (z+6)^2 = \frac{81}{2}$, which we recognize as an equation of a sphere with center $(2, 0, -6)$ and

radius $\sqrt{\frac{81}{2}} = 9/\sqrt{2}$.

**17.** (a) If the midpoint of the line segment from $P_1(x_1, y_1, z_1)$ to $P_2(x_2, y_2, z_2)$ is $Q = \left( \dfrac{x_1 + x_2}{2}, \dfrac{y_1 + y_2}{2}, \dfrac{z_1 + z_2}{2} \right)$,

then the distances $|P_1Q|$ and $|QP_2|$ are equal, and each is half of $|P_1P_2|$. We verify that this is the case:

$$|P_1P_2| = \sqrt{(x_2 - x_1)^2 + (y_2 - y_1)^2 + (z_2 - z_1)^2}$$

$$|P_1Q| = \sqrt{\left[\tfrac{1}{2}(x_1 + x_2) - x_1\right]^2 + \left[\tfrac{1}{2}(y_1 + y_2) - y_1\right]^2 + \left[\tfrac{1}{2}(z_1 + z_2) - z_1\right]^2}$$

$$= \sqrt{\left(\tfrac{1}{2}x_2 - \tfrac{1}{2}x_1\right)^2 + \left(\tfrac{1}{2}y_2 - \tfrac{1}{2}y_1\right)^2 + \left(\tfrac{1}{2}z_2 - \tfrac{1}{2}z_1\right)^2}$$

$$= \sqrt{\left(\tfrac{1}{2}\right)^2 \left[(x_2 - x_1)^2 + (y_2 - y_1)^2 + (z_2 - z_1)^2\right]} = \tfrac{1}{2}\sqrt{(x_2 - x_1)^2 + (y_2 - y_1)^2 + (z_2 - z_1)^2}$$

$$= \tfrac{1}{2}|P_1P_2|$$

$$|QP_2| = \sqrt{\left[x_2 - \tfrac{1}{2}(x_1 + x_2)\right]^2 + \left[y_2 - \tfrac{1}{2}(y_1 + y_2)\right]^2 + \left[z_2 - \tfrac{1}{2}(z_1 + z_2)\right]^2}$$

$$= \sqrt{\left(\tfrac{1}{2}x_2 - \tfrac{1}{2}x_1\right)^2 + \left(\tfrac{1}{2}y_2 - \tfrac{1}{2}y_1\right)^2 + \left(\tfrac{1}{2}z_2 - \tfrac{1}{2}z_1\right)^2} = \sqrt{\left(\tfrac{1}{2}\right)^2\left[(x_2 - x_1)^2 + (y_2 - y_1)^2 + (z_2 - z_1)^2\right]}$$

$$= \tfrac{1}{2}\sqrt{(x_2 - x_1)^2 + (y_2 - y_1)^2 + (z_2 - z_1)^2} = \tfrac{1}{2}|P_1P_2|$$

So $Q$ is indeed the midpoint of $P_1P_2$.

(b) By part (a), the midpoints of sides $AB$, $BC$ and $CA$ are $P_1\left(-\tfrac{1}{2}, 1, 4\right)$, $P_2\left(1, \tfrac{1}{2}, 5\right)$ and $P_3\left(\tfrac{5}{2}, \tfrac{3}{2}, 4\right)$. (Recall that a median of a triangle is a line segment from a vertex to the midpoint of the opposite side.) Then the lengths of the medians are:

$$|AP_2| = \sqrt{0^2 + \left(\tfrac{1}{2} - 2\right)^2 + (5 - 3)^2} = \sqrt{\tfrac{9}{4} + 4} = \sqrt{\tfrac{25}{4}} = \tfrac{5}{2}$$

$$|BP_3| = \sqrt{\left(\tfrac{5}{2} + 2\right)^2 + \left(\tfrac{3}{2}\right)^2 + (4 - 5)^2} = \sqrt{\tfrac{81}{4} + \tfrac{9}{4} + 1} = \sqrt{\tfrac{94}{4}} = \tfrac{1}{2}\sqrt{94}$$

$$|CP_1| = \sqrt{\left(-\tfrac{1}{2} - 4\right)^2 + (1 - 1)^2 + (4 - 5)^2} = \sqrt{\tfrac{81}{4} + 1} = \tfrac{1}{2}\sqrt{85}$$

**19.** (a) Since the sphere touches the $xy$-plane, its radius is the distance from its center, $(2, -3, 6)$, to the $xy$-plane, namely 6.

Therefore $r = 6$ and an equation of the sphere is $(x - 2)^2 + (y + 3)^2 + (z - 6)^2 = 6^3 = 36$.

(b) The radius of this sphere is the distance from its center $(2, -3, 6)$ to the $yz$-plane, which is 2. Therefore, an equation is

$(x - 2)^2 + (y + 3)^2 + (z - 6)^2 = 4$.

(c) Here the radius is the distance from the center $(2, -3, 6)$ to the $xz$-plane, which is 3. Therefore, an equation is

$(x - 2)^2 + (y + 3)^2 + (z - 6)^2 = 9$.

**21.** The equation $x = 5$ represents a plane parallel to the $yz$-plane and 5 units in front of it.

**23.** The inequality $y < 8$ represents a half-space consisting of all points to the left of the plane $y = 8$.

**25.** The inequality $0 \le z \le 6$ represents all points on or between the horizontal planes $z = 0$ (the $xy$-plane) and $z = 6$.

**27.** Because $z = -1$, all points in the region must lie in the horizontal plane $z = -1$. In addition, $x^2 + y^2 = 4$, so the region

consists of all points that lie on a circle with radius 2 and center on the $z$-axis that is contained in the plane $z = -1$.

**29.** The inequality $x^2 + y^2 + z^2 \le 3$ is equivalent to $\sqrt{x^2 + y^2 + z^2} \le \sqrt{3}$, so the region consists of those points whose distance

from the origin is at most $\sqrt{3}$. This is the set of all points on or inside the sphere with radius $\sqrt{3}$ and center $(0, 0, 0)$.

**31.** Here $x^2 + z^2 \le 9$ or equivalently $\sqrt{x^2 + z^2} \le 3$ which describes the set of all points in $\mathbb{R}^3$ whose distance from the $y$-axis is

at most 3. Thus, the inequality represents the region consisting of all points on or inside a circular cylinder of radius 3 with

axis the $y$-axis.

**33.** This describes all points whose $x$-coordinate is between 0 and 5, that is, $0 < x < 5$.

**35.** This describes a region all of whose points have a distance to the origin which is greater than $r$, but smaller than $R$. So

inequalities describing the region are $r < \sqrt{x^2 + y^2 + z^2} < R$, or $r^2 < x^2 + y^2 + z^2 < R^2$.

**37.** (a) To find the $x$- and $y$-coordinates of the point $P$, we project it onto $L_2$ and

project the resulting point $Q$ onto the $x$- and $y$-axes. To find the

$z$-coordinate, we project $P$ onto either the $xz$-plane or the $yz$-plane

(using our knowledge of its $x$- or $y$-coordinate) and then project the

resulting point onto the $z$-axis. (Or, we could draw a line parallel to $QO$

from $P$ to the $z$-axis.) The coordinates of $P$ are $(2, 1, 4)$.

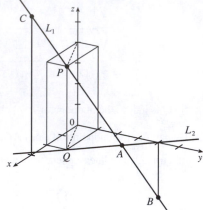

(b) $A$ is the intersection of $L_1$ and $L_2$, $B$ is directly below the

$y$-intercept of $L_2$, and $C$ is directly above the $x$-intercept of $L_2$.

**39.** We need to find a set of points $\{P(x, y, z) \,|\, |AP| = |BP|\}$.

$$\sqrt{(x+1)^2 + (y-5)^2 + (z-3)^2} = \sqrt{(x-6)^2 + (y-2)^2 + (z+2)^2} \quad \Rightarrow$$

$$(x+1)^2 + (y-5) + (z-3)^2 = (x-6)^2 + (y-2)^2 + (z+2)^2 \quad \Rightarrow$$

$$x^2 + 2x + 1 + y^2 - 10y + 25 + z^2 - 6z + 9 = x^2 - 12x + 36 + y^2 - 4y + 4 + z^2 + 4z + 4 \quad \Rightarrow \quad 14x - 6y - 10z = 9.$$

Thus the set of points is a plane perpendicular to the line segment joining $A$ and $B$ (since this plane must contain the

perpendicular bisector of the line segment $AB$).

**41.** The sphere $x^2 + y^2 + z^2 = 4$ has center $(0, 0, 0)$ and radius 2. Completing squares in $x^2 - 4x + y^2 - 4y + z^2 - 4z = -11$

gives $(x^2 - 4x + 4) + (y^2 - 4y + 4) + (z^2 - 4z + 4) = -11 + 4 + 4 + 4 \quad \Rightarrow \quad (x-2)^2 + (y-2)^2 + (z-2)^2 = 1$,

so this is the sphere with center $(2, 2, 2)$ and radius 1. The (shortest) distance between the spheres is measured along

the line segment connecting their centers. The distance between $(0, 0, 0)$ and $(2, 2, 2)$ is

$\sqrt{(2-0)^2 + (2-0)^2 + (2-0)^2} = \sqrt{12} = 2\sqrt{3}$, and subtracting the radius of each circle, the distance between the

spheres is $2\sqrt{3} - 2 - 1 = 2\sqrt{3} - 3$.

## 9.2   Vectors

**1.** (a) The cost of a theater ticket is a scalar, because it has only magnitude.

(b) The current in a river is a vector, because it has both magnitude (the speed of the current) and direction at any given
location.

(c) If we assume that the initial path is linear, the initial flight path from Houston to Dallas is a vector, because it has both
magnitude (distance) and direction.

(d) The population of the world is a scalar, because it has only magnitude.

**3.** Vectors are equal when they share the same length and direction (but not necessarily location). Using the symmetry of the

parallelogram as a guide, we see that $\overrightarrow{AB} = \overrightarrow{DC}$, $\overrightarrow{DA} = \overrightarrow{CB}$, $\overrightarrow{DE} = \overrightarrow{EB}$, and $\overrightarrow{EA} = \overrightarrow{CE}$.

**5.** (a)        (b)        (c)        (d)

**7.** $\mathbf{a} = \langle 2 - (-1), 2 - 3 \rangle = \langle 3, -1 \rangle$

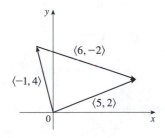

**9.** $\mathbf{a} = \langle 2 - 0, 3 - 3, -1 - 1 \rangle = \langle 2, 0, -2 \rangle$

**11.** $\langle -1, 4 \rangle + \langle 6, -2 \rangle = \langle -1 + 6, 4 + (-2) \rangle = \langle 5, 2 \rangle$

**13.** $\langle 0, 1, 2 \rangle + \langle 0, 0, -3 \rangle = \langle 0 + 0, 1 + 0, 2 + (-3) \rangle$

$$= \langle 0, 1, -1 \rangle$$

**15.** $\mathbf{a} + \mathbf{b} = \langle 5 + (-3), -12 + (-6) \rangle = \langle 2, -18 \rangle$

$2\mathbf{a} + 3\mathbf{b} = \langle 10, -24 \rangle + \langle -9, -18 \rangle = \langle 1, -42 \rangle$

$|\mathbf{a}| = \sqrt{5^2 + (-12)^2} = \sqrt{169} = 13$

$|\mathbf{a} - \mathbf{b}| = |\langle 5 - (-3), -12 - (-6) \rangle| = |\langle 8, -6 \rangle| = \sqrt{8^2 + (-6)^2} = \sqrt{100} = 10$

**17.** $\mathbf{a} + \mathbf{b} = (\mathbf{i} + 2\mathbf{j} - 3\mathbf{k}) + (-2\mathbf{i} - \mathbf{j} + 5\mathbf{k}) = -\mathbf{i} + \mathbf{j} + 2\mathbf{k}$

$2\mathbf{a} + 3\mathbf{b} = 2(\mathbf{i} + 2\mathbf{j} - 3\mathbf{k}) + 3(-2\mathbf{i} - \mathbf{j} + 5\mathbf{k}) = 2\mathbf{i} + 4\mathbf{j} - 6\mathbf{k} - 6\mathbf{i} - 3\mathbf{j} + 15\mathbf{k} = -4\mathbf{i} + \mathbf{j} + 9\mathbf{k}$

$|\mathbf{a}| = \sqrt{1^2 + 2^2 + (-3)^2} = \sqrt{14}$

$|\mathbf{a} - \mathbf{b}| = |(\mathbf{i} + 2\mathbf{j} - 3\mathbf{k}) - (-2\mathbf{i} - \mathbf{j} + 5\mathbf{k})| = |3\mathbf{i} + 3\mathbf{j} - 8\mathbf{k}| = \sqrt{3^2 + 3^2 + (-8)^2} = \sqrt{82}$

**19.** The vector $-3\mathbf{i} + 7\mathbf{j}$ has length $|-3\mathbf{i} + 7\mathbf{j}| = \sqrt{(-3)^2 + 7^2} = \sqrt{58}$, so by Equation 4 the unit vector with the same

direction is $\dfrac{1}{\sqrt{58}}(-3\mathbf{i} + 7\mathbf{j}) = -\dfrac{3}{\sqrt{58}}\mathbf{i} + \dfrac{7}{\sqrt{58}}\mathbf{j}$.

**21.** The vector $8\mathbf{i} - \mathbf{j} + 4\mathbf{k}$ has length $|8\mathbf{i} - \mathbf{j} + 4\mathbf{k}| = \sqrt{8^2 + (-1)^2 + 4^2} = \sqrt{81} = 9$, so by Equation 4 the unit vector with

the same direction is $\frac{1}{9}(8\mathbf{i} - \mathbf{j} + 4\mathbf{k}) = \frac{8}{9}\mathbf{i} - \frac{1}{9}\mathbf{j} + \frac{4}{9}\mathbf{k}$.

**23.** From the figure, we see that the $x$-component of $\mathbf{v}$ is

$v_1 = |\mathbf{v}| \cos(\pi/3) = 4 \cdot \frac{1}{2} = 2$ and the $y$-component is

$v_2 = |\mathbf{v}| \sin(\pi/3) = 4 \cdot \frac{\sqrt{3}}{2} = 2\sqrt{3}$. Thus

$\mathbf{v} = \langle v_1, v_2 \rangle = \langle 2, 2\sqrt{3} \rangle$.

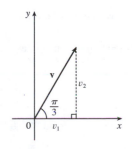

**25.** The velocity vector **v** makes an angle of $40°$ with the horizontal and
has magnitude equal to the speed at which the football was thrown.
From the figure, we see that the horizontal component of **v** is
$|\mathbf{v}| \cos 40° = 60 \cos 40° \approx 45.96$ ft/s and the vertical component
is $|\mathbf{v}| \sin 40° = 60 \sin 40° \approx 38.57$ ft/s.

**27.** The given force vectors can be expressed in terms of their horizontal and vertical components as $-300\,\mathbf{i}$ and
$200 \cos 60°\,\mathbf{i} + 200 \sin 60°\,\mathbf{j} = 200(\frac{1}{2})\,\mathbf{i} + 200\left(\frac{\sqrt{3}}{2}\right)\mathbf{j} = 100\,\mathbf{i} + 100\sqrt{3}\,\mathbf{j}$. The resultant force **F** is the sum of
these two vectors: $\mathbf{F} = (-300 + 100)\,\mathbf{i} + (0 + 100\sqrt{3})\,\mathbf{j} = -200\,\mathbf{i} + 100\sqrt{3}\,\mathbf{j}$. Then we have

$$|\mathbf{F}| \approx \sqrt{(-200)^2 + (100\sqrt{3})^2} = \sqrt{70{,}000} = 100\sqrt{7} \approx 264.6 \text{ N. Let } \theta \text{ be the angle } \mathbf{F} \text{ makes with the}$$

positive $x$-axis. Then $\tan \theta = \dfrac{100\sqrt{3}}{-200} = -\dfrac{\sqrt{3}}{2}$ and the terminal point of **F** lies in the second quadrant, so

$$\theta = \tan^{-1}\left(-\frac{\sqrt{3}}{2}\right) + 180° \approx -40.9° + 180° = 139.1°.$$

**29.** With respect to the water's surface, the woman's velocity is the vector sum of the velocity of the ship with respect
to the water, and the woman's velocity with respect to the ship. If we let north be the positive $y$-direction, then
$\mathbf{v} = \langle 0, 22 \rangle + \langle -3, 0 \rangle = \langle -3, 22 \rangle$. The woman's speed is $|\mathbf{v}| = \sqrt{9 + 484} \approx 22.2$ mi/h. The vector **v** makes an angle $\theta$
with the east, where $\theta = \tan^{-1}\left(\frac{22}{-3}\right) \approx 98°$. Therefore, the woman's direction is about $N(98 - 90)°W = N8°W$.

**31.** Let $\mathbf{T}_1$ and $\mathbf{T}_2$ represent the tension vectors in each side of the
clothesline as shown in the figure. $\mathbf{T}_1$ and $\mathbf{T}_2$ have equal vertical
components and opposite horizontal components, so we can write

$\mathbf{T}_1 = -a\,\mathbf{i} + b\,\mathbf{j}$ and $\mathbf{T}_2 = a\,\mathbf{i} + b\,\mathbf{j}$ $[a, b > 0]$. By similar triangles, $\dfrac{b}{a} = \dfrac{0.08}{4} \quad \Rightarrow \quad a = 50b$. The force due to gravity
acting on the shirt has magnitude $0.8g \approx (0.8)(9.8) = 7.84$ N, hence we have $\mathbf{w} = -7.84\,\mathbf{j}$. The resultant $\mathbf{T}_1 + \mathbf{T}_2$
of the tensile forces counterbalances **w**, so $\mathbf{T}_1 + \mathbf{T}_2 = -\mathbf{w} \quad \Rightarrow \quad (-a\,\mathbf{i} + b\,\mathbf{j}) + (a\,\mathbf{i} + b\,\mathbf{j}) = 7.84\,\mathbf{j} \quad \Rightarrow$
$(-50b\,\mathbf{i} + b\,\mathbf{j}) + (50b\,\mathbf{i} + b\,\mathbf{j}) = 2b\,\mathbf{j} = 7.84\,\mathbf{j} \quad \Rightarrow \quad b = \frac{7.84}{2} = 3.92$ and $a = 50b = 196$. Thus the tensions are
$\mathbf{T}_1 = -a\,\mathbf{i} + b\,\mathbf{j} = -196\,\mathbf{i} + 3.92\,\mathbf{j}$ and $\mathbf{T}_2 = a\,\mathbf{i} + b\,\mathbf{j} = 196\,\mathbf{i} + 3.92\,\mathbf{j}$.

Alternatively, we can find the value of $\theta$ and proceed as in Example 7.

**33.** The slope of the tangent line to the graph of $y = x^2$ at the point $(2, 4)$ is

$$\left.\frac{dy}{dx}\right|_{x=2} = 2x\bigg|_{x=2} = 4$$

and a parallel vector is $\mathbf{i} + 4\,\mathbf{j}$ which has length $|\mathbf{i} + 4\,\mathbf{j}| = \sqrt{1^2 + 4^2} = \sqrt{17}$, so unit vectors parallel to the tangent line are
$\pm\frac{1}{\sqrt{17}}(\mathbf{i} + 4\,\mathbf{j})$.

**35.** (a), (b)

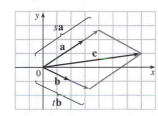

(c) From the sketch, we estimate that $s \approx 1.3$ and $t \approx 1.6$.

(d) $\mathbf{c} = s\,\mathbf{a} + t\,\mathbf{b} \iff 7 = 3s + 2t$ and $1 = 2s - t$.

Solving these equations gives $s = \frac{9}{7}$ and $t = \frac{11}{7}$.

**37.**

Let $\mathbf{a} = \langle a_1, a_2, a_3 \rangle$, as shown in the figure. Since $|\mathbf{a}| = 1$ and triangle $ABO$ is a right triangle, we have $\cos 60^\circ = \dfrac{a_1}{1} \implies a_1 = \cos 60^\circ$. Similarly, triangle $ACO$ is a right triangle, so $a_2 = \cos 72^\circ$. Finally, since $|\mathbf{a}| = 1$ we have

$$\sqrt{(\cos 60^\circ)^2 + (\cos 72^\circ)^2 + a_3^2} = 1 \implies a_3^2 = 1 - (\cos 60^\circ)^2 - (\cos 72^\circ)^2 \implies$$

$$a_3 = \sqrt{1 - (\cos 60^\circ)^2 - (\cos 72^\circ)^2}. \text{ Thus}$$

$$\mathbf{a} = \left\langle \cos 60^\circ, \cos 72^\circ, \sqrt{1 - (\cos 60^\circ)^2 - (\cos 72^\circ)^2} \right\rangle \approx \langle 0.50, 0.31, 0.81 \rangle.$$

**39.** $|\mathbf{r} - \mathbf{r}_0|$ is the distance between the points $(x, y, z)$ and $(x_0, y_0, z_0)$, so the set of points is a sphere with radius 1 and center $(x_0, y_0, z_0)$.

*Alternate method:* $|\mathbf{r} - \mathbf{r}_0| = 1 \iff \sqrt{(x - x_0)^2 + (y - y_0)^2 + (z - z_0)^2} = 1 \iff$

$(x - x_0)^2 + (y - y_0)^2 + (z - z_0)^2 = 1$, which is the equation of a sphere with radius 1 and center $(x_0, y_0, z_0)$.

**41.** $\mathbf{a} + (\mathbf{b} + \mathbf{c}) = \langle a_1, a_2 \rangle + (\langle b_1, b_2 \rangle + \langle c_1, c_2 \rangle) = \langle a_1, a_2 \rangle + \langle b_1 + c_1, b_2 + c_2 \rangle$

$$= \langle a_1 + b_1 + c_1, a_2 + b_2 + c_2 \rangle = \langle (a_1 + b_1) + c_1, (a_2 + b_2) + c_2 \rangle$$

$$= \langle a_1 + b_1, a_2 + b_2 \rangle + \langle c_1, c_2 \rangle = (\langle a_1, a_2 \rangle + \langle b_1, b_2 \rangle) + \langle c_1, c_2 \rangle$$

$$= (\mathbf{a} + \mathbf{b}) + \mathbf{c}$$

**43.** Consider triangle $ABC$, where $D$ and $E$ are the midpoints of $AB$ and $BC$. We know that $\overrightarrow{AB} + \overrightarrow{BC} = \overrightarrow{AC}$   **(1)**   and

$\overrightarrow{DB} + \overrightarrow{BE} = \overrightarrow{DE}$   **(2)**.   However, $\overrightarrow{DB} = \frac{1}{2}\overrightarrow{AB}$, and $\overrightarrow{BE} = \frac{1}{2}\overrightarrow{BC}$. Substituting these expressions for $\overrightarrow{DB}$ and $\overrightarrow{BE}$ into

**(2)** gives $\frac{1}{2}\overrightarrow{AB} + \frac{1}{2}\overrightarrow{BC} = \overrightarrow{DE}$. Comparing this with **(1)** gives $\overrightarrow{DE} = \frac{1}{2}\overrightarrow{AC}$. Therefore $\overrightarrow{AC}$ and $\overrightarrow{DE}$ are parallel and $\left|\overrightarrow{DE}\right| = \frac{1}{2}\left|\overrightarrow{AC}\right|$.

## 9.3    The Dot Product

**1.** (a) $\mathbf{a} \cdot \mathbf{b}$ is a scalar, and the dot product is defined only for vectors, so $(\mathbf{a} \cdot \mathbf{b}) \cdot \mathbf{c}$ has no meaning.

(b) $(\mathbf{a} \cdot \mathbf{b})\,\mathbf{c}$ is a scalar multiple of a vector, so it does have meaning.

(c) Both $|\mathbf{a}|$ and $\mathbf{b} \cdot \mathbf{c}$ are scalars, so $|\mathbf{a}|\,(\mathbf{b} \cdot \mathbf{c})$ is an ordinary product of real numbers, and has meaning.

(d) Both $\mathbf{a}$ and $\mathbf{b} + \mathbf{c}$ are vectors, so the dot product $\mathbf{a} \cdot (\mathbf{b} + \mathbf{c})$ has meaning.

(e) $\mathbf{a} \cdot \mathbf{b}$ is a scalar, but $\mathbf{c}$ is a vector, and so the two quantities cannot be added and $\mathbf{a} \cdot \mathbf{b} + \mathbf{c}$ has no meaning.

(f) $|\mathbf{a}|$ is a scalar, and the dot product is defined only for vectors, so $|\mathbf{a}| \cdot (\mathbf{b} + \mathbf{c})$ has no meaning.

**3.** By the definition of the dot product, $\mathbf{a} \cdot \mathbf{b} = |\mathbf{a}| \, |\mathbf{b}| \cos \theta = (6)(5) \cos \frac{2\pi}{3} = 30 \left(-\frac{1}{2}\right) = -15$.

**5.** $\mathbf{a} \cdot \mathbf{b} = \left\langle -2, \frac{1}{3} \right\rangle \cdot \langle -5, 12 \rangle = (-2)(-5) + \left(\frac{1}{3}\right)(12) = 10 + 4 = 14$

**7.** $\mathbf{a} \cdot \mathbf{b} = \left\langle 4, 1, \frac{1}{4} \right\rangle \cdot \langle 6, -3, -8 \rangle = (4)(6) + (1)(-3) + \left(\frac{1}{4}\right)(-8) = 19$

**9.** $\mathbf{a} \cdot \mathbf{b} = (2\,\mathbf{i} + \mathbf{j}) \cdot (\mathbf{i} - \mathbf{j} + \mathbf{k}) = (2)(1) + (1)(-1) + (0)(1) = 1$

**11.** $\mathbf{u}$, $\mathbf{v}$, and $\mathbf{w}$ are all unit vectors, so the triangle is an equilateral triangle. Thus the angle between $\mathbf{u}$ and $\mathbf{v}$ is $60°$ and

$\mathbf{u} \cdot \mathbf{v} = |\mathbf{u}| \, |\mathbf{v}| \cos 60° = (1)(1)\left(\frac{1}{2}\right) = \frac{1}{2}$. If $\mathbf{w}$ is moved so it has the same initial point as $\mathbf{u}$, we can see that the angle

between them is $120°$ and we have $\mathbf{u} \cdot \mathbf{w} = |\mathbf{u}| \, |\mathbf{w}| \cos 120° = (1)(1)\left(-\frac{1}{2}\right) = -\frac{1}{2}$.

**13.** (a) $\mathbf{i} \cdot \mathbf{j} = \langle 1, 0, 0 \rangle \cdot \langle 0, 1, 0 \rangle = (1)(0) + (0)(1) + (0)(0) = 0$. Similarly, $\mathbf{j} \cdot \mathbf{k} = (0)(0) + (1)(0) + (0)(1) = 0$ and

$\mathbf{k} \cdot \mathbf{i} = (0)(1) + (0)(0) + (1)(0) = 0$.

*Another method:* Because $\mathbf{i}$, $\mathbf{j}$, and $\mathbf{k}$ are mutually perpendicular, the cosine factor in each dot product is $\cos \frac{\pi}{2} = 0$.

(b) By Property 1 of the dot product, $\mathbf{i} \cdot \mathbf{i} = |\mathbf{i}|^2 = 1^2 = 1$ since $\mathbf{i}$ is a unit vector. Similarly, $\mathbf{j} \cdot \mathbf{j} = |\mathbf{j}|^2 = 1$ and

$\mathbf{k} \cdot \mathbf{k} = |\mathbf{k}|^2 = 1$.

**15.** $|\mathbf{a}| = \sqrt{(-8)^2 + 6^2} = 10$, $|\mathbf{b}| = \sqrt{\left(\sqrt{7}\right)^2 + 3^2} = 4$, and $\mathbf{a} \cdot \mathbf{b} = (-8)\left(\sqrt{7}\right) + (6)(3) = 18 - 8\sqrt{7}$. From the

definition of the dot product, we have $\cos \theta = \dfrac{\mathbf{a} \cdot \mathbf{b}}{|\mathbf{a}| \, |\mathbf{b}|} = \dfrac{18 - 8\sqrt{7}}{10 \cdot 4} = \dfrac{9 - 4\sqrt{7}}{20}$. So the angle between $\mathbf{a}$ and $\mathbf{b}$ is

$\theta = \cos^{-1}\left(\dfrac{9 - 4\sqrt{7}}{20}\right) \approx 95°$.

**17.** $|\mathbf{a}| = \sqrt{0^2 + 1^2 + 1^2} = \sqrt{2}$, $|\mathbf{b}| = \sqrt{1^2 + 2^2 + (-3)^2} = \sqrt{14}$, and $\mathbf{a} \cdot \mathbf{b} = (0)(1) + (1)(2) + (1)(-3) = -1$.

Then $\cos \theta = \dfrac{\mathbf{a} \cdot \mathbf{b}}{|\mathbf{a}| \, |\mathbf{b}|} = \dfrac{-1}{\sqrt{2} \cdot \sqrt{14}} = \dfrac{-1}{2\sqrt{7}}$ and $\theta = \cos^{-1}\left(-\dfrac{1}{2\sqrt{7}}\right) \approx 101°$.

**19.** Let $a$, $b$, and $c$ be the angles at vertices $A$, $B$, and $C$ respectively.

Then $a$ is the angle between vectors $\overrightarrow{AB}$ and $\overrightarrow{AC}$, $b$ is the angle

between vectors $\overrightarrow{BA}$ and $\overrightarrow{BC}$, and $c$ is the angle between vectors

$\overrightarrow{CA}$ and $\overrightarrow{CB}$.

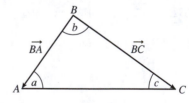

Thus $\cos a = \dfrac{\overrightarrow{AB} \cdot \overrightarrow{AC}}{\left|\overrightarrow{AB}\right| \left|\overrightarrow{AC}\right|} = \dfrac{\langle 2, 6 \rangle \cdot \langle -2, 4 \rangle}{\sqrt{2^2 + 6^2} \, \sqrt{(-2)^2 + 4^2}} = \dfrac{1}{\sqrt{40}\,\sqrt{20}}(-4 + 24) = \dfrac{20}{\sqrt{800}} = \dfrac{\sqrt{2}}{2}$ and

$a = \cos^{-1}\left(\dfrac{\sqrt{2}}{2}\right) = 45°$. Similarly, $\cos b = \dfrac{\overrightarrow{BA} \cdot \overrightarrow{BC}}{\left|\overrightarrow{BA}\right| \left|\overrightarrow{BC}\right|} = \dfrac{\langle -2, -6 \rangle \cdot \langle -4, -2 \rangle}{\sqrt{4 + 36} \, \sqrt{16 + 4}} = \dfrac{1}{\sqrt{40}\,\sqrt{20}}(8 + 12) = \dfrac{20}{\sqrt{800}} = \dfrac{\sqrt{2}}{2}$

so $b = \cos^{-1}\left(\dfrac{\sqrt{2}}{2}\right) = 45°$ and $c = 180° - (45° + 45°) = 90°$.

*Alternate solution:* Apply the Law of Cosines three times as follows: $\cos a = \dfrac{\left|\overrightarrow{BC}\right|^2 - \left|\overrightarrow{AB}\right|^2 - \left|\overrightarrow{AC}\right|^2}{2\left|\overrightarrow{AB}\right|\left|\overrightarrow{AC}\right|}$,

$$\cos b = \frac{\left|\overrightarrow{AC}\right|^2 - \left|\overrightarrow{AB}\right|^2 - \left|\overrightarrow{BC}\right|^2}{2\left|\overrightarrow{AB}\right|\left|\overrightarrow{BC}\right|}, \text{ and } \cos c = \frac{\left|\overrightarrow{AB}\right|^2 - \left|\overrightarrow{AC}\right|^2 - \left|\overrightarrow{BC}\right|^2}{2\left|\overrightarrow{AC}\right|\left|\overrightarrow{BC}\right|}.$$

**21.** (a) $\mathbf{a} \cdot \mathbf{b} = (-5)(6) + (3)(-8) + (7)(2) = -40 \neq 0$, so $\mathbf{a}$ and $\mathbf{b}$ are not orthogonal. Also, since $\mathbf{a}$ is not a scalar multiple of $\mathbf{b}$, $\mathbf{a}$ and $\mathbf{b}$ are not parallel.

(b) $\mathbf{a} \cdot \mathbf{b} = (4)(-3) + (6)(2) = 0$, so $\mathbf{a}$ and $\mathbf{b}$ are orthogonal (and not parallel).

(c) $\mathbf{a} \cdot \mathbf{b} = (-1)(3) + (2)(4) + (5)(-1) = 0$, so $\mathbf{a}$ and $\mathbf{b}$ are orthogonal (and not parallel).

(d) Because $\mathbf{a} = -\frac{2}{3}\mathbf{b}$, $\mathbf{a}$ and $\mathbf{b}$ are parallel.

**23.** $\overrightarrow{QP} = \langle -1, -3, 2 \rangle$, $\overrightarrow{QR} = \langle 4, -2, -1 \rangle$, and $\overrightarrow{QP} \cdot \overrightarrow{QR} = -4 + 6 - 2 = 0$. Thus $\overrightarrow{QP}$ and $\overrightarrow{QR}$ are orthogonal, so the angle of the triangle at vertex $Q$ is a right angle.

**25.** Let $\mathbf{a} = a_1\mathbf{i} + a_2\mathbf{j} + a_3\mathbf{k}$ be a vector orthogonal to both $\mathbf{i} + \mathbf{j}$ and $\mathbf{i} + \mathbf{k}$. Then $\mathbf{a} \cdot (\mathbf{i} + \mathbf{j}) = 0 \Leftrightarrow a_1 + a_2 = 0$ and $\mathbf{a} \cdot (\mathbf{i} + \mathbf{k}) = 0 \Leftrightarrow a_1 + a_3 = 0$, so $a_1 = -a_2 = -a_3$. Furthermore $\mathbf{a}$ is to be a unit vector, so $1 = a_1^2 + a_2^2 + a_3^2 = 3a_1^2$ implies $a_1 = \pm\frac{1}{\sqrt{3}}$. Thus $\mathbf{a} = \frac{1}{\sqrt{3}}\mathbf{i} - \frac{1}{\sqrt{3}}\mathbf{j} - \frac{1}{\sqrt{3}}\mathbf{k}$ and $\mathbf{a} = -\frac{1}{\sqrt{3}}\mathbf{i} + \frac{1}{\sqrt{3}}\mathbf{j} + \frac{1}{\sqrt{3}}\mathbf{k}$ are two such unit vectors.

**27.** The line $2x - y = 3 \Leftrightarrow y = 2x - 3$ has slope 2, so a vector parallel to the line is $\mathbf{a} = \langle 1, 2 \rangle$. The line $3x + y = 7 \Leftrightarrow y = -3x + 7$ has slope $-3$, so a vector parallel to the line is $\mathbf{b} = \langle 1, -3 \rangle$. The angle between the lines is the same as the angle $\theta$ between the vectors. Here we have $\mathbf{a} \cdot \mathbf{b} = (1)(1) + (2)(-3) = -5$, $|\mathbf{a}| = \sqrt{1^2 + 2^2} = \sqrt{5}$, and $|\mathbf{b}| = \sqrt{1^2 + (-3)^2} = \sqrt{10}$, so $\cos\theta = \frac{\mathbf{a} \cdot \mathbf{b}}{|\mathbf{a}||\mathbf{b}|} = \frac{-5}{\sqrt{5} \cdot \sqrt{10}} = \frac{-5}{5\sqrt{2}} = -\frac{1}{\sqrt{2}}$ or $-\frac{\sqrt{2}}{2}$. Thus $\theta = 135°$, and the acute angle between the lines is $180° - 135° = 45°$.

**29.** $|\mathbf{a}| = \sqrt{3^2 + (-4)^2} = 5$. The scalar projection of $\mathbf{b}$ onto $\mathbf{a}$ is $\text{comp}_\mathbf{a}\,\mathbf{b} = \frac{\mathbf{a} \cdot \mathbf{b}}{|\mathbf{a}|} = \frac{3 \cdot 5 + (-4) \cdot 0}{5} = 3$ and the vector projection of $\mathbf{b}$ onto $\mathbf{a}$ is $\text{proj}_\mathbf{a}\,\mathbf{b} = \left(\frac{\mathbf{a} \cdot \mathbf{b}}{|\mathbf{a}|}\right)\frac{\mathbf{a}}{|\mathbf{a}|} = 3 \cdot \frac{1}{5}\langle 3, -4 \rangle = \langle \frac{9}{5}, -\frac{12}{5} \rangle$.

**31.** $|\mathbf{a}| = \sqrt{4 + 1 + 16} = \sqrt{21}$ so the scalar projection of $\mathbf{b}$ onto $\mathbf{a}$ is $\text{comp}_\mathbf{a}\,\mathbf{b} = \frac{\mathbf{a} \cdot \mathbf{b}}{|\mathbf{a}|} = \frac{0 - 1 + 2}{\sqrt{21}} = \frac{1}{\sqrt{21}}$ while the vector projection of $\mathbf{b}$ onto $\mathbf{a}$ is $\text{proj}_\mathbf{a}\,\mathbf{b} = \frac{1}{\sqrt{21}}\frac{\mathbf{a}}{|\mathbf{a}|} = \frac{1}{\sqrt{21}} \cdot \frac{2\mathbf{i} - \mathbf{j} + 4\mathbf{k}}{\sqrt{21}} = \frac{1}{21}(2\mathbf{i} - \mathbf{j} + 4\mathbf{k}) = \frac{2}{21}\mathbf{i} - \frac{1}{21}\mathbf{j} + \frac{4}{21}\mathbf{k}$.

**33.** $(\text{orth}_\mathbf{a}\,\mathbf{b}) \cdot \mathbf{a} = (\mathbf{b} - \text{proj}_\mathbf{a}\,\mathbf{b}) \cdot \mathbf{a} = \mathbf{b} \cdot \mathbf{a} - (\text{proj}_\mathbf{a}\,\mathbf{b}) \cdot \mathbf{a} = \mathbf{b} \cdot \mathbf{a} - \frac{\mathbf{a} \cdot \mathbf{b}}{|\mathbf{a}|^2}\mathbf{a} \cdot \mathbf{a} = \mathbf{b} \cdot \mathbf{a} - \frac{\mathbf{a} \cdot \mathbf{b}}{|\mathbf{a}|^2}|\mathbf{a}|^2 = \mathbf{b} \cdot \mathbf{a} - \mathbf{a} \cdot \mathbf{b} = 0$. So they are orthogonal by (2).

**35.** $\text{comp}_\mathbf{a}\,\mathbf{b} = \frac{\mathbf{a} \cdot \mathbf{b}}{|\mathbf{a}|} = 2 \Leftrightarrow \mathbf{a} \cdot \mathbf{b} = 2|\mathbf{a}| = 2\sqrt{10}$. If $\mathbf{b} = \langle b_1, b_2, b_3 \rangle$, then we need $3b_1 + 0b_2 - 1b_3 = 2\sqrt{10}$. One possible solution is obtained by taking $b_1 = 0, b_2 = 0, b_3 = -2\sqrt{10}$. In general, $\mathbf{b} = \langle s, t, 3s - 2\sqrt{10} \rangle$, $s, t \in \mathbb{R}$.

**37.** The displacement vector is $\mathbf{D} = (6 - 0)\mathbf{i} + (12 - 10)\mathbf{j} + (20 - 8)\mathbf{k} = 6\mathbf{i} + 2\mathbf{j} + 12\mathbf{k}$ so, as in Example 2, the work done is $W = \mathbf{F} \cdot \mathbf{D} = (8\mathbf{i} - 6\mathbf{j} + 9\mathbf{k}) \cdot (6\mathbf{i} + 2\mathbf{j} + 12\mathbf{k}) = 48 - 12 + 108 = 144$ joules.

**39.** Here $|\mathbf{D}| = 80$ ft, $|\mathbf{F}| = 30$ lb, and $\theta = 40°$. Thus

$$W = \mathbf{F} \cdot \mathbf{D} = |\mathbf{F}|\,|\mathbf{D}|\cos\theta = (30)(80)\cos 40° = 2400\cos 40° \approx 1839 \text{ ft-lb.}$$

**41.** First note that $\mathbf{n} = \langle a, b\rangle$ is perpendicular to the line, because if $Q_1 = (a_1, b_1)$ and $Q_2 = (a_2, b_2)$ lie on the line, then

$\mathbf{n} \cdot \overrightarrow{Q_1 Q_2} = aa_2 - aa_1 + bb_2 - bb_1 = 0$, since $aa_2 + bb_2 = -c = aa_1 + bb_1$ from the equation of the line.

Let $P_2 = (x_2, y_2)$ lie on the line. Then the distance from $P_1$ to the line is the absolute value of the scalar projection

of $\overrightarrow{P_1 P_2}$ onto $\mathbf{n}$. $\quad \text{comp}_{\mathbf{n}}\left(\overrightarrow{P_1 P_2}\right) = \dfrac{|\mathbf{n} \cdot \langle x_2 - x_1, y_2 - y_1\rangle|}{|\mathbf{n}|} = \dfrac{|ax_2 - ax_1 + by_2 - by_1|}{\sqrt{a^2 + b^2}} = \dfrac{|ax_1 + by_1 + c|}{\sqrt{a^2 + b^2}}$

since $ax_2 + by_2 = -c$. The required distance is $\dfrac{|(3)(-2) + (-4)(3) + 5|}{\sqrt{3^2 + (-4)^2}} = \dfrac{13}{5}$.

**43.** For convenience, consider the unit cube positioned so that its back left corner is at the origin, and its edges lie along the coordinate axes. The diagonal of the cube that begins at the origin and ends at $(1, 1, 1)$ has vector representation $\langle 1, 1, 1\rangle$. The angle $\theta$ between this vector and the vector of the edge which also begins at the origin and runs along the $x$-axis [that is, $\langle 1, 0, 0\rangle$] is given by $\cos\theta = \dfrac{\langle 1, 1, 1\rangle \cdot \langle 1, 0, 0\rangle}{|\langle 1, 1, 1\rangle||\langle 1, 0, 0\rangle|} = \dfrac{1}{\sqrt{3}} \quad\Rightarrow\quad \theta = \cos^{-1}\left(\frac{1}{\sqrt{3}}\right) \approx 55°.$

**45.** Consider the H—C—H combination consisting of the sole carbon atom and the two hydrogen atoms that are at $(1, 0, 0)$ and $(0, 1, 0)$ (or any H—C—H combination, for that matter). Vector representations of the line segments emanating from the carbon atom and extending to these two hydrogen atoms are $\langle 1 - \frac{1}{2}, 0 - \frac{1}{2}, 0 - \frac{1}{2}\rangle = \langle \frac{1}{2}, -\frac{1}{2}, -\frac{1}{2}\rangle$ and

$\langle 0 - \frac{1}{2}, 1 - \frac{1}{2}, 0 - \frac{1}{2}\rangle = \langle -\frac{1}{2}, \frac{1}{2}, -\frac{1}{2}\rangle$. The bond angle, $\theta$, is therefore given by

$$\cos\theta = \dfrac{\langle \frac{1}{2}, -\frac{1}{2}, -\frac{1}{2}\rangle \cdot \langle -\frac{1}{2}, \frac{1}{2}, -\frac{1}{2}\rangle}{|\langle \frac{1}{2}, -\frac{1}{2}, -\frac{1}{2}\rangle||\langle -\frac{1}{2}, \frac{1}{2}, -\frac{1}{2}\rangle|} = \dfrac{-\frac{1}{4} - \frac{1}{4} + \frac{1}{4}}{\sqrt{\frac{3}{4}}\sqrt{\frac{3}{4}}} = -\dfrac{1}{3} \quad\Rightarrow\quad \theta = \cos^{-1}\left(-\frac{1}{3}\right) \approx 109.5°.$$

**47.** If $c = 0$ then $c\mathbf{a} = \mathbf{0}$, so $(c\mathbf{a}) \cdot \mathbf{b} = \mathbf{0} \cdot \mathbf{b} = 0$ by Property 5. Similarly, $\mathbf{a} \cdot (c\mathbf{b}) = \mathbf{a} \cdot \mathbf{0} = 0$, and $c(\mathbf{a} \cdot \mathbf{b}) = 0(|\mathbf{a}|\,|\mathbf{b}|\cos\theta) = 0$, thus $(c\mathbf{a}) \cdot \mathbf{b} = c(\mathbf{a} \cdot \mathbf{b}) = \mathbf{a} \cdot (c\mathbf{b})$. If $c > 0$, the angle $\theta$ between $\mathbf{a}$ and $\mathbf{b}$ coincides with the angle between $c\mathbf{a}$ and $\mathbf{b}$, so by definition of the dot product, $(c\mathbf{a}) \cdot \mathbf{b} = |c\mathbf{a}|\,|\mathbf{b}|\cos\theta = |c|\,|\mathbf{a}|\,|\mathbf{b}|\cos\theta = c\,|\mathbf{a}|\,|\mathbf{b}|\cos\theta$. Similarly, $\mathbf{a} \cdot (c\mathbf{b}) = |\mathbf{a}|\,|c\mathbf{b}|\cos\theta = |\mathbf{a}|\,|c|\,|\mathbf{b}|\cos\theta = c\,|\mathbf{a}|\,|\mathbf{b}|\cos\theta$, and $c(\mathbf{a} \cdot \mathbf{b}) = c\,|\mathbf{a}|\,|\mathbf{b}|\cos\theta$. Thus, $(c\mathbf{a}) \cdot \mathbf{b} = c(\mathbf{a} \cdot \mathbf{b}) = \mathbf{a} \cdot (c\mathbf{b})$. The case for $c < 0$ is similar. Using components, let $\mathbf{a} = \langle a_1, a_2, a_3\rangle$ and $\mathbf{b} = \langle b_1, b_2, b_3\rangle$. Then

$$(c\mathbf{a}) \cdot \mathbf{b} = \langle ca_1, ca_2, ca_3\rangle \cdot \langle b_1, b_2, b_3\rangle = (ca_1)b_1 + (ca_2)b_2 + (ca_3)b_3$$
$$= c(a_1 b_1 + a_2 b_2 + a_3 b_3) = c(\mathbf{a} \cdot \mathbf{b})$$
$$= a_1(cb_1) + a_2(cb_2) + a_3(cb_3) = \langle a_1, a_2, a_3\rangle \cdot \langle cb_1, cb_2, cb_3\rangle = \mathbf{a} \cdot (c\mathbf{b})$$

**49.** $|\mathbf{a} \cdot \mathbf{b}| = |\,|\mathbf{a}|\,|\mathbf{b}|\cos\theta| = |\mathbf{a}|\,|\mathbf{b}|\,|\cos\theta|$. Since $|\cos\theta| \le 1$, $|\mathbf{a} \cdot \mathbf{b}| = |\mathbf{a}|\,|\mathbf{b}|\,|\cos\theta| \le |\mathbf{a}|\,|\mathbf{b}|$.

*Note:* We have equality in the case of $\cos\theta = \pm 1$, so $\theta = 0$ or $\theta = \pi$, thus equality when $\mathbf{a}$ and $\mathbf{b}$ are parallel.

51. (a)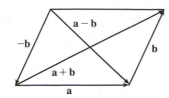

The Parallelogram Law states that the sum of the squares of the lengths of the diagonals of a parallelogram equals the sum of the squares of its (four) sides.

(b) $|\mathbf{a} + \mathbf{b}|^2 = (\mathbf{a} + \mathbf{b}) \cdot (\mathbf{a} + \mathbf{b}) = |\mathbf{a}|^2 + 2(\mathbf{a} \cdot \mathbf{b}) + |\mathbf{b}|^2$ and $|\mathbf{a} - \mathbf{b}|^2 = (\mathbf{a} - \mathbf{b}) \cdot (\mathbf{a} - \mathbf{b}) = |\mathbf{a}|^2 - 2(\mathbf{a} \cdot \mathbf{b}) + |\mathbf{b}|^2$.

Adding these two equations gives $|\mathbf{a} + \mathbf{b}|^2 + |\mathbf{a} - \mathbf{b}|^2 = 2\,|\mathbf{a}|^2 + 2\,|\mathbf{b}|^2$.

## 9.4   The Cross Product

1. (a) Since $\mathbf{b} \times \mathbf{c}$ is a vector, the dot product $\mathbf{a} \cdot (\mathbf{b} \times \mathbf{c})$ is meaningful and is a scalar.

(b) $\mathbf{b} \cdot \mathbf{c}$ is a scalar, so $\mathbf{a} \times (\mathbf{b} \cdot \mathbf{c})$ is meaningless, as the cross product is defined only for two *vectors*.

(c) Since $\mathbf{b} \times \mathbf{c}$ is a vector, the cross product $\mathbf{a} \times (\mathbf{b} \times \mathbf{c})$ is meaningful and results in another vector.

(d) $\mathbf{a} \cdot \mathbf{b}$ is a scalar, so the cross product $(\mathbf{a} \cdot \mathbf{b}) \times \mathbf{c}$ is meaningless.

(e) Since $(\mathbf{a} \cdot \mathbf{b})$ and $(\mathbf{c} \cdot \mathbf{d})$ are both scalars, the cross product $(\mathbf{a} \cdot \mathbf{b}) \times (\mathbf{c} \cdot \mathbf{d})$ is meaningless.

(f) $\mathbf{a} \times \mathbf{b}$ and $\mathbf{c} \times \mathbf{d}$ are both vectors, so the dot product $(\mathbf{a} \times \mathbf{b}) \cdot (\mathbf{c} \times \mathbf{d})$ is meaningful and is a scalar.

3. If we sketch $\mathbf{u}$ and $\mathbf{v}$ starting from the same initial point, we see that the angle between them is $60°$, so

$$|\mathbf{u} \times \mathbf{v}| = |\mathbf{u}|\,|\mathbf{v}| \sin\theta = (12)(16)\sin 60° = 192 \cdot \frac{\sqrt{3}}{2} = 96\sqrt{3}.$$

By the right-hand rule, $\mathbf{u} \times \mathbf{v}$ is directed into the page.

5. The magnitude of the torque is $|\boldsymbol{\tau}| = |\mathbf{r} \times \mathbf{F}| = |\mathbf{r}|\,|\mathbf{F}| \sin\theta = (0.18 \text{ m})(60 \text{ N})\sin(70 + 10)° = 10.8 \sin 80° \approx 10.6 \text{ N·m}.$

7. $\mathbf{a} \times \mathbf{b} = \begin{vmatrix} \mathbf{i} & \mathbf{j} & \mathbf{k} \\ 6 & 0 & -2 \\ 0 & 8 & 0 \end{vmatrix} = \begin{vmatrix} 0 & -2 \\ 8 & 0 \end{vmatrix}\mathbf{i} - \begin{vmatrix} 6 & -2 \\ 0 & 0 \end{vmatrix}\mathbf{j} + \begin{vmatrix} 6 & 0 \\ 0 & 8 \end{vmatrix}\mathbf{k}$

$\quad = [0 - (-16)]\,\mathbf{i} - (0 - 0)\,\mathbf{j} + (48 - 0)\,\mathbf{k} = 16\,\mathbf{i} + 48\,\mathbf{k}$

Now $(\mathbf{a} \times \mathbf{b}) \cdot \mathbf{a} = \langle 16, 0, 48 \rangle \cdot \langle 6, 0, -2 \rangle = 96 + 0 - 96 = 0$ and $(\mathbf{a} \times \mathbf{b}) \cdot \mathbf{b} = \langle 16, 0, 48 \rangle \cdot \langle 0, 8, 0 \rangle = 0 + 0 + 0 = 0$, so $\mathbf{a} \times \mathbf{b}$ is orthogonal to both $\mathbf{a}$ and $\mathbf{b}$.

9. $\mathbf{a} \times \mathbf{b} = \begin{vmatrix} \mathbf{i} & \mathbf{j} & \mathbf{k} \\ 1 & 3 & -2 \\ -1 & 0 & 5 \end{vmatrix} = \begin{vmatrix} 3 & -2 \\ 0 & 5 \end{vmatrix}\mathbf{i} - \begin{vmatrix} 1 & -2 \\ -1 & 5 \end{vmatrix}\mathbf{j} + \begin{vmatrix} 1 & 3 \\ -1 & 0 \end{vmatrix}\mathbf{k}$

$\quad = (15 - 0)\,\mathbf{i} - (5 - 2)\,\mathbf{j} + [0 - (-3)]\,\mathbf{k} = 15\,\mathbf{i} - 3\,\mathbf{j} + 3\,\mathbf{k}$

Since $(\mathbf{a} \times \mathbf{b}) \cdot \mathbf{a} = (15\,\mathbf{i} - 3\,\mathbf{j} + 3\,\mathbf{k}) \cdot (\mathbf{i} + 3\,\mathbf{j} - 2\,\mathbf{k}) = 15 - 9 - 6 = 0$, $\mathbf{a} \times \mathbf{b}$ is orthogonal to $\mathbf{a}$.

Since $(\mathbf{a} \times \mathbf{b}) \cdot \mathbf{b} = (15\,\mathbf{i} - 3\,\mathbf{j} + 3\,\mathbf{k}) \cdot (-\mathbf{i} + 5\,\mathbf{k}) = -15 + 0 + 15 = 0$, $\mathbf{a} \times \mathbf{b}$ is orthogonal to $\mathbf{b}$.

**11.** $\mathbf{a} \times \mathbf{b} = \begin{vmatrix} \mathbf{i} & \mathbf{j} & \mathbf{k} \\ 1 & -1 & -1 \\ \frac{1}{2} & 1 & \frac{1}{2} \end{vmatrix} = \begin{vmatrix} -1 & -1 \\ 1 & \frac{1}{2} \end{vmatrix} \mathbf{i} - \begin{vmatrix} 1 & -1 \\ \frac{1}{2} & \frac{1}{2} \end{vmatrix} \mathbf{j} + \begin{vmatrix} 1 & -1 \\ \frac{1}{2} & 1 \end{vmatrix} \mathbf{k}$

$\qquad = \left[ -\frac{1}{2} - (-1) \right] \mathbf{i} - \left[ \frac{1}{2} - (-\frac{1}{2}) \right] \mathbf{j} + \left[ 1 - (-\frac{1}{2}) \right] \mathbf{k} = \frac{1}{2} \mathbf{i} - \mathbf{j} + \frac{3}{2} \mathbf{k}$

Now $(\mathbf{a} \times \mathbf{b}) \cdot \mathbf{a} = \left( \frac{1}{2} \mathbf{i} - \mathbf{j} + \frac{3}{2} \mathbf{k} \right) \cdot (\mathbf{i} - \mathbf{j} - \mathbf{k}) = \frac{1}{2} + 1 - \frac{3}{2} = 0$ and

$(\mathbf{a} \times \mathbf{b}) \cdot \mathbf{b} = \left( \frac{1}{2} \mathbf{i} - \mathbf{j} + \frac{3}{2} \mathbf{k} \right) \cdot \left( \frac{1}{2} \mathbf{i} + \mathbf{j} + \frac{1}{2} \mathbf{k} \right) = \frac{1}{4} - 1 + \frac{3}{4} = 0$, so $\mathbf{a} \times \mathbf{b}$ is orthogonal to both $\mathbf{a}$ and $\mathbf{b}$.

**13.** $\mathbf{a} \times \mathbf{b} = \begin{vmatrix} \mathbf{i} & \mathbf{j} & \mathbf{k} \\ t & t^2 & t^3 \\ 1 & 2t & 3t^2 \end{vmatrix} = \begin{vmatrix} t^2 & t^3 \\ 2t & 3t^2 \end{vmatrix} \mathbf{i} - \begin{vmatrix} t & t^3 \\ 1 & 3t^2 \end{vmatrix} \mathbf{j} + \begin{vmatrix} t & t^2 \\ 1 & 2t \end{vmatrix} \mathbf{k}$

$\qquad = (3t^4 - 2t^4) \mathbf{i} - (3t^3 - t^3) \mathbf{j} + (2t^2 - t^2) \mathbf{k} = t^4 \mathbf{i} - 2t^3 \mathbf{j} + t^2 \mathbf{k}$

Since $(\mathbf{a} \times \mathbf{b}) \cdot \mathbf{a} = \langle t^4, -2t^3, t^2 \rangle \cdot \langle t, t^2, t^3 \rangle = t^5 - 2t^5 + t^5 = 0$, $\mathbf{a} \times \mathbf{b}$ is orthogonal to $\mathbf{a}$.

Since $(\mathbf{a} \times \mathbf{b}) \cdot \mathbf{b} = \langle t^4, -2t^3, t^2 \rangle \cdot \langle 1, 2t, 3t^2 \rangle = t^4 - 4t^4 + 3t^4 = 0$, $\mathbf{a} \times \mathbf{b}$ is orthogonal to $\mathbf{b}$.

**15.** By Example 2 , $\mathbf{i} \times \mathbf{j} = \mathbf{k}$, so $(\mathbf{i} \times \mathbf{j}) \times \mathbf{k} = \mathbf{k} \times \mathbf{k} = \mathbf{0}$ [by the margin note on page 655].

**17.** $(\mathbf{j} - \mathbf{k}) \times (\mathbf{k} - \mathbf{i}) = (\mathbf{j} - \mathbf{k}) \times \mathbf{k} + (\mathbf{j} - \mathbf{k}) \times (-\mathbf{i})$  by Property 3

$\qquad = \mathbf{j} \times \mathbf{k} + (-\mathbf{k}) \times \mathbf{k} + \mathbf{j} \times (-\mathbf{i}) + (-\mathbf{k}) \times (-\mathbf{i})$  by Property 4

$\qquad = (\mathbf{j} \times \mathbf{k}) + (-1)(\mathbf{k} \times \mathbf{k}) + (-1)(\mathbf{j} \times \mathbf{i}) + (-1)^2(\mathbf{k} \times \mathbf{i})$  by Property 2

$\qquad = \mathbf{i} + (-1)\,\mathbf{0} + (-1)(-\mathbf{k}) + \mathbf{j} = \mathbf{i} + \mathbf{j} + \mathbf{k}$  by Example 2 and the margin note on page 655

**19.** We know that the cross product of two vectors is orthogonal to both. So we calculate

$\langle 1, -1, 1 \rangle \times \langle 0, 4, 4 \rangle = \begin{vmatrix} \mathbf{i} & \mathbf{j} & \mathbf{k} \\ 1 & -1 & 1 \\ 0 & 4 & 4 \end{vmatrix} = \begin{vmatrix} -1 & 1 \\ 4 & 4 \end{vmatrix} \mathbf{i} - \begin{vmatrix} 1 & 1 \\ 0 & 4 \end{vmatrix} \mathbf{j} + \begin{vmatrix} 1 & -1 \\ 0 & 4 \end{vmatrix} \mathbf{k} = -8\,\mathbf{i} - 4\,\mathbf{j} + 4\,\mathbf{k}.$

So two unit vectors orthogonal to both are $\pm \dfrac{\langle -8, -4, 4 \rangle}{\sqrt{64 + 16 + 16}} = \pm \dfrac{\langle -8, -4, 4 \rangle}{4\sqrt{6}}$, that is, $\left\langle -\frac{2}{\sqrt{6}}, -\frac{1}{\sqrt{6}}, \frac{1}{\sqrt{6}} \right\rangle$

and $\left\langle \frac{2}{\sqrt{6}}, \frac{1}{\sqrt{6}}, -\frac{1}{\sqrt{6}} \right\rangle$.

**21.** By plotting the vertices, we can see that the parallelogram is determined by the

vectors $\overrightarrow{AB} = \langle 2, 3 \rangle$ and $\overrightarrow{AD} = \langle 4, -2 \rangle$. We know that the area of the parallelogram

determined by two vectors is equal to the length of the cross product of these vectors.

In order to compute the cross product, we consider the vector $\overrightarrow{AB}$ as the three-

dimensional vector $\langle 2, 3, 0 \rangle$ (and similarly for $\overrightarrow{AD}$), and then the area of

parallelogram $ABCD$ is

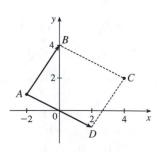

$$\left| \overrightarrow{AB} \times \overrightarrow{AD} \right| = \begin{Vmatrix} \mathbf{i} & \mathbf{j} & \mathbf{k} \\ 2 & 3 & 0 \\ 4 & -2 & 0 \end{Vmatrix} = |(0)\,\mathbf{i} - (0)\,\mathbf{j} + (-4 - 12)\,\mathbf{k}| = |-16\,\mathbf{k}| = 16$$

**23.** (a) $\overrightarrow{PQ} = \langle 4, 3, -2 \rangle$ and $\overrightarrow{PR} = \langle 5, 5, 1 \rangle$, so a vector orthogonal to the plane through $P$, $Q$, and $R$ is

$$\overrightarrow{PQ} \times \overrightarrow{PR} = \langle (3)(1) - (-2)(5), (-2)(5) - (4)(1), (4)(5) - (3)(5) \rangle = \langle 13, -14, 5 \rangle \text{ [or any scalar mutiple thereof].}$$

(b) The area of the parallelogram determined by $\overrightarrow{PQ}$ and $\overrightarrow{PR}$ is

$$\left| \overrightarrow{PQ} \times \overrightarrow{PR} \right| = |\langle 13, -14, 5 \rangle| = \sqrt{13^2 + (-14)^2 + 5^2} = \sqrt{390}, \text{ so the area of triangle } PQR \text{ is } \tfrac{1}{2}\sqrt{390}.$$

**25.** Using the notation of (1), $\mathbf{r} = \langle 0, 0.3, 0 \rangle$ and $\mathbf{F}$ has direction $\langle 0, 3, -4 \rangle$. The angle $\theta$ between them can be determined by

$$\cos\theta = \frac{\langle 0, 0.3, 0 \rangle \cdot \langle 0, 3, -4 \rangle}{|\langle 0, 0.3, 0 \rangle| \, |\langle 0, 3, -4 \rangle|} \quad \Rightarrow \quad \cos\theta = \frac{0.9}{(0.3)(5)} \quad \Rightarrow \quad \cos\theta = 0.6 \quad \Rightarrow \quad \theta \approx 53.1°. \text{ Then } |\boldsymbol{\tau}| = |\mathbf{r}| \, |\mathbf{F}| \sin\theta \quad \Rightarrow$$

$$100 = 0.3 \, |\mathbf{F}| \sin 53.1° \quad \Rightarrow \quad |\mathbf{F}| \approx 417 \text{ N}.$$

**27.** We know that the volume of the parallelepiped determined by $\mathbf{a}$, $\mathbf{b}$, and $\mathbf{c}$ is the magnitude of their scalar triple product, which

is $\mathbf{a} \cdot (\mathbf{b} \times \mathbf{c}) = \begin{vmatrix} 6 & 3 & -1 \\ 0 & 1 & 2 \\ 4 & -2 & 5 \end{vmatrix} = 6 \begin{vmatrix} 1 & 2 \\ -2 & 5 \end{vmatrix} - 3 \begin{vmatrix} 0 & 2 \\ 4 & 5 \end{vmatrix} + (-1) \begin{vmatrix} 0 & 1 \\ 4 & -2 \end{vmatrix} = 6(5+4) - 3(0-8) - (0-4) = 82.$

Thus the volume of the parallelepiped is 82 cubic units.

**29.** $\mathbf{a} = \overrightarrow{PQ} = \langle 2, 1, 1 \rangle$, $\mathbf{b} = \overrightarrow{PR} = \langle 1, -1, 2 \rangle$, and $\mathbf{c} = \overrightarrow{PS} = \langle 0, -2, 3 \rangle$.

$$\mathbf{a} \cdot (\mathbf{b} \times \mathbf{c}) = \begin{vmatrix} 2 & 1 & 1 \\ 1 & -1 & 2 \\ 0 & -2 & 3 \end{vmatrix} = 2 \begin{vmatrix} -1 & 2 \\ -2 & 3 \end{vmatrix} - 1 \begin{vmatrix} 1 & 2 \\ 0 & 3 \end{vmatrix} + 1 \begin{vmatrix} 1 & -1 \\ 0 & -2 \end{vmatrix} = 2 - 3 - 2 = -3,$$

so the volume of the parallelepiped is 3 cubic units.

**31.** $\mathbf{u} \cdot (\mathbf{v} \times \mathbf{w}) = \begin{vmatrix} 1 & 5 & -2 \\ 3 & -1 & 0 \\ 5 & 9 & -4 \end{vmatrix} = 1 \begin{vmatrix} -1 & 0 \\ 9 & -4 \end{vmatrix} - 5 \begin{vmatrix} 3 & 0 \\ 5 & -4 \end{vmatrix} + (-2) \begin{vmatrix} 3 & -1 \\ 5 & 9 \end{vmatrix} = 4 + 60 - 64 = 0$, which says that the volume

of the parallelepiped determined by $\mathbf{u}$, $\mathbf{v}$ and $\mathbf{w}$ is 0, and thus these three vectors are coplanar.

**33.** (a)

The distance between a point and a line is the length of the perpendicular from the point to the line, here $\left| \overrightarrow{PS} \right| = d$. But referring to triangle $PQS$,

$d = \left| \overrightarrow{PS} \right| = \left| \overrightarrow{QP} \right| \sin\theta = |\mathbf{b}| \sin\theta$. But $\theta$ is the angle between $\overrightarrow{QP} = \mathbf{b}$

and $\overrightarrow{QR} = \mathbf{a}$. Thus by the definition of the cross product, $\sin\theta = \dfrac{|\mathbf{a} \times \mathbf{b}|}{|\mathbf{a}| \, |\mathbf{b}|}$

and so $d = |\mathbf{b}| \sin\theta = \dfrac{|\mathbf{b}| \, |\mathbf{a} \times \mathbf{b}|}{|\mathbf{a}| \, |\mathbf{b}|} = \dfrac{|\mathbf{a} \times \mathbf{b}|}{|\mathbf{a}|}.$

(b) $\mathbf{a} = \overrightarrow{QR} = \langle -1, -2, -1 \rangle$ and $\mathbf{b} = \overrightarrow{QP} = \langle 1, -5, -7 \rangle$. Then

$$\mathbf{a} \times \mathbf{b} = \langle (-2)(-7) - (-1)(-5), (-1)(1) - (-1)(-7), (-1)(-5) - (-2)(1) \rangle = \langle 9, -8, 7 \rangle.$$

Thus the distance is $d = \dfrac{|\mathbf{a} \times \mathbf{b}|}{|\mathbf{a}|} = \dfrac{1}{\sqrt{6}} \sqrt{81 + 64 + 49} = \sqrt{\dfrac{194}{6}} = \sqrt{\dfrac{97}{3}}.$

**35.** $(\mathbf{a} - \mathbf{b}) \times (\mathbf{a} + \mathbf{b}) = (\mathbf{a} - \mathbf{b}) \times \mathbf{a} + (\mathbf{a} - \mathbf{b}) \times \mathbf{b}$      by Property 3 of the cross product

$\qquad\qquad = \mathbf{a} \times \mathbf{a} + (-\mathbf{b}) \times \mathbf{a} + \mathbf{a} \times \mathbf{b} + (-\mathbf{b}) \times \mathbf{b}$      by Property 4

$\qquad\qquad = (\mathbf{a} \times \mathbf{a}) - (\mathbf{b} \times \mathbf{a}) + (\mathbf{a} \times \mathbf{b}) - (\mathbf{b} \times \mathbf{b})$      by Property 2 (with $c = -1$)

$\qquad\qquad = \mathbf{0} - (\mathbf{b} \times \mathbf{a}) + (\mathbf{a} \times \mathbf{b}) - \mathbf{0}$      by the margin note on page 655

$\qquad\qquad = (\mathbf{a} \times \mathbf{b}) + (\mathbf{a} \times \mathbf{b})$      by Property 1

$\qquad\qquad = 2(\mathbf{a} \times \mathbf{b})$

**37.** $\mathbf{a} \times (\mathbf{b} \times \mathbf{c}) + \mathbf{b} \times (\mathbf{c} \times \mathbf{a}) + \mathbf{c} \times (\mathbf{a} \times \mathbf{b})$

$\qquad\qquad = [(\mathbf{a} \cdot \mathbf{c})\mathbf{b} - (\mathbf{a} \cdot \mathbf{b})\mathbf{c}] + [(\mathbf{b} \cdot \mathbf{a})\mathbf{c} - (\mathbf{b} \cdot \mathbf{c})\mathbf{a}] + [(\mathbf{c} \cdot \mathbf{b})\mathbf{a} - (\mathbf{c} \cdot \mathbf{a})\mathbf{b}]$      by Exercise 36

$\qquad\qquad = (\mathbf{a} \cdot \mathbf{c})\mathbf{b} - (\mathbf{a} \cdot \mathbf{b})\mathbf{c} + (\mathbf{a} \cdot \mathbf{b})\mathbf{c} - (\mathbf{b} \cdot \mathbf{c})\mathbf{a} + (\mathbf{b} \cdot \mathbf{c})\mathbf{a} - (\mathbf{a} \cdot \mathbf{c})\mathbf{b} = \mathbf{0}$

**39.** (a) No. If $\mathbf{a} \cdot \mathbf{b} = \mathbf{a} \cdot \mathbf{c}$, then $\mathbf{a} \cdot (\mathbf{b} - \mathbf{c}) = 0$, so $\mathbf{a}$ is perpendicular to $\mathbf{b} - \mathbf{c}$, which can happen if $\mathbf{b} \neq \mathbf{c}$. For example,

     let $\mathbf{a} = \langle 1, 1, 1 \rangle$, $\mathbf{b} = \langle 1, 0, 0 \rangle$ and $\mathbf{c} = \langle 0, 1, 0 \rangle$.

(b) No. If $\mathbf{a} \times \mathbf{b} = \mathbf{a} \times \mathbf{c}$ then $\mathbf{a} \times (\mathbf{b} - \mathbf{c}) = \mathbf{0}$, which implies that $\mathbf{a}$ is parallel to $\mathbf{b} - \mathbf{c}$, which of course can happen

     if $\mathbf{b} \neq \mathbf{c}$.

(c) Yes. Since $\mathbf{a} \cdot \mathbf{c} = \mathbf{a} \cdot \mathbf{b}$, $\mathbf{a}$ is perpendicular to $\mathbf{b} - \mathbf{c}$, by part (a). From part (b), $\mathbf{a}$ is also parallel to $\mathbf{b} - \mathbf{c}$. Thus since

     $\mathbf{a} \neq \mathbf{0}$ but is both parallel and perpendicular to $\mathbf{b} - \mathbf{c}$, we have $\mathbf{b} - \mathbf{c} = \mathbf{0}$, so $\mathbf{b} = \mathbf{c}$.

**41.** (a) $\mathbf{u} \cdot \mathbf{r} = \mathbf{v} \cdot \mathbf{r} \iff \mathbf{u} \cdot \mathbf{r} - \mathbf{v} \cdot \mathbf{r} = 0 \iff (\mathbf{u} - \mathbf{v}) \cdot \mathbf{r} = 0$ [by Properties 2, 3, and 4 of the dot product]. This is true

     for all vectors $\mathbf{r}$ so in particular it is true for $\mathbf{r} = \mathbf{u} - \mathbf{v}$ which gives $(\mathbf{u} - \mathbf{v}) \cdot (\mathbf{u} - \mathbf{v}) = 0 \implies |\mathbf{u} - \mathbf{v}|^2 = 0$

     [by Property 1 of the dot product] $\implies \mathbf{u} - \mathbf{v} = \mathbf{0} \implies \mathbf{u} = \mathbf{v}$.

(b) Let $\mathbf{r}$ be any vector in $V_3$. Then

$\qquad\qquad [\mathbf{a} \times (\mathbf{b} + \mathbf{c})] \cdot \mathbf{r} = \mathbf{r} \cdot [\mathbf{a} \times (\mathbf{b} + \mathbf{c})]$      by Property 2 of the dot product

$\qquad\qquad = (\mathbf{r} \times \mathbf{a}) \cdot (\mathbf{b} + \mathbf{c})$      by Equation 6

$\qquad\qquad = (\mathbf{r} \times \mathbf{a}) \cdot \mathbf{b} + (\mathbf{r} \times \mathbf{a}) \cdot \mathbf{c}$      by Property 3 of the dot product

$\qquad\qquad = \mathbf{r} \cdot (\mathbf{a} \times \mathbf{b}) + \mathbf{r} \cdot (\mathbf{a} \times \mathbf{c})$      by Equation 6

$\qquad\qquad = \mathbf{r} \cdot [(\mathbf{a} \times \mathbf{b}) + (\mathbf{a} \times \mathbf{c})]$      by Property 3 of the dot product

$\qquad\qquad = [\mathbf{a} \times \mathbf{b} + \mathbf{a} \times \mathbf{c}] \cdot \mathbf{r}$      by Property 2 of the dot product

     By part (a), we must have $\mathbf{a} \times (\mathbf{b} + \mathbf{c}) = \mathbf{a} \times \mathbf{b} + \mathbf{a} \times \mathbf{c}$.

## 9.5    Equations of Lines and Planes

**1.** (a) True; each of the first two lines has a direction vector parallel to the direction vector of the third line, so these vectors are

     each scalar multiples of the third direction vector. Then the first two direction vectors are also scalar multiples of each

     other, so these vectors, and hence the two lines, are parallel.

(b) False; for example, the $x$- and $y$-axes are both perpendicular to the $z$-axis, yet the $x$- and $y$-axes are not parallel.

(c) True; each of the first two planes has a normal vector parallel to the normal vector of the third plane, so these two normal vectors are parallel to each other and the planes are parallel.

(d) False; for example, the $xy$- and $yz$-planes are not parallel, yet they are both perpendicular to the $xz$-plane.

(e) False; the $x$- and $y$-axes are not parallel, yet they are both parallel to the plane $z = 1$.

(f) True; if each line is perpendicular to a plane, then the lines' direction vectors are both parallel to a normal vector for the plane. Thus, the direction vectors are parallel to each other and the lines are parallel.

(g) False; the planes $y = 1$ and $z = 1$ are not parallel, yet they are both parallel to the $x$-axis.

(h) True; if each plane is perpendicular to a line, then any normal vector for each plane is parallel to a direction vector for the line. Thus, the normal vectors are parallel to each other and the planes are parallel.

(i) True; see Figure 9 and the accompanying discussion.

(j) False; they can be skew, as in Example 3 .

(k) True. Consider any normal vector for the plane and any direction vector for the line. If the normal vector is perpendicular to the direction vector, the line and plane are parallel. Otherwise, the vectors meet at an angle $\theta$, $0° \le \theta < 90°$, and the line will intersect the plane at an angle $90° - \theta$.

**3.** For this line, we have $\mathbf{r}_0 = 2\,\mathbf{i} + 2.4\,\mathbf{j} + 3.5\,\mathbf{k}$ and $\mathbf{v} = 3\,\mathbf{i} + 2\,\mathbf{j} - \mathbf{k}$, so a vector equation is $\mathbf{r} = \mathbf{r}_0 + t\,\mathbf{v} = (2\,\mathbf{i} + 2.4\,\mathbf{j} + 3.5\,\mathbf{k}) + t(3\,\mathbf{i} + 2\,\mathbf{j} - \mathbf{k}) = (2 + 3t)\,\mathbf{i} + (2.4 + 2t)\,\mathbf{j} + (3.5 - t)\,\mathbf{k}$ and parametric equations are $x = 2 + 3t$, $y = 2.4 + 2t$, $z = 3.5 - t$.

**5.** A line perpendicular to the given plane has the same direction as a normal vector to the plane, such as $\mathbf{n} = \langle 1, 3, 1 \rangle$. So $\mathbf{r}_0 = \mathbf{i} + 6\,\mathbf{k}$, and we can take $\mathbf{v} = \mathbf{i} + 3\,\mathbf{j} + \mathbf{k}$. Then a vector equation is $\mathbf{r} = (\mathbf{i} + 6\,\mathbf{k}) + t(\mathbf{i} + 3\,\mathbf{j} + \mathbf{k}) = (1 + t)\,\mathbf{i} + 3t\,\mathbf{j} + (6 + t)\,\mathbf{k}$, and parametric equations are $x = 1 + t$, $y = 3t$, $z = 6 + t$.

**7.** $\mathbf{v} = \left\langle 2 - 0, 1 - \frac{1}{2}, -3 - 1 \right\rangle = \left\langle 2, \frac{1}{2}, -4 \right\rangle$, and letting $P_0 = (2, 1, -3)$, parametric equations are $x = 2 + 2t$, $y = 1 + \frac{1}{2}t$, $z = -3 - 4t$, while symmetric equations are $\dfrac{x - 2}{2} = \dfrac{y - 1}{1/2} = \dfrac{z + 3}{-4}$ or $\dfrac{x - 2}{2} = 2y - 2 = \dfrac{z + 3}{-4}$.

**9.** The line has direction $\mathbf{v} = \langle 1, 2, 1 \rangle$. Letting $P_0 = (1, -1, 1)$, parametric equations are $x = 1 + t$, $y = -1 + 2t$, $z = 1 + t$ and symmetric equations are $x - 1 = \dfrac{y + 1}{2} = z - 1$.

**11.** Direction vectors of the lines are $\mathbf{v}_1 = \langle -2 - (-4), 0 - (-6), -3 - 1 \rangle = \langle 2, 6, -4 \rangle$ and $\mathbf{v}_2 = \langle 5 - 10, 3 - 18, 14 - 4 \rangle = \langle -5, -15, 10 \rangle$, and since $\mathbf{v}_2 = -\frac{5}{2}\mathbf{v}_1$, the direction vectors and thus the lines are parallel.

**13.** (a) The line passes through the point $(1, -5, 6)$ and a direction vector for the line is $\langle -1, 2, -3 \rangle$, so symmetric equations for the line are $\dfrac{x - 1}{-1} = \dfrac{y + 5}{2} = \dfrac{z - 6}{-3}$.

(b) The line intersects the $xy$-plane when $z = 0$, so we need $\dfrac{x - 1}{-1} = \dfrac{y + 5}{2} = \dfrac{0 - 6}{-3}$ or $\dfrac{x - 1}{-1} = 2 \;\Rightarrow\; x = -1$, $\dfrac{y + 5}{2} = 2 \;\Rightarrow\; y = -1$. Thus the point of intersection with the $xy$-plane is $(-1, -1, 0)$. Similarly for the $yz$-plane, we need $x = 0 \;\Rightarrow\; 1 = \dfrac{y + 5}{2} = \dfrac{z - 6}{-3} \;\Rightarrow\; y = -3$, $z = 3$. Thus the line intersects the $yz$-plane at $(0, -3, 3)$. For the $xz$-plane, we need $y = 0$ $\Rightarrow$ $\dfrac{x-1}{-1} = \dfrac{5}{2} = \dfrac{z-6}{-3}$ $\Rightarrow$ $x = -\frac{3}{2}, z = -\frac{3}{2}$. So the line intersects the $xz$-plane

at $\left(-\frac{3}{2}, 0, -\frac{3}{2}\right)$.

**15.** From Equation 4 , the line segment from $\mathbf{r}_0 = 2\mathbf{i} - \mathbf{j} + 4\mathbf{k}$ to $\mathbf{r}_1 = 4\mathbf{i} + 6\mathbf{j} + \mathbf{k}$ is

$\mathbf{r}(t) = (1-t)\,\mathbf{r}_0 + t\,\mathbf{r}_1 = (1-t)(2\mathbf{i} - \mathbf{j} + 4\mathbf{k}) + t(4\mathbf{i} + 6\mathbf{j} + \mathbf{k}) = (2\mathbf{i} - \mathbf{j} + 4\mathbf{k}) + t(2\mathbf{i} + 7\mathbf{j} - 3\mathbf{k}), 0 \le t \le 1.$

**17.** Since the direction vectors are $\mathbf{v}_1 = \langle -6, 9, -3 \rangle$ and $\mathbf{v}_2 = \langle 2, -3, 1 \rangle$, we have $\mathbf{v}_1 = -3\mathbf{v}_2$ so the lines are parallel.

**19.** Since the direction vectors $\langle 1, 2, 3 \rangle$ and $\langle -4, -3, 2 \rangle$ are not scalar multiples of each other, the lines are not parallel, so we

check to see if the lines intersect. The parametric equations of the lines are $L_1$: $x = t, y = 1 + 2t, z = 2 + 3t$ and $L_2$:

$x = 3 - 4s, y = 2 - 3s, z = 1 + 2s$. For the lines to intersect, we must be able to find one value of $t$ and one value of $s$ that

produce the same point from the respective parametric equations. Thus we need to satisfy the following three equations:

$t = 3 - 4s, 1 + 2t = 2 - 3s, 2 + 3t = 1 + 2s$. Solving the first two equations we get $t = -1, s = 1$ and checking, we see

that these values don't satisfy the third equation. Thus the lines aren't parallel and don't intersect, so they must be skew lines.

**21.** Since the plane is perpendicular to the vector $\langle -2, 1, 5 \rangle$, we can take $\langle -2, 1, 5 \rangle$ as a normal vector to the plane.

$(6, 3, 2)$ is a point on the plane, so setting $a = -2, b = 1, c = 5$ and $x_0 = 6, y_0 = 3, z_0 = 2$ in Equation 7 gives

$-2(x - 6) + 1(y - 3) + 5(z - 2) = 0$ or $-2x + y + 5z = 1$ to be an equation of the plane.

**23.** Since the two planes are parallel, they will have the same normal vectors. So we can take $\mathbf{n} = \langle 3, 0, -7 \rangle$, and an equation of

the plane is $3(x - 4) + 0[y - (-2)] - 7(z - 3) = 0$ or $3x - 7z = -9$.

**25.** Here the vectors $\mathbf{a} = \langle 1 - 0, 0 - 1, 1 - 1 \rangle = \langle 1, -1, 0 \rangle$ and $\mathbf{b} = \langle 1 - 0, 1 - 1, 0 - 1 \rangle = \langle 1, 0, -1 \rangle$ lie in the plane, so

$\mathbf{a} \times \mathbf{b}$ is a normal vector to the plane. Thus, we can take $\mathbf{n} = \mathbf{a} \times \mathbf{b} = \langle 1 - 0, 0 + 1, 0 + 1 \rangle = \langle 1, 1, 1 \rangle$. If $P_0$ is the point

$(0, 1, 1)$, an equation of the plane is $1(x - 0) + 1(y - 1) + 1(z - 1) = 0$ or $x + y + z = 2$.

**27.** If we first find two nonparallel vectors in the plane, their cross product will be a normal vector to the plane. Since the given

line lies in the plane, its direction vector $\mathbf{a} = \langle -2, 5, 4 \rangle$ is one vector in the plane. We can verify that the given point $(6, 0, -2)$

does not lie on this line, so to find another nonparallel vector $\mathbf{b}$ which lies in the plane, we can pick any point on the line and

find a vector connecting the points. If we put $t = 0$, we see that $(4, 3, 7)$ is on the line, so

$\mathbf{b} = \langle 6 - 4, 0 - 3, -2 - 7 \rangle = \langle 2, -3, -9 \rangle$ and $\mathbf{n} = \mathbf{a} \times \mathbf{b} = \langle -45 + 12, 8 - 18, 6 - 10 \rangle = \langle -33, -10, -4 \rangle$. Thus, an

equation of the plane is $-33(x - 6) - 10(y - 0) - 4[z - (-2)] = 0$ or $33x + 10y + 4z = 190$.

**29.** A direction vector for the line of intersection is $\mathbf{a} = \mathbf{n}_1 \times \mathbf{n}_2 = \langle 1, 1, -1 \rangle \times \langle 2, -1, 3 \rangle = \langle 2, -5, -3 \rangle$, and $\mathbf{a}$ is parallel to the

desired plane. Another vector parallel to the plane is the vector connecting any point on the line of intersection to the given

point $(-1, 2, 1)$ in the plane. Setting $x = 0$, the equations of the planes reduce to $y - z = 2$ and $-y + 3z = 1$ with

simultaneous solution $y = \frac{7}{2}$ and $z = \frac{3}{2}$. So a point on the line is $\left(0, \frac{7}{2}, \frac{3}{2}\right)$ and another vector parallel to the plane is

$\langle -1, -\frac{3}{2}, -\frac{1}{2} \rangle$. Then a normal vector to the plane is $\mathbf{n} = \langle 2, -5, -3 \rangle \times \langle -1, -\frac{3}{2}, -\frac{1}{2} \rangle = \langle -2, 4, -8 \rangle$ and an equation of

the plane is $-2(x + 1) + 4(y - 2) - 8(z - 1) = 0$ or $x - 2y + 4z = -1$.

**31.** If a plane is perpendicular to two other planes, its normal vector is perpendicular to the normal vectors of the other two planes. Thus $\langle 2, 1, -2 \rangle \times \langle 1, 0, 3 \rangle = \langle 3 - 0, -2 - 6, 0 - 1 \rangle = \langle 3, -8, -1 \rangle$ is a normal vector to the desired plane. The point $(1, 5, 1)$ lies on the plane, so an equation is $3(x - 1) - 8(y - 5) - (z - 1) = 0$ or $3x - 8y - z = -38$.

**33.** To find the $x$-intercept we set $y = z = 0$ in the equation $2x + 5y + z = 10$ and obtain $2x = 10 \Rightarrow x = 5$ so the $x$-intercept is $(5, 0, 0)$. When $x = z = 0$ we get $5y = 10 \Rightarrow y = 2$, so the $y$-intercept is $(0, 2, 0)$. Setting $x = y = 0$ gives $z = 10$, so the $z$-intercept is $(0, 0, 10)$ and we graph the portion of the plane that lies in the first octant.

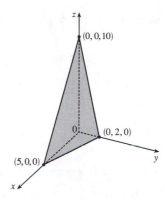

**35.** Setting $y = z = 0$ in the equation $6x - 3y + 4z = 6$ gives $6x = 6 \Rightarrow x = 1$, when $x = z = 0$ we have $-3y = 6 \Rightarrow y = -2$, and $x = y = 0$ implies $4z = 6 \Rightarrow z = \frac{3}{2}$, so the intercepts are $(1, 0, 0)$, $(0, -2, 0)$, and $(0, 0, \frac{3}{2})$. The figure shows the portion of the plane cut off by the coordinate planes.

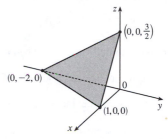

**37.** Substitute the parametric equations of the line into the equation of the plane: $(3 - t) - (2 + t) + 2(5t) = 9 \Rightarrow 8t = 8 \Rightarrow t = 1$. Therefore, the point of intersection of the line and the plane is given by $x = 3 - 1 = 2$, $y = 2 + 1 = 3$, and $z = 5(1) = 5$, that is, the point $(2, 3, 5)$.

**39.** Normal vectors for the planes are $\mathbf{n}_1 = \langle 1, 4, -3 \rangle$ and $\mathbf{n}_2 = \langle -3, 6, 7 \rangle$, so the normals (and thus the planes) aren't parallel. But $\mathbf{n}_1 \cdot \mathbf{n}_2 = -3 + 24 - 21 = 0$, so the normals (and thus the planes) are perpendicular.

**41.** Normal vectors for the planes are $\mathbf{n}_1 = \langle 1, 1, 1 \rangle$ and $\mathbf{n}_2 = \langle 1, -1, 1 \rangle$. The normals are not parallel, so neither are the planes. Furthermore, $\mathbf{n}_1 \cdot \mathbf{n}_2 = 1 - 1 + 1 = 1 \neq 0$, so the planes aren't perpendicular. The angle between them is given by

$$\cos \theta = \frac{\mathbf{n}_1 \cdot \mathbf{n}_2}{|\mathbf{n}_1|\,|\mathbf{n}_2|} = \frac{1}{\sqrt{3}\,\sqrt{3}} = \frac{1}{3} \Rightarrow \theta = \cos^{-1}\left(\tfrac{1}{3}\right) \approx 70.5°.$$

**43.** (a) To find a point on the line of intersection, set one of the variables equal to a constant, say $z = 0$. (This will fail if the line of intersection does not cross the $xy$-plane; in that case, try setting $x$ or $y$ equal to 0.) The equations of the two planes reduce to $x + y = 1$ and $x + 2y = 1$. Solving these two equations gives $x = 1$, $y = 0$. Thus a point on the line is $(1, 0, 0)$.

A vector $\mathbf{v}$ in the direction of this intersecting line is perpendicular to the normal vectors of both planes, so we can take $\mathbf{v} = \mathbf{n}_1 \times \mathbf{n}_2 = \langle 1, 1, 1 \rangle \times \langle 1, 2, 2 \rangle = \langle 2 - 2, 1 - 2, 2 - 1 \rangle = \langle 0, -1, 1 \rangle$. By Equations 2, parametric equations for the line are $x = 1$, $y = -t$, $z = t$.

(b) The angle between the planes satisfies $\cos \theta = \dfrac{\mathbf{n}_1 \cdot \mathbf{n}_2}{|\mathbf{n}_1|\,|\mathbf{n}_2|} = \dfrac{1 + 2 + 2}{\sqrt{3}\,\sqrt{9}} = \dfrac{5}{3\sqrt{3}}$. Therefore $\theta = \cos^{-1}\left(\dfrac{5}{3\sqrt{3}}\right) \approx 15.8°.$

**45.** Setting $z = 0$, the equations of the two planes become $5x - 2y = 1$ and $4x + y = 6$. Solving these two equations gives $x = 1$, $y = 2$ so a point on the line of intersection is $(1, 2, 0)$. A vector $\mathbf{v}$ in the direction of this intersecting line is perpendicular to the normal vectors of both planes. So we can use $\mathbf{v} = \mathbf{n}_1 \times \mathbf{n}_2 = \langle 5, -2, -2 \rangle \times \langle 4, 1, 1 \rangle = \langle 0, -13, 13 \rangle$ or equivalently we can take $\mathbf{v} = \langle 0, -1, 1 \rangle$, and symmetric equations for the line are $x = 1$, $\dfrac{y - 2}{-1} = \dfrac{z}{1}$ or $x = 1$, $y - 2 = -z$.

**47.** The plane contains the points $(a, 0, 0)$, $(0, b, 0)$ and $(0, 0, c)$. Thus the vectors $\mathbf{a} = \langle -a, b, 0 \rangle$ and $\mathbf{b} = \langle -a, 0, c \rangle$ lie in the plane, and $\mathbf{n} = \mathbf{a} \times \mathbf{b} = \langle bc - 0, 0 + ac, 0 + ab \rangle = \langle bc, ac, ab \rangle$ is a normal vector to the plane. The equation of the plane is therefore $bcx + acy + abz = abc + 0 + 0$ or $bcx + acy + abz = abc$. Notice that if $a \neq 0$, $b \neq 0$ and $c \neq 0$ then we can rewrite the equation as $\dfrac{x}{a} + \dfrac{y}{b} + \dfrac{z}{c} = 1$. This is a good equation to remember!

**49.** Two vectors which are perpendicular to the required line are the normal of the given plane, $\langle 1, 1, 1 \rangle$, and a direction vector for the given line, $\langle 1, -1, 2 \rangle$. So a direction vector for the required line is $\langle 1, 1, 1 \rangle \times \langle 1, -1, 2 \rangle = \langle 3, -1, -2 \rangle$. Thus $L$ is given by $\langle x, y, z \rangle = \langle 0, 1, 2 \rangle + t \langle 3, -1, -2 \rangle$, or in parametric form, $x = 3t$, $y = 1 - t$, $z = 2 - 2t$.

**51.** Let $P_i$ have normal vector $\mathbf{n}_i$. Then $\mathbf{n}_1 = \langle 3, 6, -3 \rangle$, $\mathbf{n}_2 = \langle 4, -12, 8 \rangle$, $\mathbf{n}_3 = \langle 3, -9, 6 \rangle$, $\mathbf{n}_4 = \langle 1, 2, -1 \rangle$. Now $\mathbf{n}_1 = 3\mathbf{n}_4$, so $\mathbf{n}_1$ and $\mathbf{n}_4$ are parallel, and hence $P_1$ and $P_4$ are parallel; similarly $P_2$ and $P_3$ are parallel because $\mathbf{n}_2 = \frac{4}{3}\mathbf{n}_3$. However, $\mathbf{n}_1$ and $\mathbf{n}_2$ are not parallel (so not all four planes are parallel). Notice that the point $(2, 0, 0)$ lies on both $P_1$ and on $P_4$, so these two planes are identical. The point $\left(\frac{5}{4}, 0, 0\right)$ lies on $P_2$ but not on $P_3$, so these are different planes.

**53.** Let $Q = (1, 3, 4)$ and $R = (2, 1, 1)$, points on the line corresponding to $t = 0$ and $t = 1$. Let $P = (4, 1, -2)$. Then $\mathbf{a} = \overrightarrow{QR} = \langle 1, -2, -3 \rangle$, $\mathbf{b} = \overrightarrow{QP} = \langle 3, -2, -6 \rangle$. The distance is

$$d = \frac{|\mathbf{a} \times \mathbf{b}|}{|\mathbf{a}|} = \frac{|\langle 1, -2, -3 \rangle \times \langle 3, -2, -6 \rangle|}{|\langle 1, -2, -3 \rangle|} = \frac{|\langle 6, -3, 4 \rangle|}{|\langle 1, -2, -3 \rangle|} = \frac{\sqrt{6^2 + (-3)^2 + 4^2}}{\sqrt{1^2 + (-2)^2 + (-3)^2}} = \frac{\sqrt{61}}{\sqrt{14}} = \sqrt{\frac{61}{14}}.$$

**55.** By Equation 9, the distance is $D = \dfrac{|ax_1 + by_1 + cz_1 + d|}{\sqrt{a^2 + b^2 + c^2}} = \dfrac{|3(1) + 2(-2) + 6(4) - 5|}{\sqrt{3^2 + 2^2 + 6^2}} = \dfrac{|18|}{\sqrt{49}} = \dfrac{18}{7}.$

**57.** Put $y = z = 0$ in the equation of the first plane to get the point $(2, 0, 0)$ on the plane. Because the planes are parallel, the distance $D$ between them is the distance from $(2, 0, 0)$ to the second plane. By Equation 9,

$$D = \frac{|4(2) - 6(0) + 2(0) - 3|}{\sqrt{4^2 + (-6)^2 + (2)^2}} = \frac{5}{\sqrt{56}} = \frac{5}{2\sqrt{14}} \text{ or } \frac{5\sqrt{14}}{28}.$$

**59.** The distance between two parallel planes is the same as the distance between a point on one of the planes and the other plane. Let $P_0 = (x_0, y_0, z_0)$ be a point on the plane given by $ax + by + cz + d_1 = 0$. Then $ax_0 + by_0 + cz_0 + d_1 = 0$ and the distance between $P_0$ and the plane given by $ax + by + cz + d_2 = 0$ is, from Equation 9,

$$D = \frac{|ax_0 + by_0 + cz_0 + d_2|}{\sqrt{a^2 + b^2 + c^2}} = \frac{|-d_1 + d_2|}{\sqrt{a^2 + b^2 + c^2}} = \frac{|d_1 - d_2|}{\sqrt{a^2 + b^2 + c^2}}.$$

**61.** $L_1: x = y = z \Rightarrow x = y$ **(1)**. $L_2: x + 1 = y/2 = z/3 \Rightarrow x + 1 = y/2$ **(2)**. The solution of **(1)** and **(2)** is $x = y = -2$. However, when $x = -2$, $x = z \Rightarrow z = -2$, but $x + 1 = z/3 \Rightarrow z = -3$, a contradiction. Hence the lines do not intersect. For $L_1$, $\mathbf{v}_1 = \langle 1, 1, 1 \rangle$, and for $L_2$, $\mathbf{v}_2 = \langle 1, 2, 3 \rangle$, so the lines are not parallel. Thus the lines are skew

lines. If two lines are skew, they can be viewed as lying in two parallel planes and so the distance between the skew lines

would be the same as the distance between these parallel planes. The common normal vector to the planes must be

perpendicular to both $\langle 1, 1, 1 \rangle$ and $\langle 1, 2, 3 \rangle$, the direction vectors of the two lines. So set

$\mathbf{n} = \langle 1, 1, 1 \rangle \times \langle 1, 2, 3 \rangle = \langle 3 - 2, -3 + 1, 2 - 1 \rangle = \langle 1, -2, 1 \rangle$. From above, we know that $(-2, -2, -2)$ and $(-2, -2, -3)$

are points of $L_1$ and $L_2$ respectively. So in the notation of Equation 8, $1(-2) - 2(-2) + 1(-2) + d_1 = 0 \quad \Rightarrow \quad d_1 = 0$ and

$1(-2) - 2(-2) + 1(-3) + d_2 = 0 \quad \Rightarrow \quad d_2 = 1$.

By Exercise 59, the distance between these two skew lines is $D = \dfrac{|0 - 1|}{\sqrt{1 + 4 + 1}} = \dfrac{1}{\sqrt{6}}$.

*Alternate solution (without reference to planes):* A vector which is perpendicular to both of the lines is

$\mathbf{n} = \langle 1, 1, 1 \rangle \times \langle 1, 2, 3 \rangle = \langle 1, -2, 1 \rangle$. Pick any point on each of the lines, say $(-2, -2, -2)$ and $(-2, -2, -3)$, and form the

vector $\mathbf{b} = \langle 0, 0, 1 \rangle$ connecting the two points. The distance between the two skew lines is the absolute value of the scalar

projection of $\mathbf{b}$ along $\mathbf{n}$, that is, $D = \dfrac{|\mathbf{n} \cdot \mathbf{b}|}{|\mathbf{n}|} = \dfrac{|1 \cdot 0 - 2 \cdot 0 + 1 \cdot 1|}{\sqrt{1 + 4 + 1}} = \dfrac{1}{\sqrt{6}}$.

**63.** If $a \neq 0$, then $ax + by + cz + d = 0 \quad \Rightarrow \quad a(x + d/a) + b(y - 0) + c(z - 0) = 0$ which by (7) is the scalar equation of the

plane through the point $(-d/a, 0, 0)$ with normal vector $\langle a, b, c \rangle$. Similarly, if $b \neq 0$ (or if $c \neq 0$) the equation of the plane can

be rewritten as $a(x - 0) + b(y + d/b) + c(z - 0) = 0$ [or as $a(x - 0) + b(y - 0) + c(z + d/c) = 0$] which by (7) is the

scalar equation of a plane through the point $(0, -d/b, 0)$ [or the point $(0, 0, -d/c)$] with normal vector $\langle a, b, c \rangle$.

## 9.6   Functions and Surfaces

**1.** (a) According to Table 1, $f(40, 15) = 25$, which means that if a 40-knot wind has been blowing in the open sea for 15 hours,

it will create waves with estimated heights of 25 feet.

(b) $h = f(30, t)$ means we fix $v$ at 30 and allow $t$ to vary, resulting in a function of one variable. Thus here, $h = f(30, t)$

gives the wave heights produced by 30-knot winds blowing for $t$ hours. From the table (look at the row corresponding to

$v = 30$), the function increases but at a declining rate as $t$ increases. In fact, the function values appear to be approaching a

limiting value of approximately 19, which suggests that 30-knot winds cannot produce waves higher than about 19 feet.

(c) $h = f(v, 30)$ means we fix $t$ at 30, again giving a function of one variable. So, $h = f(v, 30)$ gives the wave heights

produced by winds of speed $v$ blowing for 30 hours. From the table (look at the column corresponding to $t = 30$), the

function appears to increase at an increasing rate, with no apparent limiting value. This suggests that faster winds (lasting

30 hours) always create higher waves.

**3.** (a) $g(2, -1) = \cos(2 + 2(-1)) = \cos(0) = 1$

(b) $x + 2y$ is defined for all choices of values for $x$ and $y$ and the cosine function is defined for all input values, so the domain

of $g$ is $\mathbb{R}^2$.

(c) The range of the cosine function is $[-1, 1]$ and $x + 2y$ generates all possible input values for the cosine function, so the

range of $\cos(x + 2y)$ is $[-1, 1]$.

**5.** $\sqrt{y - x^2}$ is defined only when $y - x^2 \geq 0$, or $y \geq x^2$.

In addition, $f$ is not defined if $1 - x^2 = 0$ ⇔

$x = \pm 1$. Thus the domain of $f$ is

$\{(x, y) \mid y \geq x^2,\ x \neq \pm 1\}$.

**7.** $\sqrt{1 - x^2}$ is defined only when $1 - x^2 \geq 0$, or

$x^2 \leq 1$ ⇔ $-1 \leq x \leq 1$, and $\sqrt{1 - y^2}$ is defined

only when $1 - y^2 \geq 0$, or $y^2 \leq 1$ ⇔ $-1 \leq y \leq 1$.

Thus the domain of $f$ is

$\{(x, y) \mid -1 \leq x \leq 1,\ -1 \leq y \leq 1\}$.

**9.** $z = 3$, a horizontal plane through the point $(0, 0, 3)$.

**11.** $z = 6 - 3x - 2y$ or $3x + 2y + z = 6$, a plane with

intercepts 2, 3, and 6.

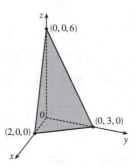

**13.** $z = y^2 + 1$, a parabolic cylinder

**15.** All six graphs have different traces in the planes $x = 0$ and $y = 0$, so we investigate these for each function.

(a) $f(x, y) = |x| + |y|$. The trace in $x = 0$ is $z = |y|$, and in $y = 0$ is $z = |x|$, so it must be graph VI.

(b) $f(x, y) = |xy|$. The trace in $x = 0$ is $z = 0$, and in $y = 0$ is $z = 0$, so it must be graph V.

(c) $f(x, y) = \dfrac{1}{1 + x^2 + y^2}$. The trace in $x = 0$ is $z = \dfrac{1}{1 + y^2}$, and in $y = 0$ is $z = \dfrac{1}{1 + x^2}$. In addition, we can see that $f$ is

close to 0 for large values of $x$ and $y$, so this is graph I.

(d) $f(x, y) = (x^2 - y^2)^2$. The trace in $x = 0$ is $z = y^4$, and in $y = 0$ is $z = x^4$. Both graph II and graph IV seem plausible;

notice the trace in $z = 0$ is $0 = (x^2 - y^2)^2$ ⇒ $y = \pm x$, so it must be graph IV.

(e) $f(x, y) = (x - y)^2$. The trace in $x = 0$ is $z = y^2$, and in $y = 0$ is $z = x^2$. Both graph II and graph IV seem plausible; notice the trace in $z = 0$ is $0 = (x - y)^2$ $\Rightarrow$ $y = x$, so it must be graph II.

(f) $f(x, y) = \sin(|x| + |y|)$. The trace in $x = 0$ is $z = \sin|y|$, and in $y = 0$ is $z = \sin|x|$. In addition, notice that the oscillating nature of the graph is characteristic of trigonometric functions. So this is graph III.

**17.** The equation of the graph is $z = \sqrt{4x^2 + y^2}$ or equivalently $4x^2 + y^2 = z^2$,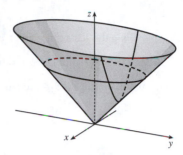
$z \geq 0$. Traces in $x = k$ are $z^2 - y^2 = 4k^2$, $z \geq 0$, a family of hyperbolas where
we have only the upper branch. Traces in $y = k$ are $z^2 - 4x^2 = k^2$, $z \geq 0$,
again a family of half-hyperbolas. Traces in $z = k$, $k \geq 0$, are $4x^2 + y^2 = k^2$
or $x^2 + \dfrac{y^2}{4} = \dfrac{k^2}{4}$, a family of ellipses. Note that the original equation can be
written as $x^2 + \dfrac{y^2}{4} = \dfrac{z^2}{4}$, $z \geq 0$, which we recognize as the upper half of an
elliptical cone.

**19.** $y = z^2 - x^2$. The traces in $x = k$ are the parabolas $y = z^2 - k^2$; the traces in
$y = k$ are $k = z^2 - x^2$, which are hyperbolas (note the hyperbolas are oriented
differently for $k > 0$ than for $k < 0$); and the traces in $z = k$ are the parabolas
$y = k^2 - x^2$. Thus, $\dfrac{y}{1} = \dfrac{z^2}{1^2} - \dfrac{x^2}{1^2}$ is a hyperbolic paraboloid.

**21.** Completing squares in $y$ and $z$ gives $4x^2 + (y - 2)^2 + 4(z - 3)^2 = 4$ or $x^2 + \dfrac{(y - 2)^2}{4} + (z - 3)^2 = 1$, an ellipsoid with
center $(0, 2, 3)$.

**23.** (a) In $\mathbb{R}^2$, $x^2 + y^2 = 1$ represents a circle of radius 1 centered at the origin.

(b) In $\mathbb{R}^3$, the equation doesn't involve $z$, which means that any horizontal plane $z = k$ intersects the surface in a circle
$x^2 + y^2 = 1$, $z = k$. Thus the surface is a circular cylinder, made up of infinitely many shifted copies of the circle
$x^2 + y^2 = 1$, with axis the $z$-axis.

(c) In $\mathbb{R}^3$, $x^2 + z^2 = 1$ also represents a circular cylinder of radius 1, this time with axis the $y$-axis.

**25.** (a) The traces of $x^2 + y^2 - z^2 = 1$ in $x = k$ are $y^2 - z^2 = 1 - k^2$, a family of hyperbolas. (Note that the hyperbolas are
oriented differently for $-1 < k < 1$ than for $k < -1$ or $k > 1$.) The traces in $y = k$ are $x^2 - z^2 = 1 - k^2$, a similar
family of hyperbolas. The traces in $z = k$ are $x^2 + y^2 = 1 + k^2$, a family of circles. For $k = 0$, the trace in the
$xy$-plane, the circle is of radius 1. As $|k|$ increases, so does the radius of the circle. This behavior, combined with the
hyperbolic vertical traces, gives the graph of the hyperboloid of one sheet in Table 2.

(b) The shape of the surface is unchanged, but the hyperboloid is rotated so that its axis is the $y$-axis. Traces in $y = k$ are circles, while traces in $x = k$ and $z = k$ are hyperbolas.

(c) Completing the square in $y$ gives $x^2 + (y+1)^2 - z^2 = 1$. The surface is a hyperboloid identical to the one in part (a) but shifted one unit in the negative $y$-direction.

27. Graph III has these traces. One indication is found by noting that the higher $z$-values occur for negative values of $y$ in the traces in $x = 1$ and $x = 2$, and for positive values of $x$ in the traces in $y = -1$ and $y = -2$. Thus the graph should have a "hill" over the fourth quadrant of the $xy$-plane. Similarly, we should expect a "valley" corresponding to the second quadrant of the $xy$-plane.

29. $f(x, y) = xye^{x+2y-9x^2-9y^2}$

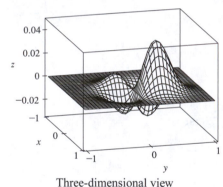

Three-dimensional view       Front view

It does appear that the function has a maximum value, at the higher of the two "hilltops." From the front view graph, we can estimate the maximum value to be approximately 0.044. Both hilltops could be considered local maximum points, as the values of $f$ there are larger than at the neighboring points. Similarly, the two "valley bottoms" visible in the graph can be considered local minimum points, as all the neighboring points give greater values of $f$. (And $f$ achieves a minimum value at the lower valley dip.)

31.

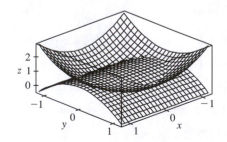

The curve of intersection looks like a bent ellipse. The projection of this curve onto the $xy$-plane is the set of points $(x, y, 0)$ which satisfy $x^2 + y^2 = 1 - y^2 \iff x^2 + 2y^2 = 1 \iff$

$$x^2 + \frac{y^2}{\left(1/\sqrt{2}\right)^2} = 1. \text{ This is an equation of an ellipse.}$$

**33.** If $(a, b, c)$ satisfies $z = y^2 - x^2$, then $c = b^2 - a^2$. $L_1 \colon x = a + t, \, y = b + t, \, z = c + 2(b - a)t$,

$L_2 \colon x = a + t, \, y = b - t, \, z = c - 2(b + a)t$. Substitute the parametric equations of $L_1$ into the equation

of the hyperbolic paraboloid in order to find the points of intersection: $z = y^2 - x^2 \quad \Rightarrow$

$c + 2(b - a)t = (b + t)^2 - (a + t)^2 = b^2 - a^2 + 2(b - a)t \quad \Rightarrow \quad c = b^2 - a^2$. As this is true for all values of $t$,

$L_1$ lies on $z = y^2 - x^2$. Performing similar operations with $L_2$ gives: $z = y^2 - x^2 \quad \Rightarrow$

$c - 2(b + a)t = (b - t)^2 - (a + t)^2 = b^2 - a^2 - 2(b + a)t \quad \Rightarrow \quad c = b^2 - a^2$. This tells us that all of $L_2$ also lies on

$z = y^2 - x^2$.

## 9.7 Cylindrical and Spherical Coordinates

**1.** See Figure 2 and the accompanying discussion on page 682; see the paragraph preceding Example 2 on page 683.

**3. (a)**

$x = 2 \cos \dfrac{\pi}{4} = \sqrt{2}, \, y = 2 \sin \dfrac{\pi}{4} = \sqrt{2}, \, z = 1$,

so the point is $\left(\sqrt{2}, \sqrt{2}, 1\right)$ in rectangular coordinates.

**(b)**

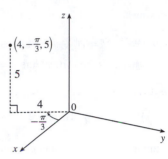

$x = 4 \cos\left(-\dfrac{\pi}{3}\right) = 2, \, y = 4 \sin\left(-\dfrac{\pi}{3}\right) = -2\sqrt{3}$,

and $z = 5$, so the point is $\left(2, -2\sqrt{3}, 5\right)$ in rectangular

coordinates.

**5. (a)** $r^2 = x^2 + y^2 = 1^2 + (-1)^2 = 2$ so $r = \sqrt{2}$; $\tan\theta = \dfrac{y}{x} = \dfrac{-1}{1} = -1$ and the point $(1, -1)$ is in the fourth quadrant of

the $xy$-plane, so $\theta = \dfrac{7\pi}{4} + 2n\pi$; $z = 4$. Thus, one set of cylindrical coordinates is $\left(\sqrt{2}, \dfrac{7\pi}{4}, 4\right)$.

**(b)** $r^2 = (-1)^2 + \left(-\sqrt{3}\right)^2 = 4$ so $r = 2$; $\tan\theta = \dfrac{-\sqrt{3}}{-1} = \sqrt{3}$ and the point $\left(-1, -\sqrt{3}\right)$ is in the third quadrant of the

$xy$-plane, so $\theta = \dfrac{4\pi}{3} + 2n\pi$; $z = 2$. Thus, one set of cylindrical coordinates is $\left(2, \dfrac{4\pi}{3}, 2\right)$.

**7. (a)**

$x = \rho \sin\phi \cos\theta = (1) \sin 0 \cos 0 = 0$,

$y = \rho \sin\phi \sin\theta = (1) \sin 0 \sin 0 = 0$, and

$z = \rho \cos\phi = (1) \cos 0 = 1$ so the point is

$(0, 0, 1)$ in rectangular coordinates.

**(b)**

$x = 2 \sin \dfrac{\pi}{4} \cos \dfrac{\pi}{3} = \dfrac{\sqrt{2}}{2}, \, y = 2 \sin \dfrac{\pi}{4} \sin \dfrac{\pi}{3} = \dfrac{\sqrt{6}}{2}$,

$z = 2 \cos \dfrac{\pi}{4} = \sqrt{2}$ so the point is $\left(\dfrac{\sqrt{2}}{2}, \dfrac{\sqrt{6}}{2}, \sqrt{2}\right)$ in

rectangular coordinates.

**9.** (a) $\rho = \sqrt{x^2 + y^2 + z^2} = \sqrt{1 + 3 + 12} = 4$, $\cos\phi = \dfrac{z}{\rho} = \dfrac{2\sqrt{3}}{4} = \dfrac{\sqrt{3}}{2}$ $\Rightarrow$ $\phi = \dfrac{\pi}{6}$, and

$\cos\theta = \dfrac{x}{\rho\sin\phi} = \dfrac{1}{4\sin(\pi/6)} = \dfrac{1}{2}$ $\Rightarrow$ $\theta = \dfrac{\pi}{3}$ [since $y > 0$]. Thus spherical coordinates are $\left(4, \dfrac{\pi}{3}, \dfrac{\pi}{6}\right)$.

(b) $\rho = \sqrt{0 + 1 + 1} = \sqrt{2}$, $\cos\phi = \dfrac{-1}{\sqrt{2}}$ $\Rightarrow$ $\phi = \dfrac{3\pi}{4}$, and $\cos\theta = \dfrac{0}{\sqrt{2}\sin(3\pi/4)} = 0$ $\Rightarrow$ $\theta = \dfrac{3\pi}{2}$ [since $y < 0$].

Thus spherical coordinates are $\left(\sqrt{2}, \dfrac{3\pi}{2}, \dfrac{3\pi}{4}\right)$.

**11.** Since $\theta = \frac{\pi}{4}$ but $r$ and $z$ may vary, the surface is a vertical half-plane including the $z$-axis and intersecting the $xy$-plane in the half-line $y = x$, $x \geq 0$.

**13.** Since $\phi = \frac{\pi}{3}$, the surface is the top half of the right circular cone with vertex at the origin and axis the positive $z$-axis.

**15.** $z = 4 - r^2 = 4 - (x^2 + y^2)$ or $4 - x^2 - y^2$, so the surface is a circular paraboloid with vertex $(0, 0, 4)$, axis the $z$-axis, and opening downward.

**17.** $r = 2\cos\theta$ $\Rightarrow$ $r^2 = x^2 + y^2 = 2r\cos\theta = 2x$ $\Leftrightarrow$ $(x-1)^2 + y^2 = 1$, which is the equation of a circular cylinder with radius 1, whose axis is the vertical line $x = 1$, $y = 0$, $z = z$.

**19.** $\rho = \sin\theta\sin\phi$ $\Rightarrow$ $\rho^2 = \rho\sin\theta\sin\phi$ $\Leftrightarrow$ $x^2 + y^2 + z^2 = y$ $\Leftrightarrow$ $x^2 + y^2 - y + \frac{1}{4} + z^2 = \frac{1}{4}$ $\Leftrightarrow$

$x^2 + (y - \frac{1}{2})^2 + z^2 = \frac{1}{4}$. Therefore, the surface is a sphere of radius $\frac{1}{2}$ centered at $(0, \frac{1}{2}, 0)$.

**21.** (a) Substituting $x^2 + y^2 = r^2$ and $y = r\sin\theta$, the equation $x^2 + y^2 = 2y$ becomes $r^2 = 2r\sin\theta$ or $r = 2\sin\theta$.

(b) $\rho^2\sin^2\phi(\cos^2\theta + \sin^2\theta) = 2\rho\sin\phi\sin\theta$ or $\rho\sin^2\phi = 2\sin\phi\sin\theta$ or $\rho\sin\phi = 2\sin\theta$.

**23.** (a) Substituting $x = r\cos\theta$ and $y = r\sin\theta$, the equation $3x + 2y + z = 6$ becomes $3r\cos\theta + 2r\sin\theta + z = 6$ or $z = 6 - r(3\cos\theta + 2\sin\theta)$.

(b) $x = \rho\sin\phi\cos\theta$, $y = \rho\sin\phi\sin\theta$, and $z = \rho\cos\phi$, so the equation becomes

$3\rho\sin\phi\cos\theta + 2\rho\sin\phi\sin\theta + \rho\cos\phi = 6$ or $\rho(3\sin\phi\cos\theta + 2\sin\phi\sin\theta + \cos\phi) = 6$.

**25.**

$0 \leq r \leq 2$ and $0 \leq z \leq 1$ describe a solid circular cylinder with radius 2, axis the $z$-axis, and height 1, but $-\pi/2 \leq \theta \leq \pi/2$ restricts the solid to the first and fourth quadrants of the $xy$-plane, so we have a half-cylinder.

**27.** $\rho = 2$ represents a sphere of radius 2, centered at the origin, so $\rho \leq 2$ is this sphere and its interior. $0 \leq \phi \leq \frac{\pi}{2}$ restricts the solid to that portion of the region that lies on or above the $xy$-plane, and $0 \leq \theta \leq \frac{\pi}{2}$ further restricts the solid to the first octant. Thus the solid is the portion in the first octant of the solid ball centered at the origin with radius 2.

**29.** $\rho \le 1$ represents the solid sphere of radius 1 centered at the origin.

$\frac{3\pi}{4} \le \phi \le \pi$ restricts the solid to that portion on or below the cone $\phi = \frac{3\pi}{4}$.

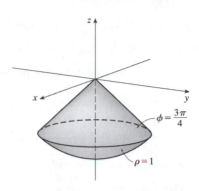

**31.** We can position the cylindrical shell vertically so that its axis coincides with the $z$-axis and its base lies in the $xy$-plane. If we use centimeters as the unit of measurement, then cylindrical coordinates conveniently describe the shell as $6 \le r \le 7$, $0 \le \theta \le 2\pi, 0 \le z \le 20$.

**33.** $z \ge \sqrt{x^2 + y^2}$ because the solid lies above the cone. Squaring both sides of this inequality gives $z^2 \ge x^2 + y^2 \Rightarrow$ $2z^2 \ge x^2 + y^2 + z^2 = \rho^2 \Rightarrow z^2 = \rho^2 \cos^2 \phi \ge \frac{1}{2}\rho^2 \Rightarrow \cos^2 \phi \ge \frac{1}{2}$. The cone opens upward so that the inequality is $\cos \phi \ge \frac{1}{\sqrt{2}}$, or equivalently $0 \le \phi \le \frac{\pi}{4}$. In spherical coordinates the sphere $z = x^2 + y^2 + z^2$ is $\rho \cos \phi = \rho^2 \Rightarrow$ $\rho = \cos \phi$. $0 \le \rho \le \cos \phi$ because the solid lies below the sphere. The solid can therefore be described as the region in spherical coordinates satisfying $0 \le \rho \le \cos \phi, 0 \le \phi \le \frac{\pi}{4}$.

**35.** In cylindrical coordinates, the equation of the cylinder is $r = 3, 0 \le z \le 10$.

The hemisphere is the upper part of the sphere radius 3, center $(0, 0, 10)$, equation $r^2 + (z - 10)^2 = 3^2, z \ge 10$. In Maple, we can use the `coords=cylindrical` option in a regular `plot3d` command. In Mathematica, we can use `RevolutionPlot3D` or `ParametricPlot3D`.

# 9 Review

## CONCEPT CHECK

**1.** A scalar is a real number, while a vector is a quantity that has both a real-valued magnitude and a direction.

**2.** To add two vectors geometrically, we can use either the Triangle Law or the Parallelogram Law, as illustrated in Figures 3 and 4 in Section 9.2. Algebraically, we add the corresponding components of the vectors.

**3.** For $c > 0$, $c\mathbf{a}$ is a vector with the same direction as $\mathbf{a}$ and length $c$ times the length of $\mathbf{a}$. If $c < 0$, $c\mathbf{a}$ points in the opposite direction as $\mathbf{a}$ and has length $|c|$ times the length of $\mathbf{a}$. (See Figures 7 and 15 in Section 9.2.) Algebraically, to find $c\mathbf{a}$ we multiply each component of $\mathbf{a}$ by $c$.

**4.** See (1) in Section 9.2.

**5.** See the definition on page 649 and the boxed equation on page 650.

6. The dot product can be used to determine the work done moving an object given the force and displacement vectors. The dot product can also be used to find the angle between two vectors and the scalar projection of one vector onto another. In particular, the dot product can determine if two vectors are orthogonal.

7. See the boxed equations on page 652 as well as Figures 5 and 6 and the accompanying discussion on pages 651–52.

8. See the definition on page 655; use either (2) or (4) in Section 9.4.

9. The cross product can be used to determine torque if the force and position vectors are known. In addition, the cross product can be used to create a vector orthogonal to two given vectors as well as to determine if two vectors are parallel. The cross product can also be used to find the area of a parallelogram determined by two vectors.

10. (a) The area of the parallelogram determined by $\mathbf{a}$ and $\mathbf{b}$ is the length of the cross product: $|\mathbf{a} \times \mathbf{b}|$.

    (b) The volume of the parallelepiped determined by $\mathbf{a}$, $\mathbf{b}$, and $\mathbf{c}$ is the magnitude of their scalar triple product: $|\mathbf{a} \cdot (\mathbf{b} \times \mathbf{c})|$.

11. If an equation of the plane is known, it can be written as $ax + by + cz + d = 0$. A normal vector, which is perpendicular to the plane, is $\langle a, b, c \rangle$ (or any scalar multiple of $\langle a, b, c \rangle$). If an equation is not known, we can use points on the plane to find two non-parallel vectors which lie in the plane. The cross product of these vectors is a vector perpendicular to the plane.

12. The angle between two intersecting planes is defined as the acute angle between their normal vectors. We can find this angle using the definition of the dot product on page 649.

13. See (1), (2), and (3) in Section 9.5.

14. See (5), (6), and (7) in Section 9.5.

15. (a) Two (nonzero) vectors are parallel if and only if one is a scalar multiple of the other. In addition, two nonzero vectors are parallel if and only if their cross product is $\mathbf{0}$.

    (b) Two vectors are perpendicular if and only if their dot product is 0.

    (c) Two planes are parallel if and only if their normal vectors are parallel.

16. (a) Determine the vectors $\overrightarrow{PQ} = \langle a_1, a_2, a_3 \rangle$ and $\overrightarrow{PR} = \langle b_1, b_2, b_3 \rangle$. If there is a scalar $t$ such that $\langle a_1, a_2, a_3 \rangle = t \langle b_1, b_2, b_3 \rangle$, then the vectors are parallel and the points must all lie on the same line.

    Alternatively, if $\overrightarrow{PQ} \times \overrightarrow{PR} = \mathbf{0}$, then $\overrightarrow{PQ}$ and $\overrightarrow{PR}$ are parallel, so $P$, $Q$, and $R$ are collinear.

    Thirdly, an algebraic method is to determine an equation of the line joining two of the points, and then check whether or not the third point satisfies this equation.

    (b) Find the vectors $\overrightarrow{PQ} = \mathbf{a}$, $\overrightarrow{PR} = \mathbf{b}$, $\overrightarrow{PS} = \mathbf{c}$. $\mathbf{a} \times \mathbf{b}$ is normal to the plane formed by $P$, $Q$ and $R$, and so $S$ lies on this plane if $\mathbf{a} \times \mathbf{b}$ and $\mathbf{c}$ are orthogonal, that is, if $(\mathbf{a} \times \mathbf{b}) \cdot \mathbf{c} = 0$. (Or use the reasoning in Example 6 in Section 9.4.)

    Alternatively, find an equation for the plane determined by three of the points and check whether or not the fourth point satisfies this equation.

**17.** (a) See Exercise 9.4.33.

(b) See Example 8 in Section 9.5.

(c) See Example 10 in Section 9.5.

**18.** One method of graphing a function of two variables is to first find traces (see Example 6 in Section 9.6 and the discussion preceding it).

**19.** See Table 2 in Section 9.6.

**20.** (a) See (1) and the discussion accompanying Figure 4 in Section 9.7.

(b) See (3) and Figures 7–9, and the accompanying discussion, in Section 9.7.

## TRUE-FALSE QUIZ

**1.** True, by Property 2 of the dot product. (See page 651.)

**3.** True. If $\theta$ is the angle between $\mathbf{u}$ and $\mathbf{v}$, then by the definition of the cross product,

$|\mathbf{u} \times \mathbf{v}| = |\mathbf{u}|\,|\mathbf{v}|\sin\theta = |\mathbf{v}|\,|\mathbf{u}|\sin\theta = |\mathbf{v} \times \mathbf{u}|$.

(Or, by Properties 1 and 2 of the cross product, $|\mathbf{u} \times \mathbf{v}| = |-\mathbf{v} \times \mathbf{u}| = |-1|\,|\mathbf{v} \times \mathbf{u}| = |\mathbf{v} \times \mathbf{u}|$.)

**5.** Property 2 of the cross product tells us that this is true.

**7.** This is true by (6) in Section 9.4.

**9.** This is true because $\mathbf{u} \times \mathbf{v}$ is orthogonal to $\mathbf{u}$ (see page 655), and the dot product of two orthogonal vectors is 0.

**11.** If $|\mathbf{u}| = 1$, $|\mathbf{v}| = 1$ and $\theta$ is the angle between these two vectors (so $0 \le \theta \le \pi$), then by the definition of the cross product,

$|\mathbf{u} \times \mathbf{v}| = |\mathbf{u}|\,|\mathbf{v}|\sin\theta = \sin\theta$, which is equal to 1 if and only if $\theta = \frac{\pi}{2}$ (that is, if and only if the two vectors are orthogonal).

Therefore, the assertion that the cross product of two unit vectors is a unit vector is false.

**13.** This is false. In $\mathbb{R}^2$, $x^2 + y^2 = 1$ represents a circle, but $\{(x, y, z) \mid x^2 + y^2 = 1\}$ represents a *three-dimensional surface*, namely, a circular cylinder with axis the $z$-axis.

**15.** False. For example, $\mathbf{i} \cdot \mathbf{j} = 0$ but $\mathbf{i} \ne \mathbf{0}$ and $\mathbf{j} \ne \mathbf{0}$.

**17.** This is true. If $\mathbf{u}$ and $\mathbf{v}$ are both nonzero, then by (2) in Section 9.3, $\mathbf{u} \cdot \mathbf{v} = 0$ implies that $\mathbf{u}$ and $\mathbf{v}$ are orthogonal. But $\mathbf{u} \times \mathbf{v} = \mathbf{0}$ implies that $\mathbf{u}$ and $\mathbf{v}$ are parallel (see the boxed statement preceding Example 1 on page 655). Two nonzero vectors can't be both parallel and orthogonal, so at least one of $\mathbf{u}$, $\mathbf{v}$ must be $\mathbf{0}$.

## EXERCISES

**1.** (a) The radius of the sphere is the distance between the points $(-1, 2, 1)$ and $(6, -2, 3)$, namely,

$\sqrt{[6 - (-1)]^2 + (-2 - 2)^2 + (3 - 1)^2} = \sqrt{69}$. By the formula for an equation of a sphere (see page 637), an equation of the sphere with center $(-1, 2, 1)$ and radius $\sqrt{69}$ is $(x + 1)^2 + (y - 2)^2 + (z - 1)^2 = 69$.

(b) The intersection of this sphere with the $yz$-plane is the set of points on the sphere whose $x$-coordinate is 0. Putting $x = 0$ into the equation, we have $(y - 2)^2 + (z - 1)^2 = 68, x = 0$ which represents a circle in the $yz$-plane with center $(0, 2, 1)$ and radius $\sqrt{68}$.

(c) Completing squares gives $(x - 4)^2 + (y + 1)^2 + (z + 3)^2 = -1 + 16 + 1 + 9 = 25$. Thus the sphere is centered at $(4, -1, -3)$ and has radius 5.

**3.** $\mathbf{u} \cdot \mathbf{v} = |\mathbf{u}|\,|\mathbf{v}| \cos 45° = (2)(3)\frac{\sqrt{2}}{2} = 3\sqrt{2}$.   $|\mathbf{u} \times \mathbf{v}| = |\mathbf{u}|\,|\mathbf{v}| \sin 45° = (2)(3)\frac{\sqrt{2}}{2} = 3\sqrt{2}$.

By the right-hand rule, $\mathbf{u} \times \mathbf{v}$ is directed out of the page.

**5.** For the two vectors to be orthogonal, we need $\langle 3, 2, x \rangle \cdot \langle 2x, 4, x \rangle = 0$   $\Leftrightarrow$   $(3)(2x) + (2)(4) + (x)(x) = 0$   $\Leftrightarrow$

$x^2 + 6x + 8 = 0$   $\Leftrightarrow$   $(x + 2)(x + 4) = 0$   $\Leftrightarrow$   $x = -2$ or $x = -4$.

**7.** (a) $(\mathbf{u} \times \mathbf{v}) \cdot \mathbf{w} = \mathbf{u} \cdot (\mathbf{v} \times \mathbf{w}) = 2$

(b) $\mathbf{u} \cdot (\mathbf{w} \times \mathbf{v}) = \mathbf{u} \cdot [-(\mathbf{v} \times \mathbf{w})] = -\mathbf{u} \cdot (\mathbf{v} \times \mathbf{w}) = -2$

(c) $\mathbf{v} \cdot (\mathbf{u} \times \mathbf{w}) = (\mathbf{v} \times \mathbf{u}) \cdot \mathbf{w} = -(\mathbf{u} \times \mathbf{v}) \cdot \mathbf{w} = -2$

(d) $(\mathbf{u} \times \mathbf{v}) \cdot \mathbf{v} = \mathbf{u} \cdot (\mathbf{v} \times \mathbf{v}) = \mathbf{u} \cdot \mathbf{0} = 0$

**9.** For simplicity, consider a unit cube positioned with its back left corner at the origin. Vector representations of the diagonals joining the points $(0, 0, 0)$ to $(1, 1, 1)$ and $(1, 0, 0)$ to $(0, 1, 1)$ are $\langle 1, 1, 1 \rangle$ and $\langle -1, 1, 1 \rangle$. Let $\theta$ be the angle between these two vectors.   $\langle 1, 1, 1 \rangle \cdot \langle -1, 1, 1 \rangle = -1 + 1 + 1 = 1 = |\langle 1, 1, 1 \rangle|\,|\langle -1, 1, 1 \rangle| \cos \theta = 3 \cos \theta$   $\Rightarrow$   $\cos \theta = \frac{1}{3}$   $\Rightarrow$

$\theta = \cos^{-1}\left(\frac{1}{3}\right) \approx 71°$.

**11.** $\overrightarrow{AB} = \langle 1, 0, -1 \rangle$, $\overrightarrow{AC} = \langle 0, 4, 3 \rangle$, so

(a) a vector perpendicular to the plane is $\overrightarrow{AB} \times \overrightarrow{AC} = \langle 0 + 4, -(3 + 0), 4 - 0 \rangle = \langle 4, -3, 4 \rangle$.

(b) $\frac{1}{2}\left|\overrightarrow{AB} \times \overrightarrow{AC}\right| = \frac{1}{2}\sqrt{16 + 9 + 16} = \frac{\sqrt{41}}{2}$.

**13.** Let $F_1$ be the magnitude of the force directed $20°$ away from the direction of shore, and let $F_2$ be the magnitude of the other force. Separating these forces into components parallel to the direction of the resultant force and perpendicular to it gives

$F_1 \cos 20° + F_2 \cos 30° = 255$ **(1)**, and $F_1 \sin 20° - F_2 \sin 30° = 0$   $\Rightarrow$   $F_1 = F_2 \dfrac{\sin 30°}{\sin 20°}$ **(2)**. Substituting **(2)**

into **(1)** gives $F_2(\sin 30° \cot 20° + \cos 30°) = 255$   $\Rightarrow$   $F_2 \approx 114$ N. Substituting this into **(2)** gives $F_1 \approx 166$ N.

**15.** The line has direction $\mathbf{v} = \langle -3, 2, 3 \rangle$. Letting $P_0 = (4, -1, 2)$, parametric equations are

$x = 4 - 3t$, $y = -1 + 2t$, $z = 2 + 3t$.

**17.** A direction vector for the line is a normal vector for the plane, $\mathbf{n} = \langle 2, -1, 5 \rangle$, and parametric equations for the line are

$x = -2 + 2t$, $y = 2 - t$, $z = 4 + 5t$.

**19.** Here the vectors $\mathbf{a} = \langle 4 - 3, 0 - (-1), 2 - 1 \rangle = \langle 1, 1, 1 \rangle$ and $\mathbf{b} = \langle 6 - 3, 3 - (-1), 1 - 1 \rangle = \langle 3, 4, 0 \rangle$ lie in the plane, so $\mathbf{n} = \mathbf{a} \times \mathbf{b} = \langle -4, 3, 1 \rangle$ is a normal vector to the plane and an equation of the plane is

$-4(x - 3) + 3(y - (-1)) + 1(z - 1) = 0$ or $-4x + 3y + z = -14$.

**21.** $\mathbf{n}_1 = \langle 1, 0, -1 \rangle$ and $\mathbf{n}_2 = \langle 0, 1, 2 \rangle$. Setting $z = 0$, it is easy to see that $(1, 3, 0)$ is a point on the line of intersection of $x - z = 1$ and $y + 2z = 3$. The direction of this line is $\mathbf{v}_1 = \mathbf{n}_1 \times \mathbf{n}_2 = \langle 1, -2, 1 \rangle$. A second vector parallel to the desired plane is $\mathbf{v}_2 = \langle 1, 1, -2 \rangle$, since it is perpendicular to $x + y - 2z = 1$. Therefore, the normal of the plane in question is $\mathbf{n} = \mathbf{v}_1 \times \mathbf{v}_2 = \langle 4 - 1, 1 + 2, 1 + 2 \rangle = 3\langle 1, 1, 1 \rangle$. Taking $(x_0, y_0, z_0) = (1, 3, 0)$, the equation we are looking for is $(x - 1) + (y - 3) + z = 0 \iff x + y + z = 4$.

**23.** Since the direction vectors $\langle 2, 3, 4 \rangle$ and $\langle 6, -1, 2 \rangle$ aren't parallel, neither are the lines. For the lines to intersect, the three equations $1 + 2t = -1 + 6s$, $2 + 3t = 3 - s$, $3 + 4t = -5 + 2s$ must be satisfied simultaneously. Solving the first two equations gives $t = \frac{1}{5}$, $s = \frac{2}{5}$ and checking we see these values don't satisfy the third equation. Thus the lines aren't parallel and they don't intersect, so they must be skew.

**25.** (a) By Exercise 9.5.59, $D = \dfrac{|-2 - (-24)|}{\sqrt{3^2 + 1^2 + (-4)^2}} = \dfrac{22}{\sqrt{26}}$.

(b) Use the formula proven in Exercise 9.4.33(a). In the notation used in that exercise, $\mathbf{a}$ is just the direction of the line; that is, $\mathbf{a} = \langle 1, -1, 2 \rangle$. A point on the line is $(1, 2, -1)$ (setting $t = 0$), and therefore $\mathbf{b} = \langle 1 - 0, 2 - 0, -1 - 0 \rangle = \langle 1, 2, -1 \rangle$.

Hence $d = \dfrac{|\mathbf{a} \times \mathbf{b}|}{|\mathbf{a}|} = \dfrac{|\langle 1, -1, 2 \rangle \times \langle 1, 2, -1 \rangle|}{\sqrt{1 + 1 + 4}} = \dfrac{|\langle -3, 3, 3 \rangle|}{\sqrt{6}} = \sqrt{\dfrac{27}{6}} = \dfrac{3}{\sqrt{2}}$.

**27.** $\ln(x - y^2)$ is defined only when $x - y^2 > 0$, or $x > y^2$, and $x$ is defined for all real numbers, so the domain of the product $x \ln(x - y^2)$ is $\{(x, y) \mid x > y^2\}$.

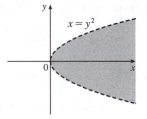

**29.** The graph is the plane $z = 6 - 2x - 3y \implies 2x + 3y + z = 6$. The intercepts with the coordinate axes are $(3, 0, 0)$, $(0, 2, 0)$, and $(0, 0, 6)$ which enable us to sketch the portion of the plane that lies in the first octant.

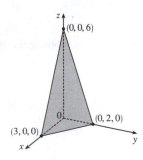

**31.** The equation is $z = 4 - x^2 - 4y^2$. The traces in $x = k$ are $z = 4 - k^2 - 4y^2$, a family of parabolas opening downward, as are the traces in $y = k$, $z = 4 - 4k^2 - x^2$. The traces in $z = k$ are $x^2 + 4y^2 = 4 - k$, a family of ellipses, so the surface is an elliptic paraboloid.

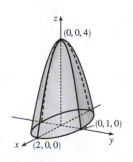

**33.** An equivalent equation is $\dfrac{x^2}{(1/2)^2} + y^2 + z^2 = 1$, an ellipsoid centered at

the origin with intercepts $\pm\frac{1}{2}$, $\pm 1$, and $\pm 1$.

**35.** $y^2 + z^2 = 1$ is the equation of a circular cylinder with axis the $x$-axis.

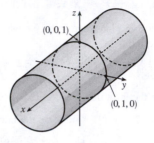

**37.** $x = r\cos\theta = 2\sqrt{3}\cos\frac{\pi}{3} = 2\sqrt{3}\cdot\frac{1}{2} = \sqrt{3}$, $y = r\sin\theta = 2\sqrt{3}\sin\frac{\pi}{3} = 2\sqrt{3}\cdot\frac{\sqrt{3}}{2} = 3$, $z = 2$, so in rectangular

coordinates the point is $\left(\sqrt{3}, 3, 2\right)$. $\rho = \sqrt{r^2 + z^2} = \sqrt{12+4} = 4$, $\theta = \frac{\pi}{3}$, and $\cos\phi = \frac{z}{\rho} = \frac{1}{2}$, so $\phi = \frac{\pi}{3}$ and spherical

coordinates are $\left(4, \frac{\pi}{3}, \frac{\pi}{3}\right)$.

**39.** $x = \rho\sin\phi\cos\theta = 8\sin\frac{\pi}{6}\cos\frac{\pi}{4} = 8\cdot\frac{1}{2}\cdot\frac{\sqrt{2}}{2} = 2\sqrt{2}$, $y = \rho\sin\phi\sin\theta = 8\sin\frac{\pi}{6}\sin\frac{\pi}{4} = 2\sqrt{2}$, and

$z = \rho\cos\phi = 8\cos\frac{\pi}{6} = 8\cdot\frac{\sqrt{3}}{2} = 4\sqrt{3}$. Thus rectangular coordinates for the point are $\left(2\sqrt{2}, 2\sqrt{2}, 4\sqrt{3}\right)$.

$r^2 = x^2 + y^2 = 8 + 8 = 16$ $\Rightarrow$ $r = 4$, $\theta = \frac{\pi}{4}$, and $z = 4\sqrt{3}$, so cylindrical coordinates are $\left(4, \frac{\pi}{4}, 4\sqrt{3}\right)$.

**41.** $x^2 + y^2 + z^2 = 4$. In cylindrical coordinates, this becomes $r^2 + z^2 = 4$. In spherical coordinates, it becomes $\rho^2 = 4$ or

$\rho = 2$.

**43.** The resulting surface is a circular paraboloid with equation $z = 4x^2 + 4y^2$. Changing to cylindrical coordinates we have

$z = 4\left(x^2 + y^2\right) = 4r^2$.

# ☐ FOCUS ON PROBLEM SOLVING

1. Since three-dimensional situations are often difficult to visualize and work with, let us first try to find an analogous problem in two dimensions. The analogue of a cube is a square and the analogue of a sphere is a circle. Thus a similar problem in two dimensions is the following: if five circles with the same radius $r$ are contained in a square of side 1 m so that the circles touch each other and four of the circles touch two sides of the square, find $r$.

The diagonal of the square is $\sqrt{2}$. The diagonal is also $4r + 2x$. But $x$ is the diagonal of a smaller square of side $r$. Therefore

$$x = \sqrt{2}\,r \quad \Rightarrow \quad \sqrt{2} = 4r + 2x = 4r + 2\sqrt{2}\,r = (4 + 2\sqrt{2})r \quad \Rightarrow \quad r = \frac{\sqrt{2}}{4 + 2\sqrt{2}}.$$

Let's use these ideas to solve the original three-dimensional problem. The diagonal of the cube is $\sqrt{1^2 + 1^2 + 1^2} = \sqrt{3}$. The diagonal of the cube is also $4r + 2x$ where $x$ is the diagonal of a smaller cube with edge $r$. Therefore

$$x = \sqrt{r^2 + r^2 + r^2} = \sqrt{3}\,r \quad \Rightarrow \quad \sqrt{3} = 4r + 2x = 4r + 2\sqrt{3}\,r = (4 + 2\sqrt{3})r. \text{ Thus } r = \frac{\sqrt{3}}{4 + 2\sqrt{3}} = \frac{2\sqrt{3} - 3}{2}.$$

The radius of each ball is $\left(\sqrt{3} - \frac{3}{2}\right)$ m.

3. (a) We find the line of intersection $L$ as in Example 9.5.7(b). Observe that the point $(-1, c, c)$ lies on both planes. Now since $L$ lies in both planes, it is perpendicular to both of the normal vectors $\mathbf{n}_1$ and $\mathbf{n}_2$, and thus parallel to their cross product

$$\mathbf{n}_1 \times \mathbf{n}_2 = \begin{vmatrix} \mathbf{i} & \mathbf{j} & \mathbf{k} \\ c & 1 & 1 \\ 1 & -c & c \end{vmatrix} = \langle 2c, -c^2 + 1, -c^2 - 1 \rangle. \text{ So symmetric equations of } L \text{ can be written as}$$

$$\frac{x + 1}{-2c} = \frac{y - c}{c^2 - 1} = \frac{z - c}{c^2 + 1}, \text{ provided that } c \neq 0, \pm 1.$$

If $c = 0$, then the two planes are given by $y + z = 0$ and $x = -1$, so symmetric equations of $L$ are $x = -1$, $y = -z$. If $c = -1$, then the two planes are given by $-x + y + z = -1$ and $x + y + z = -1$, and they intersect in the line $x = 0$, $y = -z - 1$. If $c = 1$, then the two planes are given by $x + y + z = 1$ and $x - y + z = 1$, and they intersect in the line $y = 0$, $x = 1 - z$.

(b) If we set $z = t$ in the symmetric equations and solve for $x$ and $y$ separately, we get $x + 1 = \dfrac{(t - c)(-2c)}{c^2 + 1}$,

$$y - c = \frac{(t - c)(c^2 - 1)}{c^2 + 1} \quad \Rightarrow \quad x = \frac{-2ct + (c^2 - 1)}{c^2 + 1}, \; y = \frac{(c^2 - 1)t + 2c}{c^2 + 1}. \text{ Eliminating } c \text{ from these equations, we}$$

have $x^2 + y^2 = t^2 + 1$. So the curve traced out by $L$ in the plane $z = t$ is a circle with center at $(0, 0, t)$ and radius $\sqrt{t^2 + 1}$.

(c) The area of a horizontal cross-section of the solid is $A(z) = \pi(z^2 + 1)$, so $V = \int_0^1 A(z)\,dz = \pi\left[\frac{1}{3}z^3 + z\right]_0^1 = \frac{4\pi}{3}$.

**5.** $\mathbf{v}_3 = \text{proj}_{\mathbf{v}_1}\mathbf{v}_2 = \dfrac{\mathbf{v}_1 \cdot \mathbf{v}_2}{|\mathbf{v}_1|^2}\,\mathbf{v}_1 = \dfrac{5}{2^2}\,\mathbf{v}_1$ so $|\mathbf{v}_3| = \dfrac{5}{2^2}\,|\mathbf{v}_1| = \dfrac{5}{2}$,

$\mathbf{v}_4 = \text{proj}_{\mathbf{v}_2}\mathbf{v}_3 = \dfrac{\mathbf{v}_2 \cdot \mathbf{v}_3}{|\mathbf{v}_2|^2}\,\mathbf{v}_2 = \dfrac{\mathbf{v}_2 \cdot \frac{5}{2^2}\mathbf{v}_1}{|\mathbf{v}_2|^2}\,\mathbf{v}_2 = \dfrac{5}{2^2 \cdot 3^2}(\mathbf{v}_1 \cdot \mathbf{v}_2)\,\mathbf{v}_2 = \dfrac{5^2}{2^2 \cdot 3^2}\,\mathbf{v}_2 \;\Rightarrow\; |\mathbf{v}_4| = \dfrac{5^2}{2^2 \cdot 3^2}\,|\mathbf{v}_2| = \dfrac{5^2}{2^2 \cdot 3}$,

$\mathbf{v}_5 = \text{proj}_{\mathbf{v}_3}\mathbf{v}_4 = \dfrac{\mathbf{v}_3 \cdot \mathbf{v}_4}{|\mathbf{v}_3|^2}\,\mathbf{v}_3 = \dfrac{\frac{5}{2^2}\mathbf{v}_1 \cdot \frac{5^2}{2^2\,3^2}\mathbf{v}_2}{\left(\frac{5}{2}\right)^2}\left(\dfrac{5}{2^2}\,\mathbf{v}_1\right) = \dfrac{5^2}{2^4 \cdot 3^2}(\mathbf{v}_1 \cdot \mathbf{v}_2)\,\mathbf{v}_1 = \dfrac{5^3}{2^4 \cdot 3^2}\,\mathbf{v}_1 \;\Rightarrow$

$|\mathbf{v}_5| = \dfrac{5^3}{2^4 \cdot 3^2}\,|\mathbf{v}_1| = \dfrac{5^3}{2^3 \cdot 3^2}$. Similarly, $|\mathbf{v}_6| = \dfrac{5^4}{2^4 \cdot 3^3}$, $|\mathbf{v}_7| = \dfrac{5^5}{2^5 \cdot 3^4}$, and in general, $|\mathbf{v}_n| = \dfrac{5^{n-2}}{2^{n-2} \cdot 3^{n-3}} = 3\left(\dfrac{5}{6}\right)^{n-2}$.

Thus

$$\sum_{n=1}^{\infty} |\mathbf{v}_n| = |\mathbf{v}_1| + |\mathbf{v}_2| + \sum_{n=3}^{\infty} 3\left(\tfrac{5}{6}\right)^{n-2} = 2 + 3 + \sum_{n=1}^{\infty} 3\left(\tfrac{5}{6}\right)^{n}$$

$$= 5 + \sum_{n=1}^{\infty} \tfrac{5}{2}\left(\tfrac{5}{6}\right)^{n-1} = 5 + \frac{\frac{5}{2}}{1 - \frac{5}{6}} \quad \text{[sum of a geometric series]} \quad = 5 + 15 = 20$$

**7.** (a) When $\theta = \theta_s$, the block is not moving, so the sum of the forces on the block must be **0**, thus $\mathbf{N} + \mathbf{F} + \mathbf{W} = \mathbf{0}$. This relationship is illustrated geometrically in the figure. Since the vectors form a right triangle, we have

$$\tan(\theta_s) = \frac{|\mathbf{F}|}{|\mathbf{N}|} = \frac{\mu_s n}{n} = \mu_s.$$

(b) We place the block at the origin and sketch the force vectors acting on the block, including the additional horizontal force **H**, with initial points at the origin. We then rotate this system so that **F** lies along the positive $x$-axis and the inclined plane is parallel to the $x$-axis. (See the following figure.)

$|\mathbf{F}|$ is maximal, so $|\mathbf{F}| = \mu_s n$ for $\theta > \theta_s$. Then the vectors, in terms of components parallel and perpendicular to the inclined plane, are

$$\mathbf{N} = n\,\mathbf{j} \qquad \mathbf{F} = (\mu_s n)\,\mathbf{i}$$

$$\mathbf{W} = (-mg\sin\theta)\,\mathbf{i} + (-mg\cos\theta)\,\mathbf{j} \qquad \mathbf{H} = (h_{\min}\cos\theta)\,\mathbf{i} + (-h_{\min}\sin\theta)\,\mathbf{j}$$

Equating components, we have

$$\mu_s n - mg \sin \theta + h_{\min} \cos \theta = 0 \quad \Rightarrow \quad h_{\min} \cos \theta + \mu_s n = mg \sin \theta \tag{1}$$

$$n - mg \cos \theta - h_{\min} \sin \theta = 0 \quad \Rightarrow \quad h_{\min} \sin \theta + mg \cos \theta = n \tag{2}$$

(c) Since **(2)** is solved for $n$, we substitute into **(1)**:

$$h_{\min} \cos \theta + \mu_s (h_{\min} \sin \theta + mg \cos \theta) = mg \sin \theta \quad \Rightarrow$$

$$h_{\min} \cos \theta + h_{\min} \mu_s \sin \theta = mg \sin \theta - mg \mu_s \cos \theta \quad \Rightarrow$$

$$h_{\min} = mg \left( \frac{\sin \theta - \mu_s \cos \theta}{\cos \theta + \mu_s \sin \theta} \right) = mg \left( \frac{\tan \theta - \mu_s}{1 + \mu_s \tan \theta} \right)$$

From part (a) we know $\mu_s = \tan \theta_s$, so this becomes $h_{\min} = mg \left( \dfrac{\tan \theta - \tan \theta_s}{1 + \tan \theta_s \tan \theta} \right)$ and using a trigonometric identity,

this is $mg \tan(\theta - \theta_s)$ as desired.

Note for $\theta = \theta_s$, $h_{\min} = mg \tan 0 = 0$, which makes sense since the block is at rest for $\theta_s$, thus no additional force $\mathbf{H}$ is necessary to prevent it from moving. As $\theta$ increases, the factor $\tan(\theta - \theta_s)$, and hence the value of $h_{\min}$, increases slowly for small values of $\theta - \theta_s$ but much more rapidly as $\theta - \theta_s$ becomes significant. This seems reasonable, as the steeper the inclined plane, the less the horizontal components of the various forces affect the movement of the block, so we would need a much larger magnitude of horizontal force to keep the block motionless. If we allow $\theta \to 90°$, corresponding to the inclined plane being placed vertically, the value of $h_{\min}$ is quite large; this is to be expected, as it takes a great amount of horizontal force to keep an object from moving vertically. In fact, without friction (so $\theta_s = 0$), we would have $\theta \to 90° \quad \Rightarrow \quad h_{\min} \to \infty$, and it would be impossible to keep the block from slipping.

(d) Since $h_{\max}$ is the largest value of $h$ that keeps the block from slipping, the force of friction is keeping the block from moving *up* the inclined plane; thus, $\mathbf{F}$ is directed *down* the plane. Our system of forces is similar to that in part (b), then, except that we have $\mathbf{F} = -(\mu_s n)\,\mathbf{i}$. (Note that $|\mathbf{F}|$ is again maximal.) Following our procedure in parts (b) and (c), we equate components:

$$-\mu_s n - mg \sin \theta + h_{\max} \cos \theta = 0 \quad \Rightarrow \quad h_{\max} \cos \theta - \mu_s n = mg \sin \theta$$

$$n - mg \cos \theta - h_{\max} \sin \theta = 0 \quad \Rightarrow \quad h_{\max} \sin \theta + mg \cos \theta = n$$

Then substituting,

$$h_{\max} \cos \theta - \mu_s (h_{\max} \sin \theta + mg \cos \theta) = mg \sin \theta \quad \Rightarrow$$

$$h_{\max} \cos \theta - h_{\max} \mu_s \sin \theta = mg \sin \theta + mg \mu_s \cos \theta \quad \Rightarrow$$

$$h_{\max} = mg \left( \frac{\sin \theta + \mu_s \cos \theta}{\cos \theta - \mu_s \sin \theta} \right) = mg \left( \frac{\tan \theta + \mu_s}{1 - \mu_s \tan \theta} \right)$$

$$= mg \left( \frac{\tan \theta + \tan \theta_s}{1 - \tan \theta_s \tan \theta} \right) = mg \tan(\theta + \theta_s)$$

We would expect $h_{\max}$ to increase as $\theta$ increases, with similar behavior as we established for $h_{\min}$, but with $h_{\max}$ values always larger than $h_{\min}$. We can see that this is the case if we graph $h_{\max}$ as a function of $\theta$, as the curve is the graph of $h_{\min}$ translated $2\theta_s$ to the left, so the equation does seem reasonable. Notice that the equation predicts $h_{\max} \to \infty$ as $\theta \to (90° - \theta_s)$. In fact, as $h_{\max}$ increases, the normal force increases as well. When $(90° - \theta_s) \leq \theta \leq 90°$, the horizontal force is completely counteracted by the sum of the normal and frictional forces, so no part of the horizontal force contributes to moving the block up the plane no matter how large its magnitude.

# 10 ☐ VECTOR FUNCTIONS

## 10.1  Vector Functions and Space Curves

1. The component functions $\sqrt{4 - t^2}$, $e^{-3t}$, and $\ln(t + 1)$ are all defined when $4 - t^2 \geq 0 \;\Rightarrow\; -2 \leq t \leq 2$ and

   $t + 1 > 0 \;\Rightarrow\; t > -1$, so the domain of $\mathbf{r}$ is $(-1, 2]$.

3. $\displaystyle\lim_{t\to\infty} \frac{1 + t^2}{1 - t^2} = \lim_{t\to\infty} \frac{(1/t^2) + 1}{(1/t^2) - 1} = \frac{0 + 1}{0 - 1} = -1$, $\displaystyle\lim_{t\to\infty} \tan^{-1} t = \frac{\pi}{2}$, $\displaystyle\lim_{t\to\infty} \frac{1 - e^{-2t}}{t} = \lim_{t\to\infty} \frac{1}{t} - \frac{1}{te^{2t}} = 0 - 0 = 0$. Thus

   $\displaystyle\lim_{t\to\infty} \left\langle \frac{1 + t^2}{1 - t^2}, \tan^{-1} t, \frac{1 - e^{-2t}}{t} \right\rangle = \left\langle -1, \frac{\pi}{2}, 0 \right\rangle$.

5. The corresponding parametric equations for this curve are $x = \sin t$, $y = t$.

   We can make a table of values, or we can eliminate the parameter: $t = y \;\Rightarrow\;$

   $x = \sin y$, with $y \in \mathbb{R}$. By comparing different values of $t$, we find the direction in

   which $t$ increases as indicated in the graph.

7. The corresponding parametric equations are $x = t$, $y = 2 - t$, $z = 2t$, which are

   parametric equations of a line through the point $(0, 2, 0)$ and with direction vector

   $\langle 1, -1, 2 \rangle$.

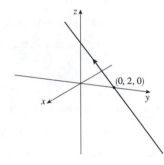

9. The corresponding parametric equations are $x = 1$, $y = \cos t$, $z = 2 \sin t$.

   Eliminating the parameter in $y$ and $z$ gives $y^2 + (z/2)^2 = \cos^2 t + \sin^2 t = 1$

   or $y^2 + z^2/4 = 1$. Since $x = 1$, the curve is an ellipse centered at $(1, 0, 0)$ in

   the plane $x = 1$.

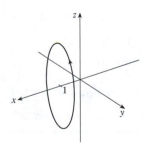

83

**11.** The parametric equations are $x = t^2$, $y = t^4$, $z = t^6$. These are positive

for $t \neq 0$ and $0$ when $t = 0$. So the curve lies entirely in the first octant.

The projection of the graph onto the $xy$-plane is $y = x^2$, $y > 0$, a half parabola.

Onto the $xz$-plane $z = x^3$, $z > 0$, a half cubic, and the $yz$-plane, $y^3 = z^2$.

**13.** The projection of the curve onto the $xy$-plane is given by $\mathbf{r}(t) = \langle t, \sin t, 0 \rangle$   [we use 0 for the $z$-component]   whose graph

is the curve $y = \sin x$, $z = 0$. Similarly, the projection onto the $xz$-plane is $\mathbf{r}(t) = \langle t, 0, 2\cos t \rangle$, whose graph is the cosine

wave $z = 2\cos x$, $y = 0$, and the projection onto the $yz$-plane is $\mathbf{r}(t) = \langle 0, \sin t, 2\cos t \rangle$ whose graph is the ellipse

$y^2 + \frac{1}{4}z^2 = 1$, $x = 0$.

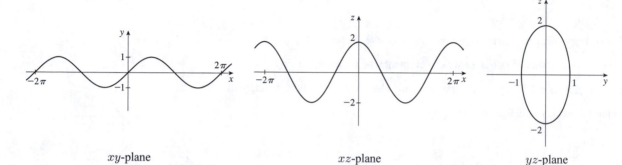

| $xy$-plane | $xz$-plane | $yz$-plane |

From the projection onto the $yz$-plane we see that the curve lies on an elliptical

cylinder with axis the $x$-axis. The other two projections show that the curve

oscillates both vertically and horizontally as we move in the $x$-direction,

suggesting that the curve is an elliptical helix that spirals along the cylinder.

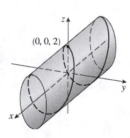

**15.** Taking $\mathbf{r}_0 = \langle 0, 0, 0 \rangle$ and $\mathbf{r}_1 = \langle 1, 2, 3 \rangle$, we have from Equation 9.5.4

$\mathbf{r}(t) = (1 - t)\,\mathbf{r}_0 + t\,\mathbf{r}_1 = (1 - t)\,\langle 0, 0, 0 \rangle + t\,\langle 1, 2, 3 \rangle$, $0 \leq t \leq 1$  or  $\mathbf{r}(t) = \langle t, 2t, 3t \rangle$, $0 \leq t \leq 1$.

Parametric equations are $x = t$, $y = 2t$, $z = 3t$, $0 \leq t \leq 1$.

**17.** Taking $\mathbf{r}_0 = \langle 1, -1, 2 \rangle$ and $\mathbf{r}_1 = \langle 4, 1, 7 \rangle$, we have

$\mathbf{r}(t) = (1 - t)\,\mathbf{r}_0 + t\,\mathbf{r}_1 = (1 - t)\,\langle 1, -1, 2 \rangle + t\,\langle 4, 1, 7 \rangle$, $0 \leq t \leq 1$  or  $\mathbf{r}(t) = \langle 1 + 3t, -1 + 2t, 2 + 5t \rangle$, $0 \leq t \leq 1$.

Parametric equations are $x = 1 + 3t$, $y = -1 + 2t$, $z = 2 + 5t$, $0 \leq t \leq 1$.

**19.** $x = t\cos t$, $y = t$, $z = t\sin t$, $t \geq 0$.   At any point $(x, y, z)$ on the curve, $x^2 + z^2 = t^2\cos^2 t + t^2\sin^2 t = t^2 = y^2$ so the

curve lies on the circular cone $x^2 + z^2 = y^2$ with axis the $y$-axis. Also notice that $y \geq 0$; the graph is II.

**21.** $x = t$, $y = 1/(1 + t^2)$, $z = t^2$.   At any point on the curve we have $z = x^2$, so the curve lies on a parabolic cylinder parallel

to the $y$-axis. Notice that $0 < y \leq 1$ and $z \geq 0$. Also the curve passes through $(0, 1, 0)$ when $t = 0$ and $y \to 0$, $z \to \infty$ as

$t \to \pm\infty$, so the graph must be V.

**23.** $x = \cos 8t$, $y = \sin 8t$, $z = e^{0.8t}$, $t \geq 0$.  $x^2 + y^2 = \cos^2 8t + \sin^2 8t = 1$, so the curve lies on a circular cylinder with

axis the $z$-axis. A point $(x, y, z)$ on the curve lies directly above the point $(x, y, 0)$, which moves counterclockwise around the

unit circle in the $xy$-plane as $t$ increases. The curve starts at $(1, 0, 1)$, when $t = 0$, and $z \to \infty$ (at an increasing rate) as

$t \to \infty$, so the graph is IV.

**25.** If $x = t \cos t$, $y = t \sin t$, $z = t$, then $x^2 + y^2 = t^2 \cos^2 t + t^2 \sin^2 t = t^2 = z^2$,

so the curve lies on the cone $z^2 = x^2 + y^2$. Since $z = t$, the curve is a spiral on

this cone.

**27.** Parametric equations for the curve are $x = t$, $y = 0$, $z = 2t - t^2$. Substituting into the equation of the paraboloid

gives $2t - t^2 = t^2$  $\Rightarrow$  $2t = 2t^2$  $\Rightarrow$  $t = 0, 1$. Since $\mathbf{r}(0) = \mathbf{0}$ and $\mathbf{r}(1) = \mathbf{i} + \mathbf{k}$, the points of intersection

are $(0, 0, 0)$ and $(1, 0, 1)$.

**29.** $\mathbf{r}(t) = \langle \cos t \sin 2t, \sin t \sin 2t, \cos 2t \rangle$.

We include both a regular plot and a plot

showing a tube of radius 0.08 around the

curve.

**31.** $\mathbf{r}(t) = \langle \cos 2t, \cos 3t, \cos 4t \rangle$

**33.**

$x = (1 + \cos 16t) \cos t$, $y = (1 + \cos 16t) \sin t$, $z = 1 + \cos 16t$. At any

point on the graph,

$x^2 + y^2 = (1 + \cos 16t)^2 \cos^2 t + (1 + \cos 16t)^2 \sin^2 t$

$= (1 + \cos 16t)^2 = z^2$, so the graph lies on the cone $x^2 + y^2 = z^2$.

From the graph at left, we see that this curve looks like the projection of a

leaved two-dimensional curve onto a cone.

**35.** If $t = -1$, then $x = 1$, $y = 4$, $z = 0$, so the curve passes through the point $(1, 4, 0)$. If $t = 3$, then $x = 9$, $y = -8$, $z = 28$, so the curve passes through the point $(9, -8, 28)$. For the point $(4, 7, -6)$ to be on the curve, we require $y = 1 - 3t = 7 \Rightarrow t = -2$. But then $z = 1 + (-2)^3 = -7 \neq -6$, so $(4, 7, -6)$ is not on the curve.

**37.** Both equations are solved for $z$, so we can substitute to eliminate $z$: $\sqrt{x^2 + y^2} = 1 + y \Rightarrow x^2 + y^2 = 1 + 2y + y^2 \Rightarrow x^2 = 1 + 2y \Rightarrow y = \frac{1}{2}(x^2 - 1)$. We can form parametric equations for the curve $C$ of intersection by choosing a parameter $x = t$, then $y = \frac{1}{2}(t^2 - 1)$ and $z = 1 + y = 1 + \frac{1}{2}(t^2 - 1) = \frac{1}{2}(t^2 + 1)$. Thus a vector function representing $C$ is $\mathbf{r}(t) = t\,\mathbf{i} + \frac{1}{2}(t^2 - 1)\,\mathbf{j} + \frac{1}{2}(t^2 + 1)\,\mathbf{k}$.

**39.** The projection of the curve $C$ of intersection onto the $xy$-plane is the circle $x^2 + y^2 = 1$, $z = 0$, so we can write $x = \cos t$, $y = \sin t$, $0 \leq t \leq 2\pi$. Since $C$ also lies on the surface $z = x^2 - y^2$, we have $z = x^2 - y^2 = \cos^2 t - \sin^2 t$ or $\cos 2t$. Thus parametric equations for $C$ are $x = \cos t$, $y = \sin t$, $z = \cos 2t$, $0 \leq t \leq 2\pi$, and the corresponding vector function is $\mathbf{r}(t) = \cos t\,\mathbf{i} + \sin t\,\mathbf{j} + \cos 2t\,\mathbf{k}$, $0 \leq t \leq 2\pi$.

**41.**

The projection of the curve $C$ of intersection onto the $xy$-plane is the circle $x^2 + y^2 = 4$, $z = 0$. Then we can write $x = 2\cos t$, $y = 2\sin t$, $0 \leq t \leq 2\pi$. Since $C$ also lies on the surface $z = x^2$, we have $z = x^2 = (2\cos t)^2 = 4\cos^2 t$. Then parametric equations for $C$ are $x = 2\cos t$, $y = 2\sin t$, $z = 4\cos^2 t$, $0 \leq t \leq 2\pi$.

**43.** For the particles to collide, we require $\mathbf{r}_1(t) = \mathbf{r}_2(t) \iff \langle t^2, 7t - 12, t^2 \rangle = \langle 4t - 3, t^2, 5t - 6 \rangle$. Equating components gives $t^2 = 4t - 3$, $7t - 12 = t^2$, and $t^2 = 5t - 6$. From the first equation, $t^2 - 4t + 3 = 0 \iff (t - 3)(t - 1) = 0$ so $t = 1$ or $t = 3$. $t = 1$ does not satisfy the other two equations, but $t = 3$ does. The particles collide when $t = 3$, at the point $(9, 9, 9)$.

**45.** Let $\mathbf{u}(t) = \langle u_1(t), u_2(t), u_3(t) \rangle$ and $\mathbf{v}(t) = \langle v_1(t), v_2(t), v_3(t) \rangle$. In each part of this problem the basic procedure is to use Equation 1 and then analyze the individual component functions using the limit properties we have already developed for real-valued functions.

(a) $\displaystyle \lim_{t \to a} \mathbf{u}(t) + \lim_{t \to a} \mathbf{v}(t) = \left\langle \lim_{t \to a} u_1(t), \lim_{t \to a} u_2(t), \lim_{t \to a} u_3(t) \right\rangle + \left\langle \lim_{t \to a} v_1(t), \lim_{t \to a} v_2(t), \lim_{t \to a} v_3(t) \right\rangle$ and the limits of these component functions must each exist since the vector functions both possess limits as $t \to a$. Then adding the two vectors and using the addition property of limits for real-valued functions, we have that

$$\lim_{t \to a} \mathbf{u}(t) + \lim_{t \to a} \mathbf{v}(t) = \left\langle \lim_{t \to a} u_1(t) + \lim_{t \to a} v_1(t), \lim_{t \to a} u_2(t) + \lim_{t \to a} v_2(t), \lim_{t \to a} u_3(t) + \lim_{t \to a} v_3(t) \right\rangle$$

$$= \left\langle \lim_{t \to a} [u_1(t) + v_1(t)], \lim_{t \to a} [u_2(t) + v_2(t)], \lim_{t \to a} [u_3(t) + v_3(t)] \right\rangle$$

$$= \lim_{t \to a} \langle u_1(t) + v_1(t), u_2(t) + v_2(t), u_3(t) + v_3(t) \rangle \qquad \text{[using (1) backward]}$$

$$= \lim_{t \to a} [\mathbf{u}(t) + \mathbf{v}(t)]$$

(b) $\displaystyle\lim_{t\to a} c\mathbf{u}(t) = \lim_{t\to a} \langle cu_1(t), cu_2(t), cu_3(t)\rangle = \left\langle \lim_{t\to a} cu_1(t), \lim_{t\to a} cu_2(t), \lim_{t\to a} cu_3(t)\right\rangle$

$\qquad\qquad = \left\langle c\lim_{t\to a} u_1(t), c\lim_{t\to a} u_2(t), c\lim_{t\to a} u_3(t)\right\rangle = c\left\langle \lim_{t\to a} u_1(t), \lim_{t\to a} u_2(t), \lim_{t\to a} u_3(t)\right\rangle$

$\qquad\qquad = c\lim_{t\to a} \langle u_1(t), u_2(t), u_3(t)\rangle = c\lim_{t\to a} \mathbf{u}(t)$

(c) $\displaystyle\lim_{t\to a}\mathbf{u}(t)\cdot\lim_{t\to a}\mathbf{v}(t) = \left\langle \lim_{t\to a} u_1(t), \lim_{t\to a} u_2(t), \lim_{t\to a} u_3(t)\right\rangle\cdot\left\langle \lim_{t\to a} v_1(t), \lim_{t\to a} v_2(t), \lim_{t\to a} v_3(t)\right\rangle$

$\qquad\qquad = \left[\lim_{t\to a} u_1(t)\right]\left[\lim_{t\to a} v_1(t)\right] + \left[\lim_{t\to a} u_2(t)\right]\left[\lim_{t\to a} v_2(t)\right] + \left[\lim_{t\to a} u_3(t)\right]\left[\lim_{t\to a} v_3(t)\right]$

$\qquad\qquad = \lim_{t\to a} u_1(t)v_1(t) + \lim_{t\to a} u_2(t)v_2(t) + \lim_{t\to a} u_3(t)v_3(t)$

$\qquad\qquad = \lim_{t\to a}\left[u_1(t)v_1(t) + u_2(t)v_2(t) + u_3(t)v_3(t)\right] = \lim_{t\to a}\left[\mathbf{u}(t)\cdot\mathbf{v}(t)\right]$

(d) $\displaystyle\lim_{t\to a}\mathbf{u}(t)\times\lim_{t\to a}\mathbf{v}(t) = \left\langle \lim_{t\to a} u_1(t), \lim_{t\to a} u_2(t), \lim_{t\to a} u_3(t)\right\rangle\times\left\langle \lim_{t\to a} v_1(t), \lim_{t\to a} v_2(t), \lim_{t\to a} v_3(t)\right\rangle$

$\qquad\qquad = \left\langle\left[\lim_{t\to a} u_2(t)\right]\left[\lim_{t\to a} v_3(t)\right] - \left[\lim_{t\to a} u_3(t)\right]\left[\lim_{t\to a} v_2(t)\right],\right.$

$\qquad\qquad\qquad \left[\lim_{t\to a} u_3(t)\right]\left[\lim_{t\to a} v_1(t)\right] - \left[\lim_{t\to a} u_1(t)\right]\left[\lim_{t\to a} v_3(t)\right],$

$\qquad\qquad\qquad \left.\left[\lim_{t\to a} u_1(t)\right]\left[\lim_{t\to a} v_2(t)\right] - \left[\lim_{t\to a} u_2(t)\right]\left[\lim_{t\to a} v_1(t)\right]\right\rangle$

$\qquad\qquad = \left\langle \lim_{t\to a}\left[u_2(t)v_3(t) - u_3(t)v_2(t)\right], \lim_{t\to a}\left[u_3(t)v_1(t) - u_1(t)v_3(t)\right],\right.$

$\qquad\qquad\qquad \left.\lim_{t\to a}\left[u_1(t)v_2(t) - u_2(t)v_1(t)\right]\right\rangle$

$\qquad\qquad = \lim_{t\to a}\langle u_2(t)v_3(t) - u_3(t)v_2(t), u_3(t)v_1(t) - u_1(t)v_3(t), u_1(t)v_2(t) - u_2(t)v_1(t)\rangle$

$\qquad\qquad = \lim_{t\to a}\left[\mathbf{u}(t)\times\mathbf{v}(t)\right]$

## 10.2  Derivatives and Integrals of Vector Functions

**1.** (a)

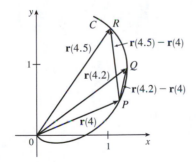

(b) $\dfrac{\mathbf{r}(4.5) - \mathbf{r}(4)}{0.5} = 2[\mathbf{r}(4.5) - \mathbf{r}(4)]$, so we draw a vector in the same

direction but with twice the length of the vector $\mathbf{r}(4.5) - \mathbf{r}(4)$.

$\dfrac{\mathbf{r}(4.2) - \mathbf{r}(4)}{0.2} = 5[\mathbf{r}(4.2) - \mathbf{r}(4)]$, so we draw a vector in the same

direction but with 5 times the length of the vector $\mathbf{r}(4.2) - \mathbf{r}(4)$.

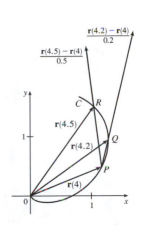

(c) By Definition 1, $\mathbf{r}'(4) = \displaystyle\lim_{h\to 0}\dfrac{\mathbf{r}(4+h) - \mathbf{r}(4)}{h}$.  $\mathbf{T}(4) = \dfrac{\mathbf{r}'(4)}{|\mathbf{r}'(4)|}$.

(d) $\mathbf{T}(4)$ is a unit vector in the same direction as $\mathbf{r}'(4)$, that is, parallel to the tangent line to the curve at $\mathbf{r}(4)$ with length 1.

**3.** Since $(x+2)^2 = t^2 = y - 1 \;\Rightarrow$

    $y = (x+2)^2 - 1$, the curve is a

    parabola.

(a), (c)

(b) $\mathbf{r}'(t) = \langle 1, 2t \rangle$,

    $\mathbf{r}'(-1) = \langle 1, -2 \rangle$

**5.** $x = \sin t, \;\; y = 2 \cos t$ so

    $x^2 + (y/2)^2 = 1$ and the curve is

    an ellipse.

(a), (c)

(b) $\mathbf{r}'(t) = \cos t\,\mathbf{i} - 2 \sin t\,\mathbf{j}$,

    $\mathbf{r}'\!\left(\dfrac{\pi}{4}\right) = \dfrac{\sqrt{2}}{2}\,\mathbf{i} - \sqrt{2}\,\mathbf{j}$

**7.** Since $y = e^{3t} = (e^t)^3 = x^3$, the

    curve is part of a cubic cuve. Note

    that here, $x > 0$.

(a), (c)

(b) $\mathbf{r}'(t) = e^t\,\mathbf{i} + 3e^{3t}\,\mathbf{j}$,

    $\mathbf{r}'(0) = \mathbf{i} + 3\mathbf{j}$

**9.** $\mathbf{r}'(t) = \left\langle \dfrac{d}{dt}\,[t \sin t], \dfrac{d}{dt}\,[t^2], \dfrac{d}{dt}\,[t \cos 2t] \right\rangle = \langle t \cos t + \sin t, 2t, t(-\sin 2t) \cdot 2 + \cos 2t \rangle$

    $= \langle t \cos t + \sin t, 2t, \cos 2t - 2t \sin 2t \rangle$

**11.** $\mathbf{r}(t) = e^{t^2}\,\mathbf{i} - \mathbf{j} + \ln(1 + 3t)\,\mathbf{k} \;\;\Rightarrow\;\; \mathbf{r}'(t) = 2te^{t^2}\,\mathbf{i} + \dfrac{3}{1 + 3t}\,\mathbf{k}$

**13.** $\mathbf{r}'(t) = \mathbf{0} + \mathbf{b} + 2t\,\mathbf{c} = \mathbf{b} + 2t\,\mathbf{c}$ by Formulas 1 and 3 of Theorem 3.

**15.** $\mathbf{r}'(t) = \langle -te^{-t} + e^{-t}, 2/(1 + t^2), 2e^t \rangle \;\;\Rightarrow\;\; \mathbf{r}'(0) = \langle 1, 2, 2 \rangle$. So $|\mathbf{r}'(0)| = \sqrt{1^2 + 2^2 + 2^2} = \sqrt{9} = 3$ and

    $\mathbf{T}(0) = \dfrac{\mathbf{r}'(0)}{|\mathbf{r}'(0)|} = \tfrac{1}{3}\langle 1, 2, 2 \rangle = \langle \tfrac{1}{3}, \tfrac{2}{3}, \tfrac{2}{3} \rangle$.

**17.** $\mathbf{r}'(t) = -\sin t\,\mathbf{i} + 3\,\mathbf{j} + 4\cos 2t\,\mathbf{k} \quad\Rightarrow\quad \mathbf{r}'(0) = 3\,\mathbf{j} + 4\,\mathbf{k}$. Thus

$$\mathbf{T}(0) = \frac{\mathbf{r}'(0)}{|\mathbf{r}'(0)|} = \frac{1}{\sqrt{0^2 + 3^2 + 4^2}}\,(3\,\mathbf{j} + 4\,\mathbf{k}) = \tfrac{1}{5}(3\,\mathbf{j} + 4\,\mathbf{k}) = \tfrac{3}{5}\,\mathbf{j} + \tfrac{4}{5}\,\mathbf{k}.$$

**19.** $\mathbf{r}(t) = \langle t, e^t, te^t \rangle \quad\Rightarrow\quad \mathbf{r}'(t) = \langle 1, e^t, (t+1)e^t \rangle$. Then $\mathbf{r}'(0) = \langle 1, 1, 1 \rangle$ and $|\mathbf{r}'(0)| = \sqrt{1^2 + 1^2 + 1^2} = \sqrt{3}$, so

$$\mathbf{T}(0) = \frac{\mathbf{r}'(0)}{|\mathbf{r}'(0)|} = \tfrac{1}{\sqrt{3}}\langle 1, 1, 1 \rangle = \left\langle \tfrac{1}{\sqrt{3}}, \tfrac{1}{\sqrt{3}}, \tfrac{1}{\sqrt{3}} \right\rangle. \quad \mathbf{r}''(t) = \langle 0, e^t, (t+2)e^t \rangle, \text{ so}$$

$$\mathbf{r}'(t) \cdot \mathbf{r}''(t) = \langle 1, e^t, (t+1)e^t \rangle \cdot \langle 0, e^t, (t+2)e^t \rangle = (1)(0) + (e^t)(e^t) + ((t+1)e^t)((t+2)e^t)$$

$$= 0 + e^{2t} + (t+1)(t+2)e^{2t} = (t^2 + 3t + 3)e^{2t}$$

**21.** The vector equation for the curve is $\mathbf{r}(t) = \langle 1 + 2\sqrt{t}, t^3 - t, t^3 + t \rangle$, so $\mathbf{r}'(t) = \langle 1/\sqrt{t}, 3t^2 - 1, 3t^2 + 1 \rangle$. The point $(3, 0, 2)$ corresponds to $t = 1$, so the tangent vector there is $\mathbf{r}'(1) = \langle 1, 2, 4 \rangle$. Thus, the tangent line goes through the point $(3, 0, 2)$ and is parallel to the vector $\langle 1, 2, 4 \rangle$. Parametric equations are $x = 3 + t$, $y = 2t$, $z = 2 + 4t$.

**23.** The vector equation for the curve is $\mathbf{r}(t) = \langle e^{-t}\cos t, e^{-t}\sin t, e^{-t} \rangle$, so

$$\mathbf{r}'(t) = \langle e^{-t}(-\sin t) + (\cos t)(-e^{-t}), e^{-t}\cos t + (\sin t)(-e^{-t}), (-e^{-t}) \rangle$$

$$= \langle -e^{-t}(\cos t + \sin t), e^{-t}(\cos t - \sin t), -e^{-t} \rangle$$

The point $(1, 0, 1)$ corresponds to $t = 0$, so the tangent vector there is

$\mathbf{r}'(0) = \langle -e^0(\cos 0 + \sin 0), e^0(\cos 0 - \sin 0), -e^0 \rangle = \langle -1, 1, -1 \rangle$. Thus, the tangent line is parallel to the vector $\langle -1, 1, -1 \rangle$ and parametric equations are $x = 1 + (-1)t = 1 - t$, $y = 0 + 1 \cdot t = t$, $z = 1 + (-1)t = 1 - t$.

**25.** First we parametrize the curve $C$ of intersection. The projection of $C$ onto the $xy$-plane is contained in the circle $x^2 + y^2 = 25$, $z = 0$, so we can write $x = 5\cos t$, $y = 5\sin t$. $C$ also lies on the cylinder $y^2 + z^2 = 20$, and $z \geq 0$ near the point $(3, 4, 2)$, so we can write $z = \sqrt{20 - y^2} = \sqrt{20 - 25\sin^2 t}$. A vector equation then for $C$ is

$$\mathbf{r}(t) = \left\langle 5\cos t, 5\sin t, \sqrt{20 - 25\sin^2 t} \right\rangle \quad\Rightarrow\quad \mathbf{r}'(t) = \left\langle -5\sin t, 5\cos t, \tfrac{1}{2}(20 - 25\sin^2 t)^{-1/2}(-50\sin t\cos t) \right\rangle.$$

The point $(3, 4, 2)$ corresponds to $t = \cos^{-1}\left(\tfrac{3}{5}\right)$, so the tangent vector there is

$$\mathbf{r}'\left(\cos^{-1}\left(\tfrac{3}{5}\right)\right) = \left\langle -5\left(\tfrac{4}{5}\right), 5\left(\tfrac{3}{5}\right), \tfrac{1}{2}\left(20 - 25\left(\tfrac{4}{5}\right)^2\right)^{-1/2}\left(-50\left(\tfrac{4}{5}\right)\left(\tfrac{3}{5}\right)\right) \right\rangle = \langle -4, 3, -6 \rangle.$$

The tangent line is parallel to this vector and passes through $(3, 4, 2)$, so a vector equation for the line is $\mathbf{r}(t) = (3 - 4t)\mathbf{i} + (4 + 3t)\mathbf{j} + (2 - 6t)\mathbf{k}$.

**27.** $\mathbf{r}(t) = \langle t, e^{-t}, 2t - t^2 \rangle \quad\Rightarrow\quad \mathbf{r}'(t) = \langle 1, -e^{-t}, 2 - 2t \rangle$. At $(0, 1, 0)$, $t = 0$ and $\mathbf{r}'(0) = \langle 1, -1, 2 \rangle$. Thus, parametric equations of the tangent line are $x = t$, $y = 1 - t$, $z = 2t$.

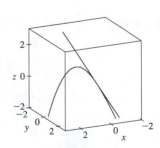

**29.** $\mathbf{r}(t) = \langle t\cos t, t, t\sin t \rangle \quad \Rightarrow \quad \mathbf{r}'(t) = \langle \cos t - t\sin t, 1, t\cos t + \sin t \rangle.$

At $(-\pi, \pi, 0)$, $t = \pi$ and $\mathbf{r}'(\pi) = \langle -1, 1, -\pi \rangle$. Thus, parametric equations

of the tangent line are $x = -\pi - t$, $y = \pi + t$, $z = -\pi t$.

**31.** The angle of intersection of the two curves is the angle between the two tangent vectors to the curves at the point of

intersection. Since $\mathbf{r}_1'(t) = \langle 1, 2t, 3t^2 \rangle$ and $t = 0$ at $(0, 0, 0)$, $\mathbf{r}_1'(0) = \langle 1, 0, 0 \rangle$ is a tangent vector to $\mathbf{r}_1$ at $(0, 0, 0)$. Similarly,

$\mathbf{r}_2'(t) = \langle \cos t, 2\cos 2t, 1 \rangle$ and since $\mathbf{r}_2(0) = \langle 0, 0, 0 \rangle$, $\mathbf{r}_2'(0) = \langle 1, 2, 1 \rangle$ is a tangent vector to $\mathbf{r}_2$ at $(0, 0, 0)$. If $\theta$ is the angle

between these two tangent vectors, then $\cos\theta = \frac{1}{\sqrt{1}\sqrt{6}} \langle 1, 0, 0 \rangle \cdot \langle 1, 2, 1 \rangle = \frac{1}{\sqrt{6}}$ and $\theta = \cos^{-1}\left(\frac{1}{\sqrt{6}}\right) \approx 66°$.

**33.** $\int_0^1 (16t^3\,\mathbf{i} - 9t^2\,\mathbf{j} + 25t^4\,\mathbf{k})\,dt = \left(\int_0^1 16t^3\,dt\right)\mathbf{i} - \left(\int_0^1 9t^2\,dt\right)\mathbf{j} + \left(\int_0^1 25t^4\,dt\right)\mathbf{k}$

$$= \left[4t^4\right]_0^1 \mathbf{i} - \left[3t^3\right]_0^1 \mathbf{j} + \left[5t^5\right]_0^1 \mathbf{k} = 4\,\mathbf{i} - 3\,\mathbf{j} + 5\,\mathbf{k}$$

**35.** $\int_0^{\pi/2} (3\sin^2 t \cos t\,\mathbf{i} + 3\sin t \cos^2 t\,\mathbf{j} + 2\sin t \cos t\,\mathbf{k})\,dt$

$$= \left(\int_0^{\pi/2} 3\sin^2 t \cos t\,dt\right)\mathbf{i} + \left(\int_0^{\pi/2} 3\sin t \cos^2 t\,dt\right)\mathbf{j} + \left(\int_0^{\pi/2} 2\sin t \cos t\,dt\right)\mathbf{k}$$

$$= \left[\sin^3 t\right]_0^{\pi/2}\mathbf{i} + \left[-\cos^3 t\right]_0^{\pi/2}\mathbf{j} + \left[\sin^2 t\right]_0^{\pi/2}\mathbf{k} = (1-0)\,\mathbf{i} + (0+1)\,\mathbf{j} + (1-0)\,\mathbf{k} = \mathbf{i} + \mathbf{j} + \mathbf{k}$$

**37.** $\int (\sec^2 t\,\mathbf{i} + t(t^2+1)^3\,\mathbf{j} + t^2\ln t\,\mathbf{k})\,dt = \left(\int \sec^2 t\,dt\right)\mathbf{i} + \left(\int t(t^2+1)^3\,dt\right)\mathbf{j} + \left(\int t^2\ln t\,dt\right)\mathbf{k}$

$$= \tan t\,\mathbf{i} + \tfrac{1}{8}(t^2+1)^4\,\mathbf{j} + \left(\tfrac{1}{3}t^3\ln t - \tfrac{1}{9}t^3\right)\mathbf{k} + \mathbf{C},$$

where $\mathbf{C}$ is a vector constant of integration. [For the $z$-component, integrate by parts with $u = \ln t$, $dv = t^2\,dt$.]

**39.** $\mathbf{r}'(t) = 2t\,\mathbf{i} + 3t^2\,\mathbf{j} + \sqrt{t}\,\mathbf{k} \quad \Rightarrow \quad \mathbf{r}(t) = t^2\,\mathbf{i} + t^3\,\mathbf{j} + \tfrac{2}{3}t^{3/2}\,\mathbf{k} + \mathbf{C},$ where $\mathbf{C}$ is a constant vector.

But $\mathbf{i} + \mathbf{j} = \mathbf{r}(1) = \mathbf{i} + \mathbf{j} + \tfrac{2}{3}\mathbf{k} + \mathbf{C}$. Thus $\mathbf{C} = -\tfrac{2}{3}\mathbf{k}$ and $\mathbf{r}(t) = t^2\,\mathbf{i} + t^3\,\mathbf{j} + \left(\tfrac{2}{3}t^{3/2} - \tfrac{2}{3}\right)\mathbf{k}.$

For Exercises 41–44, let $\mathbf{u}(t) = \langle u_1(t), u_2(t), u_3(t) \rangle$ and $\mathbf{v}(t) = \langle v_1(t), v_2(t), v_3(t) \rangle$. In each of these exercises, the procedure is to apply Theorem 2 so that the corresponding properties of derivatives of real-valued functions can be used.

**41.** $\dfrac{d}{dt}\,[\mathbf{u}(t) + \mathbf{v}(t)] = \dfrac{d}{dt}\,\langle u_1(t) + v_1(t), u_2(t) + v_2(t), u_3(t) + v_3(t) \rangle$

$$= \left\langle \dfrac{d}{dt}\,[u_1(t) + v_1(t)], \dfrac{d}{dt}\,[u_2(t) + v_2(t)], \dfrac{d}{dt}\,[u_3(t) + v_3(t)] \right\rangle$$

$$= \langle u_1'(t) + v_1'(t), u_2'(t) + v_2'(t), u_3'(t) + v_3'(t) \rangle$$

$$= \langle u_1'(t), u_2'(t), u_3'(t) \rangle + \langle v_1'(t), v_2'(t), v_3'(t) \rangle = \mathbf{u}'(t) + \mathbf{v}'(t)$$

**43.** $\dfrac{d}{dt}\left[\mathbf{u}(t) \times \mathbf{v}(t)\right] = \dfrac{d}{dt}\langle u_2(t)v_3(t) - u_3(t)v_2(t),\, u_3(t)v_1(t) - u_1(t)v_3(t),\, u_1(t)v_2(t) - u_2(t)v_1(t)\rangle$

$$= \langle u_2'v_3(t) + u_2(t)v_3'(t) - u_3'(t)v_2(t) - u_3(t)v_2'(t),$$

$$u_3'(t)v_1(t) + u_3(t)v_1'(t) - u_1'(t)v_3(t) - u_1(t)v_3'(t),$$

$$u_1'(t)v_2(t) + u_1(t)v_2'(t) - u_2'(t)v_1(t) - u_2(t)v_1'(t)\rangle$$

$$= \langle u_2'(t)v_3(t) - u_3'(t)v_2(t),\, u_3'(t)v_1(t) - u_1'(t)v_3(t),\, u_1'(t)v_2(t) - u_2'(t)v_1(t)\rangle$$

$$+ \langle u_2(t)v_3'(t) - u_3(t)v_2'(t),\, u_3(t)v_1'(t) - u_1(t)v_3'(t),\, u_1(t)v_2'(t) - u_2(t)v_1'(t)\rangle$$

$$= \mathbf{u}'(t) \times \mathbf{v}(t) + \mathbf{u}(t) \times \mathbf{v}'(t)$$

*Alternate solution:* Let $\mathbf{r}(t) = \mathbf{u}(t) \times \mathbf{v}(t)$. Then

$$\mathbf{r}(t+h) - \mathbf{r}(t) = [\mathbf{u}(t+h) \times \mathbf{v}(t+h)] - [\mathbf{u}(t) \times \mathbf{v}(t)]$$

$$= [\mathbf{u}(t+h) \times \mathbf{v}(t+h)] - [\mathbf{u}(t) \times \mathbf{v}(t)] + [\mathbf{u}(t+h) \times \mathbf{v}(t)] - [\mathbf{u}(t+h) \times \mathbf{v}(t)]$$

$$= \mathbf{u}(t+h) \times [\mathbf{v}(t+h) - \mathbf{v}(t)] + [\mathbf{u}(t+h) - \mathbf{u}(t)] \times \mathbf{v}(t)$$

(Be careful of the order of the cross product.) Dividing through by $h$ and taking the limit as $h \to 0$ we have

$$\mathbf{r}'(t) = \lim_{h \to 0} \frac{\mathbf{u}(t+h) \times [\mathbf{v}(t+h) - \mathbf{v}(t)]}{h} + \lim_{h \to 0} \frac{[\mathbf{u}(t+h) - \mathbf{u}(t)] \times \mathbf{v}(t)}{h} = \mathbf{u}(t) \times \mathbf{v}'(t) + \mathbf{u}'(t) \times \mathbf{v}(t)$$

by Exercise 10.1.45(a) and Definition 1.

**45.** $\dfrac{d}{dt}\left[\mathbf{u}(t) \cdot \mathbf{v}(t)\right] = \mathbf{u}'(t) \cdot \mathbf{v}(t) + \mathbf{u}(t) \cdot \mathbf{v}'(t)$    [by Formula 4 of Theorem 3]

$$= \langle \cos t, -\sin t, 1 \rangle \cdot \langle t, \cos t, \sin t \rangle + \langle \sin t, \cos t, t \rangle \cdot \langle 1, -\sin t, \cos t \rangle$$

$$= t\cos t - \cos t \sin t + \sin t + \sin t - \cos t \sin t + t\cos t$$

$$= 2t\cos t + 2\sin t - 2\cos t \sin t$$

**47.** By Formula 4 of Theorem 3, $f'(t) = \mathbf{u}'(t) \cdot \mathbf{v}(t) + \mathbf{u}(t) \cdot \mathbf{v}'(t)$, and $\mathbf{v}'(t) = \langle 1, 2t, 3t^2 \rangle$, so

$$f'(2) = \mathbf{u}'(2) \cdot \mathbf{v}(2) + \mathbf{u}(2) \cdot \mathbf{v}'(2) = \langle 3, 0, 4 \rangle \cdot \langle 2, 4, 8 \rangle + \langle 1, 2, -1 \rangle \cdot \langle 1, 4, 12 \rangle = 6 + 0 + 32 + 1 + 8 - 12 = 35.$$

**49.** $\dfrac{d}{dt}\left[\mathbf{r}(t) \times \mathbf{r}'(t)\right] = \mathbf{r}'(t) \times \mathbf{r}'(t) + \mathbf{r}(t) \times \mathbf{r}''(t)$ by Formula 5 of Theorem 3. But $\mathbf{r}'(t) \times \mathbf{r}'(t) = \mathbf{0}$ (by the margin note on

page 655). Thus, $\dfrac{d}{dt}\left[\mathbf{r}(t) \times \mathbf{r}'(t)\right] = \mathbf{r}(t) \times \mathbf{r}''(t)$.

**51.** $\dfrac{d}{dt}|\mathbf{r}(t)| = \dfrac{d}{dt}\left[\mathbf{r}(t) \cdot \mathbf{r}(t)\right]^{1/2} = \frac{1}{2}[\mathbf{r}(t) \cdot \mathbf{r}(t)]^{-1/2}\,[2\mathbf{r}(t) \cdot \mathbf{r}'(t)] = \dfrac{1}{|\mathbf{r}(t)|}\,\mathbf{r}(t) \cdot \mathbf{r}'(t)$

**53.** Since $\mathbf{u}(t) = \mathbf{r}(t) \cdot [\mathbf{r}'(t) \times \mathbf{r}''(t)]$,

$$\mathbf{u}'(t) = \mathbf{r}'(t) \cdot [\mathbf{r}'(t) \times \mathbf{r}''(t)] + \mathbf{r}(t) \cdot \dfrac{d}{dt}[\mathbf{r}'(t) \times \mathbf{r}''(t)]$$

$$= 0 + \mathbf{r}(t) \cdot [\mathbf{r}''(t) \times \mathbf{r}''(t) + \mathbf{r}'(t) \times \mathbf{r}'''(t)] \qquad [\text{since } \mathbf{r}'(t) \perp \mathbf{r}'(t) \times \mathbf{r}''(t)]$$

$$= \mathbf{r}(t) \cdot [\mathbf{r}'(t) \times \mathbf{r}'''(t)] \qquad [\text{since } \mathbf{r}''(t) \times \mathbf{r}''(t) = \mathbf{0}]$$

## 10.3  Arc Length and Curvature

**1.** $\mathbf{r}(t) = \langle 2\sin t, 5t, 2\cos t \rangle$  $\Rightarrow$  $\mathbf{r}'(t) = \langle 2\cos t, 5, -2\sin t \rangle$  $\Rightarrow$  $|\mathbf{r}'(t)| = \sqrt{(2\cos t)^2 + 5^2 + (-2\sin t)^2} = \sqrt{29}$.

Then using Formula 3, we have $L = \int_{-10}^{10} |\mathbf{r}'(t)|\, dt = \int_{-10}^{10} \sqrt{29}\, dt = \sqrt{29}\, t\Big]_{-10}^{10} = 20\sqrt{29}$.

**3.** $\mathbf{r}(t) = \sqrt{2}\, t\, \mathbf{i} + e^t \mathbf{j} + e^{-t} \mathbf{k}$  $\Rightarrow$  $\mathbf{r}'(t) = \sqrt{2}\, \mathbf{i} + e^t \mathbf{j} - e^{-t} \mathbf{k}$  $\Rightarrow$

$|\mathbf{r}'(t)| = \sqrt{\left(\sqrt{2}\right)^2 + (e^t)^2 + (-e^{-t})^2} = \sqrt{2 + e^{2t} + e^{-2t}} = \sqrt{(e^t + e^{-t})^2} = e^t + e^{-t}$  [since $e^t + e^{-t} > 0$].

Then $L = \int_0^1 |\mathbf{r}'(t)|\, dt = \int_0^1 (e^t + e^{-t})\, dt = \left[e^t - e^{-t}\right]_0^1 = e - e^{-1}$.

**5.** $\mathbf{r}(t) = \mathbf{i} + t^2 \mathbf{j} + t^3 \mathbf{k}$  $\Rightarrow$  $\mathbf{r}'(t) = 2t\, \mathbf{j} + 3t^2 \mathbf{k}$  $\Rightarrow$  $|\mathbf{r}'(t)| = \sqrt{4t^2 + 9t^4} = t\sqrt{4 + 9t^2}$  [since $t \geq 0$].

Then $L = \int_0^1 |\mathbf{r}'(t)|\, dt = \int_0^1 t\sqrt{4 + 9t^2}\, dt = \frac{1}{18} \cdot \frac{2}{3}(4 + 9t^2)^{3/2}\Big]_0^1 = \frac{1}{27}(13^{3/2} - 4^{3/2}) = \frac{1}{27}(13^{3/2} - 8)$.

**7.** $\mathbf{r}(t) = \langle \sqrt{t}, t, t^2 \rangle$  $\Rightarrow$  $\mathbf{r}'(t) = \left\langle \dfrac{1}{2\sqrt{t}}, 1, 2t \right\rangle$  $\Rightarrow$  $|\mathbf{r}'(t)| = \sqrt{\left(\dfrac{1}{2\sqrt{t}}\right)^2 + 1^2 + (2t)^2} = \sqrt{\dfrac{1}{4t} + 1 + 4t^2}$, so

$L = \int_1^4 |\mathbf{r}'(t)|\, dt = \int_1^4 \sqrt{\dfrac{1}{4t} + 1 + 4t^2}\, dt \approx 15.3841$.

**9.** $\mathbf{r}(t) = \langle \sin t, \cos t, \tan t \rangle$  $\Rightarrow$  $\mathbf{r}'(t) = \langle \cos t, -\sin t, \sec^2 t \rangle$  $\Rightarrow$

$|\mathbf{r}'(t)| = \sqrt{\cos^2 t + (-\sin t)^2 + (\sec^2 t)^2} = \sqrt{1 + \sec^4 t}$ and $L = \int_0^{\pi/4} |\mathbf{r}'(t)|\, dt = \int_0^{\pi/4} \sqrt{1 + \sec^4 t}\, dt \approx 1.2780$.

**11.** The projection of the curve $C$ onto the $xy$-plane is the curve $x^2 = 2y$ or $y = \frac{1}{2}x^2$, $z = 0$. Then we can choose the parameter

$x = t$  $\Rightarrow$  $y = \frac{1}{2}t^2$. Since $C$ also lies on the surface $3z = xy$, we have $z = \frac{1}{3}xy = \frac{1}{3}(t)(\frac{1}{2}t^2) = \frac{1}{6}t^3$. Then parametric

equations for $C$ are $x = t$, $y = \frac{1}{2}t^2$, $z = \frac{1}{6}t^3$ and the corresponding vector equation is $\mathbf{r}(t) = \langle t, \frac{1}{2}t^2, \frac{1}{6}t^3 \rangle$. The origin

corresponds to $t = 0$ and the point $(6, 18, 36)$ corresponds to $t = 6$, so

$$L = \int_0^6 |\mathbf{r}'(t)|\, dt = \int_0^6 |\langle 1, t, \tfrac{1}{2}t^2 \rangle|\, dt = \int_0^6 \sqrt{1^2 + t^2 + \left(\tfrac{1}{2}t^2\right)^2}\, dt = \int_0^6 \sqrt{1 + t^2 + \tfrac{1}{4}t^4}\, dt$$

$$= \int_0^6 \sqrt{(1 + \tfrac{1}{2}t^2)^2}\, dt = \int_0^6 (1 + \tfrac{1}{2}t^2)\, dt = \left[t + \tfrac{1}{6}t^3\right]_0^6 = 6 + 36 = 42$$

**13.** $\mathbf{r}(t) = 2t\, \mathbf{i} + (1 - 3t)\, \mathbf{j} + (5 + 4t)\, \mathbf{k}$  $\Rightarrow$  $\mathbf{r}'(t) = 2\mathbf{i} - 3\mathbf{j} + 4\mathbf{k}$ and $\frac{ds}{dt} = |\mathbf{r}'(t)| = \sqrt{4 + 9 + 16} = \sqrt{29}$. Then

$s = s(t) = \int_0^t |\mathbf{r}'(u)|\, du = \int_0^t \sqrt{29}\, du = \sqrt{29}\, t$. Therefore, $t = \frac{1}{\sqrt{29}}s$, and substituting for $t$ in the original equation, we

have $\mathbf{r}(t(s)) = \frac{2}{\sqrt{29}}s\, \mathbf{i} + \left(1 - \frac{3}{\sqrt{29}}s\right)\mathbf{j} + \left(5 + \frac{4}{\sqrt{29}}s\right)\mathbf{k}$.

**15.** Here $\mathbf{r}(t) = \langle 3\sin t, 4t, 3\cos t \rangle$, so $\mathbf{r}'(t) = \langle 3\cos t, 4, -3\sin t \rangle$ and $|\mathbf{r}'(t)| = \sqrt{9\cos^2 t + 16 + 9\sin^2 t} = \sqrt{25} = 5$.

The point $(0, 0, 3)$ corresponds to $t = 0$, so the arc length function beginning at $(0, 0, 3)$ and measuring in the positive

direction is given by $s(t) = \int_0^t |\mathbf{r}'(u)|\, du = \int_0^t 5\, du = 5t$.  $s(t) = 5$  $\Rightarrow$  $5t = 5$  $\Rightarrow$  $t = 1$, thus your location after

moving 5 units along the curve is $(3\sin 1, 4, 3\cos 1)$.

**17. (a)** $\mathbf{r}(t) = \langle 2\sin t, 5t, 2\cos t \rangle$ $\Rightarrow$ $\mathbf{r}'(t) = \langle 2\cos t, 5, -2\sin t \rangle$ $\Rightarrow$ $|\mathbf{r}'(t)| = \sqrt{4\cos^2 t + 25 + 4\sin^2 t} = \sqrt{29}$.

Then $\mathbf{T}(t) = \dfrac{\mathbf{r}'(t)}{|\mathbf{r}'(t)|} = \dfrac{1}{\sqrt{29}}\langle 2\cos t, 5, -2\sin t \rangle$ or $\left\langle \frac{2}{\sqrt{29}}\cos t, \frac{5}{\sqrt{29}}, -\frac{2}{\sqrt{29}}\sin t \right\rangle$.

$\mathbf{T}'(t) = \frac{1}{\sqrt{29}}\langle -2\sin t, 0, -2\cos t \rangle$ $\Rightarrow$ $|\mathbf{T}'(t)| = \frac{1}{\sqrt{29}}\sqrt{4\sin^2 t + 0 + 4\cos^2 t} = \frac{2}{\sqrt{29}}$. Thus

$\mathbf{N}(t) = \dfrac{\mathbf{T}'(t)}{|\mathbf{T}'(t)|} = \dfrac{1/\sqrt{29}}{2/\sqrt{29}}\langle -2\sin t, 0, -2\cos t \rangle = \langle -\sin t, 0, -\cos t \rangle$.

**(b)** $\kappa(t) = \dfrac{|\mathbf{T}'(t)|}{|\mathbf{r}'(t)|} = \dfrac{2/\sqrt{29}}{\sqrt{29}} = \dfrac{2}{29}$

**19. (a)** $\mathbf{r}(t) = \langle \sqrt{2}\,t, e^t, e^{-t} \rangle$ $\Rightarrow$ $\mathbf{r}'(t) = \langle \sqrt{2}, e^t, -e^{-t} \rangle$ $\Rightarrow$ $|\mathbf{r}'(t)| = \sqrt{2 + e^{2t} + e^{-2t}} = \sqrt{(e^t + e^{-t})^2} = e^t + e^{-t}$.

Then

$\mathbf{T}(t) = \dfrac{\mathbf{r}'(t)}{|\mathbf{r}'(t)|} = \dfrac{1}{e^t + e^{-t}}\langle \sqrt{2}, e^t, -e^{-t} \rangle = \dfrac{1}{e^{2t} + 1}\langle \sqrt{2}\,e^t, e^{2t}, -1 \rangle$ $\left[\text{after multiplying by } \dfrac{e^t}{e^t}\right]$ and

$\mathbf{T}'(t) = \dfrac{1}{e^{2t} + 1}\langle \sqrt{2}\,e^t, 2e^{2t}, 0 \rangle - \dfrac{2e^{2t}}{(e^{2t} + 1)^2}\langle \sqrt{2}\,e^t, e^{2t}, -1 \rangle$

$= \dfrac{1}{(e^{2t} + 1)^2}\left[(e^{2t} + 1)\langle \sqrt{2}\,e^t, 2e^{2t}, 0 \rangle - 2e^{2t}\langle \sqrt{2}\,e^t, e^{2t}, -1 \rangle\right] = \dfrac{1}{(e^{2t} + 1)^2}\langle \sqrt{2}\,e^t\left(1 - e^{2t}\right), 2e^{2t}, 2e^{2t} \rangle$

Then

$$|\mathbf{T}'(t)| = \dfrac{1}{(e^{2t} + 1)^2}\sqrt{2e^{2t}(1 - 2e^{2t} + e^{4t}) + 4e^{4t} + 4e^{4t}} = \dfrac{1}{(e^{2t} + 1)^2}\sqrt{2e^{2t}(1 + 2e^{2t} + e^{4t})}$$

$$= \dfrac{1}{(e^{2t} + 1)^2}\sqrt{2e^{2t}\left(1 + e^{2t}\right)^2} = \dfrac{\sqrt{2}\,e^t(1 + e^{2t})}{(e^{2t} + 1)^2} = \dfrac{\sqrt{2}\,e^t}{e^{2t} + 1}$$

Therefore

$$\mathbf{N}(t) = \dfrac{\mathbf{T}'(t)}{|\mathbf{T}'(t)|} = \dfrac{e^{2t} + 1}{\sqrt{2}\,e^t}\dfrac{1}{(e^{2t} + 1)^2}\langle \sqrt{2}\,e^t(1 - e^{2t}), 2e^{2t}, 2e^{2t} \rangle$$

$$= \dfrac{1}{\sqrt{2}\,e^t(e^{2t} + 1)}\langle \sqrt{2}\,e^t(1 - e^{2t}), 2e^{2t}, 2e^{2t} \rangle = \dfrac{1}{e^{2t} + 1}\langle 1 - e^{2t}, \sqrt{2}\,e^t, \sqrt{2}\,e^t \rangle$$

**(b)** $\kappa(t) = \dfrac{|\mathbf{T}'(t)|}{|\mathbf{r}'(t)|} = \dfrac{\sqrt{2}\,e^t}{e^{2t} + 1}\cdot\dfrac{1}{e^t + e^{-t}} = \dfrac{\sqrt{2}\,e^t}{e^{3t} + 2e^t + e^{-t}} = \dfrac{\sqrt{2}\,e^{2t}}{e^{4t} + 2e^{2t} + 1} = \dfrac{\sqrt{2}\,e^{2t}}{(e^{2t} + 1)^2}$

**21.** $\mathbf{r}(t) = t^3\,\mathbf{j} + t^2\,\mathbf{k}$ $\Rightarrow$ $\mathbf{r}'(t) = 3t^2\,\mathbf{j} + 2t\,\mathbf{k}$, $\mathbf{r}''(t) = 6t\,\mathbf{j} + 2\,\mathbf{k}$, $|\mathbf{r}'(t)| = \sqrt{0^2 + (3t^2)^2 + (2t)^2} = \sqrt{9t^4 + 4t^2}$,

$\mathbf{r}'(t) \times \mathbf{r}''(t) = -6t^2\,\mathbf{i}$, $|\mathbf{r}'(t) \times \mathbf{r}''(t)| = 6t^2$. Then $\kappa(t) = \dfrac{|\mathbf{r}'(t) \times \mathbf{r}''(t)|}{|\mathbf{r}'(t)|^3} = \dfrac{6t^2}{\left(\sqrt{9t^4 + 4t^2}\right)^3} = \dfrac{6t^2}{(9t^4 + 4t^2)^{3/2}}$.

**23.** $\mathbf{r}(t) = 3t\,\mathbf{i} + 4\sin t\,\mathbf{j} + 4\cos t\,\mathbf{k}$ $\Rightarrow$ $\mathbf{r}'(t) = 3\,\mathbf{i} + 4\cos t\,\mathbf{j} - 4\sin t\,\mathbf{k}$, $\mathbf{r}''(t) = -4\sin t\,\mathbf{j} - 4\cos t\,\mathbf{k}$,

$|\mathbf{r}'(t)| = \sqrt{9 + 16\cos^2 t + 16\sin^2 t} = \sqrt{9 + 16} = 5$, $\mathbf{r}'(t) \times \mathbf{r}''(t) = -16\,\mathbf{i} + 12\cos t\,\mathbf{j} - 12\sin t\,\mathbf{k}$,

$|\mathbf{r}'(t) \times \mathbf{r}''(t)| = \sqrt{256 + 144\cos^2 t + 144\sin^2 t} = \sqrt{400} = 20$. Then $\kappa(t) = \dfrac{|\mathbf{r}'(t) \times \mathbf{r}''(t)|}{|\mathbf{r}'(t)|^3} = \dfrac{20}{5^3} = \dfrac{4}{25}$.

**25.** $\mathbf{r}(t) = \langle t, t^2, t^3 \rangle \Rightarrow \mathbf{r}'(t) = \langle 1, 2t, 3t^2 \rangle$. The point $(1, 1, 1)$ corresponds to $t = 1$, and $\mathbf{r}'(1) = \langle 1, 2, 3 \rangle \Rightarrow$

$|\mathbf{r}'(1)| = \sqrt{1 + 4 + 9} = \sqrt{14}$. $\mathbf{r}''(t) = \langle 0, 2, 6t \rangle \Rightarrow \mathbf{r}''(1) = \langle 0, 2, 6 \rangle$. $\mathbf{r}'(1) \times \mathbf{r}''(1) = \langle 6, -6, 2 \rangle$, so

$|\mathbf{r}'(1) \times \mathbf{r}''(1)| = \sqrt{36 + 36 + 4} = \sqrt{76}$. Then $\kappa(1) = \dfrac{|\mathbf{r}'(1) \times \mathbf{r}''(1)|}{|\mathbf{r}'(1)|^3} = \dfrac{\sqrt{76}}{\sqrt{14}^3} = \dfrac{1}{7}\sqrt{\dfrac{19}{14}}$.

**27.** $f(x) = x^4$, $f'(x) = 4x^3$, $f''(x) = 12x^2$, $\kappa(x) = \dfrac{|f''(x)|}{[1 + (f'(x))^2]^{3/2}} = \dfrac{|12x^2|}{[1 + (4x^3)^2]^{3/2}} = \dfrac{12x^2}{(1 + 16x^6)^{3/2}}$

**29.** $f(x) = xe^x$, $f'(x) = xe^x + e^x$, $f''(x) = xe^x + 2e^x$,

$\kappa(x) = \dfrac{|f''(x)|}{[1 + (f'(x))^2]^{3/2}} = \dfrac{|xe^x + 2e^x|}{[1 + (xe^x + e^x)^2]^{3/2}} = \dfrac{|x + 2|\, e^x}{[1 + (xe^x + e^x)^2]^{3/2}}$

**31.** Since $y' = y'' = e^x$, the curvature is $\kappa(x) = \dfrac{|y''(x)|}{[1 + (y'(x))^2]^{3/2}} = \dfrac{e^x}{(1 + e^{2x})^{3/2}} = e^x(1 + e^{2x})^{-3/2}$.

To find the maximum curvature, we first find the critical numbers of $\kappa(x)$:

$\kappa'(x) = e^x(1 + e^{2x})^{-3/2} + e^x\left(-\frac{3}{2}\right)(1 + e^{2x})^{-5/2}(2e^{2x}) = e^x \dfrac{1 + e^{2x} - 3e^{2x}}{(1 + e^{2x})^{5/2}} = e^x \dfrac{1 - 2e^{2x}}{(1 + e^{2x})^{5/2}}$.

$\kappa'(x) = 0$ when $1 - 2e^{2x} = 0$, so $e^{2x} = \frac{1}{2}$ or $x = -\frac{1}{2}\ln 2$. And since $1 - 2e^{2x} > 0$ for $x < -\frac{1}{2}\ln 2$ and $1 - 2e^{2x} < 0$

for $x > -\frac{1}{2}\ln 2$, the maximum curvature is attained at the point $\left(-\frac{1}{2}\ln 2, e^{(-\ln 2)/2}\right) = \left(-\frac{1}{2}\ln 2, \frac{1}{\sqrt{2}}\right)$.

Since $\lim\limits_{x \to \infty} e^x(1 + e^{2x})^{-3/2} = 0$, $\kappa(x)$ approaches 0 as $x \to \infty$.

**33.** (a) $C$ appears to be changing direction more quickly at $P$ than $Q$, so we would expect the curvature to be greater at $P$.

(b) First we sketch approximate osculating circles at $P$ and $Q$. Using the

axes scale as a guide, we measure the radius of the osculating circle

at $P$ to be approximately 0.8 units, thus $\rho = \dfrac{1}{\kappa} \Rightarrow$

$\kappa = \dfrac{1}{\rho} \approx \dfrac{1}{0.8} \approx 1.3$. Similarly, we estimate the radius of the

osculating circle at $Q$ to be 1.4 units, so $\kappa = \dfrac{1}{\rho} \approx \dfrac{1}{1.4} \approx 0.7$.

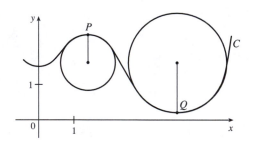

**35.** $y = x^{-2} \Rightarrow y' = -2x^{-3}$, $y'' = 6x^{-4}$, and

$\kappa(x) = \dfrac{|y''|}{[1 + (y')^2]^{3/2}} = \dfrac{|6x^{-4}|}{[1 + (-2x^{-3})^2]^{3/2}} = \dfrac{6}{x^4(1 + 4x^{-6})^{3/2}}$.

The appearance of the two humps in this graph is perhaps a little surprising, but it is

explained by the fact that $y = x^{-2}$ increases asymptotically at the origin from both

directions, and so its graph has very little bend there. [Note that $\kappa(0)$ is undefined.]

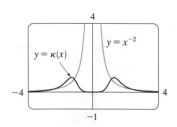

**37.** Notice that the curve $b$ has two inflection points at which the graph appears almost straight. We would expect the curvature to be 0 or nearly 0 at these values, but the curve $a$ isn't near 0 there. Thus, $a$ must be the graph of $y = f(x)$ rather than the graph of curvature, and $b$ is the graph of $y = \kappa(x)$.

**39.** Using a CAS, we find (after simplifying)

$\kappa(t) = \dfrac{6\sqrt{4\cos^2 t - 12\cos t + 13}}{(17 - 12\cos t)^{3/2}}$. (To compute cross

products in Maple, use the `VectorCalculus` or

`LinearAlgebra` package and the `CrossProduct(a,b)`

command; in Mathematica, use `Cross[a,b]`.) Curvature is

largest at integer multiples of $2\pi$.

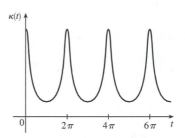

**41.** $x = t^2 \Rightarrow \dot{x} = 2t \Rightarrow \ddot{x} = 2, \quad y = t^3 \Rightarrow \dot{y} = 3t^2 \Rightarrow \ddot{y} = 6t$.

Then $\kappa(t) = \dfrac{|\dot{x}\ddot{y} - \dot{y}\ddot{x}|}{[\dot{x}^2 + \dot{y}^2]^{3/2}} = \dfrac{|(2t)(6t) - (3t^2)(2)|}{[(2t)^2 + (3t^2)^2]^{3/2}} = \dfrac{|12t^2 - 6t^2|}{(4t^2 + 9t^4)^{3/2}} = \dfrac{6t^2}{(4t^2 + 9t^4)^{3/2}}$.

**43.** $x = e^t \cos t \Rightarrow \dot{x} = e^t(\cos t - \sin t) \Rightarrow \ddot{x} = e^t(-\sin t - \cos t) + e^t(\cos t - \sin t) = -2e^t \sin t$,

$y = e^t \sin t \Rightarrow \dot{y} = e^t(\cos t + \sin t) \Rightarrow \ddot{y} = e^t(-\sin t + \cos t) + e^t(\cos t + \sin t) = 2e^t \cos t$. Then

$\kappa(t) = \dfrac{|\dot{x}\ddot{y} - \dot{y}\ddot{x}|}{[\dot{x}^2 + \dot{y}^2]^{3/2}} = \dfrac{|e^t(\cos t - \sin t)(2e^t \cos t) - e^t(\cos t + \sin t)(-2e^t \sin t)|}{([e^t(\cos t - \sin t)]^2 + [e^t(\cos t + \sin t)]^2)^{3/2}}$

$= \dfrac{|2e^{2t}(\cos^2 t - \sin t \cos t + \sin t \cos t + \sin^2 t)|}{[e^{2t}(\cos^2 t - 2\cos t \sin t + \sin^2 t + \cos^2 t + 2\cos t \sin t + \sin^2 t)]^{3/2}} = \dfrac{|2e^{2t}(1)|}{[e^{2t}(1+1)]^{3/2}} = \dfrac{2e^{2t}}{e^{3t}(2)^{3/2}} = \dfrac{1}{\sqrt{2}\, e^t}$

**45.** $(1, \frac{2}{3}, 1)$ corresponds to $t = 1$.   $\mathbf{T}(t) = \dfrac{\mathbf{r}'(t)}{|\mathbf{r}'(t)|} = \dfrac{\langle 2t, 2t^2, 1\rangle}{\sqrt{4t^2 + 4t^4 + 1}} = \dfrac{\langle 2t, 2t^2, 1\rangle}{2t^2 + 1}$, so $\mathbf{T}(1) = \langle \frac{2}{3}, \frac{2}{3}, \frac{1}{3}\rangle$.

$\mathbf{T}'(t) = -4t(2t^2 + 1)^{-2}\langle 2t, 2t^2, 1\rangle + (2t^2 + 1)^{-1}\langle 2, 4t, 0\rangle$   [by Formula 3 of Theorem 10.2.3 ]

$= (2t^2 + 1)^{-2}\langle -8t^2 + 4t^2 + 2, -8t^3 + 8t^3 + 4t, -4t\rangle = 2(2t^2 + 1)^{-2}\langle 1 - 2t^2, 2t, -2t\rangle$

$\mathbf{N}(t) = \dfrac{\mathbf{T}'(t)}{|\mathbf{T}'(t)|} = \dfrac{2(2t^2 + 1)^{-2}\langle 1 - 2t^2, 2t, -2t\rangle}{2(2t^2 + 1)^{-2}\sqrt{(1 - 2t^2)^2 + (2t)^2 + (-2t)^2}} = \dfrac{\langle 1 - 2t^2, 2t, -2t\rangle}{\sqrt{1 - 4t^2 + 4t^4 + 8t^2}} = \dfrac{\langle 1 - 2t^2, 2t, -2t\rangle}{1 + 2t^2}$

$\mathbf{N}(1) = \langle -\frac{1}{3}, \frac{2}{3}, -\frac{2}{3}\rangle$ and $\mathbf{B}(1) = \mathbf{T}(1) \times \mathbf{N}(1) = \langle -\frac{4}{9} - \frac{2}{9}, -\left(-\frac{4}{9} + \frac{1}{9}\right), \frac{4}{9} + \frac{2}{9}\rangle = \langle -\frac{2}{3}, \frac{1}{3}, \frac{2}{3}\rangle$.

**47.** $(0, \pi, -2)$ corresponds to $t = \pi$.   $\mathbf{r}(t) = \langle 2\sin 3t, t, 2\cos 3t\rangle \Rightarrow$

$\mathbf{T}(t) = \dfrac{\mathbf{r}'(t)}{|\mathbf{r}'(t)|} = \dfrac{\langle 6\cos 3t, 1, -6\sin 3t\rangle}{\sqrt{36\cos^2 3t + 1 + 36\sin^2 3t}} = \dfrac{1}{\sqrt{37}}\langle 6\cos 3t, 1, -6\sin 3t\rangle$.

$\mathbf{T}(\pi) = \frac{1}{\sqrt{37}}\langle -6, 1, 0\rangle$ is a normal vector for the normal plane, and so $\langle -6, 1, 0\rangle$ is also normal. Thus an equation for the plane is $-6(x - 0) + 1(y - \pi) + 0(z + 2) = 0$ or $y - 6x = \pi$.

$\mathbf{T}'(t) = \frac{1}{\sqrt{37}}\langle -18\sin 3t, 0, -18\cos 3t\rangle \Rightarrow |\mathbf{T}'(t)| = \dfrac{\sqrt{18^2\sin^2 3t + 18^2\cos^2 3t}}{\sqrt{37}} = \dfrac{18}{\sqrt{37}} \Rightarrow$

$\mathbf{N}(t) = \dfrac{\mathbf{T}'(t)}{|\mathbf{T}'(t)|} = \langle -\sin 3t, 0, -\cos 3t \rangle$. So $\mathbf{N}(\pi) = \langle 0, 0, 1 \rangle$ and $\mathbf{B}(\pi) = \frac{1}{\sqrt{37}} \langle -6, 1, 0 \rangle \times \langle 0, 0, 1 \rangle = \frac{1}{\sqrt{37}} \langle 1, 6, 0 \rangle$.

Since $\mathbf{B}(\pi)$ is a normal to the osculating plane, so is $\langle 1, 6, 0 \rangle$.

An equation for the plane is $1(x - 0) + 6(y - \pi) + 0(z + 2) = 0$ or $x + 6y = 6\pi$.

**49.** The ellipse is given by the parametric equations $x = 2\cos t$, $y = 3\sin t$, so using the result from Exercise 40,

$$\kappa(t) = \frac{|\dot{x}\ddot{y} - \ddot{x}\dot{y}|}{[\dot{x}^2 + \dot{y}^2]^{3/2}} = \frac{|(-2\sin t)(-3\sin t) - (3\cos t)(-2\cos t)|}{(4\sin^2 t + 9\cos^2 t)^{3/2}} = \frac{6}{(4\sin^2 t + 9\cos^2 t)^{3/2}}.$$

At $(2, 0)$, $t = 0$. Now $\kappa(0) = \frac{6}{27} = \frac{2}{9}$, so the radius of the osculating circle is

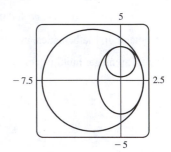

$1/\kappa(0) = \frac{9}{2}$ and its center is $\left(-\frac{5}{2}, 0\right)$. Its equation is therefore $\left(x + \frac{5}{2}\right)^2 + y^2 = \frac{81}{4}$.

At $(0, 3)$, $t = \frac{\pi}{2}$, and $\kappa\left(\frac{\pi}{2}\right) = \frac{6}{8} = \frac{3}{4}$. So the radius of the osculating circle is $\frac{4}{3}$ and

its center is $\left(0, \frac{5}{3}\right)$. Hence its equation is $x^2 + \left(y - \frac{5}{3}\right)^2 = \frac{16}{9}$.

**51.** The tangent vector is normal to the normal plane, and the vector $\langle 6, 6, -8 \rangle$ is normal to the given plane.

But $\mathbf{T}(t) \parallel \mathbf{r}'(t)$ and $\langle 6, 6, -8 \rangle \parallel \langle 3, 3, -4 \rangle$, so we need to find $t$ such that $\mathbf{r}'(t) \parallel \langle 3, 3, -4 \rangle$.

$\mathbf{r}(t) = \langle t^3, 3t, t^4 \rangle \Rightarrow \mathbf{r}'(t) = \langle 3t^2, 3, 4t^3 \rangle \parallel \langle 3, 3, -4 \rangle$ when $t = -1$. So the planes are parallel at the point $(-1, -3, 1)$.

**53.** First we parametrize the curve of intersection. We can choose $y = t$; then $x = y^2 = t^2$ and $z = x^2 = t^4$, and the curve is

given by $\mathbf{r}(t) = \langle t^2, t, t^4 \rangle$. $\mathbf{r}'(t) = \langle 2t, 1, 4t^3 \rangle$ and the point $(1, 1, 1)$ corresponds to $t = 1$, so

$\mathbf{r}'(1) = \langle 2, 1, 4 \rangle$ is a normal vector for the normal plane. Thus an equation of the normal plane is

$2(x - 1) + 1(y - 1) + 4(z - 1) = 0$ or $2x + y + 4z = 7$. $\mathbf{T}(t) = \dfrac{\mathbf{r}'(t)}{|\mathbf{r}'(t)|} = \dfrac{1}{\sqrt{4t^2 + 1 + 16t^6}} \langle 2t, 1, 4t^3 \rangle$ and

$\mathbf{T}'(t) = -\frac{1}{2}(4t^2 + 1 + 16t^6)^{-3/2}(8t + 96t^5) \langle 2t, 1, 4t^3 \rangle + (4t^2 + 1 + 16t^6)^{-1/2} \langle 2, 0, 12t^2 \rangle$. A normal vector for

the osculating plane is $\mathbf{B}(1) = \mathbf{T}(1) \times \mathbf{N}(1)$, but $\mathbf{r}'(1) = \langle 2, 1, 4 \rangle$ is parallel to $\mathbf{T}(1)$ and

$\mathbf{T}'(1) = -\frac{1}{2}(21)^{-3/2}(104)\langle 2, 1, 4 \rangle + (21)^{-1/2}\langle 2, 0, 12 \rangle = \frac{2}{21\sqrt{21}} \langle -31, -26, 22 \rangle$ is parallel to $\mathbf{N}(1)$ as is $\langle -31, -26, 22 \rangle$,

so $\langle 2, 1, 4 \rangle \times \langle -31, -26, 22 \rangle = \langle 126, -168, -21 \rangle$ is normal to the osculating plane. Thus an equation for the osculating

plane is $126(x - 1) - 168(y - 1) - 21(z - 1) = 0$ or $6x - 8y - z = -3$.

**55.** $\kappa = \left| \dfrac{d\mathbf{T}}{ds} \right| = \left| \dfrac{d\mathbf{T}/dt}{ds/dt} \right| = \dfrac{|d\mathbf{T}/dt|}{ds/dt}$ and $\mathbf{N} = \dfrac{d\mathbf{T}/dt}{|d\mathbf{T}/dt|}$, so $\kappa\mathbf{N} = \dfrac{\left|\dfrac{d\mathbf{T}}{dt}\right| \dfrac{d\mathbf{T}}{dt}}{\left|\dfrac{d\mathbf{T}}{dt}\right| \dfrac{ds}{dt}} = \dfrac{d\mathbf{T}/dt}{ds/dt} = \dfrac{d\mathbf{T}}{ds}$ by the Chain Rule.

**57.** (a) $|\mathbf{B}| = 1 \Rightarrow \mathbf{B} \cdot \mathbf{B} = 1 \Rightarrow \dfrac{d}{ds}(\mathbf{B} \cdot \mathbf{B}) = 0 \Rightarrow 2\dfrac{d\mathbf{B}}{ds} \cdot \mathbf{B} = 0 \Rightarrow \dfrac{d\mathbf{B}}{ds} \perp \mathbf{B}$

(b) $\mathbf{B} = \mathbf{T} \times \mathbf{N}$ $\Rightarrow$

$$\frac{d\mathbf{B}}{ds} = \frac{d}{ds}(\mathbf{T} \times \mathbf{N}) = \frac{d}{dt}(\mathbf{T} \times \mathbf{N})\frac{1}{ds/dt} = \frac{d}{dt}(\mathbf{T} \times \mathbf{N})\frac{1}{|\mathbf{r}'(t)|} = [(\mathbf{T}' \times \mathbf{N}) + (\mathbf{T} \times \mathbf{N}')]\frac{1}{|\mathbf{r}'(t)|}$$

$$= \left[\left(\mathbf{T}' \times \frac{\mathbf{T}'}{|\mathbf{T}'|}\right) + (\mathbf{T} \times \mathbf{N}')\right]\frac{1}{|\mathbf{r}'(t)|} = \frac{\mathbf{T} \times \mathbf{N}'}{|\mathbf{r}'(t)|} \quad \Rightarrow \quad \frac{d\mathbf{B}}{ds} \perp \mathbf{T}$$

(c) $\mathbf{B} = \mathbf{T} \times \mathbf{N}$ $\Rightarrow$ $\mathbf{T} \perp \mathbf{N}, \mathbf{B} \perp \mathbf{T}$ and $\mathbf{B} \perp \mathbf{N}$. So $\mathbf{B}, \mathbf{T}$ and $\mathbf{N}$ form an orthogonal set of vectors in the three-dimensional space $\mathbb{R}^3$. From parts (a) and (b), $d\mathbf{B}/ds$ is perpendicular to both $\mathbf{B}$ and $\mathbf{T}$, so $d\mathbf{B}/ds$ is parallel to $\mathbf{N}$. Therefore, $d\mathbf{B}/ds = -\tau(s)\mathbf{N}$, where $\tau(s)$ is a scalar.

(d) Since $\mathbf{B} = \mathbf{T} \times \mathbf{N}, \mathbf{T} \perp \mathbf{N}$ and both $\mathbf{T}$ and $\mathbf{N}$ are unit vectors, $\mathbf{B}$ is a unit vector mutually perpendicular to both $\mathbf{T}$ and $\mathbf{N}$. For a plane curve, $\mathbf{T}$ and $\mathbf{N}$ always lie in the plane of the curve, so that $\mathbf{B}$ is a constant unit vector always perpendicular to the plane. Thus $d\mathbf{B}/ds = \mathbf{0}$, but $d\mathbf{B}/ds = -\tau(s)\mathbf{N}$ and $\mathbf{N} \neq \mathbf{0}$, so $\tau(s) = 0$.

**59.** (a) $\mathbf{r}' = s'\,\mathbf{T}$ $\Rightarrow$ $\mathbf{r}'' = s''\,\mathbf{T} + s'\,\mathbf{T}' = s''\,\mathbf{T} + s'\frac{d\mathbf{T}}{ds}s' = s''\,\mathbf{T} + \kappa(s')^2\,\mathbf{N}$ by the first Serret-Frenet formula.

(b) Using part (a), we have

$$\mathbf{r}' \times \mathbf{r}'' = (s'\,\mathbf{T}) \times [s''\,\mathbf{T} + \kappa(s')^2\,\mathbf{N}]$$

$$= [(s'\,\mathbf{T}) \times (s''\,\mathbf{T})] + [(s'\mathbf{T}) \times (\kappa(s')^2\,\mathbf{N})] \quad \text{[by Property 3 of the cross product]}$$

$$= (s's'')(\mathbf{T} \times \mathbf{T}) + \kappa(s')^3(\mathbf{T} \times \mathbf{N}) = \mathbf{0} + \kappa(s')^3\,\mathbf{B} = \kappa(s')^3\,\mathbf{B}$$

(c) Using part (a), we have

$$\mathbf{r}''' = [s''\,\mathbf{T} + \kappa(s')^2\,\mathbf{N}]' = s'''\,\mathbf{T} + s''\,\mathbf{T}' + \kappa'(s')^2\,\mathbf{N} + 2\kappa s's''\,\mathbf{N} + \kappa(s')^2\,\mathbf{N}'$$

$$= s'''\,\mathbf{T} + s''\frac{d\mathbf{T}}{ds}s' + \kappa'(s')^2\,\mathbf{N} + 2\kappa s's''\,\mathbf{N} + \kappa(s')^2\frac{d\mathbf{N}}{ds}s'$$

$$= s'''\,\mathbf{T} + s''s'\kappa\,\mathbf{N} + \kappa'(s')^2\,\mathbf{N} + 2\kappa s's''\,\mathbf{N} + \kappa(s')^3(-\kappa\,\mathbf{T} + \tau\,\mathbf{B}) \quad \text{[by the second formula]}$$

$$= [s''' - \kappa^2(s')^3]\,\mathbf{T} + [3\kappa s's'' + \kappa'(s')^2]\,\mathbf{N} + \kappa\tau(s')^3\,\mathbf{B}$$

(d) Using parts (b) and (c) and the facts that $\mathbf{B} \cdot \mathbf{T} = 0, \mathbf{B} \cdot \mathbf{N} = 0$, and $\mathbf{B} \cdot \mathbf{B} = 1$, we get

$$\frac{(\mathbf{r}' \times \mathbf{r}'') \cdot \mathbf{r}'''}{|\mathbf{r}' \times \mathbf{r}''|^2} = \frac{\kappa(s')^3\,\mathbf{B} \cdot \{[s''' - \kappa^2(s')^3]\,\mathbf{T} + [3\kappa s's'' + \kappa'(s')^2]\,\mathbf{N} + \kappa\tau(s')^3\,\mathbf{B}\}}{|\kappa(s')^3\,\mathbf{B}|^2} = \frac{\kappa(s')^3\kappa\tau(s')^3}{[\kappa(s')^3]^2} = \tau.$$

**61.** For one helix, the vector equation is $\mathbf{r}(t) = \langle 10\cos t, 10\sin t, 34t/(2\pi) \rangle$ (measuring in angstroms), because the radius of each helix is 10 angstroms, and $z$ increases by 34 angstroms for each increase of $2\pi$ in $t$. Using the arc length formula, letting $t$ go from 0 to $2.9 \times 10^8 \times 2\pi$, we find the approximate length of each helix to be

$$L = \int_0^{2.9 \times 10^8 \times 2\pi} |\mathbf{r}'(t)|\, dt = \int_0^{2.9 \times 10^8 \times 2\pi} \sqrt{(-10\sin t)^2 + (10\cos t)^2 + \left(\frac{34}{2\pi}\right)^2}\, dt = \sqrt{100 + \left(\frac{34}{2\pi}\right)^2}\, t \Big]_0^{2.9 \times 10^8 \times 2\pi}$$

$$= 2.9 \times 10^8 \times 2\pi \sqrt{100 + \left(\frac{34}{2\pi}\right)^2} \approx 2.07 \times 10^{10}\text{ Å — more than two meters!}$$

## 10.4 Motion in Space: Velocity and Acceleration

**1.** (a) If $\mathbf{r}(t) = x(t)\,\mathbf{i} + y(t)\,\mathbf{j} + z(t)\,\mathbf{k}$ is the position vector of the particle at time $t$, then the average velocity over the time interval $[0, 1]$ is

$$\mathbf{v}_{\text{ave}} = \frac{\mathbf{r}(1) - \mathbf{r}(0)}{1 - 0} = \frac{(4.5\,\mathbf{i} + 6.0\,\mathbf{j} + 3.0\,\mathbf{k}) - (2.7\,\mathbf{i} + 9.8\,\mathbf{j} + 3.7\,\mathbf{k})}{1} = 1.8\,\mathbf{i} - 3.8\,\mathbf{j} - 0.7\,\mathbf{k}.$$ Similarly, over the other

intervals we have

$$[0.5, 1]: \quad \mathbf{v}_{\text{ave}} = \frac{\mathbf{r}(1) - \mathbf{r}(0.5)}{1 - 0.5} = \frac{(4.5\,\mathbf{i} + 6.0\,\mathbf{j} + 3.0\,\mathbf{k}) - (3.5\,\mathbf{i} + 7.2\,\mathbf{j} + 3.3\,\mathbf{k})}{0.5} = 2.0\,\mathbf{i} - 2.4\,\mathbf{j} - 0.6\,\mathbf{k}$$

$$[1, 2]: \quad \mathbf{v}_{\text{ave}} = \frac{\mathbf{r}(2) - \mathbf{r}(1)}{2 - 1} = \frac{(7.3\,\mathbf{i} + 7.8\,\mathbf{j} + 2.7\,\mathbf{k}) - (4.5\,\mathbf{i} + 6.0\,\mathbf{j} + 3.0\,\mathbf{k})}{1} = 2.8\,\mathbf{i} + 1.8\,\mathbf{j} - 0.3\,\mathbf{k}$$

$$[1, 1.5]: \quad \mathbf{v}_{\text{ave}} = \frac{\mathbf{r}(1.5) - \mathbf{r}(1)}{1.5 - 1} = \frac{(5.9\,\mathbf{i} + 6.4\,\mathbf{j} + 2.8\,\mathbf{k}) - (4.5\,\mathbf{i} + 6.0\,\mathbf{j} + 3.0\,\mathbf{k})}{0.5} = 2.8\,\mathbf{i} + 0.8\,\mathbf{j} - 0.4\,\mathbf{k}$$

(b) We can estimate the velocity at $t = 1$ by averaging the average velocities over the time intervals $[0.5, 1]$ and $[1, 1.5]$:

$$\mathbf{v}(1) \approx \tfrac{1}{2}[(2\,\mathbf{i} - 2.4\,\mathbf{j} - 0.6\,\mathbf{k}) + (2.8\,\mathbf{i} + 0.8\,\mathbf{j} - 0.4\,\mathbf{k})] = 2.4\,\mathbf{i} - 0.8\,\mathbf{j} - 0.5\,\mathbf{k}.$$ Then the speed is

$$|\mathbf{v}(1)| \approx \sqrt{(2.4)^2 + (-0.8)^2 + (-0.5)^2} \approx 2.58.$$

**3.** $\mathbf{r}(t) = \left\langle -\tfrac{1}{2}t^2, t \right\rangle \quad \Rightarrow$      At $t = 2$:

$\mathbf{v}(t) = \mathbf{r}'(t) = \langle -t, 1 \rangle$      $\mathbf{v}(2) = \langle -2, 1 \rangle$

$\mathbf{a}(t) = \mathbf{r}''(t) = \langle -1, 0 \rangle$      $\mathbf{a}(2) = \langle -1, 0 \rangle$

$|\mathbf{v}(t)| = \sqrt{t^2 + 1}$

**5.** $(t) = 3\cos t\,\mathbf{i} + 2\sin t\,\mathbf{j} \quad \Rightarrow$      At $t = \pi/3$:

$\mathbf{v}(t) = -3\sin t\,\mathbf{i} + 2\cos t\,\mathbf{j}$      $\mathbf{v}\left(\tfrac{\pi}{3}\right) = -\tfrac{3\sqrt{3}}{2}\,\mathbf{i} + \mathbf{j}$

$\mathbf{a}(t) = -3\cos t\,\mathbf{i} - 2\sin t\,\mathbf{j}$      $\mathbf{a}\left(\tfrac{\pi}{3}\right) = -\tfrac{3}{2}\,\mathbf{i} - \sqrt{3}\,\mathbf{j}$

$|\mathbf{v}(t)| = \sqrt{9\sin^2 t + 4\cos^2 t} = \sqrt{4 + 5\sin^2 t}$

Notice that $x^2/9 + y^2/4 = \sin^2 t + \cos^2 t = 1$, so the path is an ellipse.

**7.** $\mathbf{r}(t) = t\,\mathbf{i} + t^2\,\mathbf{j} + 2\,\mathbf{k} \quad \Rightarrow$      At $t = 1$:

$\mathbf{v}(t) = \mathbf{i} + 2t\,\mathbf{j}$      $\mathbf{v}(1) = \mathbf{i} + 2\,\mathbf{j}$

$\mathbf{a}(t) = 2\,\mathbf{j}$      $\mathbf{a}(1) = 2\,\mathbf{j}$

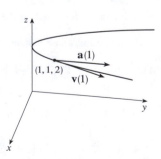

$|\mathbf{v}(t)| = \sqrt{1 + 4t^2}$

Here $x = t$, $y = t^2 \quad \Rightarrow \quad y = x^2$ and $z = 2$, so the path of the particle is a

parabola in the plane $z = 2$.

**9.** $\mathbf{r}(t) = \sqrt{2}\,t\,\mathbf{i} + e^t\,\mathbf{j} + e^{-t}\,\mathbf{k}$   $\Rightarrow$   $\mathbf{v}(t) = \mathbf{r}'(t) = \sqrt{2}\,\mathbf{i} + e^t\,\mathbf{j} - e^{-t}\,\mathbf{k}$,   $\mathbf{a}(t) = \mathbf{v}'(t) = e^t\,\mathbf{j} + e^{-t}\,\mathbf{k}$,

$|\mathbf{v}(t)| = \sqrt{2 + e^{2t} + e^{-2t}} = \sqrt{(e^t + e^{-t})^2} = e^t + e^{-t}$.

**11.** $\mathbf{r}(t) = e^t\langle \cos t, \sin t, t \rangle$   $\Rightarrow$

$\mathbf{v}(t) = \mathbf{r}'(t) = e^t\langle \cos t, \sin t, t \rangle + e^t\langle -\sin t, \cos t, 1 \rangle = e^t\langle \cos t - \sin t, \sin t + \cos t, t + 1 \rangle$

$\mathbf{a}(t) = \mathbf{v}'(t) = e^t\langle \cos t - \sin t - \sin t - \cos t, \sin t + \cos t + \cos t - \sin t, t + 1 + 1 \rangle$

$\qquad = e^t\langle -2\sin t, 2\cos t, t + 2 \rangle$

$|\mathbf{v}(t)| = e^t\sqrt{\cos^2 t + \sin^2 t - 2\cos t \sin t + \sin^2 t + \cos^2 t + 2\sin t \cos t + t^2 + 2t + 1}$

$\qquad = e^t\sqrt{t^2 + 2t + 3}$

**13.** $\mathbf{a}(t) = \mathbf{i} + 2\mathbf{j}$   $\Rightarrow$   $\mathbf{v}(t) = \int \mathbf{a}(t)\,dt = \int (\mathbf{i} + 2\mathbf{j})\,dt = t\,\mathbf{i} + 2t\,\mathbf{j} + \mathbf{C}$ and $\mathbf{k} = \mathbf{v}(0) = \mathbf{C}$,

so $\mathbf{C} = \mathbf{k}$ and $\mathbf{v}(t) = t\,\mathbf{i} + 2t\,\mathbf{j} + \mathbf{k}$.   $\mathbf{r}(t) = \int \mathbf{v}(t)\,dt = \int (t\,\mathbf{i} + 2t\,\mathbf{j} + \mathbf{k})\,dt = \frac{1}{2}t^2\,\mathbf{i} + t^2\,\mathbf{j} + t\,\mathbf{k} + \mathbf{D}$.

But $\mathbf{i} = \mathbf{r}(0) = \mathbf{D}$, so $\mathbf{D} = \mathbf{i}$ and $\mathbf{r}(t) = \left(\frac{1}{2}t^2 + 1\right)\mathbf{i} + t^2\,\mathbf{j} + t\,\mathbf{k}$.

**15.** (a) $\mathbf{a}(t) = 2t\,\mathbf{i} + \sin t\,\mathbf{j} + \cos 2t\,\mathbf{k}$   $\Rightarrow$     (b)

$\mathbf{v}(t) = \int (2t\,\mathbf{i} + \sin t\,\mathbf{j} + \cos 2t\,\mathbf{k})\,dt = t^2\,\mathbf{i} - \cos t\,\mathbf{j} + \frac{1}{2}\sin 2t\,\mathbf{k} + \mathbf{C}$

and $\mathbf{i} = \mathbf{v}(0) = -\mathbf{j} + \mathbf{C}$, so $\mathbf{C} = \mathbf{i} + \mathbf{j}$

and $\mathbf{v}(t) = (t^2 + 1)\,\mathbf{i} + (1 - \cos t)\,\mathbf{j} + \frac{1}{2}\sin 2t\,\mathbf{k}$.

$\mathbf{r}(t) = \int [(t^2 + 1)\,\mathbf{i} + (1 - \cos t)\,\mathbf{j} + \frac{1}{2}\sin 2t\,\mathbf{k}]\,dt$

$\qquad = \left(\frac{1}{3}t^3 + t\right)\mathbf{i} + (t - \sin t)\,\mathbf{j} - \frac{1}{4}\cos 2t\,\mathbf{k} + \mathbf{D}$

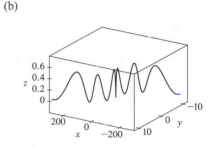

But $\mathbf{j} = \mathbf{r}(0) = -\frac{1}{4}\mathbf{k} + \mathbf{D}$, so $\mathbf{D} = \mathbf{j} + \frac{1}{4}\mathbf{k}$ and $\mathbf{r}(t) = \left(\frac{1}{3}t^3 + t\right)\mathbf{i} + (t - \sin t + 1)\,\mathbf{j} + \left(\frac{1}{4} - \frac{1}{4}\cos 2t\right)\mathbf{k}$.

**17.** $\mathbf{r}(t) = \langle t^2, 5t, t^2 - 16t \rangle$   $\Rightarrow$   $\mathbf{v}(t) = \langle 2t, 5, 2t - 16 \rangle$, $|\mathbf{v}(t)| = \sqrt{4t^2 + 25 + 4t^2 - 64t + 256} = \sqrt{8t^2 - 64t + 281}$

and $\dfrac{d}{dt}|\mathbf{v}(t)| = \frac{1}{2}(8t^2 - 64t + 281)^{-1/2}(16t - 64)$. This is zero if and only if the numerator is zero, that is,

$16t - 64 = 0$ or $t = 4$. Since $\dfrac{d}{dt}|\mathbf{v}(t)| < 0$ for $t < 4$ and $\dfrac{d}{dt}|\mathbf{v}(t)| > 0$ for $t > 4$, the minimum speed of $\sqrt{153}$ is attained

at $t = 4$ units of time.

**19.** $|\mathbf{F}(t)| = 20$ N in the direction of the positive $z$-axis, so $\mathbf{F}(t) = 20\,\mathbf{k}$. Also $m = 4$ kg, $\mathbf{r}(0) = \mathbf{0}$ and $\mathbf{v}(0) = \mathbf{i} - \mathbf{j}$.

Since $20\,\mathbf{k} = \mathbf{F}(t) = 4\,\mathbf{a}(t)$, $\mathbf{a}(t) = 5\,\mathbf{k}$. Then $\mathbf{v}(t) = 5t\,\mathbf{k} + \mathbf{c}_1$ where $\mathbf{c}_1 = \mathbf{i} - \mathbf{j}$ so $\mathbf{v}(t) = \mathbf{i} - \mathbf{j} + 5t\,\mathbf{k}$ and the

speed is $|\mathbf{v}(t)| = \sqrt{1 + 1 + 25t^2} = \sqrt{25t^2 + 2}$. Also $\mathbf{r}(t) = t\,\mathbf{i} - t\,\mathbf{j} + \frac{5}{2}t^2\,\mathbf{k} + \mathbf{c}_2$ and $\mathbf{0} = \mathbf{r}(0)$, so $\mathbf{c}_2 = \mathbf{0}$

and $\mathbf{r}(t) = t\,\mathbf{i} - t\,\mathbf{j} + \frac{5}{2}t^2\,\mathbf{k}$.

**21.** $|\mathbf{v}(0)| = 200$ m/s and, since the angle of elevation is $60°$, a unit vector in the direction of the velocity is

$(\cos 60°)\mathbf{i} + (\sin 60°)\mathbf{j} = \frac{1}{2}\mathbf{i} + \frac{\sqrt{3}}{2}\mathbf{j}$. Thus $\mathbf{v}(0) = 200\left(\frac{1}{2}\mathbf{i} + \frac{\sqrt{3}}{2}\mathbf{j}\right) = 100\,\mathbf{i} + 100\sqrt{3}\,\mathbf{j}$ and if we set up the axes so that

the projectile starts at the origin, then $\mathbf{r}(0) = \mathbf{0}$. Ignoring air resistance, the only force is that due to gravity, so

$\mathbf{F}(t) = m\mathbf{a}(t) = -mg\,\mathbf{j}$ where $g \approx 9.8$ m/s$^2$. Thus $\mathbf{a}(t) = -9.8\,\mathbf{j}$ and, integrating, we have $\mathbf{v}(t) = -9.8t\,\mathbf{j} + \mathbf{C}$. But

$100\,\mathbf{i} + 100\,\sqrt{3}\,\mathbf{j} = \mathbf{v}(0) = \mathbf{C}$, so $\mathbf{v}(t) = 100\,\mathbf{i} + \left(100\,\sqrt{3} - 9.8t\right)\mathbf{j}$ and then (integrating again)

$\mathbf{r}(t) = 100\,t\,\mathbf{i} + \left(100\,\sqrt{3}\,t - 4.9t^2\right)\mathbf{j} + \mathbf{D}$ where $\mathbf{0} = \mathbf{r}(0) = \mathbf{D}$. Thus the position function of the projectile is

$\mathbf{r}(t) = 100\,t\,\mathbf{i} + \left(100\,\sqrt{3}\,t - 4.9t^2\right)\mathbf{j}$.

(a) Parametric equations for the projectile are $x(t) = 100t$, $y(t) = 100\,\sqrt{3}\,t - 4.9t^2$. The projectile reaches the ground when

$y(t) = 0$ (and $t > 0$) $\Rightarrow$ $100\,\sqrt{3}\,t - 4.9t^2 = t(100\,\sqrt{3} - 4.9t) = 0$ $\Rightarrow$ $t = \frac{100\sqrt{3}}{4.9} \approx 35.3$ s. So the range is

$x\left(\frac{100\sqrt{3}}{4.9}\right) = 100\left(\frac{100\sqrt{3}}{4.9}\right) \approx 3535$ m.

(b) The maximum height is reached when $y(t)$ has a critical number (or equivalently, when the vertical component

of velocity is 0): $y'(t) = 0$ $\Rightarrow$ $100\,\sqrt{3} - 9.8t = 0$ $\Rightarrow$ $t = \frac{100\sqrt{3}}{9.8} \approx 17.7$ s. Thus the maximum height is

$y\left(\frac{100\sqrt{3}}{9.8}\right) = 100\,\sqrt{3}\left(\frac{100\sqrt{3}}{9.8}\right) - 4.9\left(\frac{100\sqrt{3}}{9.8}\right)^2 \approx 1531$ m.

(c) From part (a), impact occurs at $t = \frac{100\sqrt{3}}{4.9}$ s. Thus, the velocity at impact is

$\mathbf{v}\left(\frac{100\sqrt{3}}{4.9}\right) = 100\,\mathbf{i} + \left[100\,\sqrt{3} - 9.8\left(\frac{100\sqrt{3}}{4.9}\right)\right]\mathbf{j} = 100\,\mathbf{i} - 100\,\sqrt{3}\,\mathbf{j}$ and the speed is

$\left|\mathbf{v}\left(\frac{100\sqrt{3}}{4.9}\right)\right| = \sqrt{10{,}000 + 30{,}000} = 200$ m/s.

**23.** As in Example 5, $\mathbf{r}(t) = (v_0 \cos 45°)t\,\mathbf{i} + \left[(v_0 \sin 45°)t - \frac{1}{2}gt^2\right]\mathbf{j} = \frac{1}{2}\left[v_0\sqrt{2}\,t\,\mathbf{i} + \left(v_0\sqrt{2}\,t - gt^2\right)\mathbf{j}\right]$. The ball lands when

$y = 0$ (and $t > 0$) $\Rightarrow$ $t = \frac{v_0\sqrt{2}}{g}$ s. Now since it lands 90 m away, $90 = x = \frac{1}{2}v_0\,\sqrt{2}\,\frac{v_0\,\sqrt{2}}{g}$ or $v_0^2 = 90g$ and the initial

velocity is $v_0 = \sqrt{90g} \approx 30$ m/s.

**25.** Let $\alpha$ be the angle of elevation. Then $v_0 = 150$ m/s and from Example 5, the horizontal distance traveled by the projectile is

$d = \frac{v_0^2 \sin 2\alpha}{g}$. Thus $\frac{150^2 \sin 2\alpha}{g} = 800$ $\Rightarrow$ $\sin 2\alpha = \frac{800g}{150^2} \approx 0.3484$ $\Rightarrow$ $2\alpha \approx 20.4°$ or $180 - 20.4 = 159.6°$.

Two angles of elevation then are $\alpha \approx 10.2°$ and $\alpha \approx 79.8°$.

**27.** Place the catapult at the origin and assume the catapult is 100 meters from the city, so the city lies between $(100, 0)$

and $(600, 0)$. The initial speed is $v_0 = 80$ m/s and let $\theta$ be the angle the catapult is set at. As in Example 5, the trajectory of

the catapulted rock is given by $\mathbf{r}(t) = (80 \cos \theta)t\,\mathbf{i} + \left[(80 \sin \theta)t - 4.9t^2\right]\mathbf{j}$. The top of the near city wall is at $(100, 15)$,

which the rock will hit when $(80 \cos \theta)\,t = 100$ $\Rightarrow$ $t = \dfrac{5}{4 \cos \theta}$ and $(80 \sin \theta)t - 4.9t^2 = 15$ $\Rightarrow$

$80 \sin \theta \cdot \dfrac{5}{4 \cos \theta} - 4.9\left(\dfrac{5}{4 \cos \theta}\right)^2 = 15$ $\Rightarrow$ $100 \tan \theta - 7.65625 \sec^2 \theta = 15$. Replacing $\sec^2 \theta$ with $\tan^2 \theta + 1$ gives

$7.65625 \tan^2 \theta - 100 \tan \theta + 22.62625 = 0$. Using the quadratic formula, we have $\tan \theta \approx 0.230324,\ 12.8309$ $\Rightarrow$

$\theta \approx 13.0°, 85.5°$. So for $13.0° < \theta < 85.5°$, the rock will land beyond the near city wall. The base of the far wall is located at $(600, 0)$ which the rock hits if $(80 \cos \theta)t = 600 \quad \Rightarrow \quad t = \dfrac{15}{2 \cos \theta}$ and $(80 \sin \theta)t - 4.9t^2 = 0 \quad \Rightarrow$

$80 \sin \theta \cdot \dfrac{15}{2 \cos \theta} - 4.9 \left( \dfrac{15}{2 \cos \theta} \right)^2 = 0 \quad \Rightarrow \quad 600 \tan \theta - 275.625 \sec^2 \theta = 0 \quad \Rightarrow$

$275.625 \tan^2 \theta - 600 \tan \theta + 275.625 = 0$. Solutions are $\tan \theta \approx 0.658678,\ 1.51819 \quad \Rightarrow \quad \theta \approx 33.4°,\ 56.6°$. Thus the

rock lands beyond the enclosed city ground for $33.4° < \theta < 56.6°$, and the angles that allow the rock to land on city ground

are $13.0° < \theta < 33.4°,\ 56.6° < \theta < 85.5°$. If you consider that the rock can hit the far wall and bounce back into the city, we

calculate the angles that cause the rock to hit the top of the wall at $(600, 15)$: $(80 \cos \theta)t = 600 \quad \Rightarrow \quad t = \dfrac{15}{2 \cos \theta}$ and

$(80 \sin \theta)t - 4.9t^2 = 15 \quad \Rightarrow \quad 600 \tan \theta - 275.625 \sec^2 \theta = 15 \quad \Rightarrow \quad 275.625 \tan^2 \theta - 600 \tan \theta + 290.625 = 0$.

Solutions are $\tan \theta \approx 0.727506,\ 1.44936 \quad \Rightarrow \quad \theta \approx 36.0°,\ 55.4°$, so the catapult should be set with angle $\theta$ where

$13.0° < \theta < 36.0°,\ 55.4° < \theta < 85.5°$.

**29.** (a) After $t$ seconds, the boat will be $5t$ meters west of point $A$. The velocity

of the water at that location is $\frac{3}{400}(5t)(40 - 5t)\,\mathbf{j}$. The velocity of the

boat in still water is $5\,\mathbf{i}$, so the resultant velocity of the boat is

$\mathbf{v}(t) = 5\,\mathbf{i} + \frac{3}{400}(5t)(40 - 5t)\,\mathbf{j} = 5\mathbf{i} + \left( \frac{3}{2}t - \frac{3}{16}t^2 \right)\mathbf{j}$. Integrating, we obtain

$\mathbf{r}(t) = 5t\,\mathbf{i} + \left( \frac{3}{4}t^2 - \frac{1}{16}t^3 \right)\mathbf{j} + \mathbf{C}$. If we place the origin at $A$ (and consider $\mathbf{j}$

to coincide with the northern direction) then $\mathbf{r}(0) = \mathbf{0} \quad \Rightarrow \quad \mathbf{C} = \mathbf{0}$ and we have $\mathbf{r}(t) = 5t\,\mathbf{i} + \left( \frac{3}{4}t^2 - \frac{1}{16}t^3 \right)\mathbf{j}$. The boat

reaches the east bank after 8 s, and it is located at $\mathbf{r}(8) = 5(8)\mathbf{i} + \left( \frac{3}{4}(8)^2 - \frac{1}{16}(8)^3 \right)\mathbf{j} = 40\,\mathbf{i} + 16\,\mathbf{j}$. Thus the boat is 16 m

downstream.

(b) Let $\alpha$ be the angle north of east that the boat heads. Then the velocity of the boat in still water is given by

$5(\cos \alpha)\,\mathbf{i} + 5(\sin \alpha)\,\mathbf{j}$. At $t$ seconds, the boat is $5(\cos \alpha)t$ meters from the west bank, at which point the velocity

of the water is $\frac{3}{400}[5(\cos \alpha)t][40 - 5(\cos \alpha)t]\,\mathbf{j}$. The resultant velocity of the boat is given by

$\mathbf{v}(t) = 5(\cos \alpha)\,\mathbf{i} + \left[ 5 \sin \alpha + \frac{3}{400}(5t \cos \alpha)(40 - 5t \cos \alpha) \right]\mathbf{j} = (5 \cos \alpha)\,\mathbf{i} + \left( 5 \sin \alpha + \frac{3}{2}t \cos \alpha - \frac{3}{16}t^2 \cos^2 \alpha \right)\mathbf{j}$.

Integrating, $\mathbf{r}(t) = (5t \cos \alpha)\,\mathbf{i} + \left( 5t \sin \alpha + \frac{3}{4}t^2 \cos \alpha - \frac{1}{16}t^3 \cos^2 \alpha \right)\mathbf{j}$ (where we have again placed

the origin at $A$). The boat will reach the east bank when $5t \cos \alpha = 40 \quad \Rightarrow \quad t = \dfrac{40}{5 \cos \alpha} = \dfrac{8}{\cos \alpha}$.

In order to land at point $B(40, 0)$ we need $5t \sin \alpha + \frac{3}{4}t^2 \cos \alpha - \frac{1}{16}t^3 \cos^2 \alpha = 0 \quad \Rightarrow$

$5 \left( \dfrac{8}{\cos \alpha} \right) \sin \alpha + \frac{3}{4} \left( \dfrac{8}{\cos \alpha} \right)^2 \cos \alpha - \frac{1}{16} \left( \dfrac{8}{\cos \alpha} \right)^3 \cos^2 \alpha = 0 \quad \Rightarrow \quad \dfrac{1}{\cos \alpha}(40 \sin \alpha + 48 - 32) = 0 \quad \Rightarrow$

$40 \sin \alpha + 16 = 0 \quad \Rightarrow \quad \sin \alpha = -\frac{2}{5}$. Thus $\alpha = \sin^{-1} \left( -\frac{2}{5} \right) \approx -23.6°$, so the boat should head $23.6°$ south of

east (upstream). The path does seem realistic. The boat initially heads

upstream to counteract the effect of the current. Near the center of the river,

the current is stronger and the boat is pushed downstream. When the boat

nears the eastern bank, the current is slower and the boat is able to progress

upstream to arrive at point $B$.

**31.** If $\mathbf{r}'(t) = \mathbf{c} \times \mathbf{r}(t)$ then $\mathbf{r}'(t)$ is perpendicular to both $\mathbf{c}$ and $\mathbf{r}(t)$. Remember that $\mathbf{r}'(t)$ points in the direction of motion, so if $\mathbf{r}'(t)$ is always perpendicular to $\mathbf{c}$, the path of the particle must lie in a plane perpendicular to $\mathbf{c}$. But $\mathbf{r}'(t)$ is also perpendicular to the position vector $\mathbf{r}(t)$ which confines the path to a sphere centered at the origin. Considering both restrictions, the path must be contained in a circle that lies in a plane perpendicular to $\mathbf{c}$, and the circle is centered on a line through the origin in the direction of $\mathbf{c}$.

**33.** $\mathbf{r}(t) = t^3\,\mathbf{i} + t^2\,\mathbf{j} \;\Rightarrow\; \mathbf{r}'(t) = 3t^2\,\mathbf{i} + 2t\,\mathbf{j}, \quad |\mathbf{r}'(t)| = \sqrt{(3t^2)^2 + (2t)^2} = \sqrt{9t^4 + 4t^2},$

$\mathbf{r}''(t) = 6t\,\mathbf{i} + 2\,\mathbf{j}, \quad \mathbf{r}'(t) \times \mathbf{r}''(t) = -6t^2\,\mathbf{k}, \quad |\mathbf{r}'(t) \times \mathbf{r}''(t)| = 6t^2.$ Then Equation 9 gives

$$a_T = \frac{\mathbf{r}'(t) \cdot \mathbf{r}''(t)}{|\mathbf{r}'(t)|} = \frac{(3t^2)(6t) + (2t)(2)}{\sqrt{9t^4 + 4t^2}} = \frac{18t^3 + 4t}{\sqrt{9t^4 + 4t^2}}$$

$\left[\text{or by Equation 8, } a_T = v' = \dfrac{d}{dt}\sqrt{9t^4 + 4t^2} = (18t^3 + 4t)/\sqrt{9t^4 + 4t^2}\right]$ and Equation 10 gives

$$a_N = \frac{|\mathbf{r}'(t) \times \mathbf{r}''(t)|}{|\mathbf{r}'(t)|} = \frac{6t^2}{\sqrt{9t^4 + 4t^2}}.$$

**35.** $\mathbf{r}(t) = \cos t\,\mathbf{i} + \sin t\,\mathbf{j} + t\,\mathbf{k} \;\Rightarrow\; \mathbf{r}'(t) = -\sin t\,\mathbf{i} + \cos t\,\mathbf{j} + \mathbf{k}, \quad |\mathbf{r}'(t)| = \sqrt{\sin^2 t + \cos^2 t + 1} = \sqrt{2},$

$\mathbf{r}''(t) = -\cos t\,\mathbf{i} - \sin t\,\mathbf{j}, \quad \mathbf{r}'(t) \times \mathbf{r}''(t) = \sin t\,\mathbf{i} - \cos t\,\mathbf{j} + \mathbf{k}.$

Then $a_T = \dfrac{\mathbf{r}'(t) \cdot \mathbf{r}''(t)}{|\mathbf{r}'(t)|} = \dfrac{\sin t \cos t - \sin t \cos t}{\sqrt{2}} = 0$ and $a_N = \dfrac{|\mathbf{r}'(t) \times \mathbf{r}''(t)|}{|\mathbf{r}'(t)|} = \dfrac{\sqrt{\sin^2 t + \cos^2 t + 1}}{\sqrt{2}} = \dfrac{\sqrt{2}}{\sqrt{2}} = 1.$

**37.** The tangential component of $\mathbf{a}$ is the length of the projection of $\mathbf{a}$ onto $\mathbf{T}$, so we sketch the scalar projection of $\mathbf{a}$ in the tangential direction to the curve and estimate its length to be 4.5 (using the fact that $\mathbf{a}$ has length 10 as a guide). Similarly, the normal component of $\mathbf{a}$ is the length of the projection of $\mathbf{a}$ onto $\mathbf{N}$, so we sketch the scalar projection of $\mathbf{a}$ in the normal direction to the curve and estimate its length to be 9.0. Thus $a_T \approx 4.5$ cm/s$^2$ and $a_N \approx 9.0$ cm/s$^2$.

**39.** If the engines are turned off at time $t$, then the spacecraft will continue to travel in the direction of $\mathbf{v}(t)$, so we need a $t$ such that for some scalar $s > 0$, $\mathbf{r}(t) + s\,\mathbf{v}(t) = \langle 6, 4, 9 \rangle$. $\quad \mathbf{v}(t) = \mathbf{r}'(t) = \mathbf{i} + \dfrac{1}{t}\,\mathbf{j} + \dfrac{8t}{(t^2 + 1)^2}\,\mathbf{k} \;\Rightarrow$

$\mathbf{r}(t) + s\,\mathbf{v}(t) = \left\langle 3 + t + s, 2 + \ln t + \dfrac{s}{t}, 7 - \dfrac{4}{t^2 + 1} + \dfrac{8st}{(t^2+1)^2} \right\rangle \;\Rightarrow\; 3 + t + s = 6 \;\Rightarrow\; s = 3 - t,$

so $7 - \dfrac{4}{t^2 + 1} + \dfrac{8(3 - t)t}{(t^2 + 1)^2} = 9 \;\Leftrightarrow\; \dfrac{24t - 12t^2 - 4}{(t^2 + 1)^2} = 2 \;\Leftrightarrow\; t^4 + 8t^2 - 12t + 3 = 0.$

It is easily seen that $t = 1$ is a root of this polynomial. Also $2 + \ln 1 + \dfrac{3 - 1}{1} = 4$, so $t = 1$ is the desired solution.

## 10.5 Parametric Surfaces

**1.** $P(7, 10, 4)$ lies on the parametric surface $\mathbf{r}(u, v) = \langle 2u + 3v, 1 + 5u - v, 2 + u + v \rangle$ if and only if there are values for $u$

and $v$ where $2u + 3v = 7$, $1 + 5u - v = 10$, and $2 + u + v = 4$. But solving the first two equations simultaneously gives

$u = 2$, $v = 1$ and these values do not satisfy the third equation, so $P$ does not lie on the surface.

$Q(5, 22, 5)$ lies on the surface if $2u + 3v = 5$, $1 + 5u - v = 22$, and $2 + u + v = 5$ for some values of $u$ and $v$. Solving the

first two equations simultaneously gives $u = 4$, $v = -1$ and these values satisfy the third equation, so $Q$ lies on the surface.

**3.** $\mathbf{r}(u, v) = (u + v)\mathbf{i} + (3 - v)\mathbf{j} + (1 + 4u + 5v)\mathbf{k} = \langle 0, 3, 1 \rangle + u \langle 1, 0, 4 \rangle + v \langle 1, -1, 5 \rangle$. From Example 3, we recognize

this as a vector equation of a plane through the point $(0, 3, 1)$ and containing vectors $\mathbf{a} = \langle 1, 0, 4 \rangle$ and $\mathbf{b} = \langle 1, -1, 5 \rangle$. If we

wish to find a more conventional equation for the plane, a normal vector to the plane is $\mathbf{a} \times \mathbf{b} = \begin{vmatrix} \mathbf{i} & \mathbf{j} & \mathbf{k} \\ 1 & 0 & 4 \\ 1 & -1 & 5 \end{vmatrix} = 4\mathbf{i} - \mathbf{j} - \mathbf{k}$

and an equation of the plane is $4(x - 0) - (y - 3) - (z - 1) = 0$ or $4x - y - z = -4$.

**5.** $\mathbf{r}(s, t) = \langle s, t, t^2 - s^2 \rangle$, so the corresponding parametric equations for the surface are $x = s$, $y = t$, $z = t^2 - s^2$. For any

point $(x, y, z)$ on the surface, we have $z = y^2 - x^2$. With no restrictions on the parameters, the surface is $z = y^2 - x^2$, which

we recognize as a hyperbolic paraboloid.

**7.** $\mathbf{r}(u, v) = \langle u^2 + 1, v^3 + 1, u + v \rangle$, $-1 \le u \le 1$, $-1 \le v \le 1$.

The surface has parametric equations $x = u^2 + 1$, $y = v^3 + 1$, $z = u + v$,

$-1 \le u \le 1, -1 \le v \le 1$. In Maple, the surface can be graphed by entering

`plot3d([u^2+1,v^3+1,u+v],u=-1..1,v=-1..1);`. In

Mathematica we use the `ParametricPlot3D` command. If we keep $u$

constant at $u_0$, $x = u_0^2 + 1$, a constant, so the corresponding grid curves must

be the curves parallel to the $yz$-plane. If $v$ is constant, we have $y = v_0^3 + 1$,

a constant, so these grid curves are the curves parallel to the $xz$-plane.

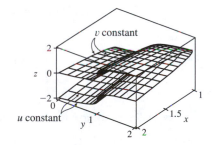

**9.** $\mathbf{r}(u, v) = \langle u \cos v, u \sin v, u^5 \rangle$.

The surface has parametric equations $x = u \cos v$, $y = u \sin v$,

$z = u^5$, $-1 \le u \le 1$, $0 \le v \le 2\pi$. Note that if $u = u_0$ is constant

then $z = u_0^5$ is constant and $x = u_0 \cos v$, $y = u_0 \sin v$ describe a

circle in $x, y$ of radius $|u_0|$, so the corresponding grid curves are

circles parallel to the $xy$-plane. If $v = v_0$, a constant, the parametric

equations become $x = u \cos v_0$, $y = u \sin v_0$, $z = u^5$. Then

$y = (\tan v_0)x$, so these are the grid curves we see that lie in vertical

planes $y = kx$ through the $z$-axis.

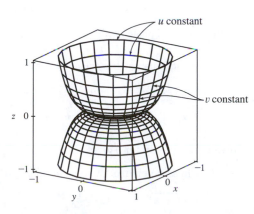

**11.** $x = \sin v,\ \ y = \cos u \sin 4v,\ \ z = \sin 2u \sin 4v,\ \ 0 \le u \le 2\pi,\ \ -\frac{\pi}{2} \le v \le \frac{\pi}{2}.$

Note that if $v = v_0$ is constant, then $x = \sin v_0$ is constant, so the corresponding grid curves must be parallel to the $yz$-plane. These are the vertically oriented grid curves we see, each shaped like a "figure-eight." When $u = u_0$ is held constant, the parametric equations become $x = \sin v,\ y = \cos u_0 \sin 4v,$ $z = \sin 2u_0 \sin 4v.$ Since $z$ is a constant multiple of $y$, the corresponding grid curves are the curves contained in planes $z = ky$ that pass through the $x$-axis.

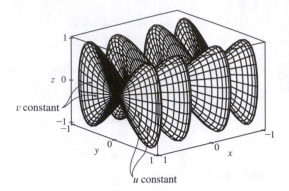

**13.** $\mathbf{r}(u, v) = u \cos v\,\mathbf{i} + u \sin v\,\mathbf{j} + v\,\mathbf{k}$. The parametric equations for the surface are $x = u \cos v,\ y = u \sin v,\ z = v$. We look at the grid curves first; if we fix $v$, then $x$ and $y$ parametrize a straight line in the plane $z = v$ which intersects the $z$-axis. If $u$ is held constant, the projection onto the $xy$-plane is circular; with $z = v$, each grid curve is a helix. The surface is a spiraling ramp, graph IV.

**15.** $\mathbf{r}(u, v) = \sin v\,\mathbf{i} + \cos u \sin 2v\,\mathbf{j} + \sin u \sin 2v\,\mathbf{k}$. Parametric equations for the surface are $x = \sin v,\ y = \cos u \sin 2v,$ $z = \sin u \sin 2v$. If $v = v_0$ is fixed, then $x = \sin v_0$ is constant, and $y = (\sin 2v_0)\cos u$ and $z = (\sin 2v_0)\sin u$ describe a circle of radius $|\sin 2v_0|$, so each corresponding grid curve is a circle contained in the vertical plane $x = \sin v_0$ parallel to the $yz$-plane. The only possible surface is graph II. The grid curves we see running lengthwise along the surface correspond to holding $u$ constant, in which case $y = (\cos u_0)\sin 2v,\ z = (\sin u_0)\sin 2v\ \Rightarrow\ z = (\tan u_0)y$, so each grid curve lies in a plane $z = ky$ that includes the $x$-axis.

**17.** $x = \cos^3 u \cos^3 v,\ y = \sin^3 u \cos^3 v,\ z = \sin^3 v$. If $v = v_0$ is held constant then $z = \sin^3 v_0$ is constant, so the corresponding grid curve lies in a horizontal plane. Several of the graphs exhibit horizontal grid curves, but the curves for this surface are neither circles nor straight lines, so graph III is the only possibility. (In fact, the horizontal grid curves here are members of the family $x = a \cos^3 u,\ y = a \sin^3 u$ and are called astroids.) The vertical grid curves we see on the surface correspond to $u = u_0$ held constant, as then we have $x = \cos^3 u_0 \cos^3 v,\ y = \sin^3 u_0 \cos^3 v$ so the corresponding grid curve lies in the vertical plane $y = (\tan^3 u_0)x$ through the $z$-axis.

**19.** From Example 3, parametric equations for the plane through the point $(1, 2, -3)$ that contains the vectors $\mathbf{a} = \langle 1, 1, -1 \rangle$ and $\mathbf{b} = \langle 1, -1, 1 \rangle$ are $x = 1 + u(1) + v(1) = 1 + u + v,\ y = 2 + u(1) + v(-1) = 2 + u - v,$ $z = -3 + u(-1) + v(1) = -3 - u + v.$

**21.** Solving the equation for $y$ gives $y^2 = 1 - x^2 + z^2\ \Rightarrow\ y = \sqrt{1 - x^2 + z^2}$. (We choose the positive root since we want the part of the hyperboloid that corresponds to $y \ge 0$.) If we let $x$ and $z$ be the parameters, parametric equations are $x = x,\ z = z,$ $y = \sqrt{1 - x^2 + z^2}.$

**23.** Since the cone intersects the sphere in the circle $x^2 + y^2 = 2$, $z = \sqrt{2}$ and we want the portion of the sphere above this, we

can parametrize the surface as $x = x$, $y = y$, $z = \sqrt{4 - x^2 - y^2}$ where $x^2 + y^2 \leq 2$.

*Alternate solution:* Using spherical coordinates, $x = 2\sin\phi\cos\theta$, $y = 2\sin\phi\sin\theta$, $z = 2\cos\phi$ where $0 \leq \phi \leq \frac{\pi}{4}$ and

$0 \leq \theta \leq 2\pi$.

**25.** Parametric equations are $x = x$, $y = 4\cos\theta$, $z = 4\sin\theta$, $0 \leq x \leq 5$, $0 \leq \theta \leq 2\pi$.

**27.** The surface appears to be a portion of a circular cylinder of radius 3 with axis the $x$-axis. An equation of the cylinder is

$y^2 + z^2 = 9$, and we can impose the restrictions $0 \leq x \leq 5$, $y \leq 0$ to obtain the portion shown. To graph the surface on a

CAS, we can use parametric equations $x = u$, $y = 3\cos v$, $z = 3\sin v$ with the parameter domain $0 \leq u \leq 5$, $\frac{\pi}{2} \leq v \leq \frac{3\pi}{2}$.

Alternatively, we can regard $x$ and $z$ as parameters. Then parametric equations are $x = x$, $z = z$, $y = -\sqrt{9 - z^2}$, where

$0 \leq x \leq 5$ and $-3 \leq z \leq 3$.

**29.** Using Equations 3, we have the parametrization $x = x$, $y = e^{-x}\cos\theta$, $z = e^{-x}\sin\theta$, $0 \leq x \leq 3$, $0 \leq \theta \leq 2\pi$.

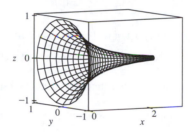

**31.** (a) $x = a\sin u\cos v$, $y = b\sin u\sin v$, $z = c\cos u$  $\Rightarrow$     (b)

$$\frac{x^2}{a^2} + \frac{y^2}{b^2} + \frac{z^2}{c^2} = (\sin u\cos v)^2 + (\sin u\sin v)^2 + (\cos u)^2$$

$$= \sin^2 u + \cos^2 u = 1$$

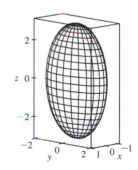

and since the ranges of $u$ and $v$ are sufficient to generate the entire graph,

the parametric equations represent an ellipsoid.

**33.** (a) Replacing $\cos u$ by $\sin u$ and $\sin u$ by $\cos u$ gives parametric equations

$x = (2 + \sin v)\sin u$, $y = (2 + \sin v)\cos u$, $z = u + \cos v$. From the graph, it

appears that the direction of the spiral is reversed. We can verify this observation by

noting that the projection of the spiral grid curves onto the $xy$-plane, given by

$x = (2 + \sin v)\sin u$, $y = (2 + \sin v)\cos u$, $z = 0$, draws a circle in the clockwise

direction for each value of $v$. The original equations, on the other hand, give circular

projections drawn in the counterclockwise direction. The equation for $z$ is identical in

both surfaces, so as $z$ increases, these grid curves spiral up in opposite directions for

the two surfaces.

(b) Replacing $\cos u$ by $\cos 2u$ and $\sin u$ by $\sin 2u$ gives parametric equations

$x = (2 + \sin v) \cos 2u$, $y = (2 + \sin v) \sin 2u$, $z = u + \cos v$. From the graph, it

appears that the number of coils in the surface doubles within the same parametric

domain. We can verify this observation by noting that the projection of the spiral grid

curves onto the $xy$-plane, given by $x = (2 + \sin v) \cos 2u$, $y = (2 + \sin v) \sin 2u$,

$z = 0$ (where $v$ is constant), complete circular revolutions for $0 \le u \le \pi$ while the

original surface requires $0 \le u \le 2\pi$ for a complete revolution. Thus, the new

surface winds around twice as fast as the original surface, and since the equation for $z$

is identical in both surfaces, we observe twice as many circular coils in the same

$z$-interval.

## 10 Review

1. A vector function is a function whose domain is a set of real numbers and whose range is a set of vectors. To find the derivative or integral, we can differentiate or integrate each component of the vector function.

2. The tip of the moving vector $\mathbf{r}(t)$ of a continuous vector function traces out a space curve.

3. The tangent vector to a smooth curve at a point $P$ with position vector $\mathbf{r}(t)$ is the vector $\mathbf{r}'(t)$. The tangent line at $P$ is the line through $P$ parallel to the tangent vector $\mathbf{r}'(t)$. The unit tangent vector is $\mathbf{T}(t) = \dfrac{\mathbf{r}'(t)}{|\mathbf{r}'(t)|}$.

4. (a)–(f) See Theorem 10.2.3.

5. Use Formula 10.3.2, or equivalently, 10.3.3.

6. (a) The curvature of a curve is $\kappa = \left| \dfrac{d\mathbf{T}}{ds} \right|$ where $\mathbf{T}$ is the unit tangent vector.

   (b) $\kappa(t) = \left| \dfrac{\mathbf{T}'(t)}{\mathbf{r}'(t)} \right|$

   (c) $\kappa(t) = \dfrac{|\mathbf{r}'(t) \times \mathbf{r}''(t)|}{|\mathbf{r}'(t)|^3}$

   (d) $\kappa(x) = \dfrac{|f''(x)|}{[1 + (f'(x))^2]^{3/2}}$

7. (a) The unit normal vector: $\mathbf{N}(t) = \dfrac{\mathbf{T}'(t)}{|\mathbf{T}'(t)|}$.  The binormal vector: $\mathbf{B}(t) = \mathbf{T}(t) \times \mathbf{N}(t)$.

   (b) See the discussion at the top of page 713.

8. (a) If $\mathbf{r}(t)$ is the position vector of the particle on the space curve, the velocity $\mathbf{v}(t) = \mathbf{r}'(t)$, the speed is given by $|\mathbf{v}(t)|$, and the acceleration $\mathbf{a}(t) = \mathbf{v}'(t) = \mathbf{r}''(t)$.

   (b) $\mathbf{a} = a_T \mathbf{T} + a_N \mathbf{N}$ where $a_T = v'$ and $a_N = \kappa v^2$.

9. See the statement of Kepler's Laws on page 722.

10. See the discussion on pages 727 and 728.

## TRUE-FALSE QUIZ

**1.** True. If we reparametrize the curve by replacing $u = t^3$, we have $\mathbf{r}(u) = u\,\mathbf{i} + 2u\,\mathbf{j} + 3u\,\mathbf{k}$, which is a line through the origin with direction vector $\mathbf{i} + 2\,\mathbf{j} + 3\,\mathbf{k}$.

**3.** False. By Formula 5 of Theorem 10.2.3, $\dfrac{d}{dt}[\mathbf{u}(t) \times \mathbf{v}(t)] = \mathbf{u}'(t) \times \mathbf{v}(t) + \mathbf{u}(t) \times \mathbf{v}'(t)$.

**5.** False. $\kappa$ is the magnitude of the rate of change of the unit tangent vector $\mathbf{T}$ with respect to arc length $s$, not with respect to $t$.

**7.** True. At an inflection point where $f$ is twice continuously differentiable we must have $f''(x) = 0$, and by Equation 10.3.11, the curvature is 0 there.

**9.** False. If $\mathbf{r}(t)$ is the position of a moving particle at time $t$ and $|\mathbf{r}(t)| = 1$ then the particle lies on the unit circle or the unit sphere, but this does not mean that the speed $|\mathbf{r}'(t)|$ must be constant. As a counterexample, let $\mathbf{r}(t) = \langle t, \sqrt{1-t^2} \rangle$, then $\mathbf{r}'(t) = \langle 1, -t/\sqrt{1-t^2} \rangle$ and $|\mathbf{r}(t)| = \sqrt{t^2 + 1 - t^2} = 1$ but $|\mathbf{r}'(t)| = \sqrt{1 + t^2/(1-t^2)} = 1/\sqrt{1-t^2}$ which is not constant.

**11.** True. See the discussion at the top of page 713.

## EXERCISES

**1.** (a) The corresponding parametric equations for the curve are $x = t$, $y = \cos \pi t$, $z = \sin \pi t$. Since $y^2 + z^2 = 1$, the curve is contained in a circular cylinder with axis the $x$-axis. Since $x = t$, the curve is a helix.

(b) $\mathbf{r}(t) = t\,\mathbf{i} + \cos \pi t\,\mathbf{j} + \sin \pi t\,\mathbf{k} \quad\Rightarrow$

$\mathbf{r}'(t) = \mathbf{i} - \pi \sin \pi t\,\mathbf{j} + \pi \cos \pi t\,\mathbf{k} \quad\Rightarrow$

$\mathbf{r}''(t) = -\pi^2 \cos \pi t\,\mathbf{j} - \pi^2 \sin \pi t\,\mathbf{k}$

**3.** The projection of the curve $C$ of intersection onto the $xy$-plane is the circle $x^2 + y^2 = 16$, $z = 0$. So we can write $x = 4\cos t$, $y = 4\sin t$, $0 \le t \le 2\pi$. From the equation of the plane, we have $z = 5 - x = 5 - 4\cos t$, so parametric equations for $C$ are $x = 4\cos t$, $y = 4\sin t$, $z = 5 - 4\cos t$, $0 \le t \le 2\pi$, and the corresponding vector function is $\mathbf{r}(t) = 4\cos t\,\mathbf{i} + 4\sin t\,\mathbf{j} + (5 - 4\cos t)\,\mathbf{k}$, $0 \le t \le 2\pi$.

**5.** $\int_0^1 (t^2\,\mathbf{i} + t\cos \pi t\,\mathbf{j} + \sin \pi t\,\mathbf{k})\,dt = \left(\int_0^1 t^2\,dt\right)\mathbf{i} + \left(\int_0^1 t\cos \pi t\,dt\right)\mathbf{j} + \left(\int_0^1 \sin \pi t\,dt\right)\mathbf{k}$

$\qquad = \left[\frac{1}{3}t^3\right]_0^1 \mathbf{i} + \left(\frac{t}{\pi}\sin \pi t\right]_0^1 - \int_0^1 \frac{1}{\pi}\sin \pi t\,dt\right)\mathbf{j} + \left[-\frac{1}{\pi}\cos \pi t\right]_0^1 \mathbf{k}$

$\qquad = \frac{1}{3}\mathbf{i} + \left[\frac{1}{\pi^2}\cos \pi t\right]_0^1 \mathbf{j} + \frac{2}{\pi}\mathbf{k} = \frac{1}{3}\mathbf{i} - \frac{2}{\pi^2}\mathbf{j} + \frac{2}{\pi}\mathbf{k}$

where we integrated by parts in the $y$-component.

**7.** $\mathbf{r}(t) = \langle t^2, t^3, t^4 \rangle \implies \mathbf{r}'(t) = \langle 2t, 3t^2, 4t^3 \rangle \implies |\mathbf{r}'(t)| = \sqrt{4t^2 + 9t^4 + 16t^6}$ and

$L = \int_0^3 |\mathbf{r}'(t)|\, dt = \int_0^3 \sqrt{4t^2 + 9t^4 + 16t^6}\, dt$. Using Simpson's Rule with $f(t) = \sqrt{4t^2 + 9t^4 + 16t^6}$ and $n = 6$ we

have $\Delta t = \frac{3-0}{6} = \frac{1}{2}$ and

$$L \approx \frac{\Delta t}{3}\left[f(0) + 4f\left(\tfrac{1}{2}\right) + 2f(1) + 4f\left(\tfrac{3}{2}\right) + 2f(2) + 4f\left(\tfrac{5}{2}\right) + f(3)\right]$$

$$= \frac{1}{6}\left[\sqrt{0 + 0 + 0} + 4 \cdot \sqrt{4\left(\tfrac{1}{2}\right)^2 + 9\left(\tfrac{1}{2}\right)^4 + 16\left(\tfrac{1}{2}\right)^6} + 2 \cdot \sqrt{4(1)^2 + 9(1)^4 + 16(1)^6}\right.$$

$$+ 4 \cdot \sqrt{4\left(\tfrac{3}{2}\right)^2 + 9\left(\tfrac{3}{2}\right)^4 + 16\left(\tfrac{3}{2}\right)^6} + 2 \cdot \sqrt{4(2)^2 + 9(2)^4 + 16(2)^6}$$

$$\left. + 4 \cdot \sqrt{4\left(\tfrac{5}{2}\right)^2 + 9\left(\tfrac{5}{2}\right)^4 + 16\left(\tfrac{5}{2}\right)^6} + \sqrt{4(3)^2 + 9(3)^4 + 16(3)^6}\right]$$

$$\approx 86.631$$

**9.** The angle of intersection of the two curves, $\theta$, is the angle between their respective tangents at the point of intersection.

For both curves the point $(1, 0, 0)$ occurs when $t = 0$.

$\mathbf{r}_1'(t) = -\sin t\,\mathbf{i} + \cos t\,\mathbf{j} + \mathbf{k} \implies \mathbf{r}_1'(0) = \mathbf{j} + \mathbf{k}$ and $\mathbf{r}_2'(t) = \mathbf{i} + 2t\,\mathbf{j} + 3t^2\,\mathbf{k} \implies \mathbf{r}_2'(0) = \mathbf{i}$.

$\mathbf{r}_1'(0) \cdot \mathbf{r}_2'(0) = (\mathbf{j} + \mathbf{k}) \cdot \mathbf{i} = 0$.   Therefore, the curves intersect in a right angle, that is, $\theta = \frac{\pi}{2}$.

**11.** (a) $\mathbf{T}(t) = \dfrac{\mathbf{r}'(t)}{|\mathbf{r}'(t)|} = \dfrac{\langle t^2, t, 1 \rangle}{|\langle t^2, t, 1 \rangle|} = \dfrac{\langle t^2, t, 1 \rangle}{\sqrt{t^4 + t^2 + 1}}$

(b) $\mathbf{T}'(t) = -\frac{1}{2}(t^4 + t^2 + 1)^{-3/2}(4t^3 + 2t)\langle t^2, t, 1 \rangle + (t^4 + t^2 + 1)^{-1/2}\langle 2t, 1, 0 \rangle$

$$= \frac{-2t^3 - t}{(t^4 + t^2 + 1)^{3/2}}\langle t^2, t, 1 \rangle + \frac{1}{(t^4 + t^2 + 1)^{1/2}}\langle 2t, 1, 0 \rangle$$

$$= \frac{\langle -2t^5 - t^3, -2t^4 - t^2, -2t^3 - t \rangle + \langle 2t^5 + 2t^3 + 2t, t^4 + t^2 + 1, 0 \rangle}{(t^4 + t^2 + 1)^{3/2}} = \frac{\langle 2t, -t^4 + 1, -2t^3 - t \rangle}{(t^4 + t^2 + 1)^{3/2}}$$

$|\mathbf{T}'(t)| = \dfrac{\sqrt{4t^2 + t^8 - 2t^4 + 1 + 4t^6 + 4t^4 + t^2}}{(t^4 + t^2 + 1)^{3/2}} = \dfrac{\sqrt{t^8 + 4t^6 + 2t^4 + 5t^2}}{(t^4 + t^2 + 1)^{3/2}}$   and   $\mathbf{N}(t) = \dfrac{\langle 2t, 1 - t^4, -2t^3 - t \rangle}{\sqrt{t^8 + 4t^6 + 2t^4 + 5t^2}}$.

(c) $\kappa(t) = \dfrac{|\mathbf{T}'(t)|}{|\mathbf{r}'(t)|} = \dfrac{\sqrt{t^8 + 4t^6 + 2t^4 + 5t^2}}{(t^4 + t^2 + 1)^2}$

**13.** $y' = 4x^3$, $y'' = 12x^2$ and $\kappa(x) = \dfrac{|y''|}{[1 + (y')^2]^{3/2}} = \dfrac{|12x^2|}{(1 + 16x^6)^{3/2}}$, so $\kappa(1) = \dfrac{12}{17^{3/2}}$.

**15.** $\mathbf{r}(t) = \langle \sin 2t, t, \cos 2t \rangle \implies \mathbf{r}'(t) = \langle 2\cos 2t, 1, -2\sin 2t \rangle \implies \mathbf{T}(t) = \frac{1}{\sqrt{5}}\langle 2\cos 2t, 1, -2\sin 2t \rangle \implies$

$\mathbf{T}'(t) = \frac{1}{\sqrt{5}}\langle -4\sin 2t, 0, -4\cos 2t \rangle \implies \mathbf{N}(t) = \langle -\sin 2t, 0, -\cos 2t \rangle$. So $\mathbf{N} = \mathbf{N}(\pi) = \langle 0, 0, -1 \rangle$ and

$\mathbf{B} = \mathbf{T} \times \mathbf{N} = \frac{1}{\sqrt{5}}\langle -1, 2, 0 \rangle$. So a normal to the osculating plane is $\langle -1, 2, 0 \rangle$ and an equation is

$-1(x - 0) + 2(y - \pi) + 0(z - 1) = 0$ or $x - 2y + 2\pi = 0$.

**17.** $\mathbf{r}(t) = t\ln t\,\mathbf{i} + t\,\mathbf{j} + e^{-t}\,\mathbf{k}$,   $\mathbf{v}(t) = \mathbf{r}'(t) = (1 + \ln t)\,\mathbf{i} + \mathbf{j} - e^{-t}\,\mathbf{k}$,

$|\mathbf{v}(t)| = \sqrt{(1 + \ln t)^2 + 1^2 + (-e^{-t})^2} = \sqrt{2 + 2\ln t + (\ln t)^2 + e^{-2t}}$,   $\mathbf{a}(t) = \mathbf{v}'(t) = \frac{1}{t}\mathbf{i} + e^{-t}\,\mathbf{k}$

**19.** We set up the axes so that the shot leaves the athlete's hand 7 ft above the origin. Then we are given $\mathbf{r}(0) = 7\mathbf{j}$,

$|\mathbf{v}(0)| = 43$ ft/s, and $\mathbf{v}(0)$ has direction given by a $45°$ angle of elevation. Then a unit vector in the direction of $\mathbf{v}(0)$ is

$\frac{1}{\sqrt{2}}(\mathbf{i} + \mathbf{j}) \quad \Rightarrow \quad \mathbf{v}(0) = \frac{43}{\sqrt{2}}(\mathbf{i} + \mathbf{j})$. Assuming air resistance is negligible, the only external force is due to gravity, so as in

Example 10.4.5 we have $\mathbf{a} = -g\mathbf{j}$ where here $g \approx 32$ ft/s². Since $\mathbf{v}'(t) = \mathbf{a}(t)$, we integrate, giving $\mathbf{v}(t) = -gt\mathbf{j} + \mathbf{C}$

where $\mathbf{C} = \mathbf{v}(0) = \frac{43}{\sqrt{2}}(\mathbf{i} + \mathbf{j}) \quad \Rightarrow \quad \mathbf{v}(t) = \frac{43}{\sqrt{2}}\mathbf{i} + \left(\frac{43}{\sqrt{2}} - gt\right)\mathbf{j}$. Since $\mathbf{r}'(t) = \mathbf{v}(t)$ we integrate again, so

$\mathbf{r}(t) = \frac{43}{\sqrt{2}}t\mathbf{i} + \left(\frac{43}{\sqrt{2}}t - \frac{1}{2}gt^2\right)\mathbf{j} + \mathbf{D}$. But $\mathbf{D} = \mathbf{r}(0) = 7\mathbf{j} \quad \Rightarrow \quad \mathbf{r}(t) = \frac{43}{\sqrt{2}}t\mathbf{i} + \left(\frac{43}{\sqrt{2}}t - \frac{1}{2}gt^2 + 7\right)\mathbf{j}$.

(a) At 2 seconds, the shot is at $\mathbf{r}(2) = \frac{43}{\sqrt{2}}(2)\mathbf{i} + \left(\frac{43}{\sqrt{2}}(2) - \frac{1}{2}g(2)^2 + 7\right)\mathbf{j} \approx 60.8\,\mathbf{i} + 3.8\,\mathbf{j}$, so the shot is about 3.8 ft above

the ground, at a horizontal distance of 60.8 ft from the athlete.

(b) The shot reaches its maximum height when the vertical component of velocity is 0: $\frac{43}{\sqrt{2}} - gt = 0 \quad \Rightarrow$

$t = \dfrac{43}{\sqrt{2}\,g} \approx 0.95$ s. Then $\mathbf{r}(0.95) \approx 28.9\,\mathbf{i} + 21.4\,\mathbf{j}$, so the maximum height is approximately 21.4 ft.

(c) The shot hits the ground when the vertical component of $\mathbf{r}(t)$ is 0, so $\frac{43}{\sqrt{2}}t - \frac{1}{2}gt^2 + 7 = 0 \quad \Rightarrow$

$-16t^2 + \frac{43}{\sqrt{2}}t + 7 = 0 \quad \Rightarrow \quad t \approx 2.11$ s. $\quad \mathbf{r}(2.11) \approx 64.2\,\mathbf{i} - 0.08\,\mathbf{j}$, thus the shot lands approximately 64.2 ft from the

athlete.

**21.** From Example 4 in Section 10.5, a parametric representation of the sphere $x^2 + y^2 + z^2 = 4$ is $x = 2\sin\phi\cos\theta$,

$y = 2\sin\phi\sin\theta$, $z = 2\cos\phi$ with $0 \leq \theta \leq 2\pi$ and $0 \leq \phi \leq \pi$. We can restrict the surface to that portion between the planes

$z = 1$ and $z = -1$ by restricting $-1 \leq z \leq 1 \quad \Rightarrow \quad -1 \leq 2\cos\phi \leq 1 \quad \Rightarrow \quad \frac{\pi}{3} \leq \phi \leq \frac{2\pi}{3}$.

**23.** By the Fundamental Theorem of Calculus, $\mathbf{r}'(t) = \left\langle \sin\left(\frac{1}{2}\pi t^2\right), \cos\left(\frac{1}{2}\pi t^2\right) \right\rangle$, $|\mathbf{r}'(t)| = 1$ and so $\mathbf{T}(t) = \mathbf{r}'(t)$.

Thus $\mathbf{T}'(t) = \pi t \left\langle \cos\left(\frac{1}{2}\pi t^2\right), -\sin\left(\frac{1}{2}\pi t^2\right) \right\rangle$ and the curvature is $\kappa = |\mathbf{T}'(t)| = \sqrt{(\pi t)^2(1)} = \pi\,|t|$.

# FOCUS ON PROBLEM SOLVING

1. (a) $\mathbf{r}(t) = R\cos\omega t\,\mathbf{i} + R\sin\omega t\,\mathbf{j} \;\Rightarrow\; \mathbf{v} = \mathbf{r}'(t) = -\omega R\sin\omega t\,\mathbf{i} + \omega R\cos\omega t\,\mathbf{j}$, so $\mathbf{r} = R(\cos\omega t\,\mathbf{i} + \sin\omega t\,\mathbf{j})$ and $\mathbf{v} = \omega R(-\sin\omega t\,\mathbf{i} + \cos\omega t\,\mathbf{j})$. $\mathbf{v}\cdot\mathbf{r} = \omega R^2(-\cos\omega t\sin\omega t + \sin\omega t\cos\omega t) = 0$, so $\mathbf{v}\perp\mathbf{r}$. Since $\mathbf{r}$ points along a radius of the circle, and $\mathbf{v}\perp\mathbf{r}$, $\mathbf{v}$ is tangent to the circle. Because it is a velocity vector, $\mathbf{v}$ points in the direction of motion.

(b) In (a), we wrote $\mathbf{v}$ in the form $\omega R\,\mathbf{u}$, where $\mathbf{u}$ is the unit vector $-\sin\omega t\,\mathbf{i} + \cos\omega t\,\mathbf{j}$. Clearly $|\mathbf{v}| = \omega R\,|\mathbf{u}| = \omega R$. At speed $\omega R$, the particle completes one revolution, a distance $2\pi R$, in time $T = \dfrac{2\pi R}{\omega R} = \dfrac{2\pi}{\omega}$.

(c) $\mathbf{a} = \dfrac{d\mathbf{v}}{dt} = -\omega^2 R\cos\omega t\,\mathbf{i} - \omega^2 R\sin\omega t\,\mathbf{j} = -\omega^2 R(\cos\omega t\,\mathbf{i} + \sin\omega t\,\mathbf{j})$, so $\mathbf{a} = -\omega^2\mathbf{r}$. This shows that $\mathbf{a}$ is proportional to $\mathbf{r}$ and points in the opposite direction (toward the origin). Also, $|\mathbf{a}| = \omega^2\,|\mathbf{r}| = \omega^2 R$.

(d) By Newton's Second Law (see Section 10.4), $\mathbf{F} = m\mathbf{a}$, so $|\mathbf{F}| = m\,|\mathbf{a}| = mR\omega^2 = \dfrac{m\,(\omega R)^2}{R} = \dfrac{m\,|\mathbf{v}|^2}{R}$.

3. (a) The projectile reaches maximum height when $0 = \dfrac{dy}{dt} = \dfrac{d}{dt}\left[(v_0\sin\alpha)t - \tfrac{1}{2}gt^2\right] = v_0\sin\alpha - gt$; that is, when

$t = \dfrac{v_0\sin\alpha}{g}$ and $y = (v_0\sin\alpha)\left(\dfrac{v_0\sin\alpha}{g}\right) - \dfrac{1}{2}g\left(\dfrac{v_0\sin\alpha}{g}\right)^2 = \dfrac{v_0^2\sin^2\alpha}{2g}$. This is the maximum height attained when the projectile is fired with an angle of elevation $\alpha$. This maximum height is largest when $\alpha = \dfrac{\pi}{2}$. In that case, $\sin\alpha = 1$ and the maximum height is $\dfrac{v_0^2}{2g}$.

(b) Let $R = v_0^2/g$. We are asked to consider the parabola $x^2 + 2Ry - R^2 = 0$ which can be rewritten as $y = -\dfrac{1}{2R}x^2 + \dfrac{R}{2}$. The points on or inside this parabola are those for which $-R \le x \le R$ and $0 \le y \le \dfrac{-1}{2R}x^2 + \dfrac{R}{2}$. When the projectile is fired at angle of elevation $\alpha$, the points $(x, y)$ along its path satisfy the relations $x = (v_0\cos\alpha)t$ and $y = (v_0\sin\alpha)t - \tfrac{1}{2}gt^2$, where $0 \le t \le (2v_0\sin\alpha)/g$ (as in Example 10.4.5). Thus

$|x| \le \left|v_0\cos\alpha\left(\dfrac{2v_0\sin\alpha}{g}\right)\right| = \left|\dfrac{v_0^2}{g}\sin 2\alpha\right| \le \left|\dfrac{v_0^2}{g}\right| = |R|$. This shows that $-R \le x \le R$.

For $t$ in the specified range, we also have $y = t\left(v_0\sin\alpha - \tfrac{1}{2}gt\right) = \tfrac{1}{2}gt\left(\dfrac{2v_0\sin\alpha}{g} - t\right) \ge 0$ and

$y = (v_0\sin\alpha)\dfrac{x}{v_0\cos\alpha} - \dfrac{g}{2}\left(\dfrac{x}{v_0\cos\alpha}\right)^2 = (\tan\alpha)x - \dfrac{g}{2v_0^2\cos^2\alpha}x^2 = -\dfrac{1}{2R\cos^2\alpha}x^2 + (\tan\alpha)x$. Thus

$$y - \left(\dfrac{-1}{2R}x^2 + \dfrac{R}{2}\right) = \dfrac{-1}{2R\cos^2\alpha}x^2 + \dfrac{1}{2R}x^2 + (\tan\alpha)x - \dfrac{R}{2}$$

$$= \dfrac{x^2}{2R}\left(1 - \dfrac{1}{\cos^2\alpha}\right) + (\tan\alpha)x - \dfrac{R}{2} = \dfrac{x^2(1 - \sec^2\alpha) + 2R(\tan\alpha)x - R^2}{2R}$$

$$= \dfrac{-(\tan^2\alpha)x^2 + 2R(\tan\alpha)x - R^2}{2R} = \dfrac{-[(\tan\alpha)x - R]^2}{2R} \le 0$$

We have shown that every target that can be hit by the projectile lies on or inside the parabola $y = -\dfrac{1}{2R}x^2 + \dfrac{R}{2}$.

Now let $(a, b)$ be any point on or inside the parabola $y = -\dfrac{1}{2R}x^2 + \dfrac{R}{2}$. Then $-R \le a \le R$ and $0 \le b \le -\dfrac{1}{2R}a^2 + \dfrac{R}{2}$.

We seek an angle $\alpha$ such that $(a, b)$ lies in the path of the projectile; that is, we wish to find an angle $\alpha$ such that

$b = -\dfrac{1}{2R\cos^2 \alpha}a^2 + (\tan \alpha)\,a$ or equivalently $b = \dfrac{-1}{2R}(\tan^2 \alpha + 1)a^2 + (\tan \alpha)\,a$. Rearranging this equation we get

$\dfrac{a^2}{2R}\tan^2 \alpha - a\tan \alpha + \left(\dfrac{a^2}{2R} + b\right) = 0$ or $a^2(\tan \alpha)^2 - 2aR(\tan \alpha) + (a^2 + 2bR) = 0 \ (*)$. This quadratic equation

for $\tan \alpha$ has real solutions exactly when the discriminant is nonnegative. Now $B^2 - 4AC \ge 0 \ \Leftrightarrow$

$(-2aR)^2 - 4a^2(a^2 + 2bR) \ge 0 \ \Leftrightarrow \ 4a^2(R^2 - a^2 - 2bR) \ge 0 \ \Leftrightarrow \ -a^2 - 2bR + R^2 \ge 0 \ \Leftrightarrow$

$b \le \dfrac{1}{2R}(R^2 - a^2) \ \Leftrightarrow \ b \le \dfrac{-1}{2R}a^2 + \dfrac{R}{2}$. This condition is satisfied since $(a, b)$ is on or inside the parabola

$y = -\dfrac{1}{2R}x^2 + \dfrac{R}{2}$. It follows that $(a, b)$ lies in the path of the projectile when $\tan \alpha$ satisfies $(*)$, that is, when

$\tan \alpha = \dfrac{2aR \pm \sqrt{4a^2(R^2 - a^2 - 2bR)}}{2a^2} = \dfrac{R \pm \sqrt{R^2 - 2bR - a^2}}{a}$.

(c)

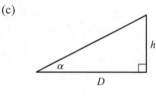

If the gun is pointed at a target with height $h$ at a distance $D$ downrange, then

$\tan \alpha = h/D$. When the projectile reaches a distance $D$ downrange (remember we are assuming that it doesn't hit the ground first), we have $D = x = (v_0 \cos \alpha)t$,

so $t = \dfrac{D}{v_0 \cos \alpha}$ and $y = (v_0 \sin \alpha)t - \tfrac{1}{2}gt^2 = D\tan \alpha - \dfrac{gD^2}{2v_0^2 \cos^2 \alpha}$.

Meanwhile, the target, whose $x$-coordinate is also $D$, has fallen from height $h$ to height

$h - \tfrac{1}{2}gt^2 = D\tan \alpha - \dfrac{gD^2}{2v_0^2 \cos^2 \alpha}$. Thus the projectile hits the target.

5. (a) $\mathbf{a} = -g\mathbf{j} \ \Rightarrow \ \mathbf{v} = \mathbf{v_0} - gt\mathbf{j} = 2\mathbf{i} - gt\mathbf{j} \ \Rightarrow \ \mathbf{s} = \mathbf{s_0} + 2t\mathbf{i} - \tfrac{1}{2}gt^2\mathbf{j} = 3.5\mathbf{j} + 2t\mathbf{i} - \tfrac{1}{2}gt^2\mathbf{j} \ \Rightarrow$

$\mathbf{s} = 2t\mathbf{i} + \left(3.5 - \tfrac{1}{2}gt^2\right)\mathbf{j}$. Therefore $y = 0$ when $t = \sqrt{7/g}$ seconds. At that instant, the ball is $2\sqrt{7/g} \approx 0.94$ ft to the

right of the table top. Its coordinates (relative to an origin on the floor directly under the table's edge) are $(0.94, 0)$. At

impact, the velocity is $\mathbf{v} = 2\mathbf{i} - \sqrt{7g}\,\mathbf{j}$, so the speed is $|\mathbf{v}| = \sqrt{4 + 7g} \approx 15$ ft/s.

(b) The slope of the curve when $t = \sqrt{\dfrac{7}{g}}$ is $\dfrac{dy}{dx} = \dfrac{dy/dt}{dx/dt} = \dfrac{-gt}{2} = \dfrac{-g\sqrt{7/g}}{2} = \dfrac{-\sqrt{7g}}{2}$. Thus $\cot \theta = \dfrac{\sqrt{7g}}{2}$

and $\theta \approx 7.6°$.

(c) From (a), $|\mathbf{v}| = \sqrt{4 + 7g}$. So the ball rebounds with speed $0.8\sqrt{4 + 7g} \approx 12.08$ ft/s at angle of inclination

$90° - \theta \approx 82.3886°$. By Example 10.4.5, the horizontal distance traveled between bounces is $d = \dfrac{v_0^2 \sin 2\alpha}{g}$, where

$v_0 \approx 12.08$ ft/s and $\alpha \approx 82.3886°$. Therefore, $d \approx 1.197$ ft. So the ball strikes the floor at about

$2\sqrt{7/g} + 1.197 \approx 2.13$ ft to the right of the table's edge.

**7.** The trajectory of the projectile is given by $\mathbf{r}(t) = (v\cos\alpha)t\,\mathbf{i} + \left[(v\sin\alpha)t - \frac{1}{2}gt^2\right]\mathbf{j}$, so

$$\mathbf{v}(t) = \mathbf{r}'(t) = v\cos\alpha\,\mathbf{i} + (v\sin\alpha - gt)\,\mathbf{j} \text{ and}$$

$$|\mathbf{v}(t)| = \sqrt{(v\cos\alpha)^2 + (v\sin\alpha - gt)^2} = \sqrt{v^2 - (2vg\sin\alpha)\,t + g^2 t^2} = \sqrt{g^2\left(t^2 - \frac{2v}{g}(\sin\alpha)\,t + \frac{v^2}{g^2}\right)}$$

$$= g\sqrt{\left(t - \frac{v}{g}\sin\alpha\right)^2 + \frac{v^2}{g^2} - \frac{v^2}{g^2}\sin^2\alpha} = g\sqrt{\left(t - \frac{v}{g}\sin\alpha\right)^2 + \frac{v^2}{g^2}\cos^2\alpha}$$

The projectile hits the ground when $(v\sin\alpha)t - \frac{1}{2}gt^2 = 0 \;\Rightarrow\; t = \frac{2v}{g}\sin\alpha$, so the distance traveled by the projectile is

$$L(\alpha) = \int_0^{(2v/g)\sin\alpha} |\mathbf{v}(t)|\,dt = \int_0^{(2v/g)\sin\alpha} g\sqrt{\left(t - \frac{v}{g}\sin\alpha\right)^2 + \frac{v^2}{g^2}\cos^2\alpha}\,dt$$

$$= g\left[\frac{t - (v/g)\sin\alpha}{2}\sqrt{\left(t - \frac{v}{g}\sin\alpha\right)^2 + \left(\frac{v}{g}\cos\alpha\right)^2}\right.$$

$$\left. + \frac{[(v/g)\cos\alpha]^2}{2}\ln\left(t - \frac{v}{g}\sin\alpha + \sqrt{\left(t - \frac{v}{g}\sin\alpha\right)^2 + \left(\frac{v}{g}\cos\alpha\right)^2}\right)\right]_0^{(2v/g)\sin\alpha}$$

[using Formula 21 in the Table of Integrals]

$$= \frac{g}{2}\left[\frac{v}{g}\sin\alpha\sqrt{\left(\frac{v}{g}\sin\alpha\right)^2 + \left(\frac{v}{g}\cos\alpha\right)^2} + \left(\frac{v}{g}\cos\alpha\right)^2\ln\left(\frac{v}{g}\sin\alpha + \sqrt{\left(\frac{v}{g}\sin\alpha\right)^2 + \left(\frac{v}{g}\cos\alpha\right)^2}\right)\right.$$

$$\left. + \frac{v}{g}\sin\alpha\sqrt{\left(\frac{v}{g}\sin\alpha\right)^2 + \left(\frac{v}{g}\cos\alpha\right)^2} - \left(\frac{v}{g}\cos\alpha\right)^2\ln\left(-\frac{v}{g}\sin\alpha + \sqrt{\left(\frac{v}{g}\sin\alpha\right)^2 + \left(\frac{v}{g}\cos\alpha\right)^2}\right)\right]$$

$$= \frac{g}{2}\left[\frac{v}{g}\sin\alpha\cdot\frac{v}{g} + \frac{v^2}{g^2}\cos^2\alpha\ln\left(\frac{v}{g}\sin\alpha + \frac{v}{g}\right) + \frac{v}{g}\sin\alpha\cdot\frac{v}{g} - \frac{v^2}{g^2}\cos^2\alpha\ln\left(-\frac{v}{g}\sin\alpha + \frac{v}{g}\right)\right]$$

$$= \frac{v^2}{g}\sin\alpha + \frac{v^2}{2g}\cos^2\alpha\ln\left(\frac{(v/g)\sin\alpha + v/g}{-(v/g)\sin\alpha + v/g}\right) = \frac{v^2}{g}\sin\alpha + \frac{v^2}{2g}\cos^2\alpha\ln\left(\frac{1 + \sin\alpha}{1 - \sin\alpha}\right)$$

We want to maximize $L(\alpha)$ for $0 \le \alpha \le \pi/2$.

$$L'(\alpha) = \frac{v^2}{g}\cos\alpha + \frac{v^2}{2g}\left[\cos^2\alpha\cdot\frac{1 - \sin\alpha}{1 + \sin\alpha}\cdot\frac{2\cos\alpha}{(1 - \sin\alpha)^2} - 2\cos\alpha\sin\alpha\ln\left(\frac{1 + \sin\alpha}{1 - \sin\alpha}\right)\right]$$

$$= \frac{v^2}{g}\cos\alpha + \frac{v^2}{2g}\left[\cos^2\alpha\cdot\frac{2}{\cos\alpha} - 2\cos\alpha\sin\alpha\ln\left(\frac{1 + \sin\alpha}{1 - \sin\alpha}\right)\right]$$

$$= \frac{v^2}{g}\cos\alpha + \frac{v^2}{g}\cos\alpha\left[1 - \sin\alpha\ln\left(\frac{1 + \sin\alpha}{1 - \sin\alpha}\right)\right] = \frac{v^2}{g}\cos\alpha\left[2 - \sin\alpha\ln\left(\frac{1 + \sin\alpha}{1 - \sin\alpha}\right)\right]$$

$L(\alpha)$ has critical points for $0 < \alpha < \pi/2$ when $L'(\alpha) = 0 \;\Rightarrow\; 2 - \sin\alpha\ln\left(\frac{1 + \sin\alpha}{1 - \sin\alpha}\right) = 0$ [since $\cos\alpha \ne 0$].

Solving by graphing (or using a CAS) gives $\alpha \approx 0.9855$. Compare values at the critical point and the endpoints:

$L(0) = 0$, $L(\pi/2) = v^2/g$, and $L(0.9855) \approx 1.20v^2/g$. Thus the distance traveled by the projectile is maximized

for $\alpha \approx 0.9855$  or  $\approx 56°$.

**9.** We can write the vector equation as $\mathbf{r}_1(t) = \mathbf{a}t^2 + \mathbf{b}t + \mathbf{c}$ where $\mathbf{a} = \langle a_1, a_2, a_3 \rangle$, $\mathbf{b} = \langle b_1, b_2, b_3 \rangle$, and $\mathbf{c} = \langle c_1, c_2, c_3 \rangle$.

We know from Example 3 in Section 10.5 that $\mathbf{r}_2(u, v) = \mathbf{r}_0 + u\mathbf{a} + v\mathbf{b}$ is the vector equation of a plane, and

$\mathbf{r}_1(t) = \mathbf{r}_2(t^2, t)$ when $\mathbf{r}_0 = \mathbf{c}$, so the curve lies in the plane $\mathbf{r}(u, v) = \mathbf{c} + u\mathbf{a} + v\mathbf{b}$. [Here we assume that $\mathbf{a}$ and $\mathbf{b}$ are

nonparallel. Otherwise, $\mathbf{r}_1(t)$ and $\mathbf{r}(u, v)$ are identical lines which lie in many planes.] We can find an equation in terms

of $x$, $y$, $z$ for the plane by observing that $\mathbf{a}$ and $\mathbf{b}$ are parallel to the plane (and not to each other), so a normal vector for the

plane is $\mathbf{a} \times \mathbf{b} = \langle a_2 b_3 - a_3 b_2, a_3 b_1 - a_1 b_3, a_1 b_2 - a_2 b_1 \rangle$. The point $(c_1, c_2, c_3)$ lies on the plane, so an equation of the

plane is

$$(a_2 b_3 - a_3 b_2)(x - c_1) + (a_3 b_1 - a_1 b_3)(y - c_2) + (a_1 b_2 - a_2 b_1)(z - c_3) = 0$$

or

$$(a_2 b_3 - a_3 b_2)x + (a_3 b_1 - a_1 b_3)y + (a_1 b_2 - a_2 b_1)z = a_2 b_3 c_1 - a_3 b_2 c_1 + a_3 b_1 c_2 - a_1 b_3 c_2 + a_1 b_2 c_3 - a_2 b_1 c_3$$

*Note:* Another approach is to show that the osculating plane at any point on the curve is the same and so the entire curve must

lie in that plane.

# 11 ☐ PARTIAL DERIVATIVES

## 11.1 Functions of Several Variables

1. (a) From Table 1, $f(-15, 40) = -27$, which means that if the temperature is $-15°$C and the wind speed is 40 km/h, then the air would feel equivalent to approximately $-27°$C without wind.

   (b) The question is asking: when the temperature is $-20°$C, what wind speed gives a wind-chill index of $-30°$C? From Table 1, the speed is 20 km/h.

   (c) The question is asking: when the wind speed is 20 km/h, what temperature gives a wind-chill index of $-49°$C? From Table 1, the temperature is $-35°$C.

   (d) The function $W = f(-5, v)$ means that we fix $T$ at $-5$ and allow $v$ to vary, resulting in a function of one variable. In other words, the function gives wind-chill index values for different wind speeds when the temperature is $-5°$C. From Table 1 (look at the row corresponding to $T = -5$), the function decreases and appears to approach a constant value as $v$ increases.

   (e) The function $W = f(T, 50)$ means that we fix $v$ at 50 and allow $T$ to vary, again giving a function of one variable. In other words, the function gives wind-chill index values for different temperatures when the wind speed is 50 km/h . From Table 1 (look at the column corresponding to $v = 50$), the function increases almost linearly as $T$ increases.

3. If the amounts of labor and capital are both doubled, we replace $L, K$ in the function with $2L, 2K$, giving

$$P(2L, 2K) = 1.01(2L)^{0.75}(2K)^{0.25} = 1.01(2^{0.75})(2^{0.25})L^{0.75}K^{0.25} = (2^1)1.01L^{0.75}K^{0.25} = 2P(L, K)$$

Thus, the production is doubled. It is also true for the general case $P(L, K) = bL^\alpha K^{1-\alpha}$:

$$P(2L, 2K) = b(2L)^\alpha(2K)^{1-\alpha} = b(2^\alpha)(2^{1-\alpha})L^\alpha K^{1-\alpha} = (2^{\alpha+1-\alpha})bL^\alpha K^{1-\alpha} = 2P(L, K).$$

5. $\ln(9 - x^2 - 9y^2)$ is defined only when $9 - x^2 - 9y^2 > 0$, or $\frac{1}{9}x^2 + y^2 < 1$. So the domain of $f$ is $\left\{(x, y) \mid \frac{1}{9}x^2 + y^2 < 1\right\}$, the interior of an ellipse.

   For this domain, the range of $g(x, y) = 9 - x^2 - 9y^2$ is $(0, 9]$, so the range of $f(x, y) = \ln(g(x, y))$ is $(-\infty, \ln 9]$.

7. (a) $f(1, 1, 1) = \sqrt{1} + \sqrt{1} + \sqrt{1} + \ln(4 - 1^2 - 1^2 - 1^2) = 3 + \ln 1 = 3$

(b) $\sqrt{x}$, $\sqrt{y}$, $\sqrt{z}$ are defined only when $x \geq 0$, $y \geq 0$, $z \geq 0$, and $\ln(4 - x^2 - y^2 - z^2)$ is defined when

$4 - x^2 - y^2 - z^2 > 0 \quad \Leftrightarrow \quad x^2 + y^2 + z^2 < 4$, thus the domain is

$\left\{ (x, y, z) \mid x^2 + y^2 + z^2 < 4,\ x \geq 0,\ y \geq 0,\ z \geq 0 \right\}$, the portion of the interior of a sphere of radius 2, centered at the

origin, that is in the first octant.

**9.** The point $(-3, 3)$ lies between the level curves with $z$-values 50 and 60. Since the point is a little closer to the level curve with

$z = 60$, we estimate that $f(-3, 3) \approx 56$. The point $(3, -2)$ appears to be just about halfway between the level curves with

$z$-values 30 and 40, so we estimate $f(3, -2) \approx 35$. The graph rises as we approach the origin, gradually from above, steeply

from below.

**11.** The point $(160, 10)$, corresponding to day 160 and a depth of 10 m, lies between the isothermals with temperature values

of 8 and $12°$C. Since the point appears to be located about three-fourths the distance from the $8°$C isothermal to the $12°$C

isothermal, we estimate the temperature at that point to be approximately $11°$C. The point $(180, 5)$ lies between the 16 and

$20°$C isothermals, very close to the $20°$C level curve, so we estimate the temperature there to be about $19.5°$C.

**13.** Near $A$, the level curves are very close together, indicating that the terrain is quite steep. At $B$, the level curves are much

farther apart, so we would expect the terrain to be much less steep than near $A$, perhaps almost flat.

**15.**

**17.**

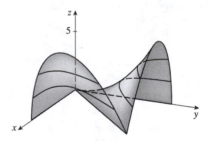

**19.** The level curves are $(y - 2x)^2 = k$ or $y = 2x \pm \sqrt{k}$,

$k \geq 0$, a family of pairs of parallel lines.

**21.** The level curves are $\sqrt{x} + y = k$ or $y = -\sqrt{x} + k$, a

family of vertical translations of the graph of the root

function $y = -\sqrt{x}$.

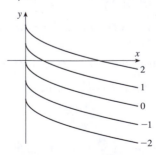

**23.** The level curves are $ye^x = k$ or $y = ke^{-x}$, a family of exponential curves.

**25.** The level curves are $\sqrt{y^2 - x^2} = k$ or $y^2 - x^2 = k^2$, $k \geq 0$. When $k = 0$ the level curve is the pair of lines $y = \pm x$. For $k > 0$, the level curves are hyperbolas with axis the $y$-axis.

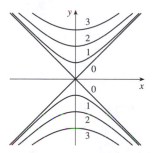

**27.** The contour map consists of the level curves $k = x^2 + 9y^2$, a family of ellipses with major axis the $x$-axis. (Or, if $k = 0$, the origin.)

The graph of $f(x, y)$ is the surface $z = x^2 + 9y^2$, an elliptic paraboloid.

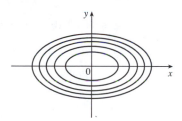

If we visualize lifting each ellipse $k = x^2 + 9y^2$ of the contour map to the plane $z = k$, we have horizontal traces that indicate the shape of the graph of $f$.

**29.** The isothermals are given by $k = 100/(1 + x^2 + 2y^2)$ or $x^2 + 2y^2 = (100 - k)/k$ $[0 < k \leq 100]$, a family of ellipses.

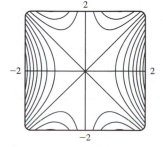

**31.** $f(x, y) = xy^2 - x^3$

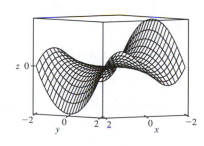

The traces parallel to the $yz$-plane (such as the left-front trace in the graph above) are parabolas; those parallel to the $xz$-plane (such as the right-front trace) are cubic curves. The surface is called a monkey saddle because a monkey sitting on the surface near the origin has places for both legs and tail to rest.

**33.** $f(x, y) = e^{-(x^2+y^2)/3}\left(\sin(x^2) + \cos(y^2)\right)$

**35.** $z = \sin(xy)$     (a) C     (b) II

Reasons: This function is periodic in both $x$ and $y$, and the function is the same when $x$ is interchanged with $y$, so its graph is symmetric about the plane $y = x$. In addition, the function is 0 along the $x$- and $y$-axes. These conditions are satisfied only by C and II.

**37.** $z = \sin(x - y)$     (a) F     (b) I

Reasons: This function is periodic in both $x$ and $y$ but is constant along the lines $y = x + k$, a condition satisfied only by F and I.

**39.** $z = (1 - x^2)(1 - y^2)$     (a) B     (b) VI

Reasons: This function is 0 along the lines $x = \pm 1$ and $y = \pm 1$. The only contour map in which this could occur is VI. Also note that the trace in the $xz$-plane is the parabola $z = 1 - x^2$ and the trace in the $yz$-plane is the parabola $z = 1 - y^2$, so the graph is B.

**41.** $k = x + 3y + 5z$ is a family of parallel planes with normal vector $\langle 1, 3, 5 \rangle$.

**43.** Equations for the level surfaces are $k = y^2 + z^2$. For $k > 0$, we have a family of circular cylinders with axis the $x$-axis and radius $\sqrt{k}$. When $k = 0$ the level surface is the $x$-axis. (There are no level surfaces for $k < 0$.)

**45.** (a) The graph of $g$ is the graph of $f$ shifted upward 2 units.

(b) The graph of $g$ is the graph of $f$ stretched vertically by a factor of 2.

(c) The graph of $g$ is the graph of $f$ reflected about the $xy$-plane.

(d) The graph of $g(x, y) = -f(x, y) + 2$ is the graph of $f$ reflected about the $xy$-plane and then shifted upward 2 units.

**47.** $f(x, y) = e^{cx^2+y^2}$. First, if $c = 0$, the graph is the cylindrical surface

$z = e^{y^2}$ (whose level curves are parallel lines). When $c > 0$, the vertical trace above the $y$-axis remains fixed while the sides of the surface in the $x$-direction "curl" upward, giving the graph a shape resembling an elliptic paraboloid. The level curves of the surface are ellipses centered at the origin.

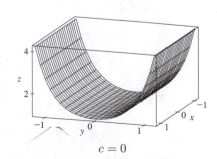

$c = 0$

For $0 < c < 1$, the ellipses have major axis the $x$-axis and the eccentricity increases as $c \to 0$.

$c = 0.5$ (level curves in increments of 1)

For $c = 1$ the level curves are circles centered at the origin.

$c = 1$ (level curves in increments of 1)

When $c > 1$, the level curves are ellipses with major axis the $y$-axis, and the eccentricity increases as $c$ increases.

$c = 2$ (level curves in increments of 4)

For values of $c < 0$, the sides of the surface in the $x$-direction curl downward and approach the $xy$-plane (while the vertical trace $x = 0$ remains fixed), giving a saddle-shaped appearance to the graph near the point $(0, 0, 1)$. The level curves consist of a family of hyperbolas. As $c$ decreases, the surface becomes flatter in the $x$-direction and the surface's approach to the curve in the trace $x = 0$ becomes steeper, as the graphs demonstrate.

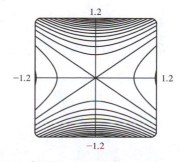

$c = -0.5$ (level curves in increments of 0.25)

$c = -2$ (level curves in increments of 0.25)

**49.** (a) $P = bL^\alpha K^{1-\alpha}$ $\Rightarrow$ $\dfrac{P}{K} = bL^\alpha K^{-\alpha}$ $\Rightarrow$ $\dfrac{P}{K} = b\left(\dfrac{L}{K}\right)^\alpha$ $\Rightarrow$ $\ln\dfrac{P}{K} = \ln\left(b\left(\dfrac{L}{K}\right)^\alpha\right)$ $\Rightarrow$

$\ln\dfrac{P}{K} = \ln b + \alpha \ln\left(\dfrac{L}{K}\right)$

(b) We list the values for $\ln(L/K)$ and $\ln(P/K)$ for the years 1899–1922. (Historically, these values were rounded to

2 decimal places.)

| Year | $x = \ln(L/K)$ | $y = \ln(P/K)$ | | Year | $x = \ln(L/K)$ | $y = \ln(P/K)$ |
|------|------|------|---|------|------|------|
| 1899 | 0 | 0 | | 1911 | −0.38 | −0.34 |
| 1900 | −0.02 | −0.06 | | 1912 | −0.38 | −0.24 |
| 1901 | −0.04 | −0.02 | | 1913 | −0.41 | −0.25 |
| 1902 | −0.04 | 0 | | 1914 | −0.47 | −0.37 |
| 1903 | −0.07 | −0.05 | | 1915 | −0.53 | −0.34 |
| 1904 | −0.13 | −0.12 | | 1916 | −0.49 | −0.28 |
| 1905 | −0.18 | −0.04 | | 1917 | −0.53 | −0.39 |
| 1906 | −0.20 | −0.07 | | 1918 | −0.60 | −0.50 |
| 1907 | −0.23 | −0.15 | | 1919 | −0.68 | −0.57 |
| 1908 | −0.41 | −0.38 | | 1920 | −0.74 | −0.57 |
| 1909 | −0.33 | −0.24 | | 1921 | −1.05 | −0.85 |
| 1910 | −0.35 | −0.27 | | 1922 | −0.98 | −0.59 |

After entering the $(x, y)$ pairs into a calculator or CAS, the resulting least squares regression line through the points is

approximately $y = 0.75136x + 0.01053$, which we round to $y = 0.75x + 0.01$.

(c) Comparing the regression line from part (b) to the equation $y = \ln b + \alpha x$ with $x = \ln(L/K)$ and $y = \ln(P/K)$, we have

$\alpha = 0.75$ and $\ln b = 0.01$ $\Rightarrow$ $b = e^{0.01} \approx 1.01$. Thus, the Cobb-Douglas production function is

$P = bL^\alpha K^{1-\alpha} = 1.01 L^{0.75} K^{0.25}$.

## 11.2   Limits and Continuity

1. In general, we can't say anything about $f(3,1)$! $\lim\limits_{(x,y)\to(3,1)} f(x,y) = 6$ means that the values of $f(x,y)$ approach 6 as

$(x,y)$ approaches, but is not equal to, $(3,1)$. If $f$ is continuous, we know that $\lim\limits_{(x,y)\to(a,b)} f(x,y) = f(a,b)$, so

$\lim\limits_{(x,y)\to(3,1)} f(x,y) = f(3,1) = 6$.

3. We make a table of values of

$$f(x,y) = \frac{x^2y^3 + x^3y^2 - 5}{2 - xy} \text{ for a set}$$

of $(x,y)$ points near the origin.

| y \ x | −0.2 | −0.1 | −0.05 | 0 | 0.05 | 0.1 | 0.2 |
|---|---|---|---|---|---|---|---|
| −0.2 | −2.551 | −2.525 | −2.513 | −2.500 | −2.488 | −2.475 | −2.451 |
| −0.1 | −2.525 | −2.513 | −2.506 | −2.500 | −2.494 | −2.488 | −2.475 |
| −0.05 | −2.513 | −2.506 | −2.503 | −2.500 | −2.497 | −2.494 | −2.488 |
| 0 | −2.500 | −2.500 | −2.500 | | −2.500 | −2.500 | −2.500 |
| 0.05 | −2.488 | −2.494 | −2.497 | −2.500 | −2.503 | −2.506 | −2.513 |
| 0.1 | −2.475 | −2.488 | −2.494 | −2.500 | −2.506 | −2.513 | −2.525 |
| 0.2 | −2.451 | −2.475 | −2.488 | −2.500 | −2.513 | −2.525 | −2.551 |

As the table shows, the values of $f(x,y)$ seem to approach $-2.5$ as $(x,y)$ approaches the origin from a variety of different

directions. This suggests that $\lim\limits_{(x,y)\to(0,0)} f(x,y) = -2.5$. Since $f$ is a rational function, it is continuous on its domain. $f$ is

defined at $(0,0)$, so we can use direct substitution to establish that $\lim\limits_{(x,y)\to(0,0)} f(x,y) = \frac{0^2 0^3 + 0^3 0^2 - 5}{2 - 0\cdot 0} = -\frac{5}{2}$, verifying

our guess.

5. $f(x,y) = 5x^3 - x^2y^2$ is a polynomial, and hence continuous, so $\lim\limits_{(x,y)\to(1,2)} f(x,y) = f(1,2) = 5(1)^3 - (1)^2(2)^2 = 1$.

7. $f(x,y) = y^4/(x^4 + 3y^4)$.   First approach $(0,0)$ along the $x$-axis. Then $f(x,0) = 0/x^4 = 0$ for $x \neq 0$, so $f(x,y) \to 0$.

Now approach $(0,0)$ along the $y$-axis. Then for $y \neq 0$, $f(0,y) = y^4/3y^4 = 1/3$, so $f(x,y) \to 1/3$. Since $f$ has two different

limits along two different lines, the limit does not exist.

9. $f(x,y) = (xy\cos y)/(3x^2 + y^2)$.   On the $x$-axis, $f(x,0) = 0$ for $x \neq 0$, so $f(x,y) \to 0$ as $(x,y) \to (0,0)$ along the

$x$-axis. Approaching $(0,0)$ along the line $y = x$, $f(x,x) = (x^2\cos x)/4x^2 = \frac{1}{4}\cos x$ for $x \neq 0$, so $f(x,y) \to \frac{1}{4}$ along this

line. Thus the limit does not exist.

11. $f(x,y) = \dfrac{xy}{\sqrt{x^2 + y^2}}$.   We can see that the limit along any line through $(0,0)$ is 0, as well as along other paths through

$(0,0)$ such as $x = y^2$ and $y = x^2$. So we suspect that the limit exists and equals 0; we use the Squeeze Theorem to prove our

assertion. $0 \leq \left| \dfrac{xy}{\sqrt{x^2 + y^2}} \right| \leq |x|$ since $|y| \leq \sqrt{x^2 + y^2}$, and $|x| \to 0$ as $(x,y) \to (0,0)$. So $\lim\limits_{(x,y)\to(0,0)} f(x,y) = 0$.

**13.** Let $f(x,y) = \dfrac{x^2 y e^y}{x^4 + 4y^2}$. Then $f(x,0) = 0$ for $x \neq 0$, so $f(x,y) \to 0$ as $(x,y) \to (0,0)$ along the $x$-axis. Approaching

$(0,0)$ along the $y$-axis or the line $y = x$ also gives a limit of $0$. But $f\left(x, x^2\right) = \dfrac{x^2 x^2 \, e^{x^2}}{x^4 + 4(x^2)^2} = \dfrac{x^4 e^{x^2}}{5x^4} = \dfrac{e^{x^2}}{5}$ for $x \neq 0$, so

$f(x,y) \to e^0/5 = \frac{1}{5}$ as $(x,y) \to (0,0)$ along the parabola $y = x^2$. Thus the limit doesn't exist.

**15.** $\displaystyle\lim_{(x,y)\to(0,0)} \frac{x^2 + y^2}{\sqrt{x^2 + y^2 + 1} - 1} = \lim_{(x,y)\to(0,0)} \frac{x^2 + y^2}{\sqrt{x^2 + y^2 + 1} - 1} \cdot \frac{\sqrt{x^2 + y^2 + 1} + 1}{\sqrt{x^2 + y^2 + 1} + 1}$

$$= \lim_{(x,y)\to(0,0)} \frac{\left(x^2 + y^2\right)\left(\sqrt{x^2 + y^2 + 1} + 1\right)}{x^2 + y^2} = \lim_{(x,y)\to(0,0)} \left(\sqrt{x^2 + y^2 + 1} + 1\right) = 2$$

**17.** $e^{-xy}$ and $\sin(\pi z/2)$ are each compositions of continuous functions, and hence continuous, so their product

$f(x,y,z) = e^{-xy} \sin(\pi z/2)$ is a continuous function. Then

$$\lim_{(x,y,z)\to(3,0,1)} f(x,y,z) = f(3,0,1) = e^{-(3)(0)} \sin(\pi \cdot 1/2) = 1.$$

**19.** $f(x,y,z) = \dfrac{xy + yz^2 + xz^2}{x^2 + y^2 + z^4}$. Then $f(x,0,0) = 0/x^2 = 0$ for $x \neq 0$, so as $(x,y,z) \to (0,0,0)$ along the $x$-axis,

$f(x,y,z) \to 0$. But $f(x,x,0) = x^2/(2x^2) = \frac{1}{2}$ for $x \neq 0$, so as $(x,y,z) \to (0,0,0)$ along the line $y = x$, $z = 0$,

$f(x,y,z) \to \frac{1}{2}$. Thus the limit doesn't exist.

**21.**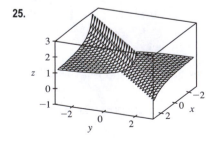
From the ridges on the graph, we see that as $(x,y) \to (0,0)$ along the
lines under the two ridges, $f(x,y)$ approaches different values. So the
limit does not exist.

**23.** $h(x,y) = g(f(x,y)) = (2x + 3y - 6)^2 + \sqrt{2x + 3y - 6}$. Since $f$ is a polynomial, it is continuous on $\mathbb{R}^2$ and $g$ is

continuous on its domain $\{t \mid t \geq 0\}$. Thus $h$ is continuous on its domain.

$D = \{(x,y) \mid 2x + 3y - 6 \geq 0\} = \{(x,y) \mid y \geq -\frac{2}{3}x + 2\}$, which consists of all points on or above the line $y = -\frac{2}{3}x + 2$.

**25.**
From the graph, it appears that $f$ is discontinuous along the line $y = x$.

If we consider $f(x,y) = e^{1/(x-y)}$ as a composition of functions,

$g(x,y) = 1/(x-y)$ is a rational function and therefore continuous except

where $x - y = 0 \;\Rightarrow\; y = x$. Since the function $h(t) = e^t$ is continuous

everywhere, the composition $h(g(x,y)) = e^{1/(x-y)} = f(x,y)$ is

continuous except along the line $y = x$, as we suspected.

**27.** $F(x,y) = \arctan\left(x + \sqrt{y}\right) = g(f(x,y))$ where $f(x,y) = x + \sqrt{y}$, continuous on its domain $\{(x,y) \mid y \geq 0\}$, and

$g(t) = \arctan t$ is continuous everywhere. Thus $F$ is continuous on its domain $\{(x,y) \mid y \geq 0\}$.

**29.** $G(x,y) = \ln(x^2 + y^2 - 4) = g(f(x,y))$ where $f(x,y) = x^2 + y^2 - 4$, continuous on $\mathbb{R}^2$, and $g(t) = \ln t$, continuous on its

domain $\{t \mid t > 0\}$. Thus $G$ is continuous on its domain $\{(x,y) \mid x^2 + y^2 - 4 > 0\} = \{(x,y) \mid x^2 + y^2 > 4\}$, the exterior

of the circle $x^2 + y^2 = 4$.

**31.** $\sqrt{y}$ is continuous on its domain $\{y \mid y \geq 0\}$ and $x^2 - y^2 + z^2$ is continuous everywhere, so $f(x,y,z) = \dfrac{\sqrt{y}}{x^2 - y^2 + z^2}$ is

continuous for $y \geq 0$ and $x^2 - y^2 + z^2 \neq 0 \;\Rightarrow\; y^2 \neq x^2 + z^2$, that is, $\{(x,y,z) \mid y \geq 0, y \neq \sqrt{x^2 + z^2}\,\}$.

**33.** $f(x,y) = \begin{cases} \dfrac{x^2 y^3}{2x^2 + y^2} & \text{if } (x,y) \neq (0,0) \\ 1 & \text{if } (x,y) = (0,0) \end{cases}$   The first piece of $f$ is a rational function defined everywhere except at the

origin, so $f$ is continuous on $\mathbb{R}^2$ except possibly at the origin. Since $x^2 \leq 2x^2 + y^2$, we have $\left|x^2 y^3/(2x^2 + y^2)\right| \leq |y^3|$. We

know that $|y^3| \to 0$ as $(x,y) \to (0,0)$. So, by the Squeeze Theorem, $\displaystyle\lim_{(x,y)\to(0,0)} f(x,y) = \lim_{(x,y)\to(0,0)} \dfrac{x^2 y^3}{2x^2 + y^2} = 0$.

But $f(0,0) = 1$, so $f$ is discontinuous at $(0,0)$. Therefore, $f$ is continuous on the set $\{(x,y) \mid (x,y) \neq (0,0)\}$.

**35.** $\displaystyle\lim_{(x,y)\to(0,0)} \frac{x^3 + y^3}{x^2 + y^2} = \lim_{r\to 0^+} \frac{(r\cos\theta)^3 + (r\sin\theta)^3}{r^2} = \lim_{r\to 0^+} (r\cos^3\theta + r\sin^3\theta) = 0$

**37.** $\displaystyle\lim_{(x,y,z)\to(0,0,0)} \frac{xyz}{x^2 + y^2 + z^2} = \lim_{\rho\to 0^+} \frac{(\rho\sin\phi\cos\theta)(\rho\sin\phi\sin\theta)(\rho\cos\phi)}{\rho^2} = \lim_{\rho\to 0^+} (\rho\sin^2\phi\cos\phi\sin\theta\cos\theta) \doteq 0$

**39.** $f(x,y) = \begin{cases} \dfrac{\sin(xy)}{xy} & \text{if } (x,y) \neq (0,0) \\ 1 & \text{if } (x,y) = (0,0) \end{cases}$

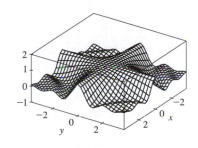

From the graph, it appears that $f$ is continuous everywhere. We know

$xy$ is continuous on $\mathbb{R}^2$ and $\sin t$ is continuous everywhere, so

$\sin(xy)$ is continuous on $\mathbb{R}^2$ and $\dfrac{\sin(xy)}{xy}$ is continuous on $\mathbb{R}^2$

except possibly where $xy = 0$. To show that $f$ is continuous at those points, consider any point $(a,b)$ in $\mathbb{R}^2$ where $ab = 0$.

Because $xy$ is continuous, $xy \to ab = 0$ as $(x,y) \to (a,b)$. If we let $t = xy$, then $t \to 0$ as $(x,y) \to (a,b)$ and

$\displaystyle\lim_{(x,y)\to(a,b)} \frac{\sin(xy)}{xy} = \lim_{t\to 0} \frac{\sin(t)}{t} = 1$ by Equation 3.3.2. Thus $\displaystyle\lim_{(x,y)\to(a,b)} f(x,y) = f(a,b)$ and $f$ is continuous on $\mathbb{R}^2$.

## 11.3   Partial Derivatives

1. (a) $\partial T/\partial x$ represents the rate of change of $T$ when we fix $y$ and $t$ and consider $T$ as a function of the single variable $x$, which

   describes how quickly the temperature changes when longitude changes but latitude and time are constant. $\partial T/\partial y$

   represents the rate of change of $T$ when we fix $x$ and $t$ and consider $T$ as a function of $y$, which describes how quickly the

   temperature changes when latitude changes but longitude and time are constant. $\partial T/\partial t$ represents the rate of change of $T$

   when we fix $x$ and $y$ and consider $T$ as a function of $t$, which describes how quickly the temperature changes over time for

   a constant longitude and latitude.

   (b) $f_x(158, 21, 9)$ represents the rate of change of temperature at longitude $158°$W, latitude $21°$N at 9:00 AM when only

   longitude varies. Since the air is warmer to the west than to the east, increasing longitude results in an increased air

   temperature, so we would expect $f_x(158, 21, 9)$ to be positive. $f_y(158, 21, 9)$ represents the rate of change of temperature

   at the same time and location when only latitude varies. Since the air is warmer to the south and cooler to the north,

   increasing latitude results in a decreased air temperature, so we would expect $f_y(158, 21, 9)$ to be negative. $f_t(158, 21, 9)$

   represents the rate of change of temperature at the same time and location when only time varies. Since typically air

   temperature increases from the morning to the afternoon as the sun warms it, we would expect $f_t(158, 21, 9)$ to be

   positive.

3. (a) By Definition 4, $f_T(-15, 30) = \lim\limits_{h \to 0} \dfrac{f(-15 + h, 30) - f(-15, 30)}{h}$, which we can approximate by considering $h = 5$

   and $h = -5$ and using the values given in the table:

   $$f_T(-15, 30) \approx \frac{f(-10, 30) - f(-15, 30)}{5} = \frac{-20 - (-26)}{5} = \frac{6}{5} = 1.2,$$

   $$f_T(-15, 30) \approx \frac{f(-20, 30) - f(-15, 30)}{-5} = \frac{-33 - (-26)}{-5} = \frac{-7}{-5} = 1.4.\ \text{Averaging these values, we estimate}$$

   $f_T(-15, 30)$ to be approximately 1.3. Thus, when the actual temperature is $-15°$C and the wind speed is 30 km/h, the

   apparent temperature rises by about $1.3°$C for every degree that the actual temperature rises.

   Similarly, $f_v(-15, 30) = \lim\limits_{h \to 0} \dfrac{f(-15, 30 + h) - f(-15, 30)}{h}$ which we can approximate by considering $h = 10$

   and $h = -10$: $f_v(-15, 30) \approx \dfrac{f(-15, 40) - f(-15, 30)}{10} = \dfrac{-27 - (-26)}{10} = \dfrac{-1}{10} = -0.1,$

   $$f_v(-15, 30) \approx \frac{f(-15, 20) - f(-15, 30)}{-10} = \frac{-24 - (-26)}{-10} = \frac{2}{-10} = -0.2.\ \text{Averaging these values, we estimate}$$

   $f_v(-15, 30)$ to be approximately $-0.15$. Thus, when the actual temperature is $-15°$C and the wind speed is 30 km/h, the

   apparent temperature decreases by about $0.15°$C for every km/h that the wind speed increases.

   (b) For a fixed wind speed $v$, the values of the wind-chill index $W$ increase as temperature $T$ increases (look at a column of

   the table), so $\dfrac{\partial W}{\partial T}$ is positive. For a fixed temperature $T$, the values of $W$ decrease (or remain constant) as $v$ increases

   (look at a row of the table), so $\dfrac{\partial W}{\partial v}$ is negative (or perhaps 0).

   (c) For fixed values of $T$, the function values $f(T, v)$ appear to become constant (or nearly constant) as $v$ increases, so the

   corresponding rate of change is 0 or near 0 as $v$ increases. This suggests that $\lim\limits_{v \to \infty} (\partial W/\partial v) = 0$.

**5.** (a) If we start at $(1, 2)$ and move in the positive $x$-direction, the graph of $f$ increases. Thus $f_x(1, 2)$ is positive.

(b) If we start at $(1, 2)$ and move in the positive $y$-direction, the graph of $f$ decreases. Thus $f_y(1, 2)$ is negative.

**7.** (a) $f_{xx} = \frac{\partial}{\partial x}(f_x)$, so $f_{xx}$ is the rate of change of $f_x$ in the $x$-direction. $f_x$ is negative at $(-1, 2)$ and if we move in the positive $x$-direction, the surface becomes less steep. Thus the values of $f_x$ are increasing and $f_{xx}(-1, 2)$ is positive.

(b) $f_{yy}$ is the rate of change of $f_y$ in the $y$-direction. $f_y$ is negative at $(-1, 2)$ and if we move in the positive $y$-direction, the surface becomes steeper. Thus the values of $f_y$ are decreasing, and $f_{yy}(-1, 2)$ is negative.

**9.** First of all, if we start at the point $(3, -3)$ and move in the positive $y$-direction, we see that both $b$ and $c$ decrease, while $a$ increases. Both $b$ and $c$ have a low point at about $(3, -1.5)$, while $a$ is 0 at this point. So $a$ is definitely the graph of $f_y$, and one of $b$ and $c$ is the graph of $f$. To see which is which, we start at the point $(-3, -1.5)$ and move in the positive $x$-direction. $b$ traces out a line with negative slope, while $c$ traces out a parabola opening downward. This tells us that $b$ is the $x$-derivative of $c$. So $c$ is the graph of $f$, $b$ is the graph of $f_x$, and $a$ is the graph of $f_y$.

**11.** $f(x, y) = 16 - 4x^2 - y^2 \Rightarrow f_x(x, y) = -8x$ and $f_y(x, y) = -2y \Rightarrow f_x(1, 2) = -8$ and $f_y(1, 2) = -4$. The graph of $f$ is the paraboloid $z = 16 - 4x^2 - y^2$ and the vertical plane $y = 2$ intersects it in the parabola $z = 12 - 4x^2$, $y = 2$ (the curve $C_1$ in the first figure). The slope of the tangent line to this parabola at $(1, 2, 8)$ is $f_x(1, 2) = -8$. Similarly the plane $x = 1$ intersects the paraboloid in the parabola $z = 12 - y^2$, $x = 1$ (the curve $C_2$ in the second figure) and the slope of the tangent line at $(1, 2, 8)$ is $f_y(1, 2) = -4$.

**13.** $f(x, y) = x^2 + y^2 + x^2y \Rightarrow f_x = 2x + 2xy, \quad f_y = 2y + x^2$

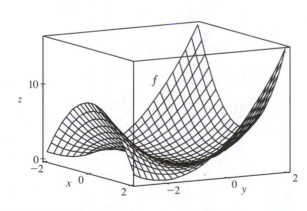

Note that the traces of $f$ in planes parallel to the $xz$-plane are parabolas which open downward for $y < -1$ and upward for $y > -1$, and the traces of $f_x$ in these planes are straight lines, which have negative slopes for $y < -1$ and positive slopes for $y > -1$. The traces of $f$ in planes parallel to the $yz$-plane are parabolas which always open upward, and the traces of $f_y$ in these planes are straight lines with positive slopes.

**15.** $f(x, y) = y^5 - 3xy \Rightarrow f_x(x, y) = 0 - 3y = -3y, f_y(x, y) = 5y^4 - 3x$

**17.** $f(x, t) = e^{-t} \cos \pi x \Rightarrow f_x(x, t) = e^{-t}(-\sin \pi x)(\pi) = -\pi e^{-t} \sin \pi x, f_t(x, t) = e^{-t}(-1) \cos \pi x = -e^{-t} \cos \pi x$

**19.** $z = (2x + 3y)^{10} \Rightarrow \dfrac{\partial z}{\partial x} = 10(2x + 3y)^9 \cdot 2 = 20(2x + 3y)^9, \dfrac{\partial z}{\partial y} = 10(2x + 3y)^9 \cdot 3 = 30(2x + 3y)^9$

**21.** $f(x, y) = \dfrac{x - y}{x + y} \Rightarrow f_x(x, y) = \dfrac{(1)(x + y) - (x - y)(1)}{(x + y)^2} = \dfrac{2y}{(x + y)^2},$

$f_y(x, y) = \dfrac{(-1)(x + y) - (x - y)(1)}{(x + y)^2} = -\dfrac{2x}{(x + y)^2}$

**23.** $w = \sin \alpha \cos \beta \Rightarrow \dfrac{\partial w}{\partial \alpha} = \cos \alpha \cos \beta, \dfrac{\partial w}{\partial \beta} = -\sin \alpha \sin \beta$

**25.** $f(r, s) = r \ln(r^2 + s^2) \Rightarrow f_r(r, s) = r \cdot \dfrac{2r}{r^2 + s^2} + \ln(r^2 + s^2) \cdot 1 = \dfrac{2r^2}{r^2 + s^2} + \ln(r^2 + s^2),$

$f_s(r, s) = r \cdot \dfrac{2s}{r^2 + s^2} + 0 = \dfrac{2rs}{r^2 + s^2}$

**27.** $u = te^{w/t} \Rightarrow \dfrac{\partial u}{\partial t} = t \cdot e^{w/t}(-wt^{-2}) + e^{w/t} \cdot 1 = e^{w/t} - \dfrac{w}{t}e^{w/t} = e^{w/t}\left(1 - \dfrac{w}{t}\right), \dfrac{\partial u}{\partial w} = te^{w/t} \cdot \dfrac{1}{t} = e^{w/t}$

**29.** $f(x, y, z) = xz - 5x^2y^3z^4 \Rightarrow f_x(x, y, z) = z - 10xy^3z^4, f_y(x, y, z) = -15x^2y^2z^4, f_z(x, y, z) = x - 20x^2y^3z^3$

**31.** $w = \ln(x + 2y + 3z) \Rightarrow \dfrac{\partial w}{\partial x} = \dfrac{1}{x + 2y + 3z}, \dfrac{\partial w}{\partial y} = \dfrac{2}{x + 2y + 3z}, \dfrac{\partial w}{\partial z} = \dfrac{3}{x + 2y + 3z}$

**33.** $u = xy \sin^{-1}(yz) \Rightarrow \dfrac{\partial u}{\partial x} = y \sin^{-1}(yz), \dfrac{\partial u}{\partial y} = xy \cdot \dfrac{1}{\sqrt{1 - (yz)^2}}(z) + \sin^{-1}(yz) \cdot x = \dfrac{xyz}{\sqrt{1 - y^2z^2}} + x \sin^{-1}(yz),$

$\dfrac{\partial u}{\partial z} = xy \cdot \dfrac{1}{\sqrt{1 - (yz)^2}}(y) = \dfrac{xy^2}{\sqrt{1 - y^2z^2}}$

**35.** $f(x, y, z, t) = xyz^2 \tan(yt) \Rightarrow f_x(x, y, z, t) = yz^2 \tan(yt),$

$f_y(x, y, z, t) = xyz^2 \cdot \sec^2(yt) \cdot t + xz^2 \tan(yt) = xyz^2t \sec^2(yt) + xz^2 \tan(yt),$

$f_z(x, y, z, t) = 2xyz \tan(yt), f_t(x, y, z, t) = xyz^2 \sec^2(yt) \cdot y = xy^2z^2 \sec^2(yt)$

**37.** $u = \sqrt{x_1^2 + x_2^2 + \cdots + x_n^2}.$ For each $i = 1, \ldots, n, u_{x_i} = \frac{1}{2}\left(x_1^2 + x_2^2 + \cdots + x_n^2\right)^{-1/2}(2x_i) = \dfrac{x_i}{\sqrt{x_1^2 + x_2^2 + \cdots + x_n^2}}.$

**39.** $f(x, y) = \ln\left(x + \sqrt{x^2 + y^2}\right) \Rightarrow$

$f_x(x, y) = \dfrac{1}{x + \sqrt{x^2 + y^2}}\left[1 + \frac{1}{2}(x^2 + y^2)^{-1/2}(2x)\right] = \dfrac{1}{x + \sqrt{x^2 + y^2}}\left(1 + \dfrac{x}{\sqrt{x^2 + y^2}}\right),$

so $f_x(3, 4) = \dfrac{1}{3 + \sqrt{3^2 + 4^2}}\left(1 + \dfrac{3}{\sqrt{3^2 + 4^2}}\right) = \frac{1}{8}\left(1 + \frac{3}{5}\right) = \frac{1}{5}.$

**41.** $f(x, y, z) = \dfrac{y}{x + y + z}$ $\Rightarrow$ $f_y(x, y, z) = \dfrac{1(x + y + z) - y(1)}{(x + y + z)^2} = \dfrac{x + z}{(x + y + z)^2}$,

so $f_y(2, 1, -1) = \dfrac{2 + (-1)}{(2 + 1 + (-1))^2} = \dfrac{1}{4}$.

**43.** $f(x, y) = xy^2 - x^3 y$ $\Rightarrow$

$$f_x(x, y) = \lim_{h \to 0} \frac{f(x + h, y) - f(x, y)}{h} = \lim_{h \to 0} \frac{(x + h)y^2 - (x + h)^3 y - (xy^2 - x^3 y)}{h}$$

$$= \lim_{h \to 0} \frac{h(y^2 - 3x^2 y - 3xyh - yh^2)}{h} = \lim_{h \to 0}(y^2 - 3x^2 y - 3xyh - yh^2) = y^2 - 3x^2 y$$

$$f_y(x, y) = \lim_{h \to 0} \frac{f(x, y + h) - f(x, y)}{h} = \lim_{h \to 0} \frac{x(y + h)^2 - x^3(y + h) - (xy^2 - x^3 y)}{h} = \lim_{h \to 0} \frac{h(2xy + xh - x^3)}{h}$$

$$= \lim_{h \to 0}(2xy + xh - x^3) = 2xy - x^3$$

**45.** $x^2 + y^2 + z^2 = 3xyz$ $\Rightarrow$ $\dfrac{\partial}{\partial x}(x^2 + y^2 + z^2) = \dfrac{\partial}{\partial x}(3xyz)$ $\Rightarrow$ $2x + 0 + 2z\dfrac{\partial z}{\partial x} = 3y\left(x\dfrac{\partial z}{\partial x} + z \cdot 1\right)$ $\Leftrightarrow$

$2z\dfrac{\partial z}{\partial x} - 3xy\dfrac{\partial z}{\partial x} = 3yz - 2x$ $\Leftrightarrow$ $(2z - 3xy)\dfrac{\partial z}{\partial x} = 3yz - 2x$, so $\dfrac{\partial z}{\partial x} = \dfrac{3yz - 2x}{2z - 3xy}$.

$\dfrac{\partial}{\partial y}(x^2 + y^2 + z^2) = \dfrac{\partial}{\partial y}(3xyz)$ $\Rightarrow$ $0 + 2y + 2z\dfrac{\partial z}{\partial y} = 3x\left(y\dfrac{\partial z}{\partial y} + z \cdot 1\right)$ $\Leftrightarrow$ $2z\dfrac{\partial z}{\partial y} - 3xy\dfrac{\partial z}{\partial y} = 3xz - 2y$ $\Leftrightarrow$

$(2z - 3xy)\dfrac{\partial z}{\partial y} = 3xz - 2y$, so $\dfrac{\partial z}{\partial y} = \dfrac{3xz - 2y}{2z - 3xy}$.

**47.** $x - z = \arctan(yz)$ $\Rightarrow$ $\dfrac{\partial}{\partial x}(x - z) = \dfrac{\partial}{\partial x}(\arctan(yz))$ $\Rightarrow$ $1 - \dfrac{\partial z}{\partial x} = \dfrac{1}{1 + (yz)^2} \cdot y\dfrac{\partial z}{\partial x}$ $\Leftrightarrow$

$1 = \left(\dfrac{y}{1 + y^2 z^2} + 1\right)\dfrac{\partial z}{\partial x}$ $\Leftrightarrow$ $1 = \left(\dfrac{y + 1 + y^2 z^2}{1 + y^2 z^2}\right)\dfrac{\partial z}{\partial x}$, so $\dfrac{\partial z}{\partial x} = \dfrac{1 + y^2 z^2}{1 + y + y^2 z^2}$.

$\dfrac{\partial}{\partial y}(x - z) = \dfrac{\partial}{\partial y}(\arctan(yz))$ $\Rightarrow$ $0 - \dfrac{\partial z}{\partial y} = \dfrac{1}{1 + (yz)^2} \cdot \left(y\dfrac{\partial z}{\partial y} + z \cdot 1\right)$ $\Leftrightarrow$

$-\dfrac{z}{1 + y^2 z^2} = \left(\dfrac{y}{1 + y^2 z^2} + 1\right)\dfrac{\partial z}{\partial y}$ $\Leftrightarrow$ $-\dfrac{z}{1 + y^2 z^2} = \left(\dfrac{y + 1 + y^2 z^2}{1 + y^2 z^2}\right)\dfrac{\partial z}{\partial y}$ $\Leftrightarrow$ $\dfrac{\partial z}{\partial y} = -\dfrac{z}{1 + y + y^2 z^2}$.

**49.** (a) $z = f(x) + g(y)$ $\Rightarrow$ $\dfrac{\partial z}{\partial x} = f'(x)$, $\dfrac{\partial z}{\partial y} = g'(y)$

(b) $z = f(x + y)$. Let $u = x + y$. Then $\dfrac{\partial z}{\partial x} = \dfrac{df}{du}\dfrac{\partial u}{\partial x} = \dfrac{df}{du}(1) = f'(u) = f'(x + y)$,

$\dfrac{\partial z}{\partial y} = \dfrac{df}{du}\dfrac{\partial u}{\partial y} = \dfrac{df}{du}(1) = f'(u) = f'(x + y)$.

**51.** $f(x, y) = x^3 y^5 + 2x^4 y$ $\Rightarrow$ $f_x(x, y) = 3x^2 y^5 + 8x^3 y$, $f_y(x, y) = 5x^3 y^4 + 2x^4$. Then $f_{xx}(x, y) = 6xy^5 + 24x^2 y$,

$f_{xy}(x, y) = 15x^2 y^4 + 8x^3$, $f_{yx}(x, y) = 15x^2 y^4 + 8x^3$, and $f_{yy}(x, y) = 20x^3 y^3$.

**53.** $w = \sqrt{u^2 + v^2} \;\Rightarrow\; w_u = \frac{1}{2}(u^2 + v^2)^{-1/2} \cdot 2u = \frac{u}{\sqrt{u^2 + v^2}}, \; w_v = \frac{1}{2}(u^2 + v^2)^{-1/2} \cdot 2v = \frac{v}{\sqrt{u^2 + v^2}}.$ Then

$$w_{uu} = \frac{1 \cdot \sqrt{u^2 + v^2} - u \cdot \frac{1}{2}(u^2 + v^2)^{-1/2}(2u)}{\left(\sqrt{u^2 + v^2}\right)^2} = \frac{\sqrt{u^2 + v^2} - u^2/\sqrt{u^2 + v^2}}{u^2 + v^2} = \frac{u^2 + v^2 - u^2}{(u^2 + v^2)^{3/2}} = \frac{v^2}{(u^2 + v^2)^{3/2}},$$

$$w_{uv} = u\left(-\frac{1}{2}\right)(u^2 + v^2)^{-3/2}(2v) = -\frac{uv}{(u^2 + v^2)^{3/2}}, \quad w_{vu} = v\left(-\frac{1}{2}\right)(u^2 + v^2)^{-3/2}(2u) = -\frac{uv}{(u^2 + v^2)^{3/2}},$$

$$w_{vv} = \frac{1 \cdot \sqrt{u^2 + v^2} - v \cdot \frac{1}{2}(u^2 + v^2)^{-1/2}(2v)}{\left(\sqrt{u^2 + v^2}\right)^2} = \frac{\sqrt{u^2 + v^2} - v^2/\sqrt{u^2 + v^2}}{u^2 + v^2} = \frac{u^2 + v^2 - v^2}{(u^2 + v^2)^{3/2}} = \frac{u^2}{(u^2 + v^2)^{3/2}}.$$

**55.** $z = \arctan\dfrac{x + y}{1 - xy} \;\Rightarrow$

$$z_x = \frac{1}{1 + \left(\frac{x+y}{1-xy}\right)^2} \cdot \frac{(1)(1 - xy) - (x + y)(-y)}{(1 - xy)^2} = \frac{1 + y^2}{(1 - xy)^2 + (x + y)^2} = \frac{1 + y^2}{1 + x^2 + y^2 + x^2 y^2}$$

$$= \frac{1 + y^2}{(1 + x^2)(1 + y^2)} = \frac{1}{1 + x^2},$$

$$z_y = \frac{1}{1 + \left(\frac{x+y}{1-xy}\right)^2} \cdot \frac{(1)(1 - xy) - (x + y)(-x)}{(1 - xy)^2} = \frac{1 + x^2}{(1 - xy)^2 + (x + y)^2} = \frac{1 + x^2}{(1 + x^2)(1 + y^2)} = \frac{1}{1 + y^2}.$$

Then $z_{xx} = -(1 + x^2)^{-2} \cdot 2x = -\dfrac{2x}{(1 + x^2)^2}, \; z_{xy} = 0, \, z_{yx} = 0, \, z_{yy} = -(1 + y^2)^{-2} \cdot 2y = -\dfrac{2y}{(1 + y^2)^2}.$

**57.** $u = xe^{xy} \;\Rightarrow\; u_x = x \cdot ye^{xy} + e^{xy} \cdot 1 = (xy + 1)e^{xy}, \; u_{xy} = (xy + 1) \cdot xe^{xy} + e^{xy} \cdot x = (x^2 y + 2x)e^{xy}$ and

$u_y = x(xe^{xy}) = x^2 e^{xy}, \; u_{yx} = x^2 \cdot ye^{xy} + e^{xy} \cdot 2x = (x^2 y + 2x)e^{xy}.$ Thus $u_{xy} = u_{yx}.$

**59.** $f(x, y) = 3xy^4 + x^3 y^2 \;\Rightarrow\; f_x = 3y^4 + 3x^2 y^2, \; f_{xx} = 6xy^2, \; f_{xxy} = 12xy$ and

$f_y = 12xy^3 + 2x^3 y, \; f_{yy} = 36xy^2 + 2x^3, \; f_{yyy} = 72xy.$

**61.** $f(x, y, z) = \cos(4x + 3y + 2z) \;\Rightarrow$

$f_x = -\sin(4x + 3y + 2z)(4) = -4\sin(4x + 3y + 2z), \; f_{xy} = -4\cos(4x + 3y + 2z)(3) = -12\cos(4x + 3y + 2z),$

$f_{xyz} = -12(-\sin(4x + 3y + 2z))(2) = 24\sin(4x + 3y + 2z)$ and

$f_y = -\sin(4x + 3y + 2z)(3) = -3\sin(4x + 3y + 2z),$

$f_{yz} = -3\cos(4x + 3y + 2z)(2) = -6\cos(4x + 3y + 2z), \; f_{yzz} = -6(-\sin(4x + 3y + 2z))(2) = 12\sin(4x + 3y + 2z).$

**63.** $u = e^{r\theta}\sin\theta \;\Rightarrow\; \dfrac{\partial u}{\partial \theta} = e^{r\theta}\cos\theta + \sin\theta \cdot e^{r\theta}(r) = e^{r\theta}(\cos\theta + r\sin\theta),$

$\dfrac{\partial^2 u}{\partial r\, \partial\theta} = e^{r\theta}(\sin\theta) + (\cos\theta + r\sin\theta)e^{r\theta}(\theta) = e^{r\theta}(\sin\theta + \theta\cos\theta + r\theta\sin\theta),$

$\dfrac{\partial^3 u}{\partial r^2\, \partial\theta} = e^{r\theta}(\theta\sin\theta) + (\sin\theta + \theta\cos\theta + r\theta\sin\theta) \cdot e^{r\theta}(\theta) = \theta e^{r\theta}(2\sin\theta + \theta\cos\theta + r\theta\sin\theta).$

**65.** Assuming that the third partial derivatives of $f$ are continuous (easily verified), we can write $f_{xzy} = f_{yxz}.$ Then

$f(x, y, z) = xy^2 z^3 + \arcsin\left(x\sqrt{z}\right) \;\Rightarrow\; f_y = 2xyz^3 + 0, \; f_{yx} = 2yz^3,$ and $f_{yxz} = 6yz^2 = f_{xzy}.$

**67.** By Definition 4, $f_x(3, 2) = \lim\limits_{h \to 0} \dfrac{f(3 + h, 2) - f(3, 2)}{h}$ which we can approximate by considering $h = 0.5$ and $h = -0.5$:

$$f_x(3, 2) \approx \frac{f(3.5, 2) - f(3, 2)}{0.5} = \frac{22.4 - 17.5}{0.5} = 9.8,\ f_x(3, 2) \approx \frac{f(2.5, 2) - f(3, 2)}{-0.5} = \frac{10.2 - 17.5}{-0.5} = 14.6.\ \text{Averaging}$$

these values, we estimate $f_x(3, 2)$ to be approximately 12.2. Similarly, $f_x(3, 2.2) = \lim\limits_{h \to 0} \dfrac{f(3 + h, 2.2) - f(3, 2.2)}{h}$ which

we can approximate by considering $h = 0.5$ and $h = -0.5$: $f_x(3, 2.2) \approx \dfrac{f(3.5, 2.2) - f(3, 2.2)}{0.5} = \dfrac{26.1 - 15.9}{0.5} = 20.4,$

$$f_x(3, 2.2) \approx \frac{f(2.5, 2.2) - f(3, 2.2)}{-0.5} = \frac{9.3 - 15.9}{-0.5} = 13.2.\ \text{Averaging these values, we have } f_x(3, 2.2) \approx 16.8.$$

To estimate $f_{xy}(3, 2)$, we first need an estimate for $f_x(3, 1.8)$:

$$f_x(3, 1.8) \approx \frac{f(3.5, 1.8) - f(3, 1.8)}{0.5} = \frac{20.0 - 18.1}{0.5} = 3.8,\ f_x(3, 1.8) \approx \frac{f(2.5, 1.8) - f(3, 1.8)}{-0.5} = \frac{12.5 - 18.1}{-0.5} = 11.2.$$

Averaging these values, we get $f_x(3, 1.8) \approx 7.5$. Now $f_{xy}(x, y) = \dfrac{\partial}{\partial y}\,[f_x(x, y)]$ and $f_x(x, y)$ is itself a function of two

variables, so Definition 4 says that $f_{xy}(x, y) = \dfrac{\partial}{\partial y}\,[f_x(x, y)] = \lim\limits_{h \to 0} \dfrac{f_x(x, y + h) - f_x(x, y)}{h} \quad \Rightarrow$

$f_{xy}(3, 2) = \lim\limits_{h \to 0} \dfrac{f_x(3, 2 + h) - f_x(3, 2)}{h}$. We can estimate this value using our previous work with $h = 0.2$ and $h = -0.2$:

$$f_{xy}(3, 2) \approx \frac{f_x(3, 2.2) - f_x(3, 2)}{0.2} = \frac{16.8 - 12.2}{0.2} = 23,\ f_{xy}(3, 2) \approx \frac{f_x(3, 1.8) - f_x(3, 2)}{-0.2} = \frac{7.5 - 12.2}{-0.2} = 23.5.$$

Averaging these values, we estimate $f_{xy}(3, 2)$ to be approximately 23.25.

**69.** $u = e^{-\alpha^2 k^2 t} \sin kx \quad \Rightarrow \quad u_x = k e^{-\alpha^2 k^2 t} \cos kx,\ u_{xx} = -k^2 e^{-\alpha^2 k^2 t} \sin kx,$ and $u_t = -\alpha^2 k^2 e^{-\alpha^2 k^2 t} \sin kx.$

Thus $\alpha^2 u_{xx} = u_t.$

**71.** $u = \dfrac{1}{\sqrt{x^2 + y^2 + z^2}} \quad \Rightarrow \quad u_x = \left(-\tfrac{1}{2}\right)(x^2 + y^2 + z^2)^{-3/2}(2x) = -x(x^2 + y^2 + z^2)^{-3/2}$ and

$$u_{xx} = -(x^2 + y^2 + z^2)^{-3/2} - x\left(-\tfrac{3}{2}\right)(x^2 + y^2 + z^2)^{-5/2}(2x) = \frac{2x^2 - y^2 - z^2}{(x^2 + y^2 + z^2)^{5/2}}.$$

By symmetry, $u_{yy} = \dfrac{2y^2 - x^2 - z^2}{(x^2 + y^2 + z^2)^{5/2}}$ and $u_{zz} = \dfrac{2z^2 - x^2 - y^2}{(x^2 + y^2 + z^2)^{5/2}}.$

Thus $u_{xx} + u_{yy} + u_{zz} = \dfrac{2x^2 - y^2 - z^2 + 2y^2 - x^2 - z^2 + 2z^2 - x^2 - y^2}{(x^2 + y^2 + z^2)^{5/2}} = 0.$

**73.** Let $v = x + at,\ w = x - at.$  Then $u_t = \dfrac{\partial[f(v) + g(w)]}{\partial t} = \dfrac{df(v)}{dv}\dfrac{\partial v}{\partial t} + \dfrac{dg(w)}{dw}\dfrac{\partial w}{\partial t} = af'(v) - ag'(w)$ and

$u_{tt} = \dfrac{\partial[af'(v) - ag'(w)]}{\partial t} = a[af''(v) + ag''(w)] = a^2[f''(v) + g''(w)].$ Similarly, by using the Chain Rule we have

$u_x = f'(v) + g'(w)$ and $u_{xx} = f''(v) + g''(w).$ Thus $u_{tt} = a^2 u_{xx}.$

**75.** $u = xe^y + ye^x \Rightarrow \dfrac{\partial u}{\partial x} = e^y + ye^x,\ \dfrac{\partial^2 u}{\partial x^2} = ye^x,\ \dfrac{\partial^3 u}{\partial x^3} = ye^x,\ \dfrac{\partial u}{\partial y} = xe^y + e^x,\ \dfrac{\partial^2 u}{\partial y^2} = xe^y,\ \dfrac{\partial^3 u}{\partial y^3} = xe^y,$

$\dfrac{\partial^3 u}{\partial x\, \partial y^2} = \dfrac{\partial}{\partial x}\left(\dfrac{\partial^2 u}{\partial y^2}\right) = \dfrac{\partial}{\partial x}\left(xe^y\right) = e^y,\quad \dfrac{\partial^2 u}{\partial x\, \partial y} = e^y + e^x \Rightarrow \dfrac{\partial^3 u}{\partial x^2\, \partial y} = e^x.$ Then

$$\dfrac{\partial^3 u}{\partial x^3} + \dfrac{\partial^3 u}{\partial y^3} = ye^x + xe^y = x\left(e^y\right) + y\left(e^x\right) = x\,\dfrac{\partial^3 u}{\partial x\, \partial y^2} + y\,\dfrac{\partial^3 u}{\partial x^2\, \partial y}$$

**77.** If we fix $K = K_0$, $P(L, K_0)$ is a function of a single variable $L$, and $\dfrac{dP}{dL} = \alpha\dfrac{P}{L}$ is a separable differential equation. Then

$\dfrac{dP}{P} = \alpha\dfrac{dL}{L} \Rightarrow \displaystyle\int \dfrac{dP}{P} = \int \alpha\dfrac{dL}{L} \Rightarrow \ln|P| = \alpha\ln|L| + C(K_0)$, where $C(K_0)$ can depend on $K_0$. Then

$|P| = e^{\alpha\ln|L| + C(K_0)}$, and since $P > 0$ and $L > 0$, we have $P = e^{\alpha\ln L}e^{C(K_0)} = e^{C(K_0)}e^{\ln L^\alpha} = C_1(K_0)L^\alpha$ where

$C_1(K_0) = e^{C(K_0)}$.

**79.** By the Chain Rule, taking the partial derivative of both sides with respect to $R_1$ gives

$\dfrac{\partial R^{-1}}{\partial R}\dfrac{\partial R}{\partial R_1} = \dfrac{\partial\left[(1/R_1) + (1/R_2) + (1/R_3)\right]}{\partial R_1}$ or $-R^{-2}\dfrac{\partial R}{\partial R_1} = -R_1^{-2}$. Thus $\dfrac{\partial R}{\partial R_1} = \dfrac{R^2}{R_1^2}$.

**81.** (a) $P = \dfrac{mRT}{V}$ so $\dfrac{\partial P}{\partial V} = \dfrac{-mRT}{V^2}$; $V = \dfrac{mRT}{P}$, so $\dfrac{\partial V}{\partial T} = \dfrac{mR}{P}$; $T = \dfrac{PV}{mR}$, so $\dfrac{\partial T}{\partial P} = \dfrac{V}{mR}$.

Thus $\dfrac{\partial P}{\partial V}\dfrac{\partial V}{\partial T}\dfrac{\partial T}{\partial P} = \dfrac{-mRT}{V^2}\dfrac{mR}{P}\dfrac{V}{mR} = \dfrac{-mRT}{PV} = -1$, since $PV = mRT$.

(b) By part (a), $PV = mRT \Rightarrow P = \dfrac{mRT}{V}$, so $\dfrac{\partial P}{\partial T} = \dfrac{mR}{V}$. Also, $PV = mRT \Rightarrow V = \dfrac{mRT}{P}$ and $\dfrac{\partial V}{\partial T} = \dfrac{mR}{P}$.

Since $T = \dfrac{PV}{mR}$, we have $T\dfrac{\partial P}{\partial T}\dfrac{\partial V}{\partial T} = \dfrac{PV}{mR}\cdot\dfrac{mR}{V}\cdot\dfrac{mR}{P} = mR$.

**83.** $\dfrac{\partial K}{\partial m} = \tfrac{1}{2}v^2,\ \dfrac{\partial K}{\partial v} = mv,\ \dfrac{\partial^2 K}{\partial v^2} = m$. Thus $\dfrac{\partial K}{\partial m}\cdot\dfrac{\partial^2 K}{\partial v^2} = \tfrac{1}{2}v^2 m = K$.

**85.** $f_x(x, y) = x + 4y \Rightarrow f_{xy}(x, y) = 4$ and $f_y(x, y) = 3x - y \Rightarrow f_{yx}(x, y) = 3$. Since $f_{xy}$ and $f_{yx}$ are continuous

everywhere but $f_{xy}(x, y) \neq f_{yx}(x, y)$, Clairaut's Theorem implies that such a function $f(x, y)$ does not exist.

**87.** By the geometry of partial derivatives, the slope of the tangent line is $f_x(1, 2)$. By implicit differentiation of

$4x^2 + 2y^2 + z^2 = 16$, we get $8x + 2z\left(\partial z/\partial x\right) = 0 \Rightarrow \partial z/\partial x = -4x/z$, so when $x = 1$ and $z = 2$ we have

$\partial z/\partial x = -2$. So the slope is $f_x(1, 2) = -2$. Thus the tangent line is given by $z - 2 = -2(x - 1),\ y = 2$. Taking the

parameter to be $t = x - 1$, we can write parametric equations for this line: $x = 1 + t,\ y = 2,\ z = 2 - 2t$.

**89.** Let $g(x) = f(x, 0) = x(x^2)^{-3/2}e^0 = x\,|x|^{-3}$. But we are using the point $(1, 0)$, so near $(1, 0)$, $g(x) = x^{-2}$. Then

$g'(x) = -2x^{-3}$ and $g'(1) = -2$, so using (1) we have $f_x(1, 0) = g'(1) = -2$.

**91.** (a)

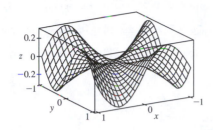

(b) For $(x, y) \neq (0, 0)$,

$$f_x(x, y) = \frac{(3x^2 y - y^3)(x^2 + y^2) - (x^3 y - xy^3)(2x)}{(x^2 + y^2)^2}$$

$$= \frac{x^4 y + 4x^2 y^3 - y^5}{(x^2 + y^2)^2}$$

and by symmetry $f_y(x, y) = \dfrac{x^5 - 4x^3 y^2 - xy^4}{(x^2 + y^2)^2}$.

(c) $f_x(0, 0) = \lim\limits_{h \to 0} \dfrac{f(h, 0) - f(0, 0)}{h} = \lim\limits_{h \to 0} \dfrac{(0/h^2) - 0}{h} = 0$ and $f_y(0, 0) = \lim\limits_{h \to 0} \dfrac{f(0, h) - f(0, 0)}{h} = 0$.

(d) By (3), $f_{xy}(0, 0) = \dfrac{\partial f_x}{\partial y} = \lim\limits_{h \to 0} \dfrac{f_x(0, h) - f_x(0, 0)}{h} = \lim\limits_{h \to 0} \dfrac{(-h^5 - 0)/h^4}{h} = -1$ while by (2),

$$f_{yx}(0, 0) = \frac{\partial f_y}{\partial x} = \lim_{h \to 0} \frac{f_y(h, 0) - f_y(0, 0)}{h} = \lim_{h \to 0} \frac{h^5 / h^4}{h} = 1.$$

(e) For $(x, y) \neq (0, 0)$, we use a CAS to compute

$$f_{xy}(x, y) = \frac{x^6 + 9x^4 y^2 - 9x^2 y^4 - y^6}{(x^2 + y^2)^3}$$

Now as $(x, y) \to (0, 0)$ along the $x$-axis, $f_{xy}(x, y) \to 1$ while as

$(x, y) \to (0, 0)$ along the $y$-axis, $f_{xy}(x, y) \to -1$. Thus $f_{xy}$ isn't

continuous at $(0, 0)$ and Clairaut's Theorem doesn't apply, so there is

no contradiction. The graphs of $f_{xy}$ and $f_{yx}$ are identical except at the

origin, where we observe the discontinuity.

## 11.4   Tangent Planes and Linear Approximations

**1.** $z = f(x, y) = 3y^2 - 2x^2 + x \quad \Rightarrow \quad f_x(x, y) = -4x + 1, \ f_y(x, y) = 6y$, so $f_x(2, -1) = -7, \ f_y(2, -1) = -6$.

By Equation 2, an equation of the tangent plane is $z - (-3) = f_x(2, -1)(x - 2) + f_y(2, -1)[y - (-1)] \quad \Rightarrow$

$z + 3 = -7(x - 2) - 6(y + 1)$  or  $z = -7x - 6y + 5$.

**3.** $z = f(x, y) = \sqrt{xy} \quad \Rightarrow \quad f_x(x, y) = \frac{1}{2}(xy)^{-1/2} \cdot y = \frac{1}{2}\sqrt{y/x}, \ f_y(x, y) = \frac{1}{2}(xy)^{-1/2} \cdot x = \frac{1}{2}\sqrt{x/y}$, so $f_x(1, 1) = \frac{1}{2}$

and $f_y(1, 1) = \frac{1}{2}$. Thus an equation of the tangent plane is $z - 1 = f_x(1, 1)(x - 1) + f_y(1, 1)(y - 1) \quad \Rightarrow$

$z - 1 = \frac{1}{2}(x - 1) + \frac{1}{2}(y - 1)$ or $x + y - 2z = 0$.

**5.** $z = f(x, y) = y \cos(x - y) \quad \Rightarrow \quad f_x = y(-\sin(x - y)(1)) = -y\sin(x - y)$,

$f_y = y(-\sin(x - y)(-1)) + \cos(x - y) = y\sin(x - y) + \cos(x - y)$, so $f_x(2, 2) = -2\sin(0) = 0$,

$f_y(2, 2) = 2\sin(0) + \cos(0) = 1$ and an equation of the tangent plane is $z - 2 = 0(x - 2) + 1(y - 2)$ or $z = y$.

**7.** $z = f(x, y) = x^2 + xy + 3y^2$, so $f_x(x, y) = 2x + y \;\Rightarrow\; f_x(1, 1) = 3$, $f_y(x, y) = x + 6y \;\Rightarrow\; f_y(1, 1) = 7$ and an

equation of the tangent plane is $z - 5 = 3(x - 1) + 7(y - 1)$ or $z = 3x + 7y - 5$. After zooming in, the surface and the

tangent plane become almost indistinguishable. (Here, the tangent plane is below the surface.) If we zoom in farther, the

surface and the tangent plane will appear to coincide.

**9.** $f(x, y) = \dfrac{xy \sin (x - y)}{1 + x^2 + y^2}$. A CAS gives $f_x(x, y) = \dfrac{y \sin (x - y) + xy \cos (x - y)}{1 + x^2 + y^2} - \dfrac{2x^2 y \sin (x - y)}{(1 + x^2 + y^2)^2}$ and

$f_y(x, y) = \dfrac{x \sin (x - y) - xy \cos (x - y)}{1 + x^2 + y^2} - \dfrac{2xy^2 \sin (x - y)}{(1 + x^2 + y^2)^2}$. We use the CAS to evaluate these at $(1, 1)$, and then

substitute the results into Equation 2 to compute an equation of the tangent plane: $z = \frac{1}{3}x - \frac{1}{3}y$. The surface and tangent

plane are shown in the first graph below. After zooming in, the surface and the tangent plane become almost indistinguishable,

as shown in the second graph. (Here, the tangent plane is shown with fewer traces than the surface.) If we zoom in farther, the

surface and the tangent plane will appear to coincide.

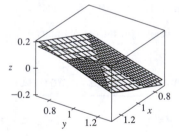

**11.** $f(x, y) = x\sqrt{y}$. The partial derivatives are $f_x(x, y) = \sqrt{y}$ and $f_y(x, y) = \dfrac{x}{2\sqrt{y}}$, so $f_x(1, 4) = 2$ and $f_y(1, 4) = \frac{1}{4}$. Both

$f_x$ and $f_y$ are continuous functions for $y > 0$, so by Theorem 8, $f$ is differentiable at $(1, 4)$. By Equation 3, the linearization of

$f$ at $(1, 4)$ is given by $L(x, y) = f(1, 4) + f_x(1, 4)(x - 1) + f_y(1, 4)(y - 4) = 2 + 2(x - 1) + \frac{1}{4}(y - 4) = 2x + \frac{1}{4}y - 1$.

**13.** $f(x, y) = \dfrac{x}{x + y}$. The partial derivatives are $f_x(x, y) = \dfrac{1(x + y) - x(1)}{(x + y)^2} = y/(x + y)^2$ and

$f_y(x, y) = x(-1)(x + y)^{-2} \cdot 1 = -x/(x + y)^2$, so $f_x(2, 1) = \frac{1}{9}$ and $f_y(2, 1) = -\frac{2}{9}$. Both $f_x$ and $f_y$ are continuous

functions for $y \neq -x$, so $f$ is differentiable at $(2, 1)$ by Theorem 8. The linearization of $f$ at $(2, 1)$ is given by

$L(x, y) = f(2, 1) + f_x(2, 1)(x - 2) + f_y(2, 1)(y - 1) = \frac{2}{3} + \frac{1}{9}(x - 2) - \frac{2}{9}(y - 1) = \frac{1}{9}x - \frac{2}{9}y + \frac{2}{3}$.

**15.** Let $f(x, y) = \dfrac{2x + 3}{4y + 1}$. Then $f_x(x, y) = \dfrac{2}{4y + 1}$ and $f_y(x, y) = (2x + 3)(-1)(4y + 1)^{-2}(4) = \dfrac{-8x - 12}{(4y + 1)^2}$. Both $f_x$ and $f_y$

are continuous functions for $y \neq -\frac{1}{4}$, so by Theorem 8, $f$ is differentiable at $(0, 0)$. We have $f_x(0, 0) = 2$, $f_y(0, 0) = -12$

and the linear approximation of $f$ at $(0, 0)$ is $f(x, y) \approx f(0, 0) + f_x(0, 0)(x - 0) + f_y(0, 0)(y - 0) = 3 + 2x - 12y$.

**17.** We can estimate $f(2.2, 4.9)$ using a linear approximation of $f$ at $(2, 5)$, given by

$f(x, y) \approx f(2, 5) + f_x(2, 5)(x - 2) + f_y(2, 5)(y - 5) = 6 + 1(x - 2) + (-1)(y - 5) = x - y + 9$. Thus

$f(2.2, 4.9) \approx 2.2 - 4.9 + 9 = 6.3$.

**19.** $f(x, y, z) = \sqrt{x^2 + y^2 + z^2} \quad \Rightarrow \quad f_x(x, y, z) = \dfrac{x}{\sqrt{x^2 + y^2 + z^2}}, \; f_y(x, y, z) = \dfrac{y}{\sqrt{x^2 + y^2 + z^2}}$, and

$f_z(x, y, z) = \dfrac{z}{\sqrt{x^2 + y^2 + z^2}}$, so $f_x(3, 2, 6) = \frac{3}{7}$, $f_y(3, 2, 6) = \frac{2}{7}$, $f_z(3, 2, 6) = \frac{6}{7}$. Then the linear approximation of $f$ at

$(3, 2, 6)$ is given by

$$f(x, y, z) \approx f(3, 2, 6) + f_x(3, 2, 6)(x - 3) + f_y(3, 2, 6)(y - 2) + f_z(3, 2, 6)(z - 6)$$
$$= 7 + \tfrac{3}{7}(x - 3) + \tfrac{2}{7}(y - 2) + \tfrac{6}{7}(z - 6) = \tfrac{3}{7}x + \tfrac{2}{7}y + \tfrac{6}{7}z$$

Thus $\sqrt{(3.02)^2 + (1.97)^2 + (5.99)^2} = f(3.02, 1.97, 5.99) \approx \frac{3}{7}(3.02) + \frac{2}{7}(1.97) + \frac{6}{7}(5.99) \approx 6.9914$.

**21.** From the table, $f(94, 80) = 127$. To estimate $f_T(94, 80)$ and $f_H(94, 80)$ we follow the procedure used in Section 11.3. Since

$f_T(94, 80) = \lim\limits_{h \to 0} \dfrac{f(94 + h, 80) - f(94, 80)}{h}$, we approximate this quantity with $h = \pm 2$ and use the values given in the

table:

$$f_T(94, 80) \approx \frac{f(96, 80) - f(94, 80)}{2} = \frac{135 - 127}{2} = 4, \quad f_T(94, 80) \approx \frac{f(92, 80) - f(94, 80)}{-2} = \frac{119 - 127}{-2} = 4$$

Averaging these values gives $f_T(94, 80) \approx 4$. Similarly, $f_H(94, 80) = \lim\limits_{h \to 0} \dfrac{f(94, 80 + h) - f(94, 80)}{h}$, so we use $h = \pm 5$:

$$f_H(94, 80) \approx \frac{f(94, 85) - f(94, 80)}{5} = \frac{132 - 127}{5} = 1, \quad f_H(94, 80) \approx \frac{f(94, 75) - f(94, 80)}{-5} = \frac{122 - 127}{-5} = 1$$

Averaging these values gives $f_H(94, 80) \approx 1$. The linear approximation, then, is

$$f(T, H) \approx f(94, 80) + f_T(94, 80)(T - 94) + f_H(94, 80)(H - 80)$$
$$\approx 127 + 4(T - 94) + 1(H - 80) \qquad [\text{or } 4T + H - 329]$$

Thus when $T = 95$ and $H = 78$, $f(95, 78) \approx 127 + 4(95 - 94) + 1(78 - 80) = 129$, so we estimate the heat index to be

approximately $129°F$.

**23.** $z = x^3 \ln(y^2) \quad \Rightarrow \quad dz = \dfrac{\partial z}{\partial x} dx + \dfrac{\partial z}{\partial y} dy = 3x^2 \ln(y^2) \, dx + x^3 \cdot \dfrac{1}{y^2}(2y) \, dy = 3x^2 \ln(y^2) \, dx + \dfrac{2x^3}{y} \, dy$

**25.** $m = p^5 q^3 \quad \Rightarrow \quad dm = \dfrac{\partial m}{\partial p} dp + \dfrac{\partial m}{\partial q} dq = 5p^4 q^3 \, dp + 3p^5 q^2 \, dq$

**27.** $R = \alpha\beta^2 \cos\gamma \quad \Rightarrow \quad dR = \dfrac{\partial R}{\partial \alpha} d\alpha + \dfrac{\partial R}{\partial \beta} d\beta + \dfrac{\partial R}{\partial \gamma} d\gamma = \beta^2 \cos\gamma \, d\alpha + 2\alpha\beta \cos\gamma \, d\beta - \alpha\beta^2 \sin\gamma \, d\gamma$

**29.** $dx = \Delta x = 0.05$, $dy = \Delta y = 0.1$, $z = 5x^2 + y^2$, $z_x = 10x$, $z_y = 2y$. Thus when $x = 1$ and $y = 2$,

$dz = z_x(1,2)\, dx + z_y(1,2)\, dy = (10)(0.05) + (4)(0.1) = 0.9$ while

$\Delta z = f(1.05, 2.1) - f(1,2) = 5(1.05)^2 + (2.1)^2 - 5 - 4 = 0.9225$.

**31.** $dA = \dfrac{\partial A}{\partial x}\, dx + \dfrac{\partial A}{\partial y}\, dy = y\, dx + x\, dy$ and $|\Delta x| \le 0.1$, $|\Delta y| \le 0.1$. We use $dx = 0.1$, $dy = 0.1$ with $x = 30$, $y = 24$; then

the maximum error in the area is about $dA = 24(0.1) + 30(0.1) = 5.4$ cm$^2$.

**33.** The volume of a can is $V = \pi r^2 h$ and $\Delta V \approx dV$ is an estimate of the amount of tin. Here $dV = 2\pi rh\, dr + \pi r^2\, dh$, so put

$dr = 0.04$, $dh = 0.08$ (0.04 on top, 0.04 on bottom) and then $\Delta V \approx dV = 2\pi(48)(0.04) + \pi(16)(0.08) \approx 16.08$ cm$^3$.

Thus the amount of tin is about 16 cm$^3$.

**35.** The errors in measurement are at most 2%, so $\left|\dfrac{\Delta w}{w}\right| \le 0.02$ and $\left|\dfrac{\Delta h}{h}\right| \le 0.02$. The relative error in the calculated surface

area is

$$\frac{\Delta S}{S} \approx \frac{dS}{S} = \frac{0.1091(0.425w^{0.425-1})h^{0.725}\, dw + 0.1091w^{0.425}(0.725h^{0.725-1})\, dh}{0.1091w^{0.425}h^{0.725}} = 0.425\frac{dw}{w} + 0.725\frac{dh}{h}$$

To estimate the maximum relative error, we use $\dfrac{dw}{w} = \left|\dfrac{\Delta w}{w}\right| = 0.02$ and $\dfrac{dh}{h} = \left|\dfrac{\Delta h}{h}\right| = 0.02$ $\Rightarrow$

$\dfrac{dS}{S} = 0.425\,(0.02) + 0.725\,(0.02) = 0.023$. Thus the maximum percentage error is approximately 2.3%.

**37.** First we find $\dfrac{\partial R}{\partial R_1}$ implicitly by taking partial derivatives of both sides with respect to $R_1$:

$$\frac{\partial}{\partial R_1}\left(\frac{1}{R}\right) = \frac{\partial\,[(1/R_1)+(1/R_2)+(1/R_3)]}{\partial R_1} \quad\Rightarrow\quad -R^{-2}\frac{\partial R}{\partial R_1} = -R_1^{-2} \quad\Rightarrow\quad \frac{\partial R}{\partial R_1} = \frac{R^2}{R_1^2}.$$ Then by symmetry,

$\dfrac{\partial R}{\partial R_2} = \dfrac{R^2}{R_2^2}$, $\dfrac{\partial R}{\partial R_3} = \dfrac{R^2}{R_3^2}$. When $R_1 = 25$, $R_2 = 40$ and $R_3 = 50$, $\dfrac{1}{R} = \dfrac{17}{200}$ $\Leftrightarrow$ $R = \frac{200}{17}$ Ω. Since the possible error

for each $R_i$ is 0.5%, the maximum error of $R$ is attained by setting $\Delta R_i = 0.005 R_i$. So

$$\Delta R \approx dR = \frac{\partial R}{\partial R_1}\Delta R_1 + \frac{\partial R}{\partial R_2}\Delta R_2 + \frac{\partial R}{\partial R_3}\Delta R_3 = (0.005)R^2\left(\frac{1}{R_1}+\frac{1}{R_2}+\frac{1}{R_3}\right) = (0.005)R = \tfrac{1}{17} \approx 0.059 \text{ Ω}.$$

**39.** $\mathbf{r}(u,v) = (u+v)\,\mathbf{i} + 3u^2\,\mathbf{j} + (u-v)\,\mathbf{k}$.

$\mathbf{r}_u = \mathbf{i} + 6u\,\mathbf{j} + \mathbf{k}$ and $\mathbf{r}_v = \mathbf{i} - \mathbf{k}$, so $\mathbf{r}_u \times \mathbf{r}_v = -6u\,\mathbf{i} + 2\,\mathbf{j} - 6u\,\mathbf{k}$.

Since the point $(2,3,0)$ corresponds to $u = 1$, $v = 1$, a normal vector

to the surface at $(2,3,0)$ is $-6\,\mathbf{i} + 2\,\mathbf{j} - 6\,\mathbf{k}$, and an equation of the

tangent plane is $-6x + 2y - 6z = -6$ or $3x - y + 3z = 3$.

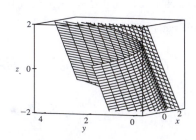

**41.** $\mathbf{r}(u,v) = u^2\,\mathbf{i} + 2u\sin v\,\mathbf{j} + u\cos v\,\mathbf{k} \;\Rightarrow\; \mathbf{r}(1,0) = (1,0,1)$.

$\mathbf{r}_u = 2u\,\mathbf{i} + 2\sin v\,\mathbf{j} + \cos v\,\mathbf{k}$ and $\mathbf{r}_v = 2u\cos v\,\mathbf{j} - u\sin v\,\mathbf{k}$,

so a normal vector to the surface at the point $(1,0,1)$ is

$\mathbf{r}_u(1,0) \times \mathbf{r}_v(1,0) = (2\,\mathbf{i} + \mathbf{k}) \times (2\,\mathbf{j}) = -2\,\mathbf{i} + 4\,\mathbf{k}$.

Thus an equation of the tangent plane at $(1,0,1)$ is

$-2(x-1) + 0(y-0) + 4(z-1) = 0$ or $-x + 2z = 1$.

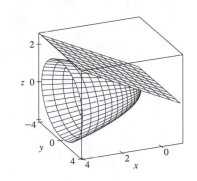

**43.** $\mathbf{r}(u,v) = u\,\mathbf{i} + \ln(uv)\,\mathbf{j} + v\,\mathbf{k} \;\Rightarrow\; \mathbf{r}_u(u,v) = \mathbf{i} + \frac{1}{u}\,\mathbf{j}$,

$\mathbf{r}_v(u,v) = \frac{1}{v}\,\mathbf{j} + \mathbf{k}$. $\mathbf{r}(1,1) = \mathbf{i} + \mathbf{k}$, so the point corresponding to

$u = 1, v = 1$ is $(1,0,1)$. A normal vector for the tangent plane is

$\mathbf{r}_u(1,1) \times \mathbf{r}_v(1,1) = (\mathbf{i} + \mathbf{j}) \times (\mathbf{j} + \mathbf{k}) = \mathbf{i} - \mathbf{j} + \mathbf{k}$, so an equation

of the tangent plane is $(x-1) - (y-0) + (z-1) = 0$

or $x - y + z = 2$.

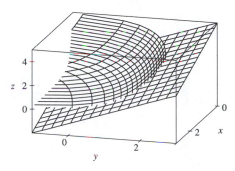

**45.** $\Delta z = f(a + \Delta x, b + \Delta y) - f(a,b) = (a + \Delta x)^2 + (b + \Delta y)^2 - (a^2 + b^2)$

$\quad = a^2 + 2a\,\Delta x + (\Delta x)^2 + b^2 + 2b\,\Delta y + (\Delta y)^2 - a^2 - b^2 = 2a\,\Delta x + (\Delta x)^2 + 2b\,\Delta y + (\Delta y)^2$

But $f_x(a,b) = 2a$ and $f_y(a,b) = 2b$ and so $\Delta z = f_x(a,b)\,\Delta x + f_y(a,b)\,\Delta y + \Delta x\,\Delta x + \Delta y\,\Delta y$, which is Definition 7

with $\varepsilon_1 = \Delta x$ and $\varepsilon_2 = \Delta y$. Hence $f$ is differentiable.

**47.** To show that $f$ is continuous at $(a,b)$ we need to show that $\displaystyle\lim_{(x,y)\to(a,b)} f(x,y) = f(a,b)$ or

equivalently $\displaystyle\lim_{(\Delta x, \Delta y)\to(0,0)} f(a + \Delta x, b + \Delta y) = f(a,b)$. Since $f$ is differentiable at $(a,b)$,

$f(a + \Delta x, b + \Delta y) - f(a,b) = \Delta z = f_x(a,b)\,\Delta x + f_y(a,b)\,\Delta y + \epsilon_1\,\Delta x + \epsilon_2\,\Delta y$, where $\epsilon_1$ and $\epsilon_2 \to 0$ as

$(\Delta x, \Delta y) \to (0,0)$. Thus $f(a + \Delta x, b + \Delta y) = f(a,b) + f_x(a,b)\,\Delta x + f_y(a,b)\,\Delta y + \varepsilon_1\,\Delta x + \varepsilon_2\,\Delta y$. Taking the limit of

both sides as $(\Delta x, \Delta y) \to (0,0)$ gives $\displaystyle\lim_{(\Delta x, \Delta y)\to(0,0)} f(a + \Delta x, b + \Delta y) = f(a,b)$. Thus $f$ is continuous at $(a,b)$.

## 11.5  The Chain Rule

**1.** $z = x^2 + y^2 + xy$, $x = \sin t$, $y = e^t \;\Rightarrow\; \dfrac{dz}{dt} = \dfrac{\partial z}{\partial x}\dfrac{dx}{dt} + \dfrac{\partial z}{\partial y}\dfrac{dy}{dt} = (2x + y)\cos t + (2y + x)e^t$

**3.** $z = \sqrt{1 + x^2 + y^2}$, $x = \ln t$, $y = \cos t \;\Rightarrow\;$

$\dfrac{dz}{dt} = \dfrac{\partial z}{\partial x}\dfrac{dx}{dt} + \dfrac{\partial z}{\partial y}\dfrac{dy}{dt} = \frac{1}{2}(1 + x^2 + y^2)^{-1/2}(2x) \cdot \dfrac{1}{t} + \frac{1}{2}(1 + x^2 + y^2)^{-1/2}(2y)(-\sin t) = \dfrac{1}{\sqrt{1 + x^2 + y^2}}\left(\dfrac{x}{t} - y\sin t\right)$

**5.** $w = xe^{y/z}$, $x = t^2$, $y = 1 - t$, $z = 1 + 2t$ $\Rightarrow$

$$\frac{dw}{dt} = \frac{\partial w}{\partial x}\frac{dx}{dt} + \frac{\partial w}{\partial y}\frac{dy}{dt} + \frac{\partial w}{\partial z}\frac{dz}{dt} = e^{y/z} \cdot 2t + xe^{y/z}\left(\frac{1}{z}\right)\cdot(-1) + xe^{y/z}\left(-\frac{y}{z^2}\right)\cdot 2 = e^{y/z}\left(2t - \frac{x}{z} - \frac{2xy}{z^2}\right)$$

**7.** $z = x^2 y^3$, $x = s\cos t$, $y = s\sin t$ $\Rightarrow$

$$\frac{\partial z}{\partial s} = \frac{\partial z}{\partial x}\frac{\partial x}{\partial s} + \frac{\partial z}{\partial y}\frac{\partial y}{\partial s} = 2xy^3\cos t + 3x^2 y^2\sin t$$

$$\frac{\partial z}{\partial t} = \frac{\partial z}{\partial x}\frac{\partial x}{\partial t} + \frac{\partial z}{\partial y}\frac{\partial y}{\partial t} = (2xy^3)(-s\sin t) + (3x^2 y^2)(s\cos t) = -2sxy^3\sin t + 3sx^2 y^2\cos t$$

**9.** $z = \sin\theta\cos\phi$, $\theta = st^2$, $\phi = s^2 t$ $\Rightarrow$

$$\frac{\partial z}{\partial s} = \frac{\partial z}{\partial\theta}\frac{\partial\theta}{\partial s} + \frac{\partial z}{\partial\phi}\frac{\partial\phi}{\partial s} = (\cos\theta\cos\phi)(t^2) + (-\sin\theta\sin\phi)(2st) = t^2\cos\theta\cos\phi - 2st\sin\theta\sin\phi$$

$$\frac{\partial z}{\partial t} = \frac{\partial z}{\partial\theta}\frac{\partial\theta}{\partial t} + \frac{\partial z}{\partial\phi}\frac{\partial\phi}{\partial t} = (\cos\theta\cos\phi)(2st) + (-\sin\theta\sin\phi)(s^2) = 2st\cos\theta\cos\phi - s^2\sin\theta\sin\phi$$

**11.** $z = e^r\cos\theta$, $r = st$, $\theta = \sqrt{s^2 + t^2}$ $\Rightarrow$

$$\frac{\partial z}{\partial s} = \frac{\partial z}{\partial r}\frac{\partial r}{\partial s} + \frac{\partial z}{\partial\theta}\frac{\partial\theta}{\partial s} = e^r\cos\theta\cdot t + e^r(-\sin\theta)\cdot\tfrac{1}{2}(s^2 + t^2)^{-1/2}(2s) = te^r\cos\theta - e^r\sin\theta\cdot\frac{s}{\sqrt{s^2 + t^2}}$$

$$= e^r\left(t\cos\theta - \frac{s}{\sqrt{s^2 + t^2}}\sin\theta\right)$$

$$\frac{\partial z}{\partial t} = \frac{\partial z}{\partial r}\frac{\partial r}{\partial t} + \frac{\partial z}{\partial\theta}\frac{\partial\theta}{\partial t} = e^r\cos\theta\cdot s + e^r(-\sin\theta)\cdot\tfrac{1}{2}(s^2 + t^2)^{-1/2}(2t) = se^r\cos\theta - e^r\sin\theta\cdot\frac{t}{\sqrt{s^2 + t^2}}$$

$$= e^r\left(s\cos\theta - \frac{t}{\sqrt{s^2 + t^2}}\sin\theta\right)$$

**13.** When $t = 3$, $x = g(3) = 2$ and $y = h(3) = 7$. By the Chain Rule (2),

$$\frac{dz}{dt} = \frac{\partial f}{\partial x}\frac{dx}{dt} + \frac{\partial f}{\partial y}\frac{dy}{dt} = f_x(2, 7)g'(3) + f_y(2, 7)\,h'(3) = (6)(5) + (-8)(-4) = 62.$$

**15.** $g(u, v) = f(x(u, v), y(u, v))$ where $x = e^u + \sin v$, $y = e^u + \cos v$ $\Rightarrow$

$\dfrac{\partial x}{\partial u} = e^u$, $\dfrac{\partial x}{\partial v} = \cos v$, $\dfrac{\partial y}{\partial u} = e^u$, $\dfrac{\partial y}{\partial v} = -\sin v$. By the Chain Rule (3), $\dfrac{\partial g}{\partial u} = \dfrac{\partial f}{\partial x}\dfrac{\partial x}{\partial u} + \dfrac{\partial f}{\partial y}\dfrac{\partial y}{\partial u}$. Then

$$g_u(0,0) = f_x(x(0,0), y(0,0))\,x_u(0,0) + f_y(x(0,0), y(0,0))\,y_u(0,0) = f_x(1,2)(e^0) + f_y(1,2)(e^0) = 2(1) + 5(1) = 7.$$

Similarly, $\dfrac{\partial g}{\partial v} = \dfrac{\partial f}{\partial x}\dfrac{\partial x}{\partial v} + \dfrac{\partial f}{\partial y}\dfrac{\partial y}{\partial v}$. Then

$$g_v(0,0) = f_x(x(0,0), y(0,0))\,x_v(0,0) + f_y(x(0,0), y(0,0))\,y_v(0,0) = f_x(1,2)(\cos 0) + f_y(1,2)(-\sin 0)$$
$$= 2(1) + 5(0) = 2$$

**17.**

$u = f(x, y)$, $x = x(r, s, t)$, $y = y(r, s, t)$ $\Rightarrow$

$$\frac{\partial u}{\partial r} = \frac{\partial u}{\partial x}\frac{\partial x}{\partial r} + \frac{\partial u}{\partial y}\frac{\partial y}{\partial r},\quad \frac{\partial u}{\partial s} = \frac{\partial u}{\partial x}\frac{\partial x}{\partial s} + \frac{\partial u}{\partial y}\frac{\partial y}{\partial s},$$

$$\frac{\partial u}{\partial t} = \frac{\partial u}{\partial x}\frac{\partial x}{\partial t} + \frac{\partial u}{\partial y}\frac{\partial y}{\partial t}$$

**19.**

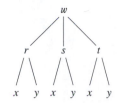

$w = f(r, s, t), \quad r = r(x, y), \quad s = s(x, y), \quad t = t(x, y) \quad \Rightarrow$

$$\frac{\partial w}{\partial x} = \frac{\partial w}{\partial r}\frac{\partial r}{\partial x} + \frac{\partial w}{\partial s}\frac{\partial s}{\partial x} + \frac{\partial w}{\partial t}\frac{\partial t}{\partial x}, \quad \frac{\partial w}{\partial y} = \frac{\partial w}{\partial r}\frac{\partial r}{\partial y} + \frac{\partial w}{\partial s}\frac{\partial s}{\partial y} + \frac{\partial w}{\partial t}\frac{\partial t}{\partial y}$$

**21.** $z = x^2 + xy^3, \ x = uv^2 + w^3, \ y = u + ve^w \quad \Rightarrow \quad \dfrac{\partial z}{\partial u} = \dfrac{\partial z}{\partial x}\dfrac{\partial x}{\partial u} + \dfrac{\partial z}{\partial y}\dfrac{\partial y}{\partial u} = (2x + y^3)(v^2) + (3xy^2)(1),$

$\dfrac{\partial z}{\partial v} = \dfrac{\partial z}{\partial x}\dfrac{\partial x}{\partial v} + \dfrac{\partial z}{\partial y}\dfrac{\partial y}{\partial v} = (2x + y^3)(2uv) + (3xy^2)(e^w), \quad \dfrac{\partial z}{\partial w} = \dfrac{\partial z}{\partial x}\dfrac{\partial x}{\partial w} + \dfrac{\partial z}{\partial y}\dfrac{\partial y}{\partial w} = (2x + y^3)(3w^2) + (3xy^2)(ve^w).$

When $u = 2$, $v = 1$, and $w = 0$, we have $x = 2$, $y = 3$, so $\dfrac{\partial z}{\partial u} = (31)(1) + (54)(1) = 85$,

$\dfrac{\partial z}{\partial v} = (31)(4) + (54)(1) = 178, \ \dfrac{\partial z}{\partial w} = (31)(0) + (54)(1) = 54.$

**23.** $R = \ln(u^2 + v^2 + w^2), \ u = x + 2y, \ v = 2x - y, \ w = 2xy \quad \Rightarrow$

$\dfrac{\partial R}{\partial x} = \dfrac{\partial R}{\partial u}\dfrac{\partial u}{\partial x} + \dfrac{\partial R}{\partial v}\dfrac{\partial v}{\partial x} + \dfrac{\partial R}{\partial w}\dfrac{\partial w}{\partial x} = \dfrac{2u}{u^2 + v^2 + w^2}\,(1) + \dfrac{2v}{u^2 + v^2 + w^2}\,(2) + \dfrac{2w}{u^2 + v^2 + w^2}\,(2y)$

$\qquad = \dfrac{2u + 4v + 4wy}{u^2 + v^2 + w^2},$

$\dfrac{\partial R}{\partial y} = \dfrac{\partial R}{\partial u}\dfrac{\partial u}{\partial y} + \dfrac{\partial R}{\partial v}\dfrac{\partial v}{\partial y} + \dfrac{\partial R}{\partial w}\dfrac{\partial w}{\partial y} = \dfrac{2u}{u^2 + v^2 + w^2}\,(2) + \dfrac{2v}{u^2 + v^2 + w^2}\,(-1) + \dfrac{2w}{u^2 + v^2 + w^2}\,(2x)$

$\qquad = \dfrac{4u - 2v + 4wx}{u^2 + v^2 + w^2}.$

When $x = y = 1$ we have $u = 3$, $v = 1$, and $w = 2$, so $\dfrac{\partial R}{\partial x} = \dfrac{9}{7}$ and $\dfrac{\partial R}{\partial y} = \dfrac{9}{7}.$

**25.** $u = x^2 + yz, \ x = pr\cos\theta, \ y = pr\sin\theta, \ z = p + r \quad \Rightarrow$

$\dfrac{\partial u}{\partial p} = \dfrac{\partial u}{\partial x}\dfrac{\partial x}{\partial p} + \dfrac{\partial u}{\partial y}\dfrac{\partial y}{\partial p} + \dfrac{\partial u}{\partial z}\dfrac{\partial z}{\partial p} = (2x)(r\cos\theta) + (z)(r\sin\theta) + (y)(1) = 2xr\cos\theta + zr\sin\theta + y,$

$\dfrac{\partial u}{\partial r} = \dfrac{\partial u}{\partial x}\dfrac{\partial x}{\partial r} + \dfrac{\partial u}{\partial y}\dfrac{\partial y}{\partial r} + \dfrac{\partial u}{\partial z}\dfrac{\partial z}{\partial r} = (2x)(p\cos\theta) + (z)(p\sin\theta) + (y)(1) = 2xp\cos\theta + zp\sin\theta + y,$

$\dfrac{\partial u}{\partial \theta} = \dfrac{\partial u}{\partial x}\dfrac{\partial x}{\partial \theta} + \dfrac{\partial u}{\partial y}\dfrac{\partial y}{\partial \theta} + \dfrac{\partial u}{\partial z}\dfrac{\partial z}{\partial \theta} = (2x)(-pr\sin\theta) + (z)(pr\cos\theta) + (y)(0) = -2xpr\sin\theta + zpr\cos\theta.$

When $p = 2$, $r = 3$, and $\theta = 0$ we have $x = 6$, $y = 0$, and $z = 5$, so $\dfrac{\partial u}{\partial p} = 36$, $\dfrac{\partial u}{\partial r} = 24$, and $\dfrac{\partial u}{\partial \theta} = 30$.

**27.** $\cos(x - y) = xe^y$, so let $F(x, y) = \cos(x - y) - xe^y = 0$.

Then $\dfrac{dy}{dx} = -\dfrac{F_x}{F_y} = -\dfrac{-\sin(x - y) - e^y}{-\sin(x - y)(-1) - xe^y} = \dfrac{\sin(x - y) + e^y}{\sin(x - y) - xe^y}.$

**29.** $x^2 + y^2 + z^2 = 3xyz$, so let $F(x, y, z) = x^2 + y^2 + z^2 - 3xyz = 0$. Then by Equations 7

$\dfrac{\partial z}{\partial x} = -\dfrac{F_x}{F_z} = -\dfrac{2x - 3yz}{2z - 3xy} = \dfrac{3yz - 2x}{2z - 3xy} \quad \text{and} \quad \dfrac{\partial z}{\partial y} = -\dfrac{F_y}{F_z} = -\dfrac{2y - 3xz}{2z - 3xy} = \dfrac{3xz - 2y}{2z - 3xy}.$

**31.** $x - z = \arctan(yz)$, so let $F(x, y, z) = x - z - \arctan(yz) = 0$. Then

$$\frac{\partial z}{\partial x} = -\frac{F_x}{F_z} = -\frac{1}{-1 - \dfrac{1}{1 + (yz)^2}(y)} = \frac{1 + y^2 z^2}{1 + y + y^2 z^2} \text{ and}$$

$$\frac{\partial z}{\partial y} = -\frac{F_y}{F_z} = -\frac{-\dfrac{1}{1 + (yz)^2}(z)}{-1 - \dfrac{1}{1 + (yz)^2}(y)} = -\frac{\dfrac{z}{1 + y^2 z^2}}{\dfrac{1 + y^2 z^2 + y}{1 + y^2 z^2}} = -\frac{z}{1 + y + y^2 z^2}.$$

**33.** Since $x$ and $y$ are each functions of $t$, $T(x, y)$ is a function of $t$, so by the Chain Rule, $\dfrac{dT}{dt} = \dfrac{\partial T}{\partial x}\dfrac{dx}{dt} + \dfrac{\partial T}{\partial y}\dfrac{dy}{dt}$. After

3 seconds, $x = \sqrt{1 + t} = \sqrt{1 + 3} = 2$, $y = 2 + \frac{1}{3}t = 2 + \frac{1}{3}(3) = 3$, $\dfrac{dx}{dt} = \dfrac{1}{2\sqrt{1 + t}} = \dfrac{1}{2\sqrt{1 + 3}} = \dfrac{1}{4}$, and $\dfrac{dy}{dt} = \dfrac{1}{3}$.

Then $\dfrac{dT}{dt} = T_x(2, 3)\dfrac{dx}{dt} + T_y(2, 3)\dfrac{dy}{dt} = 4\left(\frac{1}{4}\right) + 3\left(\frac{1}{3}\right) = 2$. Thus the temperature is rising at a rate of $2°\text{C/s}$.

**35.** $C = 1449.2 + 4.6T - 0.055T^2 + 0.00029T^3 + 0.016D$, so $\dfrac{\partial C}{\partial T} = 4.6 - 0.11T + 0.00087T^2$ and $\dfrac{\partial C}{\partial D} = 0.016$.

According to the graph, the diver is experiencing a temperature of approximately $12.5°\text{C}$ at $t = 20$ minutes, so

$\dfrac{\partial C}{\partial T} = 4.6 - 0.11(12.5) + 0.00087(12.5)^2 \approx 3.36$. By sketching tangent lines at $t = 20$ to the graphs given, we estimate

$\dfrac{dD}{dt} \approx \dfrac{1}{2}$ and $\dfrac{dT}{dt} \approx -\dfrac{1}{10}$. Then, by the Chain Rule, $\dfrac{dC}{dt} = \dfrac{\partial C}{\partial T}\dfrac{dT}{dt} + \dfrac{\partial C}{\partial D}\dfrac{dD}{dt} \approx (3.36)\left(-\frac{1}{10}\right) + (0.016)\left(\frac{1}{2}\right) \approx -0.33$.

Thus the speed of sound experienced by the diver is decreasing at a rate of approximately $0.33 \text{ m/s}$ per minute.

**37.** (a) $V = \ell w h$, so by the Chain Rule,

$$\frac{dV}{dt} = \frac{\partial V}{\partial \ell}\frac{d\ell}{dt} + \frac{\partial V}{\partial w}\frac{dw}{dt} + \frac{\partial V}{\partial h}\frac{dh}{dt} = wh\frac{d\ell}{dt} + \ell h\frac{dw}{dt} + \ell w\frac{dh}{dt} = 2 \cdot 2 \cdot 2 + 1 \cdot 2 \cdot 2 + 1 \cdot 2 \cdot (-3) = 6 \text{ m}^3/\text{s}.$$

(b) $S = 2(\ell w + \ell h + wh)$, so by the Chain Rule,

$$\frac{dS}{dt} = \frac{\partial S}{\partial \ell}\frac{d\ell}{dt} + \frac{\partial S}{\partial w}\frac{dw}{dt} + \frac{\partial S}{\partial h}\frac{dh}{dt} = 2(w + h)\frac{d\ell}{dt} + 2(\ell + h)\frac{dw}{dt} + 2(\ell + w)\frac{dh}{dt}$$
$$= 2(2 + 2)2 + 2(1 + 2)2 + 2(1 + 2)(-3) = 10 \text{ m}^2/\text{s}$$

(c) $L^2 = \ell^2 + w^2 + h^2 \Rightarrow 2L\dfrac{dL}{dt} = 2\ell\dfrac{d\ell}{dt} + 2w\dfrac{dw}{dt} + 2h\dfrac{dh}{dt} = 2(1)(2) + 2(2)(2) + 2(2)(-3) = 0 \Rightarrow$

$dL/dt = 0 \text{ m/s}$.

**39.** $\dfrac{dP}{dt} = 0.05$, $\dfrac{dT}{dt} = 0.15$, $V = 8.31\dfrac{T}{P}$ and $\dfrac{dV}{dt} = \dfrac{8.31}{P}\dfrac{dT}{dt} - 8.31\dfrac{T}{P^2}\dfrac{dP}{dt}$. Thus when $P = 20$ and $T = 320$,

$\dfrac{dV}{dt} = 8.31\left[\dfrac{0.15}{20} - \dfrac{(0.05)(320)}{400}\right] \approx -0.27 \text{ L/s}$.

**41.** Let $x$ be the length of the first side of the triangle and $y$ the length of the second side. The area $A$ of the triangle is given by

$A = \frac{1}{2}xy\sin\theta$ where $\theta$ is the angle between the two sides. Thus $A$ is a function of $x$, $y$, and $\theta$, and $x$, $y$, and $\theta$ are each in turn

functions of time $t$. We are given that $\dfrac{dx}{dt} = 3$, $\dfrac{dy}{dt} = -2$, and because $A$ is constant, $\dfrac{dA}{dt} = 0$. By the Chain Rule,

$$\frac{dA}{dt} = \frac{\partial A}{\partial x}\frac{dx}{dt} + \frac{\partial A}{\partial y}\frac{dy}{dt} + \frac{\partial A}{\partial \theta}\frac{d\theta}{dt} \quad \Rightarrow \quad \frac{dA}{dt} = \tfrac{1}{2}y\sin\theta \cdot \frac{dx}{dt} + \tfrac{1}{2}x\sin\theta \cdot \frac{dy}{dt} + \tfrac{1}{2}xy\cos\theta \cdot \frac{d\theta}{dt}. \text{ When } x = 20, \ y = 30,$$

and $\theta = \pi/6$ we have

$$0 = \tfrac{1}{2}(30)\left(\sin\tfrac{\pi}{6}\right)(3) + \tfrac{1}{2}(20)\left(\sin\tfrac{\pi}{6}\right)(-2) + \tfrac{1}{2}(20)(30)\left(\cos\tfrac{\pi}{6}\right)\frac{d\theta}{dt}$$

$$= 45 \cdot \tfrac{1}{2} - 20 \cdot \tfrac{1}{2} + 300 \cdot \frac{\sqrt{3}}{2} \cdot \frac{d\theta}{dt} = \tfrac{25}{2} + 150\sqrt{3}\,\frac{d\theta}{dt}$$

Solving for $\dfrac{d\theta}{dt}$ gives $\dfrac{d\theta}{dt} = \dfrac{-25/2}{150\sqrt{3}} = -\dfrac{1}{12\sqrt{3}}$, so the angle between the sides is decreasing at a rate of

$1/(12\sqrt{3}) \approx 0.048 \ \text{rad/s}.$

**43.** (a) By the Chain Rule, $\dfrac{\partial z}{\partial r} = \dfrac{\partial z}{\partial x}\cos\theta + \dfrac{\partial z}{\partial y}\sin\theta, \ \dfrac{\partial z}{\partial \theta} = \dfrac{\partial z}{\partial x}(-r\sin\theta) + \dfrac{\partial z}{\partial y}r\cos\theta.$

(b) $\left(\dfrac{\partial z}{\partial r}\right)^2 = \left(\dfrac{\partial z}{\partial x}\right)^2\cos^2\theta + 2\dfrac{\partial z}{\partial x}\dfrac{\partial z}{\partial y}\cos\theta\sin\theta + \left(\dfrac{\partial z}{\partial y}\right)^2\sin^2\theta,$

$\left(\dfrac{\partial z}{\partial \theta}\right)^2 = \left(\dfrac{\partial z}{\partial x}\right)^2 r^2\sin^2\theta - 2\dfrac{\partial z}{\partial x}\dfrac{\partial z}{\partial y}r^2\cos\theta\sin\theta + \left(\dfrac{\partial z}{\partial y}\right)^2 r^2\cos^2\theta.$ Thus

$$\left(\frac{\partial z}{\partial r}\right)^2 + \frac{1}{r^2}\left(\frac{\partial z}{\partial \theta}\right)^2 = \left[\left(\frac{\partial z}{\partial x}\right)^2 + \left(\frac{\partial z}{\partial y}\right)^2\right](\cos^2\theta + \sin^2\theta) = \left(\frac{\partial z}{\partial x}\right)^2 + \left(\frac{\partial z}{\partial y}\right)^2.$$

**45.** Let $u = x - y$. Then $\dfrac{\partial z}{\partial x} = \dfrac{dz}{du}\dfrac{\partial u}{\partial x} = \dfrac{dz}{du}$ and $\dfrac{\partial z}{\partial y} = \dfrac{dz}{du}(-1)$. Thus $\dfrac{\partial z}{\partial x} + \dfrac{\partial z}{\partial y} = 0.$

**47.** Let $u = x + at, \ v = x - at$. Then $z = f(u) + g(v)$, so $\partial z/\partial u = f'(u)$ and $\partial z/\partial v = g'(v)$.

Thus $\dfrac{\partial z}{\partial t} = \dfrac{\partial z}{\partial u}\dfrac{\partial u}{\partial t} + \dfrac{\partial z}{\partial v}\dfrac{\partial v}{\partial t} = af'(u) - ag'(v)$ and

$$\frac{\partial^2 z}{\partial t^2} = a\frac{\partial}{\partial t}[f'(u) - g'(v)] = a\left(\frac{df'(u)}{du}\frac{\partial u}{\partial t} - \frac{dg'(v)}{dv}\frac{\partial v}{\partial t}\right) = a^2 f''(u) + a^2 g''(v).$$

Similarly $\dfrac{\partial z}{\partial x} = f'(u) + g'(v)$ and $\dfrac{\partial^2 z}{\partial x^2} = f''(u) + g''(v)$. Thus $\dfrac{\partial^2 z}{\partial t^2} = a^2\dfrac{\partial^2 z}{\partial x^2}.$

**49.** $\dfrac{\partial z}{\partial s} = \dfrac{\partial z}{\partial x}2s + \dfrac{\partial z}{\partial y}2r$. Then

$$\frac{\partial^2 z}{\partial r\,\partial s} = \frac{\partial}{\partial r}\left(\frac{\partial z}{\partial x}2s\right) + \frac{\partial}{\partial r}\left(\frac{\partial z}{\partial y}2r\right)$$

$$= \frac{\partial^2 z}{\partial x^2}\frac{\partial x}{\partial r}2s + \frac{\partial}{\partial y}\left(\frac{\partial z}{\partial x}\right)\frac{\partial y}{\partial r}2s + \frac{\partial z}{\partial x}\frac{\partial}{\partial r}2s + \frac{\partial^2 z}{\partial y^2}\frac{\partial y}{\partial r}2r + \frac{\partial}{\partial x}\left(\frac{\partial z}{\partial y}\right)\frac{\partial x}{\partial r}2r + \frac{\partial z}{\partial y}2$$

$$= 4rs\frac{\partial^2 z}{\partial x^2} + \frac{\partial^2 z}{\partial y\,\partial x}4s^2 + 0 + 4rs\frac{\partial^2 z}{\partial y^2} + \frac{\partial^2 z}{\partial x\,\partial y}4r^2 + 2\frac{\partial z}{\partial y}$$

By the continuity of the partials, $\dfrac{\partial^2 z}{\partial r\partial s} = 4rs\dfrac{\partial^2 z}{\partial x^2} + 4rs\dfrac{\partial^2 z}{\partial y^2} + (4r^2 + 4s^2)\dfrac{\partial^2 z}{\partial x\,\partial y} + 2\dfrac{\partial z}{\partial y}.$

**51.** $\dfrac{\partial z}{\partial r} = \dfrac{\partial z}{\partial x} \cos\theta + \dfrac{\partial z}{\partial y} \sin\theta$ and $\dfrac{\partial z}{\partial \theta} = -\dfrac{\partial z}{\partial x} r \sin\theta + \dfrac{\partial z}{\partial y} r \cos\theta$. Then

$$\frac{\partial^2 z}{\partial r^2} = \cos\theta \left( \frac{\partial^2 z}{\partial x^2} \cos\theta + \frac{\partial^2 z}{\partial y\,\partial x} \sin\theta \right) + \sin\theta \left( \frac{\partial^2 z}{\partial y^2} \sin\theta + \frac{\partial^2 z}{\partial x\,\partial y} \cos\theta \right)$$

$$= \cos^2\theta \, \frac{\partial^2 z}{\partial x^2} + 2\cos\theta \sin\theta \, \frac{\partial^2 z}{\partial x\,\partial y} + \sin^2\theta \, \frac{\partial^2 z}{\partial y^2}$$

and

$$\frac{\partial^2 z}{\partial \theta^2} = -r\cos\theta \, \frac{\partial z}{\partial x} + (-r\sin\theta)\left( \frac{\partial^2 z}{\partial x^2}(-r\sin\theta) + \frac{\partial^2 z}{\partial y\,\partial x} r\cos\theta \right)$$

$$-r\sin\theta \, \frac{\partial z}{\partial y} + r\cos\theta \left( \frac{\partial^2 z}{\partial y^2} r\cos\theta + \frac{\partial^2 z}{\partial x\,\partial y}(-r\sin\theta) \right)$$

$$= -r\cos\theta \, \frac{\partial z}{\partial x} - r\sin\theta \, \frac{\partial z}{\partial y} + r^2 \sin^2\theta \, \frac{\partial^2 z}{\partial x^2} - 2r^2 \cos\theta \sin\theta \, \frac{\partial^2 z}{\partial x\,\partial y} + r^2 \cos^2\theta \, \frac{\partial^2 z}{\partial y^2}$$

Thus

$$\frac{\partial^2 z}{\partial r^2} + \frac{1}{r^2} \frac{\partial^2 z}{\partial \theta^2} + \frac{1}{r} \frac{\partial z}{\partial r} = (\cos^2\theta + \sin^2\theta) \frac{\partial^2 z}{\partial x^2} + (\sin^2\theta + \cos^2\theta) \frac{\partial^2 z}{\partial y^2}$$

$$-\frac{1}{r} \cos\theta \, \frac{\partial z}{\partial x} - \frac{1}{r} \sin\theta \, \frac{\partial z}{\partial y} + \frac{1}{r}\left( \cos\theta \, \frac{\partial z}{\partial x} + \sin\theta \, \frac{\partial z}{\partial y} \right)$$

$$= \frac{\partial^2 z}{\partial x^2} + \frac{\partial^2 z}{\partial y^2} \text{ as desired.}$$

**53.** $F(x, y, z) = 0$ is assumed to define $z$ as a function of $x$ and $y$, that is, $z = f(x, y)$. So by (7), $\dfrac{\partial z}{\partial x} = -\dfrac{F_x}{F_z}$ since $F_z \neq 0$.

Similarly, it is assumed that $F(x, y, z) = 0$ defines $x$ as a function of $y$ and $z$, that is $x = h(x, z)$. Then $F(h(y, z), y, z) = 0$

and by the Chain Rule, $F_x \dfrac{\partial x}{\partial y} + F_y \dfrac{\partial y}{\partial y} + F_z \dfrac{\partial z}{\partial y} = 0$. But $\dfrac{\partial z}{\partial y} = 0$ and $\dfrac{\partial y}{\partial y} = 1$, so $F_x \dfrac{\partial x}{\partial y} + F_y = 0 \;\Rightarrow\; \dfrac{\partial x}{\partial y} = -\dfrac{F_y}{F_x}$.

A similar calculation shows that $\dfrac{\partial y}{\partial z} = -\dfrac{F_z}{F_y}$. Thus $\dfrac{\partial z}{\partial x} \dfrac{\partial x}{\partial y} \dfrac{\partial y}{\partial z} = \left( -\dfrac{F_x}{F_z} \right)\left( -\dfrac{F_y}{F_x} \right)\left( -\dfrac{F_z}{F_y} \right) = -1$.

## 11.6 Directional Derivatives and the Gradient Vector

**1.** We can approximate the directional derivative of the pressure function at K in the direction of S by the average rate of change of pressure between the points where the red line intersects the contour lines closest to K (extend the red line slightly at the left). In the direction of S, the pressure changes from 1000 millibars to 996 millibars and we estimate the distance between these two points to be approximately 50 km (using the fact that the distance from K to S is 300 km). Then the rate of change of pressure in the direction given is approximately $\frac{996 - 1000}{50} = -0.08$ millibar/km.

**3.** $D_{\mathbf{u}} f(-20, 30) = \nabla f(-20, 30) \cdot \mathbf{u} = f_T(-20, 30)\left( \frac{1}{\sqrt{2}} \right) + f_v(-20, 30)\left( \frac{1}{\sqrt{2}} \right)$.

$f_T(-20, 30) = \lim\limits_{h \to 0} \dfrac{f(-20 + h, 30) - f(-20, 30)}{h}$, so we can approximate $f_T(-20, 30)$ by considering $h = \pm 5$ and

using the values given in the table: $f_T(-20, 30) \approx \dfrac{f(-15, 30) - f(-20, 30)}{5} = \dfrac{-26 - (-33)}{5} = 1.4$,

$f_T(-20, 30) \approx \dfrac{f(-25, 30) - f(-20, 30)}{-5} = \dfrac{-39 - (-33)}{-5} = 1.2$. Averaging these values gives $f_T(-20, 30) \approx 1.3$.

Similarly, $f_v(-20, 30) = \lim\limits_{h \to 0} \dfrac{f(-20, 30 + h) - f(-20, 30)}{h}$, so we can approximate $f_v(-20, 30)$ with $h = \pm 10$:

$$f_v(-20, 30) \approx \frac{f(-20, 40) - f(-20, 30)}{10} = \frac{-34 - (-33)}{10} = -0.1,$$

$$f_v(-20, 30) \approx \frac{f(-20, 20) - f(-20, 30)}{-10} = \frac{-30 - (-33)}{-10} = -0.3. \text{ Averaging these values gives } f_v(-20, 30) \approx -0.2.$$

Then $D_\mathbf{u} f(-20, 30) \approx 1.3 \left( \frac{1}{\sqrt{2}} \right) + (-0.2) \left( \frac{1}{\sqrt{2}} \right) \approx 0.778$.

**5.** $f(x, y) = ye^{-x} \quad \Rightarrow \quad f_x(x, y) = -ye^{-x}$ and $f_y(x, y) = e^{-x}$. If $\mathbf{u}$ is a unit vector in the direction of $\theta = 2\pi/3$, then

from Equation 6, $D_\mathbf{u} f(0, 4) = f_x(0, 4) \cos\left( \frac{2\pi}{3} \right) + f_y(0, 4) \sin\left( \frac{2\pi}{3} \right) = -4 \cdot \left( -\frac{1}{2} \right) + 1 \cdot \frac{\sqrt{3}}{2} = 2 + \frac{\sqrt{3}}{2}$.

**7.** $f(x, y) = \sin(2x + 3y)$

(a) $\nabla f(x, y) = \dfrac{\partial f}{\partial x} \mathbf{i} + \dfrac{\partial f}{\partial y} \mathbf{j} = [\cos(2x + 3y) \cdot 2] \mathbf{i} + [\cos(2x + 3y) \cdot 3] \mathbf{j} = 2 \cos(2x + 3y) \mathbf{i} + 3 \cos(2x + 3y) \mathbf{j}$

(b) $\nabla f(-6, 4) = (2 \cos 0) \mathbf{i} + (3 \cos 0) \mathbf{j} = 2 \mathbf{i} + 3 \mathbf{j}$

(c) By Equation 9, $D_\mathbf{u} f(-6, 4) = \nabla f(-6, 4) \cdot \mathbf{u} = (2 \mathbf{i} + 3 \mathbf{j}) \cdot \frac{1}{2} (\sqrt{3}\,\mathbf{i} - \mathbf{j}) = \frac{1}{2} (2\sqrt{3} - 3) = \sqrt{3} - \frac{3}{2}$.

**9.** $f(x, y, z) = xe^{2yz}$

(a) $\nabla f(x, y, z) = \langle f_x(x, y, z), f_y(x, y, z), f_z(x, y, z) \rangle = \langle e^{2yz}, 2xze^{2yz}, 2xye^{2yz} \rangle$

(b) $\nabla f(3, 0, 2) = \langle 1, 12, 0 \rangle$

(c) By Equation 14, $D_\mathbf{u} f(3, 0, 2) = \nabla f(3, 0, 2) \cdot \mathbf{u} = \langle 1, 12, 0 \rangle \cdot \langle \frac{2}{3}, -\frac{2}{3}, \frac{1}{3} \rangle = \frac{2}{3} - \frac{24}{3} + 0 = -\frac{22}{3}$.

**11.** $f(x, y) = 1 + 2x \sqrt{y} \quad \Rightarrow \quad \nabla f(x, y) = \left\langle 2 \sqrt{y}, 2x \cdot \frac{1}{2} y^{-1/2} \right\rangle = \langle 2 \sqrt{y}, x/\sqrt{y} \rangle, \nabla f(3, 4) = \langle 4, \frac{3}{2} \rangle$, and a unit vector in

the direction of $\mathbf{v}$ is $\mathbf{u} = \dfrac{1}{\sqrt{4^2 + (-3)^2}} \langle 4, -3 \rangle = \langle \frac{4}{5}, -\frac{3}{5} \rangle$, so $D_\mathbf{u} f(3, 4) = \nabla f(3, 4) \cdot \mathbf{u} = \langle 4, \frac{3}{2} \rangle \cdot \langle \frac{4}{5}, -\frac{3}{5} \rangle = \frac{23}{10}$.

**13.** $g(p, q) = p^4 - p^2 q^3 \quad \Rightarrow \quad \nabla g(p, q) = (4p^3 - 2pq^3) \mathbf{i} + (-3p^2 q^2) \mathbf{j}, \nabla g(2, 1) = 28 \mathbf{i} - 12 \mathbf{j}$, and a unit

vector in the direction of $\mathbf{v}$ is $\mathbf{u} = \dfrac{1}{\sqrt{1^2 + 3^2}} (\mathbf{i} + 3 \mathbf{j}) = \dfrac{1}{\sqrt{10}} (\mathbf{i} + 3 \mathbf{j})$, so

$$D_\mathbf{u} g(2, 1) = \nabla g(2, 1) \cdot \mathbf{u} = (28 \mathbf{i} - 12 \mathbf{j}) \cdot \frac{1}{\sqrt{10}} (\mathbf{i} + 3 \mathbf{j}) = \frac{1}{\sqrt{10}} (28 - 36) = -\frac{8}{\sqrt{10}} \text{ or } -\frac{4\sqrt{10}}{5}.$$

**15.** $f(x, y, z) = xe^y + ye^z + ze^x \quad \Rightarrow \quad \nabla f(x, y, z) = \langle e^y + ze^x, xe^y + e^z, ye^z + e^x \rangle, \nabla f(0, 0, 0) = \langle 1, 1, 1 \rangle$, and a unit

vector in the direction of $\mathbf{v}$ is $\mathbf{u} = \dfrac{1}{\sqrt{25 + 1 + 4}} \langle 5, 1, -2 \rangle = \dfrac{1}{\sqrt{30}} \langle 5, 1, -2 \rangle$, so

$$D_\mathbf{u} f(0, 0, 0) = \nabla f(0, 0, 0) \cdot \mathbf{u} = \langle 1, 1, 1 \rangle \cdot \frac{1}{\sqrt{30}} \langle 5, 1, -2 \rangle = \frac{4}{\sqrt{30}}.$$

**17.** $g(x, y, z) = (x + 2y + 3z)^{3/2} \quad \Rightarrow$

$$\nabla g(x, y, z) = \left\langle \tfrac{3}{2} (x + 2y + 3z)^{1/2} (1), \tfrac{3}{2} (x + 2y + 3z)^{1/2} (2), \tfrac{3}{2} (x + 2y + 3z)^{1/2} (3) \right\rangle$$

$$= \left\langle \tfrac{3}{2} \sqrt{x + 2y + 3z}, 3 \sqrt{x + 2y + 3z}, \tfrac{9}{2} \sqrt{x + 2y + 3z} \right\rangle, \nabla g(1, 1, 2) = \left\langle \tfrac{9}{2}, 9, \tfrac{27}{2} \right\rangle,$$

and a unit vector in the direction of $\mathbf{v} = 2\mathbf{j} - \mathbf{k}$ is $\mathbf{u} = \frac{2}{\sqrt{5}} \mathbf{j} - \frac{1}{\sqrt{5}} \mathbf{k}$, so

$$D_\mathbf{u} g(1, 1, 2) = \left\langle \tfrac{9}{2}, 9, \tfrac{27}{2} \right\rangle \cdot \left\langle 0, \tfrac{2}{\sqrt{5}}, -\tfrac{1}{\sqrt{5}} \right\rangle = \frac{18}{\sqrt{5}} - \frac{27}{2\sqrt{5}} = \frac{9}{2\sqrt{5}}.$$

**19.** $f(x,y) = \sqrt{xy} \quad \Rightarrow \quad \nabla f(x,y) = \left\langle \frac{1}{2}(xy)^{-1/2}(y), \frac{1}{2}(xy)^{-1/2}(x) \right\rangle = \left\langle \frac{y}{2\sqrt{xy}}, \frac{x}{2\sqrt{xy}} \right\rangle$, so $\nabla f(2,8) = \left\langle 1, \frac{1}{4} \right\rangle$.

The unit vector in the direction of $\overrightarrow{PQ} = \langle 5-2, 4-8 \rangle = \langle 3, -4 \rangle$ is $\mathbf{u} = \left\langle \frac{3}{5}, -\frac{4}{5} \right\rangle$, so

$D_{\mathbf{u}} f(2,8) = \nabla f(2,8) \cdot \mathbf{u} = \left\langle 1, \frac{1}{4} \right\rangle \cdot \left\langle \frac{3}{5}, -\frac{4}{5} \right\rangle = \frac{2}{5}$.

**21.** $f(x,y) = \sin(xy) \quad \Rightarrow \quad \nabla f(x,y) = \langle y\cos(xy), x\cos(xy) \rangle$, $\nabla f(1,0) = \langle 0, 1 \rangle$. Thus the maximum rate of change is $|\nabla f(1,0)| = 1$ in the direction $\langle 0, 1 \rangle$.

**23.** $f(x,y,z) = \sqrt{x^2 + y^2 + z^2} \quad \Rightarrow$

$\nabla f(x,y,z) = \left\langle \frac{1}{2}(x^2+y^2+z^2)^{-1/2} \cdot 2x, \frac{1}{2}(x^2+y^2+z^2)^{-1/2} \cdot 2y, \frac{1}{2}(x^2+y^2+z^2)^{-1/2} \cdot 2z \right\rangle$

$\qquad = \left\langle \frac{x}{\sqrt{x^2+y^2+z^2}}, \frac{y}{\sqrt{x^2+y^2+z^2}}, \frac{z}{\sqrt{x^2+y^2+z^2}} \right\rangle$,

$\nabla f(3,6,-2) = \left\langle \frac{3}{\sqrt{49}}, \frac{6}{\sqrt{49}}, \frac{-2}{\sqrt{49}} \right\rangle = \left\langle \frac{3}{7}, \frac{6}{7}, -\frac{2}{7} \right\rangle$. Thus the maximum rate of change is

$|\nabla f(3,6,-2)| = \sqrt{\left(\frac{3}{7}\right)^2 + \left(\frac{6}{7}\right)^2 + \left(-\frac{2}{7}\right)^2} = \sqrt{\frac{9+36+4}{49}} = 1$ in the direction $\left\langle \frac{3}{7}, \frac{6}{7}, -\frac{2}{7} \right\rangle$ or equivalently $\langle 3, 6, -2 \rangle$.

**25.** (a) As in the proof of Theorem 15, $D_{\mathbf{u}} f = |\nabla f| \cos\theta$. Since the minimum value of $\cos\theta$ is $-1$ occurring when $\theta = \pi$, the minimum value of $D_{\mathbf{u}} f$ is $-|\nabla f|$ occurring when $\theta = \pi$, that is when $\mathbf{u}$ is in the opposite direction of $\nabla f$ (assuming $\nabla f \neq \mathbf{0}$).

(b) $f(x,y) = x^4 y - x^2 y^3 \quad \Rightarrow \quad \nabla f(x,y) = \langle 4x^3 y - 2xy^3, x^4 - 3x^2 y^2 \rangle$, so $f$ decreases fastest at the point $(2,-3)$ in the direction $-\nabla f(2,-3) = -\langle 12, -92 \rangle = \langle -12, 92 \rangle$.

**27.** The direction of fastest change is $\nabla f(x,y) = (2x-2)\mathbf{i} + (2y-4)\mathbf{j}$, so we need to find all points $(x,y)$ where $\nabla f(x,y)$ is parallel to $\mathbf{i} + \mathbf{j} \quad \Leftrightarrow \quad (2x-2)\mathbf{i} + (2y-4)\mathbf{j} = k(\mathbf{i}+\mathbf{j}) \quad \Leftrightarrow \quad k = 2x-2$ and $k = 2y-4$. Then $2x-2 = 2y-4 \quad \Rightarrow \quad y = x+1$, so the direction of fastest change is $\mathbf{i} + \mathbf{j}$ at all points on the line $y = x+1$.

**29.** $T = \dfrac{k}{\sqrt{x^2+y^2+z^2}}$ and $120 = T(1,2,2) = \dfrac{k}{3}$ so $k = 360$.

(a) $\mathbf{u} = \dfrac{\langle 1,-1,1 \rangle}{\sqrt{3}}$,

$D_{\mathbf{u}} T(1,2,2) = \nabla T(1,2,2) \cdot \mathbf{u} = \left[ -360(x^2+y^2+z^2)^{-3/2} \langle x,y,z \rangle \right]_{(1,2,2)} \cdot \mathbf{u} = -\frac{40}{3} \langle 1,2,2 \rangle \cdot \frac{1}{\sqrt{3}} \langle 1,-1,1 \rangle = -\frac{40}{3\sqrt{3}}$

(b) From (a), $\nabla T = -360(x^2+y^2+z^2)^{-3/2} \langle x,y,z \rangle$, and since $\langle x,y,z \rangle$ is the position vector of the point $(x,y,z)$, the vector $-\langle x,y,z \rangle$, and thus $\nabla T$, always points toward the origin.

**31.** $\nabla V(x,y,z) = \langle 10x - 3y + yz, xz - 3x, xy \rangle$, $\nabla V(3,4,5) = \langle 38, 6, 12 \rangle$

(a) $D_{\mathbf{u}} V(3,4,5) = \langle 38, 6, 12 \rangle \cdot \frac{1}{\sqrt{3}} \langle 1, 1, -1 \rangle = \frac{32}{\sqrt{3}}$

(b) $\nabla V(3,4,5) = \langle 38, 6, 12 \rangle$, or equivalently, $\langle 19, 3, 6 \rangle$.

(c) $|\nabla V(3,4,5)| = \sqrt{38^2 + 6^2 + 12^2} = \sqrt{1624} = 2\sqrt{406}$

**33.** A unit vector in the direction of $\overrightarrow{AB}$ is $\mathbf{i}$ and a unit vector in the direction of $\overrightarrow{AC}$ is $\mathbf{j}$. Thus $D_{\overrightarrow{AB}} f(1,3) = f_x(1,3) = 3$ and

$D_{\overrightarrow{AC}} f(1,3) = f_y(1,3) = 26$. Therefore $\nabla f(1,3) = \langle f_x(1,3), f_y(1,3) \rangle = \langle 3, 26 \rangle$, and by definition,

$D_{\overrightarrow{AD}} f(1,3) = \nabla f \cdot \mathbf{u}$ where $\mathbf{u}$ is a unit vector in the direction of $\overrightarrow{AD}$, which is $\left\langle \frac{5}{13}, \frac{12}{13} \right\rangle$. Therefore,

$D_{\overrightarrow{AD}} f(1,3) = \langle 3, 26 \rangle \cdot \left\langle \frac{5}{13}, \frac{12}{13} \right\rangle = 3 \cdot \frac{5}{13} + 26 \cdot \frac{12}{13} = \frac{327}{13}$.

**35.** (a) $\nabla(au + bv) = \left\langle \dfrac{\partial(au + bv)}{\partial x}, \dfrac{\partial(au + bv)}{\partial y} \right\rangle = \left\langle a\dfrac{\partial u}{\partial x} + b\dfrac{\partial v}{\partial x}, a\dfrac{\partial u}{\partial y} + b\dfrac{\partial v}{\partial y} \right\rangle = a\left\langle \dfrac{\partial u}{\partial x}, \dfrac{\partial u}{\partial y} \right\rangle + b\left\langle \dfrac{\partial v}{\partial x}, \dfrac{\partial v}{\partial y} \right\rangle$

$= a\,\nabla u + b\,\nabla v$

(b) $\nabla(uv) = \left\langle v\dfrac{\partial u}{\partial x} + u\dfrac{\partial v}{\partial x}, v\dfrac{\partial u}{\partial y} + u\dfrac{\partial v}{\partial y} \right\rangle = v\left\langle \dfrac{\partial u}{\partial x}, \dfrac{\partial u}{\partial y} \right\rangle + u\left\langle \dfrac{\partial v}{\partial x}, \dfrac{\partial v}{\partial y} \right\rangle = v\,\nabla u + u\,\nabla v$

(c) $\nabla\left(\dfrac{u}{v}\right) = \left\langle \dfrac{v\dfrac{\partial u}{\partial x} - u\dfrac{\partial v}{\partial x}}{v^2}, \dfrac{v\dfrac{\partial u}{\partial y} - u\dfrac{\partial v}{\partial y}}{v^2} \right\rangle = \dfrac{v\left\langle \dfrac{\partial u}{\partial x}, \dfrac{\partial u}{\partial y} \right\rangle - u\left\langle \dfrac{\partial v}{\partial x}, \dfrac{\partial v}{\partial y} \right\rangle}{v^2} = \dfrac{v\,\nabla u - u\,\nabla v}{v^2}$

(d) $\nabla u^n = \left\langle \dfrac{\partial(u^n)}{\partial x}, \dfrac{\partial(u^n)}{\partial y} \right\rangle = \left\langle nu^{n-1}\dfrac{\partial u}{\partial x}, nu^{n-1}\dfrac{\partial u}{\partial y} \right\rangle = nu^{n-1}\,\nabla u$

**37.** $f(x,y) = x^3 + 5x^2 y + y^3 \quad \Rightarrow$

$D_{\mathbf{u}} f(x,y) = \nabla f(x,y) \cdot \mathbf{u} = \langle 3x^2 + 10xy, 5x^2 + 3y^2 \rangle \cdot \langle \frac{3}{5}, \frac{4}{5} \rangle = \frac{9}{5}x^2 + 6xy + 4x^2 + \frac{12}{5}y^2 = \frac{29}{5}x^2 + 6xy + \frac{12}{5}y^2$. Then

$D_{\mathbf{u}}^2 f(x,y) = D_{\mathbf{u}}\left[D_{\mathbf{u}} f(x,y)\right] = \nabla\left[D_{\mathbf{u}} f(x,y)\right] \cdot \mathbf{u} = \langle \frac{58}{5}x + 6y, 6x + \frac{24}{5}y \rangle \cdot \langle \frac{3}{5}, \frac{4}{5} \rangle$

$= \frac{174}{25}x + \frac{18}{5}y + \frac{24}{5}x + \frac{96}{25}y = \frac{294}{25}x + \frac{186}{25}y$

and $D_{\mathbf{u}}^2 f(2,1) = \frac{294}{25}(2) + \frac{186}{25}(1) = \frac{774}{25}$.

**39.** Let $F(x,y,z) = 2(x-2)^2 + (y-1)^2 + (z-3)^2$. Then $2(x-2)^2 + (y-1)^2 + (z-3)^2 = 10$ is a level surface of $F$.

$F_x(x,y,z) = 4(x-2) \quad \Rightarrow \quad F_x(3,3,5) = 4$, $F_y(x,y,z) = 2(y-1) \quad \Rightarrow \quad F_y(3,3,5) = 4$, and

$F_z(x,y,z) = 2(z-3) \quad \Rightarrow \quad F_z(3,3,5) = 4$.

(a) Equation 19 gives an equation of the tangent plane at $(3,3,5)$ as $4(x-3) + 4(y-3) + 4(z-5) = 0 \quad \Leftrightarrow$

$4x + 4y + 4z = 44$ or equivalently $x + y + z = 11$.

(b) By Equation 20, the normal line has symmetric equations $\dfrac{x-3}{4} = \dfrac{y-3}{4} = \dfrac{z-5}{4}$ or equivalently

$x - 3 = y - 3 = z - 5$. Corresponding parametric equations are $x = 3 + t$, $y = 3 + t$, $z = 5 + t$.

**41.** Let $F(x,y,z) = x^2 - 2y^2 + z^2 + yz$. Then $x^2 - 2y^2 + z^2 + yz = 2$ is a level surface of $F$

and $\nabla F(x,y,z) = \langle 2x, -4y + z, 2z + y \rangle$.

(a) $\nabla F(2,1,-1) = \langle 4, -5, -1 \rangle$ is a normal vector for the tangent plane at $(2,1,-1)$, so an equation of the tangent plane

is $4(x-2) - 5(y-1) - 1(z+1) = 0$ or $4x - 5y - z = 4$.

(b) The normal line has direction $\langle 4, -5, -1 \rangle$, so parametric equations are $x = 2 + 4t$, $y = 1 - 5t$, $z = -1 - t$, and

symmetric equations are $\dfrac{x-2}{4} = \dfrac{y-1}{-5} = \dfrac{z+1}{-1}$.

**43.** $F(x, y, z) = -z + xe^y \cos z$  $\Rightarrow$  $\nabla F(x, y, z) = \langle e^y \cos z, xe^y \cos z, -1 - xe^y \sin z \rangle$ and $\nabla F(1, 0, 0) = \langle 1, 1, -1 \rangle$.

(a) $1(x - 1) + 1(y - 0) - 1(z - 0) = 0$ or $x + y - z = 1$

(b) $x - 1 = y = -z$

**45.** $F(x, y, z) = xy + yz + zx$,

$\nabla F(x, y, z) = \langle y + z, x + z, y + x \rangle$,

$\nabla F(1, 1, 1) = \langle 2, 2, 2 \rangle$, so an equation of the tangent

plane is $2x + 2y + 2z = 6$ or $x + y + z = 3$, and the

normal line is given by $x - 1 = y - 1 = z - 1$ or

$x = y = z$. To graph the surface we solve for $z$:

$z = \dfrac{3 - xy}{x + y}$.

**47.** $f(x, y) = xy$  $\Rightarrow$  $\nabla f(x, y) = \langle y, x \rangle$,

$\nabla f(3, 2) = \langle 2, 3 \rangle$. $\nabla f(3, 2)$ is perpendicular to the

tangent line, so the tangent line has equation

$\nabla f(3, 2) \cdot \langle x - 3, y - 2 \rangle = 0$  $\Rightarrow$

$\langle 2, 3 \rangle \cdot \langle x - 3, x - 2 \rangle = 0$  $\Rightarrow$

$2(x - 3) + 3(y - 2) = 0$ or $2x + 3y = 12$.

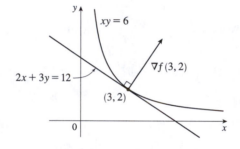

**49.** $\nabla F(x_0, y_0, z_0) = \left\langle \dfrac{2x_0}{a^2}, \dfrac{2y_0}{b^2}, \dfrac{2z_0}{c^2} \right\rangle$. Thus an equation of the tangent plane at $(x_0, y_0, z_0)$ is

$\dfrac{2x_0}{a^2} x + \dfrac{2y_0}{b^2} y + \dfrac{2z_0}{c^2} z = 2\left( \dfrac{x_0^2}{a^2} + \dfrac{y_0^2}{b^2} + \dfrac{z_0^2}{c^2} \right) = 2(1) = 2$ since $(x_0, y_0, z_0)$ is a point on the ellipsoid. Hence

$\dfrac{x_0}{a^2} x + \dfrac{y_0}{b^2} y + \dfrac{z_0}{c^2} z = 1$ is an equation of the tangent plane.

**51.** The hyperboloid $x^2 - y^2 - z^2 = 1$ is a level surface of $F(x, y, z) = x^2 - y^2 - z^2$ and $\nabla F(x, y, z) = \langle 2x, -2y, -2z \rangle$ is a

normal vector to the surface and hence a normal vector for the tangent plane at $(x, y, z)$. The tangent plane is parallel to the

plane $z = x + y$ or $x + y - z = 0$ if and only if the corresponding normal vectors are parallel, so we need a point $(x_0, y_0, z_0)$

on the hyperboloid where $\langle 2x_0, -2y_0, -2z_0 \rangle = c \langle 1, 1, -1 \rangle$ or equivalently $\langle x_0, -y_0, -z_0 \rangle = k \langle 1, 1, -1 \rangle$ for some $k \neq 0$.

Then we must have $x_0 = k$, $y_0 = -k$, $z_0 = k$ and substituting into the equation of the hyperboloid gives

$k^2 - (-k)^2 - k^2 = 1$  $\Leftrightarrow$  $-k^2 = 1$, an impossibility. Thus there is no such point on the hyperboloid.

**53.** Let $(x_0, y_0, z_0)$ be a point on the cone [other than $(0, 0, 0)$]. The cone is a level surface of $F(x, y, z) = x^2 + y^2 - z^2$ and

$\nabla F(x, y, z) = \langle 2x, 2y, -2z \rangle$, so $\nabla F(x_0, y_0, z_0) = \langle 2x_0, 2y_0, -2z_0 \rangle$ is a normal vector to the cone at this point and an

equation of the tangent plane there is $2x_0\,(x - x_0) + 2y_0\,(y - y_0) - 2z_0\,(z - z_0) = 0$ or $x_0 x + y_0 y - z_0 z = x_0^2 + y_0^2 - z_0^2$.

But $x_0^2 + y_0^2 = z_0^2$ so the tangent plane is given by $x_0 x + y_0 y - z_0 z = 0$, a plane which always contains the origin.

**55.** Let $(x_0, y_0, z_0)$ be a point on the surface. Then an equation of the tangent plane at the point is

$\dfrac{x}{2\sqrt{x_0}} + \dfrac{y}{2\sqrt{y_0}} + \dfrac{z}{2\sqrt{z_0}} = \dfrac{\sqrt{x_0} + \sqrt{y_0} + \sqrt{z_0}}{2}$. But $\sqrt{x_0} + \sqrt{y_0} + \sqrt{z_0} = \sqrt{c}$, so the equation is

$\dfrac{x}{\sqrt{x_0}} + \dfrac{y}{\sqrt{y_0}} + \dfrac{z}{\sqrt{z_0}} = \sqrt{c}$. The $x$-, $y$-, and $z$-intercepts are $\sqrt{cx_0}$, $\sqrt{cy_0}$ and $\sqrt{cz_0}$ respectively. (The $x$-intercept is found by

setting $y = z = 0$ and solving the resulting equation for $x$, and the $y$- and $z$-intercepts are found similarly.) So the sum of the

intercepts is $\sqrt{c}\left(\sqrt{x_0} + \sqrt{y_0} + \sqrt{z_0}\right) = c$, a constant.

**57.** If $f(x, y, z) = z - x^2 - y^2$ and $g(x, y, z) = 4x^2 + y^2 + z^2$, then the tangent line is perpendicular to both $\nabla f$ and $\nabla g$

at $(-1, 1, 2)$. The vector $\mathbf{v} = \nabla f \times \nabla g$ will therefore be parallel to the tangent line.

We have $\nabla f(x, y, z) = \langle -2x, -2y, 1 \rangle \;\;\Rightarrow\;\; \nabla f(-1, 1, 2) = \langle 2, -2, 1 \rangle$, and $\nabla g(x, y, z) = \langle 8x, 2y, 2z \rangle \;\;\Rightarrow$

$\nabla g(-1, 1, 2) = \langle -8, 2, 4 \rangle$. Hence $\mathbf{v} = \nabla f \times \nabla g = \begin{vmatrix} \mathbf{i} & \mathbf{j} & \mathbf{k} \\ 2 & -2 & 1 \\ -8 & 2 & 4 \end{vmatrix} = -10\,\mathbf{i} - 16\,\mathbf{j} - 12\,\mathbf{k}$.

Parametric equations are: $x = -1 - 10t, \;\; y = 1 - 16t, \;\; z = 2 - 12t$.

**59.** (a) The direction of the normal line of $F$ is given by $\nabla F$, and that of $G$ by $\nabla G$. Assuming that

$\nabla F \neq 0 \neq \nabla G$, the two normal lines are perpendicular at $P$ if $\nabla F \cdot \nabla G = 0$ at $P \;\;\Leftrightarrow$

$\langle \partial F/\partial x, \partial F/\partial y, \partial F/\partial z \rangle \cdot \langle \partial G/\partial x, \partial G/\partial y, \partial G/\partial z \rangle = 0$ at $P \;\;\Leftrightarrow\;\; F_x G_x + F_y G_y + F_z G_z = 0$ at $P$.

(b) Here $F = x^2 + y^2 - z^2$ and $G = x^2 + y^2 + z^2 - r^2$, so

$\nabla F \cdot \nabla G = \langle 2x, 2y, -2z \rangle \cdot \langle 2x, 2y, 2z \rangle = 4x^2 + 4y^2 - 4z^2 = 4F = 0$, since the point $(x, y, z)$ lies on the graph of

$F = 0$. To see that this is true without using calculus, note that $G = 0$ is the equation of a sphere centered at the origin and

$F = 0$ is the equation of a right circular cone with vertex at the origin (which is generated by lines through the origin). At

any point of intersection, the sphere's normal line (which passes through the origin) lies on the cone, and thus is

perpendicular to the cone's normal line. So the surfaces with equations $F = 0$ and $G = 0$ are everywhere orthogonal.

**61.** Let $\mathbf{u} = \langle a, b \rangle$ and $\mathbf{v} = \langle c, d \rangle$. Then we know that at the given point, $D_{\mathbf{u}}\,f = \nabla f \cdot \mathbf{u} = af_x + bf_y$ and

$D_{\mathbf{v}}\,f = \nabla f \cdot \mathbf{v} = cf_x + df_y$. But these are just two linear equations in the two unknowns $f_x$ and $f_y$, and since $\mathbf{u}$ and $\mathbf{v}$ are

not parallel, we can solve the equations to find $\nabla f = \langle f_x, f_y \rangle$ at the given point. In fact,

$\nabla f = \left\langle \dfrac{d\,D_{\mathbf{u}}\,f - b\,D_{\mathbf{v}}\,f}{ad - bc}, \dfrac{a\,D_{\mathbf{v}}\,f - c\,D_{\mathbf{u}}\,f}{ad - bc} \right\rangle$.

## 11.7 Maximum and Minimum Values

**1.** (a) First we compute $D(1, 1) = f_{xx}(1, 1) f_{yy}(1, 1) - [f_{xy}(1, 1)]^2 = (4)(2) - (1)^2 = 7$. Since $D(1, 1) > 0$ and

$f_{xx}(1, 1) > 0$, $f$ has a local minimum at $(1, 1)$ by the Second Derivatives Test.

(b) $D(1, 1) = f_{xx}(1, 1) f_{yy}(1, 1) - [f_{xy}(1, 1)]^2 = (4)(2) - (3)^2 = -1$. Since $D(1, 1) < 0$, $f$ has a saddle point at $(1, 1)$ by

the Second Derivatives Test.

**3.** In the figure, a point at approximately $(1, 1)$ is enclosed by level curves which are oval in shape and indicate that as we move

away from the point in any direction the values of $f$ are increasing. Hence we would expect a local minimum at or near $(1, 1)$.

The level curves near $(0, 0)$ resemble hyperbolas, and as we move away from the origin, the values of $f$ increase in some

directions and decrease in others, so we would expect to find a saddle point there.

To verify our predictions, we have $f(x, y) = 4 + x^3 + y^3 - 3xy$ $\Rightarrow$ $f_x(x, y) = 3x^2 - 3y$, $f_y(x, y) = 3y^2 - 3x$. We

have critical points where these partial derivatives are equal to 0: $3x^2 - 3y = 0$, $3y^2 - 3x = 0$. Substituting $y = x^2$ from the

first equation into the second equation gives $3(x^2)^2 - 3x = 0$ $\Rightarrow$ $3x(x^3 - 1) = 0$ $\Rightarrow$ $x = 0$ or $x = 1$. Then we have

two critical points, $(0, 0)$ and $(1, 1)$. The second partial derivatives are $f_{xx}(x, y) = 6x$, $f_{xy}(x, y) = -3$, and $f_{yy}(x, y) = 6y$,

so $D(x, y) = f_{xx}(x, y) f_{yy}(x, y) - [f_{xy}(x, y)]^2 = (6x)(6y) - (-3)^2 = 36xy - 9$. Then $D(0, 0) = 36(0)(0) - 9 = -9$,

and $D(1, 1) = 36(1)(1) - 9 = 27$. Since $D(0, 0) < 0$, $f$ has a saddle point at $(0, 0)$ by the Second Derivatives Test. Since

$D(1, 1) > 0$ and $f_{xx}(1, 1) > 0$, $f$ has a local minimum at $(1, 1)$.

**5.** $f(x, y) = x^2 + xy + y^2 + y$ $\Rightarrow$ $f_x = 2x + y$, $f_y = x + 2y + 1$, $f_{xx} = 2$, $f_{xy} = 1$, $f_{yy} = 2$. Then $f_x = 0$ implies

$y = -2x$, and substitution into $f_y = x + 2y + 1 = 0$ gives $x + 2(-2x) + 1 = 0$ $\Rightarrow$ $-3x = -1$ $\Rightarrow$ $x = \frac{1}{3}$.

Then $y = -\frac{2}{3}$ and the only critical point is $\left(\frac{1}{3}, -\frac{2}{3}\right)$. $D(x, y) = f_{xx} f_{yy} - (f_{xy})^2 = (2)(2) - (1)^2 = 3$, and since

$D\left(\frac{1}{3}, -\frac{2}{3}\right) = 3 > 0$ and $f_{xx}\left(\frac{1}{3}, -\frac{2}{3}\right) = 2 > 0$, $f\left(\frac{1}{3}, -\frac{2}{3}\right) = -\frac{1}{3}$ is a local minimum by the Second Derivatives Test.

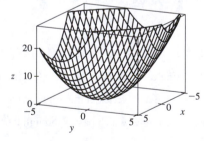

**7.** $f(x, y) = x^4 + y^4 - 4xy + 2$ $\Rightarrow$ $f_x = 4x^3 - 4y$, $f_y = 4y^3 - 4x$,

$f_{xx} = 12x^2$, $f_{xy} = -4$, $f_{yy} = 12y^2$. Then $f_x = 0$ implies $y = x^3$,

and substitution into $f_y = 0$ $\Rightarrow$ $x = y^3$ gives $x^9 - x = 0$ $\Rightarrow$

$x(x^8 - 1) = 0$ $\Rightarrow$ $x = 0$ or $x = \pm 1$. Thus the critical points are $(0, 0)$,

$(1, 1)$, and $(-1, -1)$. Now $D(0, 0) = 0 \cdot 0 - (-4)^2 = -16 < 0$,

so $(0, 0)$ is a saddle point. $D(1, 1) = (12)(12) - (-4)^2 > 0$ and

$f_{xx}(1, 1) = 12 > 0$, so $f(1, 1) = 0$ is a local minimum. $D(-1, -1) = (12)(12) - (-4)^2 > 0$ and

$f_{xx} = (-1, -1) = 12 > 0$, so $f(-1, -1) = 0$ is also a local minimum.

**9.** $f(x, y) = x^3 - 12xy + 8y^3 \quad \Rightarrow \quad f_x = 3x^2 - 12y, \ f_y = -12x + 24y^2,$

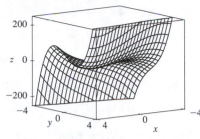

$f_{xx} = 6x, \ f_{xy} = -12, \ f_{yy} = 48y.$ Then $f_x = 0$ implies $x^2 = 4y$ and

$f_y = 0$ implies $x = 2y^2$. Substituting the second equation into the first

gives $(2y^2)^2 = 4y \quad \Rightarrow \quad 4y^4 = 4y \quad \Rightarrow \quad 4y(y^3 - 1) = 0 \quad \Rightarrow \quad y = 0$ or

$y = 1$. If $y = 0$ then $x = 0$ and if $y = 1$ then $x = 2$, so the critical points

are $(0, 0)$ and $(2, 1)$. $\quad D(0, 0) = (0)(0) - (-12)^2 = -144 < 0$, so $(0, 0)$ is a saddle point.

$D(2, 1) = (12)(48) - (-12)^2 = 432 > 0$ and $f_{xx}(2, 1) = 12 > 0$ so $f(2, 1) = -8$ is a local minimum.

**11.** $f(x, y) = e^x \cos y \quad \Rightarrow \quad f_x = e^x \cos y, \ f_y = -e^x \sin y.$

Now $f_x = 0$ implies $\cos y = 0$ or $y = \frac{\pi}{2} + n\pi$ for $n$ an integer.

But $\sin\left(\frac{\pi}{2} + n\pi\right) \neq 0$, so there are no critical points.

**13.** $f(x, y) = (x^2 + y^2)e^{y^2 - x^2} \quad \Rightarrow$

$f_x = (x^2 + y^2)e^{y^2 - x^2}(-2x) + 2xe^{y^2 - x^2} = 2xe^{y^2 - x^2}(1 - x^2 - y^2),$

$f_y = (x^2 + y^2)e^{y^2 - x^2}(2y) + 2ye^{y^2 - x^2} = 2ye^{y^2 - x^2}(1 + x^2 + y^2),$

$f_{xx} = 2xe^{y^2 - x^2}(-2x) + (1 - x^2 - y^2)\left(2x\left(-2xe^{y^2 - x^2}\right) + 2e^{y^2 - x^2}\right) = 2e^{y^2 - x^2}((1 - x^2 - y^2)(1 - 2x^2) - 2x^2),$

$f_{xy} = 2xe^{y^2 - x^2}(-2y) + 2x(2y)e^{y^2 - x^2}(1 - x^2 - y^2) = -4xye^{y^2 - x^2}(x^2 + y^2),$

$f_{yy} = 2ye^{y^2 - x^2}(2y) + (1 + x^2 + y^2)\left(2y\left(2ye^{y^2 - x^2}\right) + 2e^{y^2 - x^2}\right) = 2e^{y^2 - x^2}((1 + x^2 + y^2)(1 + 2y^2) + 2y^2).$

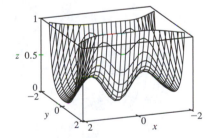

$f_y = 0$ implies $y = 0$, and substituting into $f_x = 0$ gives

$2xe^{-x^2}(1 - x^2) = 0 \quad \Rightarrow \quad x = 0$ or $x = \pm 1$. Thus the critical points are

$(0, 0)$ and $(\pm 1, 0)$. Now $D(0, 0) = (2)(2) - 0 > 0$ and $f_{xx}(0, 0) = 2 > 0$,

so $f(0, 0) = 0$ is a local minimum. $D(\pm 1, 0) = (-4e^{-1})(4e^{-1}) - 0 < 0$

so $(\pm 1, 0)$ are saddle points.

**15.** $f(x, y) = y^2 - 2y \cos x \quad \Rightarrow \quad f_x = 2y \sin x, \ f_y = 2y - 2\cos x,$

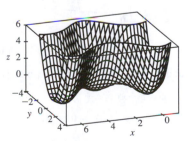

$f_{xx} = 2y \cos x, \ f_{xy} = 2\sin x, \ f_{yy} = 2.$ Then $f_x = 0$ implies $y = 0$ or

$\sin x = 0 \quad \Rightarrow \quad x = 0, \pi,$ or $2\pi$ for $-1 \leq x \leq 7$. Substituting $y = 0$ into

$f_y = 0$ gives $\cos x = 0 \quad \Rightarrow \quad x = \frac{\pi}{2}$ or $\frac{3\pi}{2}$, substituting $x = 0$ or $x = 2\pi$

into $f_y = 0$ gives $y = 1$, and substituting $x = \pi$ into $f_y = 0$ gives $y = -1$.

Thus the critical points are $(0, 1), \left(\frac{\pi}{2}, 0\right), (\pi, -1), \left(\frac{3\pi}{2}, 0\right),$ and $(2\pi, 1)$.

$D\left(\frac{\pi}{2}, 0\right) = D\left(\frac{3\pi}{2}, 0\right) = -4 < 0$ so $\left(\frac{\pi}{2}, 0\right)$ and $\left(\frac{3\pi}{2}, 0\right)$ are saddle points. $D(0, 1) = D(\pi, -1) = D(2\pi, 1) = 4 > 0$ and

$f_{xx}(0, 1) = f_{xx}(\pi, -1) = f_{xx}(2\pi, 1) = 2 > 0$, so $f(0, 1) = f(\pi, -1) = f(2\pi, 1) = -1$ are local minima.

**17.** $f(x, y) = x^2 + 4y^2 - 4xy + 2 \;\; \Rightarrow \;\; f_x = 2x - 4y$, $f_y = 8y - 4x$, $f_{xx} = 2$, $f_{xy} = -4$, $f_{yy} = 8$. Then $f_x = 0$

and $f_y = 0$ each implies $y = \frac{1}{2}x$, so all points of the form $\left(x_0, \frac{1}{2}x_0\right)$ are critical points and for each of these we have

$D\left(x_0, \frac{1}{2}x_0\right) = (2)(8) - (-4)^2 = 0$. The Second Derivatives Test gives no information, but

$f(x, y) = x^2 + 4y^2 - 4xy + 2 = (x - 2y)^2 + 2 \geq 2$ with equality if and only if $y = \frac{1}{2}x$. Thus $f\left(x_0, \frac{1}{2}x_0\right) = 2$ are all local

(and absolute) minima.

**19.** $f(x, y) = x^2 + y^2 + x^{-2}y^{-2}$

  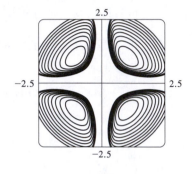

From the graphs, there appear to be local minima of about $f(1, \pm 1) = f(-1, \pm 1) \approx 3$ (and no local maxima or saddle

points). $f_x = 2x - 2x^{-3}y^{-2}$, $f_y = 2y - 2x^{-2}y^{-3}$, $f_{xx} = 2 + 6x^{-4}y^{-2}$, $f_{xy} = 4x^{-3}y^{-3}$, $f_{yy} = 2 + 6x^{-2}y^{-4}$. Then

$f_x = 0$ implies $2x^4y^2 - 2 = 0$ or $x^4y^2 = 1$ or $y^2 = x^{-4}$. Note that neither $x$ nor $y$ can be zero. Now $f_y = 0$ implies

$2x^2y^4 - 2 = 0$, and with $y^2 = x^{-4}$ this implies $2x^{-6} - 2 = 0$ or $x^6 = 1$. Thus $x = \pm 1$ and if $x = 1$, $y = \pm 1$; if $x = -1$,

$y = \pm 1$. So the critical points are $(1, 1)$, $(1, -1)$, $(-1, 1)$ and $(-1, -1)$. Now $D(1, \pm 1) = D(-1, \pm 1) = 64 - 16 > 0$ and

$f_{xx} > 0$ always, so $f(1, \pm 1) = f(-1, \pm 1) = 3$ are local minima.

**21.** $f(x, y) = \sin x + \sin y + \sin(x + y)$, $0 \leq x \leq 2\pi$, $0 \leq y \leq 2\pi$

From the graphs it appears that $f$ has a local maximum at about $(1, 1)$ with value approximately 2.6, a local minimum

at about $(5, 5)$ with value approximately $-2.6$, and a saddle point at about $(3, 3)$.

$f_x = \cos x + \cos(x + y)$, $f_y = \cos y + \cos(x + y)$, $f_{xx} = -\sin x - \sin(x + y)$, $f_{yy} = -\sin y - \sin(x + y)$,

$f_{xy} = -\sin(x + y)$. Setting $f_x = 0$ and $f_y = 0$ and subtracting gives $\cos x - \cos y = 0$ or $\cos x = \cos y$. Thus $x = y$

or $x = 2\pi - y$. If $x = y$, $f_x = 0$ becomes $\cos x + \cos 2x = 0$ or $2\cos^2 x + \cos x - 1 = 0$, a quadratic in $\cos x$. Thus

$\cos x = -1$ or $\frac{1}{2}$ and $x = \pi$, $\frac{\pi}{3}$, or $\frac{5\pi}{3}$, giving the critical points $(\pi, \pi)$, $\left(\frac{\pi}{3}, \frac{\pi}{3}\right)$ and $\left(\frac{5\pi}{3}, \frac{5\pi}{3}\right)$. Similarly if

$x = 2\pi - y$, $f_x = 0$ becomes $(\cos x) + 1 = 0$ and the resulting critical point is $(\pi, \pi)$. Now

$D(x, y) = \sin x \, \sin y + \sin x \, \sin(x + y) + \sin y \, \sin(x + y)$. So $D(\pi, \pi) = 0$ and the Second Derivatives Test doesn't apply.

However, along the line $y = x$ we have $f(x, x) = 2 \sin x + \sin 2x = 2 \sin x + 2 \sin x \, \cos x = 2 \sin x(1 + \cos x)$, and

$f(x, x) > 0$ for $0 < x < \pi$ while $f(x, x) < 0$ for $\pi < x < 2\pi$. Thus every disk with center $(\pi, \pi)$ contains points where $f$ is

positive as well as points where $f$ is negative, so the graph crosses its tangent plane $(z = 0)$ there and $(\pi, \pi)$ is a saddle point.

$D\left(\frac{\pi}{3}, \frac{\pi}{3}\right) = \frac{9}{4} > 0$ and $f_{xx}\left(\frac{\pi}{3}, \frac{\pi}{3}\right) < 0$ so $f\left(\frac{\pi}{3}, \frac{\pi}{3}\right) = \frac{3\sqrt{3}}{2}$ is a local maximum while $D\left(\frac{5\pi}{3}, \frac{5\pi}{3}\right) = \frac{9}{4} > 0$ and

$f_{xx}\left(\frac{5\pi}{3}, \frac{5\pi}{3}\right) > 0$, so $f\left(\frac{5\pi}{3}, \frac{5\pi}{3}\right) = -\frac{3\sqrt{3}}{2}$ is a local minimum.

**23.** $f(x, y) = x^4 - 5x^2 + y^2 + 3x + 2 \;\Rightarrow\; f_x(x, y) = 4x^3 - 10x + 3$ and $f_y(x, y) = 2y$. $f_y = 0 \;\Rightarrow\; y = 0$, and the graph

of $f_x$ shows that the roots of $f_x = 0$ are approximately $x = -1.714, 0.312$ and $1.402$. (Alternatively, we could have used a

calculator or a CAS to find these roots.) So to three decimal places, the critical points are $(-1.714, 0)$, $(1.402, 0)$, and

$(0.312, 0)$. Now since $f_{xx} = 12x^2 - 10$, $f_{xy} = 0$, $f_{yy} = 2$, and $D = 24x^2 - 20$, we have $D(-1.714, 0) > 0$,

$f_{xx}(-1.714, 0) > 0$, $D(1.402, 0) > 0$, $f_{xx}(1.402, 0) > 0$, and $D(0.312, 0) < 0$. Therefore $f(-1.714, 0) \approx -9.200$ and

$f(1.402, 0) \approx 0.242$ are local minima, and $(0.312, 0)$ is a saddle point. The lowest point on the graph is approximately

$(-1.714, 0, -9.200)$.

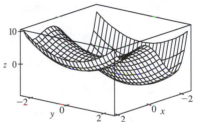

**25.** $f(x, y) = 2x + 4x^2 - y^2 + 2xy^2 - x^4 - y^4 \;\Rightarrow\; f_x(x, y) = 2 + 8x + 2y^2 - 4x^3$, $f_y(x, y) = -2y + 4xy - 4y^3$.

Now $f_y = 0 \;\Leftrightarrow\; 2y(2y^2 - 2x + 1) = 0 \;\Leftrightarrow\; y = 0$ or $y^2 = x - \frac{1}{2}$. The first of these implies that $f_x = -4x^3 + 8x + 2$,

and the second implies that $f_x = 2 + 8x + 2\left(x - \frac{1}{2}\right) - 4x^3 = -4x^3 + 10x + 1$. From the graphs, we see that the first

possibility for $f_x$ has roots at approximately $-1.267, -0.259$, and $1.526$, and the second has a root at

approximately $1.629$ (the negative roots do not give critical points, since $y^2 = x - \frac{1}{2}$ must be positive). So to three decimal

places, $f$ has critical points at $(-1.267, 0)$, $(-0.259, 0)$, $(1.526, 0)$, and $(1.629, \pm 1.063)$. Now since $f_{xx} = 8 - 12x^2$,

$f_{xy} = 4y$, $f_{yy} = 4x - 12y^2$, and $D = (8 - 12x^2)(4x - 12y^2) - 16y^2$, we have $D(-1.267, 0) > 0$, $f_{xx}(-1.267, 0) > 0$,

$D(-0.259, 0) < 0$, $D(1.526, 0) < 0$, $D(1.629, \pm 1.063) > 0$, and $f_{xx}(1.629, \pm 1.063) < 0$. Therefore, to three decimal

places, $f(-1.267, 0) \approx 1.310$ and $f(1.629, \pm 1.063) \approx 8.105$ are local maxima, and $(-0.259, 0)$ and $(1.526, 0)$ are saddle

points. The highest points on the graph are approximately $(1.629, \pm 1.063, 8.105)$.

**27.** Since $f$ is a polynomial it is continuous on $D$, so an absolute maximum and minimum exist. Here $f_x = 4$, $f_y = -5$ so there are no critical points inside $D$. Thus the absolute extrema must both occur on the boundary. Along $L_1$: $x = 0$ and $f(0, y) = 1 - 5y$ for $0 \le y \le 3$, a decreasing function in $y$, so the maximum value is $f(0,0) = 1$ and the minimum value is $f(0,3) = -14$. Along $L_2$: $y = 0$ and $f(x,0) = 1 + 4x$ for $0 \le x \le 2$, an increasing function in $x$, so the minimum value is $f(0,0) = 1$ and the maximum value is $f(2,0) = 9$. Along $L_3$: $y = -\frac{3}{2}x + 3$ and $f\left(x, -\frac{3}{2}x + 3\right) = \frac{23}{2}x - 14$ for $0 \le x \le 2$, an increasing function in $x$, so the minimum value is $f(0,3) = -14$ and the maximum value is $f(2,0) = 9$. Thus the absolute maximum of $f$ on $D$ is $f(2,0) = 9$ and the absolute minimum is $f(0,3) = -14$.

**29.** $f_x(x,y) = 2x + 2xy$, $f_y(x,y) = 2y + x^2$, and setting $f_x = f_y = 0$ gives $(0,0)$ as the only critical point in $D$, with $f(0,0) = 4$.

On $L_1$: $y = -1$, $f(x, -1) = 5$, a constant.

On $L_2$: $x = 1$, $f(1, y) = y^2 + y + 5$, a quadratic in $y$ which attains its maximum at $(1,1)$, $f(1,1) = 7$ and its minimum at $\left(1, -\frac{1}{2}\right)$, $f\left(1, -\frac{1}{2}\right) = \frac{19}{4}$.

On $L_3$: $f(x, 1) = 2x^2 + 5$ which attains its maximum at $(-1, 1)$ and $(1, 1)$ with $f(\pm 1, 1) = 7$ and its minimum at $(0, 1)$, $f(0, 1) = 5$.

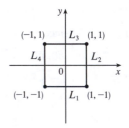

On $L_4$: $f(-1, y) = y^2 + y + 5$ with maximum at $(-1, 1)$, $f(-1, 1) = 7$ and minimum at $\left(-1, -\frac{1}{2}\right)$, $f\left(-1, -\frac{1}{2}\right) = \frac{19}{4}$. Thus the absolute maximum is attained at both $(\pm 1, 1)$ with $f(\pm 1, 1) = 7$ and the absolute minimum on $D$ is attained at $(0,0)$ with $f(0,0) = 4$.

**31.** $f_x(x,y) = 6x^2$ and $f_y(x,y) = 4y^3$. And so $f_x = 0$ and $f_y = 0$ only occur when $x = y = 0$. Hence, the only critical point inside the disk is at $x = y = 0$ where $f(0,0) = 0$. Now on the circle $x^2 + y^2 = 1$, $y^2 = 1 - x^2$ so let $g(x) = f(x,y) = 2x^3 + (1 - x^2)^2 = x^4 + 2x^3 - 2x^2 + 1, -1 \le x \le 1$. Then $g'(x) = 4x^3 + 6x^2 - 4x = 0 \implies x = 0$, $-2$, or $\frac{1}{2}$. $f(0, \pm 1) = g(0) = 1$, $f\left(\frac{1}{2}, \pm \frac{\sqrt{3}}{2}\right) = g\left(\frac{1}{2}\right) = \frac{13}{16}$, and $(-2, -3)$ is not in $D$. Checking the endpoints, we get $f(-1, 0) = g(-1) = -2$ and $f(1, 0) = g(1) = 2$. Thus the absolute maximum and minimum of $f$ on $D$ are $f(1, 0) = 2$ and $f(-1, 0) = -2$.

*Another method:* On the boundary $x^2 + y^2 = 1$ we can write $x = \cos\theta$, $y = \sin\theta$, so $f(\cos\theta, \sin\theta) = 2\cos^3\theta + \sin^4\theta$, $0 \le \theta \le 2\pi$.

**33.** $f(x,y) = -(x^2 - 1)^2 - (x^2 y - x - 1)^2 \implies f_x(x,y) = -2(x^2 - 1)(2x) - 2(x^2 y - x - 1)(2xy - 1)$ and $f_y(x,y) = -2(x^2 y - x - 1)x^2$. Setting $f_y(x,y) = 0$ gives either $x = 0$ or $x^2 y - x - 1 = 0$.

There are no critical points for $x = 0$, since $f_x(0, y) = -2$, so we set $x^2 y - x - 1 = 0 \iff y = \dfrac{x+1}{x^2}$ $[x \ne 0]$,

so $f_x\left(x, \dfrac{x+1}{x^2}\right) = -2(x^2 - 1)(2x) - 2\left(x^2 \dfrac{x+1}{x^2} - x - 1\right)\left(2x \dfrac{x+1}{x^2} - 1\right) = -4x(x^2 - 1)$. Therefore

$f_x(x, y) = f_y(x, y) = 0$ at the points $(1, 2)$ and $(-1, 0)$. To classify these critical points, we calculate

$f_{xx}(x, y) = -12x^2 - 12x^2 y^2 + 12xy + 4y + 2$, $f_{yy}(x, y) = -2x^4$,

and $f_{xy}(x, y) = -8x^3 y + 6x^2 + 4x$. In order to use the Second Derivatives

Test we calculate

$D(-1, 0) = f_{xx}(-1, 0)\, f_{yy}(-1, 0) - [f_{xy}(-1, 0)]^2 = 16 > 0$,

$f_{xx}(-1, 0) = -10 < 0$, $D(1, 2) = 16 > 0$, and $f_{xx}(1, 2) = -26 < 0$, so

both $(-1, 0)$ and $(1, 2)$ give local maxima.

**35.** Let $d$ be the distance from $(2, 1, -1)$ to any point $(x, y, z)$ on the plane $x + y - z = 1$, so

$d = \sqrt{(x-2)^2 + (y-1)^2 + (z+1)^2}$ where $z = x + y - 1$, and we minimize

$d^2 = f(x, y) = (x-2)^2 + (y-1)^2 + (x+y)^2$. Then $f_x(x, y) = 2(x-2) + 2(x+y) = 4x + 2y - 4$,

$f_y(x, y) = 2(y-1) + 2(x+y) = 2x + 4y - 2$. Solving $4x + 2y - 4 = 0$ and $2x + 4y - 2 = 0$ simultaneously gives $x = 1$,

$y = 0$. An absolute minimum exists (since there is a minimum distance from the point to the plane) and it must occur at a

critical point, so the shortest distance occurs for $x = 1$, $y = 0$ for which $d = \sqrt{(1-2)^2 + (0-1)^2 + (0+1)^2} = \sqrt{3}$.

**37.** Let $d$ be the distance from the point $(4, 2, 0)$ to any point $(x, y, z)$ on the cone, so $d = \sqrt{(x-4)^2 + (y-2)^2 + z^2}$ where

$z^2 = x^2 + y^2$, and we minimize $d^2 = (x-4)^2 + (y-2)^2 + x^2 + y^2 = f(x, y)$. Then

$f_x(x, y) = 2(x-4) + 2x = 4x - 8$, $f_y(x, y) = 2(y-2) + 2y = 4y - 4$, and the critical points occur when

$f_x = 0 \Rightarrow x = 2$, $f_y = 0 \Rightarrow y = 1$. Thus the only critical point is $(2, 1)$. An absolute minimum exists (since there

is a minimum distance from the cone to the point) which must occur at a critical point, so the points on the cone closest

to $(4, 2, 0)$ are $\left(2, 1, \pm\sqrt{5}\right)$.

**39.** $x + y + z = 100$, so maximize $f(x, y) = xy(100 - x - y)$.   $f_x = 100y - 2xy - y^2$, $f_y = 100x - x^2 - 2xy$,

$f_{xx} = -2y$, $f_{yy} = -2x$, $f_{xy} = 100 - 2x - 2y$. Then $f_x = 0$ implies $y = 0$ or $y = 100 - 2x$. Substituting $y = 0$ into

$f_y = 0$ gives $x = 0$ or $x = 100$ and substituting $y = 100 - 2x$ into $f_y = 0$ gives $3x^2 - 100x = 0$ so $x = 0$ or $\frac{100}{3}$.

Thus the critical points are $(0, 0)$, $(100, 0)$, $(0, 100)$ and $\left(\frac{100}{3}, \frac{100}{3}\right)$.

$D(0, 0) = D(100, 0) = D(0, 100) = -10{,}000$ while $D\left(\frac{100}{3}, \frac{100}{3}\right) = \frac{10{,}000}{3}$ and $f_{xx}\left(\frac{100}{3}, \frac{100}{3}\right) = -\frac{200}{3} < 0$. Thus $(0, 0)$,

$(100, 0)$ and $(0, 100)$ are saddle points whereas $f\left(\frac{100}{3}, \frac{100}{3}\right)$ is a local maximum. Thus the numbers are $x = y = z = \frac{100}{3}$.

**41.** Center the sphere at the origin so that its equation is $x^2 + y^2 + z^2 = r^2$, and orient the inscribed rectangular box so that its

edges are parallel to the coordinate axes. Any vertex of the box satisfies $x^2 + y^2 + z^2 = r^2$, so take $(x, y, z)$ to be the vertex

in the first octant. Then the box has length $2x$, width $2y$, and height $2z = 2\sqrt{r^2 - x^2 - y^2}$ with volume given by

$V(x, y) = (2x)(2y)\left(2\sqrt{r^2 - x^2 - y^2}\right) = 8xy\sqrt{r^2 - x^2 - y^2}$ for $0 < x < r$, $0 < y < r$. Then

$V_x = (8xy) \cdot \frac{1}{2}(r^2 - x^2 - y^2)^{-1/2}(-2x) + \sqrt{r^2 - x^2 - y^2} \cdot 8y = \dfrac{8y(r^2 - 2x^2 - y^2)}{\sqrt{r^2 - x^2 - y^2}}$ and $V_y = \dfrac{8x(r^2 - x^2 - 2y^2)}{\sqrt{r^2 - x^2 - y^2}}$.

Setting $V_x = 0$ gives $y = 0$ or $2x^2 + y^2 = r^2$, but $y > 0$ so only the latter solution applies. Similarly, $V_y = 0$ with $x > 0$

implies $x^2 + 2y^2 = r^2$. Substituting, we have $2x^2 + y^2 = x^2 + 2y^2 \;\Rightarrow\; x^2 = y^2 \;\Rightarrow\; y = x$. Then $x^2 + 2y^2 = r^2 \;\Rightarrow\;$

$3x^2 = r^2 \;\Rightarrow\; x = \sqrt{r^2/3} = r/\sqrt{3} = y$. Thus the only critical point is $\left(r/\sqrt{3}, r/\sqrt{3}\right)$. There must be a maximum

volume and here it must occur at a critical point, so the maximum volume occurs when $x = y = r/\sqrt{3}$ and the maximum

volume is $V\left(\frac{r}{\sqrt{3}}, \frac{r}{\sqrt{3}}\right) = 8\left(\frac{r}{\sqrt{3}}\right)\left(\frac{r}{\sqrt{3}}\right)\sqrt{r^2 - \left(\frac{r}{\sqrt{3}}\right)^2 - \left(\frac{r}{\sqrt{3}}\right)^2} = \dfrac{8}{3\sqrt{3}}\, r^3.$

**43.** Maximize $f(x, y) = \dfrac{xy}{3}(6 - x - 2y)$, then the maximum volume is $V = xyz$.

$f_x = \frac{1}{3}(6y - 2xy - y^2) = \frac{1}{3}y(6 - 2x - 2y)$ and $f_y = \frac{1}{3}x\,(6 - x - 4y)$. Setting $f_x = 0$ and $f_y = 0$ gives the critical point

$(2, 1)$ which geometrically must give a maximum. Thus the volume of the largest such box is $V = (2)(1)\left(\frac{2}{3}\right) = \frac{4}{3}$.

**45.** Let the dimensions be $x$, $y$, and $z$; then $4x + 4y + 4z = c$ and the volume is

$V = xyz = xy\left(\frac{1}{4}c - x - y\right) = \frac{1}{4}cxy - x^2y - xy^2$, $x > 0$, $y > 0$. Then $V_x = \frac{1}{4}cy - 2xy - y^2$ and $V_y = \frac{1}{4}cx - x^2 - 2xy$,

so $V_x = 0 = V_y$ when $2x + y = \frac{1}{4}c$ and $x + 2y = \frac{1}{4}c$. Solving, we get $x = \frac{1}{12}c$, $y = \frac{1}{12}c$ and $z = \frac{1}{4}c - x - y = \frac{1}{12}c$. From

the geometrical nature of the problem, this critical point must give an absolute maximum. Thus the box is a cube with edge

length $\frac{1}{12}c$.

**47.** Let the dimensions be $x$, $y$ and $z$, then minimize $xy + 2(xz + yz)$ if $xyz = 32{,}000$ cm$^3$. Then

$f(x, y) = xy + [64{,}000(x + y)/xy] = xy + 64{,}000(x^{-1} + y^{-1})$, $f_x = y - 64{,}000x^{-2}$, $f_y = x - 64{,}000y^{-2}$.

And $f_x = 0$ implies $y = 64{,}000/x^2$; substituting into $f_y = 0$ implies $x^3 = 64{,}000$ or $x = 40$ and then $y = 40$. Now

$D(x, y) = [(2)(64{,}000)]^2 x^{-3}y^{-3} - 1 > 0$ for $(40, 40)$ and $f_{xx}(40, 40) > 0$ so this is indeed a minimum. Thus the

dimensions of the box are $x = y = 40$ cm, $z = 20$ cm.

**49.** Let $x$, $y$, $z$ be the dimensions of the rectangular box. Then the volume of the box is $xyz$ and

$L = \sqrt{x^2 + y^2 + z^2} \;\Rightarrow\; L^2 = x^2 + y^2 + z^2 \;\Rightarrow\; z = \sqrt{L^2 - x^2 - y^2}.$

Substituting, we have volume $V(x, y) = xy\sqrt{L^2 - x^2 - y^2}$ $(x, y > 0)$.

$V_x = xy \cdot \frac{1}{2}(L^2 - x^2 - y^2)^{-1/2}(-2x) + y\sqrt{L^2 - x^2 - y^2} = y\sqrt{L^2 - x^2 - y^2} - \dfrac{x^2y}{\sqrt{L^2 - x^2 - y^2}},$

$V_y = x\sqrt{L^2 - x^2 - y^2} - \dfrac{xy^2}{\sqrt{L^2 - x^2 - y^2}}.$  $V_x = 0$ implies $y(L^2 - x^2 - y^2) = x^2y \;\Rightarrow\; y(L^2 - 2x^2 - y^2) = 0 \;\Rightarrow\;$

$2x^2 + y^2 = L^2$ (since $y > 0$), and $V_y = 0$ implies $x(L^2 - x^2 - y^2) = xy^2 \;\Rightarrow\; x(L^2 - x^2 - 2y^2) = 0 \;\Rightarrow\;$

$x^2 + 2y^2 = L^2$ (since $x > 0$). Substituting $y^2 = L^2 - 2x^2$ into $x^2 + 2y^2 = L^2$ gives $x^2 + 2L^2 - 4x^2 = L^2 \;\Rightarrow\;$

$3x^2 = L^2 \;\Rightarrow\; x = L/\sqrt{3}$ (since $x > 0$) and then $y = \sqrt{L^2 - 2\left(L/\sqrt{3}\right)^2} = L/\sqrt{3}.$

So the only critical point is $\left(L/\sqrt{3}, L/\sqrt{3}\right)$ which, from the geometrical nature of the problem, must give an absolute

maximum. Thus the maximum volume is $V\left(L/\sqrt{3}, L/\sqrt{3}\right) = \left(L/\sqrt{3}\right)^2\sqrt{L^2 - \left(L/\sqrt{3}\right)^2 - \left(L/\sqrt{3}\right)^2} = L^3/(3\sqrt{3})$

cubic units.

**51.** Note that here the variables are $m$ and $b$, and $f(m, b) = \sum\limits_{i=1}^{n} [y_i - (mx_i + b)]^2$. Then $f_m = \sum\limits_{i=1}^{n} -2x_i[y_i - (mx_i + b)] = 0$

implies $\sum\limits_{i=1}^{n} (x_i y_i - mx_i^2 - bx_i) = 0$ or $\sum\limits_{i=1}^{n} x_i y_i = m \sum\limits_{i=1}^{n} x_i^2 + b \sum\limits_{i=1}^{n} x_i$ and $f_b = \sum\limits_{i=1}^{n} -2[y_i - (mx_i + b)] = 0$ implies

$\sum\limits_{i=1}^{n} y_i = m \sum\limits_{i=1}^{n} x_i + \sum\limits_{i=1}^{n} b = m \left( \sum\limits_{i=1}^{n} x_i \right) + nb$. Thus we have the two desired equations.

Now $f_{mm} = \sum\limits_{i=1}^{n} 2x_i^2$, $f_{bb} = \sum\limits_{i=1}^{n} 2 = 2n$ and $f_{mb} = \sum\limits_{i=1}^{n} 2x_i$. And $f_{mm}(m, b) > 0$ always and

$D(m, b) = 4n \left( \sum\limits_{i=1}^{n} x_i^2 \right) - 4 \left( \sum\limits_{i=1}^{n} x_i \right)^2 = 4 \left[ n \left( \sum\limits_{i=1}^{n} x_i^2 \right) - \left( \sum\limits_{i=1}^{n} x_i \right)^2 \right] > 0$ always so the solutions of these two

equations do indeed minimize $\sum\limits_{i=1}^{n} d_i^2$.

## 11.8 Lagrange Multipliers

**1.** At the extreme values of $f$, the level curves of $f$ just touch the curve $g(x, y) = 8$ with a common tangent line. (See Figure 1

and the accompanying discussion.) We can observe several such occurrences on the contour map, but the level curve

$f(x, y) = c$ with the largest value of $c$ which still intersects the curve $g(x, y) = 8$ is approximately $c = 59$, and the smallest

value of $c$ corresponding to a level curve which intersects $g(x, y) = 8$ appears to be $c = 30$. Thus we estimate the maximum

value of $f$ subject to the constraint $g(x, y) = 8$ to be about 59 and the minimum to be 30.

**3.** $f(x, y) = x^2 + y^2$, $g(x, y) = xy = 1$, and $\nabla f = \lambda \nabla g \Rightarrow \langle 2x, 2y \rangle = \langle \lambda y, \lambda x \rangle$, so $2x = \lambda y$, $2y = \lambda x$, and $xy = 1$.

From the last equation, $x \neq 0$ and $y \neq 0$, so $2x = \lambda y \Rightarrow \lambda = 2x/y$. Substituting, we have $2y = (2x/y)x \Rightarrow$

$y^2 = x^2 \Rightarrow y = \pm x$. But $xy = 1$, so $x = y = \pm 1$ and the possible points for the extreme values of $f$ are $(1, 1)$ and

$(-1, -1)$. Here there is no maximum value, since the constraint $xy = 1$ allows $x$ or $y$ to become arbitrarily large, and hence

$f(x, y) = x^2 + y^2$ can be made arbitrarily large. The minimum value is $f(1, 1) = f(-1, -1) = 2$.

**5.** $f(x, y) = x^2 y$, $g(x, y) = x^2 + 2y^2 = 6 \Rightarrow \nabla f = \langle 2xy, x^2 \rangle$, $\lambda \nabla g = \langle 2\lambda x, 4\lambda y \rangle$. Then $2xy = 2\lambda x$ implies $x = 0$

or $\lambda = y$. If $x = 0$, then $x^2 = 4\lambda y$ implies $\lambda = 0$ or $y = 0$. However, if $y = 0$ then $g(x, y) = 0$, a contradiction.

So $\lambda = 0$ and then $g(x, y) = 6 \Rightarrow y = \pm\sqrt{3}$. If $\lambda = y$, then $x^2 = 4\lambda y$ implies $x^2 = 4y^2$, and so $g(x, y) = 6 \Rightarrow$

$4y^2 + 2y^2 = 6 \Rightarrow y^2 = 1 \Rightarrow y = \pm 1$. Thus $f$ has possible extreme values at the points $(0, \pm\sqrt{3})$, $(\pm 2, 1)$, and

$(\pm 2, -1)$. After evaluating $f$ at these points, we find the maximum value to be $f(\pm 2, 1) = 4$ and the minimum to be

$f(\pm 2, -1) = -4$.

**7.** $f(x, y, z) = 2x + 6y + 10z$, $g(x, y, z) = x^2 + y^2 + z^2 = 35 \Rightarrow \nabla f = \langle 2, 6, 10 \rangle$, $\lambda \nabla g = \langle 2\lambda x, 2\lambda y, 2\lambda z \rangle$. Then

$2\lambda x = 2$, $2\lambda y = 6$, $2\lambda z = 10$ imply $x = \dfrac{1}{\lambda}$, $y = \dfrac{3}{\lambda}$, and $z = \dfrac{5}{\lambda}$. But $35 = x^2 + y^2 + z^2 = \left(\dfrac{1}{\lambda}\right)^2 + \left(\dfrac{3}{\lambda}\right)^2 + \left(\dfrac{5}{\lambda}\right)^2 \Rightarrow$

$35 = \dfrac{35}{\lambda^2} \Rightarrow \lambda = \pm 1$, so $f$ has possible extreme values at the points $(1, 3, 5)$, $(-1, -3, -5)$. The maximum value of $f$ on

$x^2 + y^2 + z^2 = 35$ is $f(1, 3, 5) = 70$, and the minimum is $f(-1, -3, -5) = -70$.

**9.** $f(x, y, z) = xyz$, $g(x, y, z) = x^2 + 2y^2 + 3z^2 = 6$. $\nabla f = \lambda \nabla g$ $\Rightarrow$ $\langle yz, xz, xy \rangle = \lambda \langle 2x, 4y, 6z \rangle$. If any of $x$, $y$, or $z$ is zero then $x = y = z = 0$ which contradicts $x^2 + 2y^2 + 3z^2 = 6$. Then $\lambda = (yz)/(2x) = (xz)/(4y) = (xy)/(6z)$ or $x^2 = 2y^2$ and $z^2 = \frac{2}{3}y^2$. Thus $x^2 + 2y^2 + 3z^2 = 6$ implies $6y^2 = 6$ or $y = \pm 1$. Then the possible points are $\left( \sqrt{2}, \pm 1, \sqrt{\frac{2}{3}} \right)$, $\left( \sqrt{2}, \pm 1, -\sqrt{\frac{2}{3}} \right)$, $\left( -\sqrt{2}, \pm 1, \sqrt{\frac{2}{3}} \right)$, $\left( -\sqrt{2}, \pm 1, -\sqrt{\frac{2}{3}} \right)$. The maximum value of $f$ on the ellipsoid is $\frac{2}{\sqrt{3}}$, occurring when all coordinates are positive or exactly two are negative and the minimum is $-\frac{2}{\sqrt{3}}$ occurring when 1 or 3 of the coordinates are negative.

**11.** $f(x, y, z) = x^2 + y^2 + z^2$, $g(x, y, z) = x^4 + y^4 + z^4 = 1$ $\Rightarrow$ $\nabla f = \langle 2x, 2y, 2z \rangle$, $\lambda \nabla g = \langle 4\lambda x^3, 4\lambda y^3, 4\lambda z^3 \rangle$.

*Case 1:* If $x \neq 0$, $y \neq 0$ and $z \neq 0$, then $\nabla f = \lambda \nabla g$ implies $\lambda = 1/(2x^2) = 1/(2y^2) = 1/(2z^2)$ or $x^2 = y^2 = z^2$ and $3x^4 = 1$ or $x = \pm \frac{1}{\sqrt[4]{3}}$ giving the points $\left( \pm \frac{1}{\sqrt[4]{3}}, \frac{1}{\sqrt[4]{3}}, \frac{1}{\sqrt[4]{3}} \right)$, $\left( \pm \frac{1}{\sqrt[4]{3}}, -\frac{1}{\sqrt[4]{3}}, \frac{1}{\sqrt[4]{3}} \right)$, $\left( \pm \frac{1}{\sqrt[4]{3}}, \frac{1}{\sqrt[4]{3}}, -\frac{1}{\sqrt[4]{3}} \right)$, $\left( \pm \frac{1}{\sqrt[4]{3}}, -\frac{1}{\sqrt[4]{3}}, -\frac{1}{\sqrt[4]{3}} \right)$ all with an $f$-value of $\sqrt{3}$.

*Case 2:* If one of the variables equals zero and the other two are not zero, then the squares of the two nonzero coordinates are equal with common value $\frac{1}{\sqrt{2}}$ and corresponding $f$ value of $\sqrt{2}$.

*Case 3:* If exactly two of the variables are zero, then the third variable has value $\pm 1$ with the corresponding $f$ value of 1. Thus on $x^4 + y^4 + z^4 = 1$, the maximum value of $f$ is $\sqrt{3}$ and the minimum value is 1.

**13.** $f(x, y, z, t) = x + y + z + t$, $g(x, y, z, t) = x^2 + y^2 + z^2 + t^2 = 1$ $\Rightarrow$ $\langle 1, 1, 1, 1 \rangle = \langle 2\lambda x, 2\lambda y, 2\lambda z, 2\lambda t \rangle$, so $\lambda = 1/(2x) = 1/(2y) = 1/(2z) = 1/(2t)$ and $x = y = z = t$. But $x^2 + y^2 + z^2 + t^2 = 1$, so the possible points are $\left( \pm \frac{1}{2}, \pm \frac{1}{2}, \pm \frac{1}{2}, \pm \frac{1}{2} \right)$. Thus the maximum value of $f$ is $f\left( \frac{1}{2}, \frac{1}{2}, \frac{1}{2}, \frac{1}{2} \right) = 2$ and the minimum value is $f\left( -\frac{1}{2}, -\frac{1}{2}, -\frac{1}{2}, -\frac{1}{2} \right) = -2$.

**15.** $f(x, y, z) = x + 2y$, $g(x, y, z) = x + y + z = 1$, $h(x, y, z) = y^2 + z^2 = 4$ $\Rightarrow$ $\nabla f = \langle 1, 2, 0 \rangle$, $\lambda \nabla g = \langle \lambda, \lambda, \lambda \rangle$ and $\mu \nabla h = \langle 0, 2\mu y, 2\mu z \rangle$. Then $1 = \lambda$, $2 = \lambda + 2\mu y$ and $0 = \lambda + 2\mu z$ so $\mu y = \frac{1}{2} = -\mu z$ or $y = 1/(2\mu)$, $z = -1/(2\mu)$. Thus $x + y + z = 1$ implies $x = 1$ and $y^2 + z^2 = 4$ implies $\mu = \pm \frac{1}{2\sqrt{2}}$. Then the possible points are $\left( 1, \pm\sqrt{2}, \mp\sqrt{2} \right)$ and the maximum value is $f\left( 1, \sqrt{2}, -\sqrt{2} \right) = 1 + 2\sqrt{2}$ and the minimum value is $f\left( 1, -\sqrt{2}, \sqrt{2} \right) = 1 - 2\sqrt{2}$.

**17.** $f(x, y, z) = yz + xy$, $g(x, y, z) = xy = 1$, $h(x, y, z) = y^2 + z^2 = 1$ $\Rightarrow$ $\nabla f = \langle y, x + z, y \rangle$, $\lambda \nabla g = \langle \lambda y, \lambda x, 0 \rangle$, $\mu \nabla h = \langle 0, 2\mu y, 2\mu z \rangle$. Then $y = \lambda y$ implies $\lambda = 1$ [$y \neq 0$ since $g(x, y, z) = 1$], $x + z = \lambda x + 2\mu y$ and $y = 2\mu z$. Thus $\mu = z/(2y) = y/(2y)$ or $y^2 = z^2$, and so $y^2 + z^2 = 1$ implies $y = \pm\frac{1}{\sqrt{2}}$, $z = \pm\frac{1}{\sqrt{2}}$. Then $xy = 1$ implies $x = \pm\sqrt{2}$ and the possible points are $\left( \pm\sqrt{2}, \pm\frac{1}{\sqrt{2}}, \frac{1}{\sqrt{2}} \right)$, $\left( \pm\sqrt{2}, \pm\frac{1}{\sqrt{2}}, -\frac{1}{\sqrt{2}} \right)$. Hence the maximum of $f$ subject to the constraints is $f\left( \pm\sqrt{2}, \pm\frac{1}{\sqrt{2}}, \pm\frac{1}{\sqrt{2}} \right) = \frac{3}{2}$ and the minimum is $f\left( \pm\sqrt{2}, \pm\frac{1}{\sqrt{2}}, \mp\frac{1}{\sqrt{2}} \right) = \frac{1}{2}$.

*Note:* Since $xy = 1$ is one of the constraints we could have solved the problem by solving $f(y, z) = yz + 1$ subject to $y^2 + z^2 = 1$.

**19.** $f(x, y) = e^{-xy}$.    For the interior of the region, we find the critical points: $f_x = -ye^{-xy}$, $f_y = -xe^{-xy}$, so the only

critical point is $(0, 0)$, and $f(0, 0) = 1$. For the boundary, we use Lagrange multipliers. $g(x, y) = x^2 + 4y^2 = 1$ $\Rightarrow$

$\lambda \nabla g = \langle 2\lambda x, 8\lambda y \rangle$, so setting $\nabla f = \lambda \nabla g$ we get $-ye^{-xy} = 2\lambda x$ and $-xe^{-xy} = 8\lambda y$. The first of these gives

$e^{-xy} = -2\lambda x/y$, and then the second gives $-x(-2\lambda x/y) = 8\lambda y$ $\Rightarrow$ $x^2 = 4y^2$. Solving this last equation with the

constraint $x^2 + 4y^2 = 1$ gives $x = \pm\frac{1}{\sqrt{2}}$ and $y = \pm\frac{1}{2\sqrt{2}}$. Now $f\left(\pm\frac{1}{\sqrt{2}}, \mp\frac{1}{2\sqrt{2}}\right) = e^{1/4} \approx 1.284$ and

$f\left(\pm\frac{1}{\sqrt{2}}, \pm\frac{1}{2\sqrt{2}}\right) = e^{-1/4} \approx 0.779$. The former are the maxima on the region and the latter are the minima.

**21.** (a) $f(x, y) = x$,  $g(x, y) = y^2 + x^4 - x^3 = 0$ $\Rightarrow$ $\nabla f = \langle 1, 0 \rangle = \lambda \nabla g = \lambda \langle 4x^3 - 3x^2, 2y \rangle$. Then

$1 = \lambda(4x^3 - 3x^2)$ **(1)** and $0 = 2\lambda y$ **(2)**. We have $\lambda \neq 0$ from **(1)**, so **(2)** gives $y = 0$. Then, from the constraint equation,

$x^4 - x^3 = 0$ $\Rightarrow$ $x^3(x - 1) = 0$ $\Rightarrow$ $x = 0$ or $x = 1$. But $x = 0$ contradicts **(1)**, so the only possible extreme value

subject to the constraint is $f(1, 0) = 1$. (The question remains whether this is indeed the minimum of $f$.)

(b) The constraint is $y^2 + x^4 - x^3 = 0$ $\Leftrightarrow$ $y^2 = x^3 - x^4$. The left side is non-negative, so we must have $x^3 - x^4 \geq 0$

which is true only for $0 \leq x \leq 1$. Therefore the minimum possible value for $f(x, y) = x$ is 0 which occurs for $x = y = 0$.

However, $\lambda \nabla g(0, 0) = \lambda \langle 0 - 0, 0 \rangle = \langle 0, 0 \rangle$ and $\nabla f(0, 0) = \langle 1, 0 \rangle$, so $\nabla f(0, 0) \neq \lambda \nabla g(0, 0)$ for all values of $\lambda$.

(c) Here $\nabla g(0, 0) = \mathbf{0}$ but the method of Lagrange multipliers requires that $\nabla g \neq \mathbf{0}$ everywhere on the constraint curve.

**23.** $P(L, K) = bL^\alpha K^{1-\alpha}$, $g(L, K) = mL + nK = p$ $\Rightarrow$ $\nabla P = \langle \alpha b L^{\alpha-1} K^{1-\alpha}, (1-\alpha)bL^\alpha K^{-\alpha} \rangle$, $\lambda \nabla g = \langle \lambda m, \lambda n \rangle$.

Then $\alpha b(K/L)^{1-\alpha} = \lambda m$ and $(1-\alpha)b(L/K)^\alpha = \lambda n$ and $mL + nK = p$, so $\alpha b(K/L)^{1-\alpha}/m = (1-\alpha)b(L/K)^\alpha/n$ or

$n\alpha/[m(1-\alpha)] = (L/K)^\alpha(L/K)^{1-\alpha}$ or $L = Kn\alpha/[m(1-\alpha)]$. Substituting into $mL + nK = p$ gives $K = (1-\alpha)p/n$

and $L = \alpha p/m$ for the maximum production.

**25.** Let the sides of the rectangle be $x$ and $y$. Then $f(x, y) = xy$, $g(x, y) = 2x + 2y = p$ $\Rightarrow$ $\nabla f(x, y) = \langle y, x \rangle$,

$\lambda \nabla g = \langle 2\lambda, 2\lambda \rangle$. Then $\lambda = \frac{1}{2}y = \frac{1}{2}x$ implies $x = y$ and the rectangle with maximum area is a square with side length $\frac{1}{4}p$.

**27.** Let $f(x, y, z) = d^2 = (x - 2)^2 + (y - 1)^2 + (z + 1)^2$, then we want to minimize $f$ subject to the constraint

$g(x, y, z) = x + y - z = 1$.    $\nabla f = \lambda \nabla g$ $\Rightarrow$ $\langle 2(x - 2), 2(y - 1), 2(z + 1) \rangle = \lambda \langle 1, 1, -1 \rangle$, so $x = (\lambda + 4)/2$,

$y = (\lambda + 2)/2$, $z = -(\lambda + 2)/2$. Substituting into the constraint equation gives $\dfrac{\lambda + 4}{2} + \dfrac{\lambda + 2}{2} + \dfrac{\lambda + 2}{2} = 1$ $\Rightarrow$

$3\lambda + 8 = 2$ $\Rightarrow$ $\lambda = -2$, so $x = 1$, $y = 0$, and $z = 0$. This must correspond to a minimum, so the shortest distance is

$d = \sqrt{(1 - 2)^2 + (0 - 1)^2 + (0 + 1)^2} = \sqrt{3}$.

**29.** Let $f(x, y, z) = d^2 = (x - 4)^2 + (y - 2)^2 + z^2$. Then we want to minimize $f$ subject to the constraint

$g(x, y, z) = x^2 + y^2 - z^2 = 0$.    $\nabla f = \lambda \nabla g$ $\Rightarrow$ $\langle 2(x - 4), 2(y - 2), 2z \rangle = \langle 2\lambda x, 2\lambda y, -2\lambda z \rangle$, so $x - 4 = \lambda x$,

$y - 2 = \lambda y$, and $z = -\lambda z$. From the last equation we have $z + \lambda z = 0$ $\Rightarrow$ $z(1 + \lambda) = 0$, so either $z = 0$ or $\lambda = -1$.

But from the constraint equation we have $z = 0$ $\Rightarrow$ $x^2 + y^2 = 0$ $\Rightarrow$ $x = y = 0$ which is not possible from the first

two equations. So $\lambda = -1$ and $x - 4 = \lambda x$ $\Rightarrow$ $x = 2$, $y - 2 = \lambda y$ $\Rightarrow$ $y = 1$, and $x^2 + y^2 - z^2 = 0$ $\Rightarrow$

$4 + 1 - z^2 = 0$ $\Rightarrow$ $z = \pm\sqrt{5}$. This must correspond to a minimum, so the points on the cone closest to $(4, 2, 0)$

are $(2, 1, \pm\sqrt{5})$.

**31.** $f(x, y, z) = xyz$, $g(x, y, z) = x + y + z = 100$ $\Rightarrow$ $\nabla f = \langle yz, xz, xy \rangle = \lambda \nabla g = \langle \lambda, \lambda, \lambda \rangle$. Then $\lambda = yz = xz = xy$ implies $x = y = z = \frac{100}{3}$.

**33.** If the dimensions are $2x$, $2y$, and $2z$, then maximize $f(x, y, z) = (2x)(2y)(2z) = 8xyz$ subject to

$g(x, y, z) = x^2 + y^2 + z^2 = r^2$ ($x > 0$, $y > 0$, $z > 0$). Then $\nabla f = \lambda \nabla g$ $\Rightarrow$ $\langle 8yz, 8xz, 8xy \rangle = \lambda \langle 2x, 2y, 2z \rangle$ $\Rightarrow$

$8yz = 2\lambda x$, $8xz = 2\lambda y$, and $8xy = 2\lambda z$, so $\lambda = \dfrac{4yz}{x} = \dfrac{4xz}{y} = \dfrac{4xy}{z}$. This gives $x^2 z = y^2 z$ $\Rightarrow$ $x^2 = y^2$ (since $z \neq 0$)

and $xy^2 = xz^2$ $\Rightarrow$ $z^2 = y^2$, so $x^2 = y^2 = z^2$ $\Rightarrow$ $x = y = z$, and substituting into the constraint

equation gives $3x^2 = r^2$ $\Rightarrow$ $x = r/\sqrt{3} = y = z$. Thus the largest volume of such a box is

$$f\left(\tfrac{r}{\sqrt{3}}, \tfrac{r}{\sqrt{3}}, \tfrac{r}{\sqrt{3}}\right) = 8\left(\tfrac{r}{\sqrt{3}}\right)\left(\tfrac{r}{\sqrt{3}}\right)\left(\tfrac{r}{\sqrt{3}}\right) = \frac{8}{3\sqrt{3}} r^3.$$

**35.** $f(x, y, z) = xyz$, $g(x, y, z) = x + 2y + 3z = 6$ $\Rightarrow$ $\nabla f = \langle yz, xz, xy \rangle = \lambda \nabla g = \langle \lambda, 2\lambda, 3\lambda \rangle$.

Then $\lambda = yz = \frac{1}{2}xz = \frac{1}{3}xy$ implies $x = 2y$, $z = \frac{2}{3}y$. But $2y + 2y + 2y = 6$ so $y = 1$, $x = 2$, $z = \frac{2}{3}$ and the volume

is $V = \frac{4}{3}$.

**37.** $f(x, y, z) = xyz$, $g(x, y, z) = 4(x + y + z) = c$ $\Rightarrow$ $\nabla f = \langle yz, xz, xy \rangle$, $\lambda \nabla g = \langle 4\lambda, 4\lambda, 4\lambda \rangle$. Thus

$4\lambda = yz = xz = xy$ or $x = y = z = \frac{1}{12}c$ are the dimensions giving the maximum volume.

**39.** If the dimensions of the box are given by $x$, $y$, and $z$, then we need to find the maximum value of $f(x, y, z) = xyz$

$[x, y, z > 0]$ subject to the constraint $L = \sqrt{x^2 + y^2 + z^2}$ or $g(x, y, z) = x^2 + y^2 + z^2 = L^2$. $\nabla f = \lambda \nabla g$ $\Rightarrow$

$\langle yz, xz, xy \rangle = \lambda \langle 2x, 2y, 2z \rangle$, so $yz = 2\lambda x$ $\Rightarrow$ $\lambda = \dfrac{yz}{2x}$, $xz = 2\lambda y$ $\Rightarrow$ $\lambda = \dfrac{xz}{2y}$, and $xy = 2\lambda z$ $\Rightarrow$ $\lambda = \dfrac{xy}{2z}$.

Thus $\lambda = \dfrac{yz}{2x} = \dfrac{xz}{2y}$ $\Rightarrow$ $x^2 = y^2$ [since $z \neq 0$] $\Rightarrow$ $x = y$ and $\lambda = \dfrac{yz}{2x} = \dfrac{xy}{2z}$ $\Rightarrow$ $x = z$ [since $y \neq 0$].

Substituting into the constraint equation gives $x^2 + x^2 + x^2 = L^2$ $\Rightarrow$ $x^2 = L^2/3$ $\Rightarrow$ $x = L/\sqrt{3} = y = z$ and the

maximum volume is $\left(L/\sqrt{3}\right)^3 = L^3/\left(3\sqrt{3}\right)$.

**41.** We need to find the extreme values of $f(x, y, z) = x^2 + y^2 + z^2$ subject to the two constraints $g(x, y, z) = x + y + 2z = 2$

and $h(x, y, z) = x^2 + y^2 - z = 0$. $\nabla f = \langle 2x, 2y, 2z \rangle$, $\lambda \nabla g = \langle \lambda, \lambda, 2\lambda \rangle$ and $\mu \nabla h = \langle 2\mu x, 2\mu y, -\mu \rangle$. Thus we need

$2x = \lambda + 2\mu x$ **(1)**, $2y = \lambda + 2\mu y$ **(2)**, $2z = 2\lambda - \mu$ **(3)**, $x + y + 2z = 2$ **(4)**, and $x^2 + y^2 - z = 0$ **(5)**.

From **(1)** and **(2)**, $2(x - y) = 2\mu(x - y)$, so if $x \neq y$, $\mu = 1$. Putting this in **(3)** gives $2z = 2\lambda - 1$ or $\lambda = z + \frac{1}{2}$, but putting

$\mu = 1$ into **(1)** says $\lambda = 0$. Hence $z + \frac{1}{2} = 0$ or $z = -\frac{1}{2}$. Then **(4)** and **(5)** become $x + y - 3 = 0$ and $x^2 + y^2 + \frac{1}{2} = 0$. The

last equation cannot be true, so this case gives no solution. So we must have $x = y$. Then **(4)** and **(5)** become $2x + 2z = 2$ and

$2x^2 - z = 0$ which imply $z = 1 - x$ and $z = 2x^2$. Thus $2x^2 = 1 - x$ or $2x^2 + x - 1 = (2x - 1)(x + 1) = 0$ so $x = \frac{1}{2}$ or

$x = -1$. The two points to check are $\left(\frac{1}{2}, \frac{1}{2}, \frac{1}{2}\right)$ and $(-1, -1, 2)$: $f\left(\frac{1}{2}, \frac{1}{2}, \frac{1}{2}\right) = \frac{3}{4}$ and $f(-1, -1, 2) = 6$. Thus $\left(\frac{1}{2}, \frac{1}{2}, \frac{1}{2}\right)$ is

the point on the ellipse nearest the origin and $(-1, -1, 2)$ is the one farthest from the origin.

**43.** $f(x, y, z) = ye^{x-z}$, $g(x, y, z) = 9x^2 + 4y^2 + 36z^2 = 36$, $h(x, y, z) = xy + yz = 1$. $\nabla f = \lambda \nabla g + \mu \nabla h$ $\Rightarrow$

$\langle ye^{x-z}, e^{x-z}, -ye^{x-z} \rangle = \lambda \langle 18x, 8y, 72z \rangle + \mu \langle y, x+z, y \rangle$, so $ye^{x-z} = 18\lambda x + \mu y$, $e^{x-z} = 8\lambda y + \mu(x+z)$,

$-ye^{x-z} = 72\lambda z + \mu y$, $9x^2 + 4y^2 + 36z^2 = 36$, $xy + yz = 1$. Using a CAS to solve these 5 equations simultaneously for $x$,

$y$, $z$, $\lambda$, and $\mu$ (in Maple, use the `allvalues` command), we get 4 real-valued solutions:

$$x \approx 0.222444, \quad y \approx -2.157012, \quad z \approx -0.686049, \quad \lambda \approx -0.200401, \quad \mu \approx 2.108584$$

$$x \approx -1.951921, \quad y \approx -0.545867, \quad z \approx 0.119973, \quad \lambda \approx 0.003141, \quad \mu \approx -0.076238$$

$$x \approx 0.155142, \quad y \approx 0.904622, \quad z \approx 0.950293, \quad \lambda \approx -0.012447, \quad \mu \approx 0.489938$$

$$x \approx 1.138731, \quad y \approx 1.768057, \quad z \approx -0.573138, \quad \lambda \approx 0.317141, \quad \mu \approx 1.862675$$

Substituting these values into $f$ gives $f(0.222444, -2.157012, -0.686049) \approx -5.3506$,

$f(-1.951921, -0.545867, 0.119973) \approx -0.0688$, $f(0.155142, 0.904622, 0.950293) \approx 0.4084$,

$f(1.138731, 1.768057, -0.573138) \approx 9.7938$. Thus the maximum is approximately 9.7938, and the mininum is

approximately $-5.3506$.

**45.** (a) We wish to maximize $f(x_1, x_2, \ldots, x_n) = \sqrt[n]{x_1 x_2 \cdots x_n}$ subject to

$g(x_1, x_2, \ldots, x_n) = x_1 + x_2 + \cdots + x_n = c$ and $x_i > 0$.

$\nabla f = \left\langle \frac{1}{n}(x_1 x_2 \cdots x_n)^{\frac{1}{n}-1}(x_2 \cdots x_n), \frac{1}{n}(x_1 x_2 \cdots x_n)^{\frac{1}{n}-1}(x_1 x_3 \cdots x_n), \ldots, \frac{1}{n}(x_1 x_2 \cdots x_n)^{\frac{1}{n}-1}(x_1 \cdots x_{n-1}) \right\rangle$

and $\lambda \nabla g = \langle \lambda, \lambda, \ldots, \lambda \rangle$, so we need to solve the system of equations

$$\frac{1}{n}(x_1 x_2 \cdots x_n)^{\frac{1}{n}-1}(x_2 \cdots x_n) = \lambda \quad \Rightarrow \quad x_1^{1/n} x_2^{1/n} \cdots x_n^{1/n} = n\lambda x_1$$

$$\frac{1}{n}(x_1 x_2 \cdots x_n)^{\frac{1}{n}-1}(x_1 x_3 \cdots x_n) = \lambda \quad \Rightarrow \quad x_1^{1/n} x_2^{1/n} \cdots x_n^{1/n} = n\lambda x_2$$

$$\vdots$$

$$\frac{1}{n}(x_1 x_2 \cdots x_n)^{\frac{1}{n}-1}(x_1 \cdots x_{n-1}) = \lambda \quad \Rightarrow \quad x_1^{1/n} x_2^{1/n} \cdots x_n^{1/n} = n\lambda x_n$$

This implies $n\lambda x_1 = n\lambda x_2 = \cdots = n\lambda x_n$. Note $\lambda \neq 0$, otherwise we can't have all $x_i > 0$. Thus $x_1 = x_2 = \cdots = x_n$.

But $x_1 + x_2 + \cdots + x_n = c$ $\Rightarrow$ $nx_1 = c$ $\Rightarrow$ $x_1 = \frac{c}{n} = x_2 = x_3 = \cdots = x_n$. Then the only point where $f$ can

have an extreme value is $\left( \frac{c}{n}, \frac{c}{n}, \ldots, \frac{c}{n} \right)$. Since we can choose values for $(x_1, x_2, \ldots, x_n)$ that make $f$ as close to

zero (but not equal) as we like, $f$ has no minimum value. Thus the maximum value is

$f\left( \frac{c}{n}, \frac{c}{n}, \ldots, \frac{c}{n} \right) = \sqrt[n]{\frac{c}{n} \cdot \frac{c}{n} \cdots \frac{c}{n}} = \frac{c}{n}$.

(b) From part (a), $\frac{c}{n}$ is the maximum value of $f$. Thus $f(x_1, x_2, \ldots, x_n) = \sqrt[n]{x_1 x_2 \cdots x_n} \leq \frac{c}{n}$. But

$x_1 + x_2 + \cdots + x_n = c$, so $\sqrt[n]{x_1 x_2 \cdots x_n} \leq \dfrac{x_1 + x_2 + \cdots + x_n}{n}$. These two means are equal when $f$ attains its

maximum value $\frac{c}{n}$, but this can occur only at the point $\left( \frac{c}{n}, \frac{c}{n}, \ldots, \frac{c}{n} \right)$ we found in part (a). So the means are equal only

when $x_1 = x_2 = x_3 = \cdots = x_n = \dfrac{c}{n}$.

# 11 Review

CONCEPT CHECK

1. (a) A function $f$ of two variables is a rule that assigns to each ordered pair $(x, y)$ of real numbers in its domain a unique real number denoted by $f(x, y)$.

   (b) One way to visualize a function of two variables is by graphing it, resulting in the surface $z = f(x, y)$. Another method for visualizing a function of two variables is a contour map. The contour map consists of level curves of the function which are horizontal traces of the graph of the function projected onto the $xy$-plane.

2. A function $f$ of three variables is a rule that assigns to each ordered triple $(x, y, z)$ in its domain a unique real number $f(x, y, z)$. We can visualize a function of three variables by examining its level surfaces $f(x, y, z) = k$, where $k$ is a constant.

3. $\lim_{(x,y)\to(a,b)} f(x, y) = L$ means the values of $f(x, y)$ approach the number $L$ as the point $(x, y)$ approaches the point $(a, b)$ along any path that is within the domain of $f$. We can show that a limit at a point does not exist by finding two different paths approaching the point along which $f(x, y)$ has different limits.

4. (a) See Definition 11.2.3.

   (b) If $f$ is continuous on $\mathbb{R}^2$, its graph will appear as a surface without holes or breaks.

5. (a) See (2) and (3) in Section 11.3.

   (b) See "Interpretations of Partial Derivatives" on page 759.

   (c) To find $f_x$, regard $y$ as a constant and differentiate $f(x, y)$ with respect to $x$. To find $f_y$, regard $x$ as a constant and differentiate $f(x, y)$ with respect to $y$.

6. See the statement of Clairaut's Theorem on page 763.

7. (a) See (2) in Section 11.4

   (b) See (19) and the preceding discussion in Section 11.6.

   (c) See "Tangent Planes to Parametric Surfaces" on page 777.

8. See (3) and (4) and the accompanying discussion in Section 11.4. We can interpret the linearization of $f$ at $(a, b)$ geometrically as the linear function whose graph is the tangent plane to the graph of $f$ at $(a, b)$. Thus it is the linear function which best approximates $f$ near $(a, b)$.

9. (a) See Definition 11.4.7.

   (b) Use Theorem 11.4.8.

10. See (10) and the associated discussion in Section 11.4.

11. See (2) and (3) in Section 11.5.

12. See (7) and the preceding discussion in Section 11.5.

13. (a) See Definition 11.6.2. We can interpret it as the rate of change of $f$ at $(x_0, y_0)$ in the direction of $\mathbf{u}$. Geometrically, if $P$ is the point $(x_0, y_0, f(x_0, y_0))$ on the graph of $f$ and $C$ is the curve of intersection of the graph of $f$ with the vertical plane that passes through $P$ in the direction $\mathbf{u}$, the directional derivative of $f$ at $(x_0, y_0)$ in the direction of $\mathbf{u}$ is the slope of the tangent line to $C$ at $P$. (See Figure 5 in Section 11.6.)

(b) See Theorem 11.6.3.

14. (a) See (8) and (13) in Section 11.6.

(b) $D_{\mathbf{u}}\, f(x, y) = \nabla f(x, y) \cdot \mathbf{u}$ or $D_{\mathbf{u}}\, f(x, y, z) = \nabla f(x, y, z) \cdot \mathbf{u}$

(c) The gradient vector of a function points in the direction of maximum rate of increase of the function. On a graph of the function, the gradient points in the direction of steepest ascent.

15. (a) $f$ has a local maximum at $(a, b)$ if $f(x, y) \leq f(a, b)$ when $(x, y)$ is near $(a, b)$.

(b) $f$ has an absolute maximum at $(a, b)$ if $f(x, y) \leq f(a, b)$ for all points $(x, y)$ in the domain of $f$.

(c) $f$ has a local minimum at $(a, b)$ if $f(x, y) \geq f(a, b)$ when $(x, y)$ is near $(a, b)$.

(d) $f$ has an absolute minimum at $(a, b)$ if $f(x, y) \geq f(a, b)$ for all points $(x, y)$ in the domain of $f$.

(e) $f$ has a saddle point at $(a, b)$ if $f(a, b)$ is a local maximum in one direction but a local minimum in another.

16. (a) By Theorem 11.7.2, if $f$ has a local maximum at $(a, b)$ and the first-order partial derivatives of $f$ exist there, then
$f_x(a, b) = 0$ and $f_y(a, b) = 0$.

(b) A critical point of $f$ is a point $(a, b)$ such that $f_x(a, b) = 0$ and $f_y(a, b) = 0$ or one of these partial derivatives does not exist.

17. See (3) in Section 11.7.

18. (a) See Figure 11 and the accompanying discussion in Section 11.7.

(b) See Theorem 11.7.8.

(c) See the procedure outlined in (9) in Section 11.7.

19. See the discussion beginning on page 813; see "Two Constraints" on page 817.

## TRUE-FALSE QUIZ

1. True. $f_y(a, b) = \lim\limits_{h \to 0} \dfrac{f(a, b + h) - f(a, b)}{h}$ from Equation 11.3.3. Let $h = y - b$. As $h \to 0$, $y \to b$. Then by substituting,

we get $f_y(a, b) = \lim\limits_{y \to b} \dfrac{f(a, y) - f(a, b)}{y - b}$.

3. False. $f_{xy} = \dfrac{\partial^2 f}{\partial y\, \partial x}$.

5. False. See Example 11.2.3.

**7.** True. If $f$ has a local minimum and $f$ is differentiable at $(a, b)$ then by Theorem 11.7.2, $f_x(a, b) = 0$ and $f_y(a, b) = 0$, so
$$\nabla f(a, b) = \langle f_x(a, b), f_y(a, b) \rangle = \langle 0, 0 \rangle = \mathbf{0}.$$

**9.** False. $\nabla f(x, y) = \langle 0, 1/y \rangle$.

**11.** True. $\nabla f = \langle \cos x, \cos y \rangle$, so $|\nabla f| = \sqrt{\cos^2 x + \cos^2 y}$. But $|\cos \theta| \leq 1$, so $|\nabla f| \leq \sqrt{2}$. Now
$$D_{\mathbf{u}} f(x, y) = \nabla f \cdot \mathbf{u} = |\nabla f| \, |\mathbf{u}| \cos \theta, \text{ but } \mathbf{u} \text{ is a unit vector, so } |D_{\mathbf{u}} f(x, y)| \leq \sqrt{2} \cdot 1 \cdot 1 = \sqrt{2}.$$

## EXERCISES

**1.** $\ln(x + y + 1)$ is defined only when $x + y + 1 > 0 \quad \Leftrightarrow \quad y > -x - 1$,

so the domain of $f$ is $\{(x, y) \mid y > -x - 1\}$, all those points above the

line $y = -x - 1$.

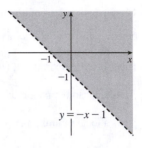

**3.** $z = f(x, y) = 1 - y^2$, a

parabolic cylinder

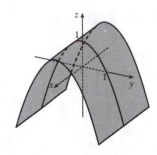

**5.** The level curves are $\sqrt{4x^2 + y^2} = k$ or

$$4x^2 + y^2 = k^2, \ k \geq 0, \text{ a family of ellipses.}$$

**7.**

**9.** $f$ is a rational function, so it is continuous on its domain. Since $f$ is defined at $(1, 1)$, we use direct substitution to evaluate the

limit: $\displaystyle \lim_{(x,y) \to (1,1)} \frac{2xy}{x^2 + 2y^2} = \frac{2(1)(1)}{1^2 + 2(1)^2} = \frac{2}{3}$.

**11.** (a) $T_x(6, 4) = \displaystyle\lim_{h \to 0} \frac{T(6 + h, 4) - T(6, 4)}{h}$, so we can approximate $T_x(6, 4)$ by considering $h = \pm 2$ and

using the values given in the table: $T_x(6, 4) \approx \dfrac{T(8, 4) - T(6, 4)}{2} = \dfrac{86 - 80}{2} = 3$,

$T_x(6, 4) \approx \dfrac{T(4, 4) - T(6, 4)}{-2} = \dfrac{72 - 80}{-2} = 4$. Averaging these values, we estimate $T_x(6, 4)$ to be approximately

$3.5°$C/m. Similarly, $T_y(6, 4) = \displaystyle\lim_{h \to 0} \frac{T(6, 4 + h) - T(6, 4)}{h}$, which we can approximate with $h = \pm 2$:

$T_y(6, 4) \approx \dfrac{T(6, 6) - T(6, 4)}{2} = \dfrac{75 - 80}{2} = -2.5$, $T_y(6, 4) \approx \dfrac{T(6, 2) - T(6, 4)}{-2} = \dfrac{87 - 80}{-2} = -3.5$. Averaging these

values, we estimate $T_y(6, 4)$ to be approximately $-3.0°$C/m.

(b) Here $\mathbf{u} = \left\langle \frac{1}{\sqrt{2}}, \frac{1}{\sqrt{2}} \right\rangle$, so by Equation 11.6.9, $D_{\mathbf{u}} T(6,4) = \nabla T(6,4) \cdot \mathbf{u} = T_x(6,4)\frac{1}{\sqrt{2}} + T_y(6,4)\frac{1}{\sqrt{2}}$. Using our

estimates from part (a), we have $D_{\mathbf{u}} T(6,4) \approx (3.5)\frac{1}{\sqrt{2}} + (-3.0)\frac{1}{\sqrt{2}} = \frac{1}{2\sqrt{2}} \approx 0.35$. This means that as we move

through the point $(6,4)$ in the direction of $\mathbf{u}$, the temperature increases at a rate of approximately $0.35°\text{C/m}$.

Alternatively, we can use Definition 11.6.2: $D_{\mathbf{u}} T(6,4) = \lim\limits_{h \to 0} \dfrac{T\left(6 + h\,\frac{1}{\sqrt{2}}, 4 + h\,\frac{1}{\sqrt{2}}\right) - T(6,4)}{h}$,

which we can estimate with $h = \pm 2\sqrt{2}$. Then $D_{\mathbf{u}} T(6,4) \approx \dfrac{T(8,6) - T(6,4)}{2\sqrt{2}} = \dfrac{80 - 80}{2\sqrt{2}} = 0$,

$D_{\mathbf{u}} T(6,4) \approx \dfrac{T(4,2) - T(6,4)}{-2\sqrt{2}} = \dfrac{74 - 80}{-2\sqrt{2}} = \dfrac{3}{\sqrt{2}}$. Averaging these values, we have $D_{\mathbf{u}} T(6,4) \approx \dfrac{3}{2\sqrt{2}} \approx 1.1°\text{C/m}$.

(c) $T_{xy}(x,y) = \dfrac{\partial}{\partial y}\left[T_x(x,y)\right] = \lim\limits_{h \to 0} \dfrac{T_x(x, y+h) - T_x(x,y)}{h}$, so $T_{xy}(6,4) = \lim\limits_{h \to 0} \dfrac{T_x(6, 4+h) - T_x(6,4)}{h}$ which we can

estimate with $h = \pm 2$. We have $T_x(6,4) \approx 3.5$ from part (a), but we will also need values for $T_x(6,6)$ and $T_x(6,2)$. If we

use $h = \pm 2$ and the values given in the table, we have

$T_x(6,6) \approx \dfrac{T(8,6) - T(6,6)}{2} = \dfrac{80 - 75}{2} = 2.5$, $T_x(6,6) \approx \dfrac{T(4,6) - T(6,6)}{-2} = \dfrac{68 - 75}{-2} = 3.5$.

Averaging these values, we estimate $T_x(6,6) \approx 3.0$. Similarly,

$T_x(6,2) \approx \dfrac{T(8,2) - T_x(6,2)}{2} = \dfrac{90 - 87}{2} = 1.5$, $T_x(6,2) \approx \dfrac{T(4,2) - T(6,2)}{-2} = \dfrac{74 - 87}{-2} = 6.5$.

Averaging these values, we estimate $T_x(6,2) \approx 4.0$. Finally, we estimate $T_{xy}(6,4)$:

$T_{xy}(6,4) \approx \dfrac{T_x(6,6) - T_x(6,4)}{2} = \dfrac{3.0 - 3.5}{2} = -0.25$, $T_{xy}(6,4) \approx \dfrac{T_x(6,2) - T_x(6,4)}{-2} = \dfrac{4.0 - 3.5}{-2} = -0.25$.

Averaging these values, we have $T_{xy}(6,4) \approx -0.25$.

**13.** $f(x,y) = \sqrt{2x + y^2}$ $\Rightarrow$ $f_x = \frac{1}{2}(2x + y^2)^{-1/2}(2) = \dfrac{1}{\sqrt{2x + y^2}}$, $f_y = \frac{1}{2}(2x + y^2)^{-1/2}(2y) = \dfrac{y}{\sqrt{2x + y^2}}$

**15.** $g(u,v) = u\tan^{-1} v$ $\Rightarrow$ $g_u = \tan^{-1} v$, $g_v = \dfrac{u}{1 + v^2}$

**17.** $T(p,q,r) = p\ln(q + e^r)$ $\Rightarrow$ $T_p = \ln(q + e^r)$, $T_q = \dfrac{p}{q + e^r}$, $T_r = \dfrac{pe^r}{q + e^r}$

**19.** $f(x,y) = 4x^3 - xy^2$ $\Rightarrow$ $f_x = 12x^2 - y^2$, $f_y = -2xy$, $f_{xx} = 24x$, $f_{yy} = -2x$, $f_{xy} = f_{yx} = -2y$

**21.** $f(x,y,z) = x^k y^l z^m$ $\Rightarrow$ $f_x = kx^{k-1}y^l z^m$, $f_y = lx^k y^{l-1}z^m$, $f_z = mx^k y^l z^{m-1}$, $f_{xx} = k(k-1)x^{k-2}y^l z^m$,

$f_{yy} = l(l-1)x^k y^{l-2}z^m$, $f_{zz} = m(m-1)x^k y^l z^{m-2}$, $f_{xy} = f_{yx} = klx^{k-1}y^{l-1}z^m$, $f_{xz} = f_{zx} = kmx^{k-1}y^l z^{m-1}$,

$f_{yz} = f_{zy} = lmx^k y^{l-1}z^{m-1}$

**23.** $z = xy + xe^{y/x}$ $\Rightarrow$ $\dfrac{\partial z}{\partial x} = y - \dfrac{y}{x}e^{y/x} + e^{y/x}$, $\dfrac{\partial z}{\partial y} = x + e^{y/x}$ and

$x\dfrac{\partial z}{\partial x} + y\dfrac{\partial z}{\partial y} = x\left(y - \dfrac{y}{x}e^{y/x} + e^{y/x}\right) + y\left(x + e^{y/x}\right) = xy - ye^{y/x} + xe^{y/x} + xy + ye^{y/x} = xy + xy + xe^{y/x} = xy + z$.

**25.** (a) $z_x = 6x + 2$ $\Rightarrow$ $z_x(1,-2) = 8$ and $z_y = -2y$ $\Rightarrow$ $z_y(1,-2) = 4$, so an equation of the tangent plane is

$z - 1 = 8(x - 1) + 4(y + 2)$ or $z = 8x + 4y + 1$.

(b) A normal vector to the tangent plane (and the surface) at $(1, -2, 1)$ is $\langle 8, 4, -1 \rangle$. Then parametric equations for the normal

line there are $x = 1 + 8t$, $y = -2 + 4t$, $z = 1 - t$, and symmetric equations are $\dfrac{x-1}{8} = \dfrac{y+2}{4} = \dfrac{z-1}{-1}$.

**27.** (a) Let $F(x, y, z) = x^2 + 2y^2 - 3z^2$. Then $F_x = 2x$, $F_y = 4y$, $F_z = -6z$, so $F_x(2, -1, 1) = 4$, $F_y(2, -1, 1) = -4$,

$F_z(2, -1, 1) = -6$. From Equation 11.6.19, an equation of the tangent plane is $4(x-2) - 4(y+1) - 6(z-1) = 0$

or, equivalently, $2x - 2y - 3z = 3$.

(b) From Equations 11.6.20, symmetric equations for the normal line are $\dfrac{x-2}{4} = \dfrac{y+1}{-4} = \dfrac{z-1}{-6}$.

**29.** (a) $\mathbf{r}(u, v) = (u + v)\,\mathbf{i} + u^2\,\mathbf{j} + v^2\,\mathbf{k}$ and the point $(3, 4, 1)$ corresponds to $u = 2$, $v = 1$. Then $\mathbf{r}_u = \mathbf{i} + 2u\,\mathbf{j}$ $\Rightarrow$

$\mathbf{r}_u(2, 1) = \mathbf{i} + 4\,\mathbf{j}$ and $\mathbf{r}_v = \mathbf{i} + 2v\,\mathbf{k}$ $\Rightarrow$ $\mathbf{r}_v(2, 1) = \mathbf{i} + 2\,\mathbf{j}$. A normal vector to the surface at $(3, 4, 1)$ is

$\mathbf{r}_u \times \mathbf{r}_v = 8\,\mathbf{i} - 2\,\mathbf{j} - 4\,\mathbf{k}$, so an equation of the tangent plane there is $8(x-3) - 2(y-4) - 4(z-1) = 0$ or equivalently

$4x - y - 2z = 6$.

(b) A direction vector for the normal line through $(3, 4, 1)$ is $8\,\mathbf{i} - 2\,\mathbf{j} - 4\,\mathbf{k}$, so a vector equation is

$\mathbf{r}(t) = (3\,\mathbf{i} + 4\,\mathbf{j} + \mathbf{k}) + t\,(8\,\mathbf{i} - 2\,\mathbf{j} - 4\,\mathbf{k})$, and the corresponding parametric equations are $x = 3 + 8t$, $y = 4 - 2t$,

$z = 1 - 4t$.

**31.** The hyperboloid is a level surface of the function $F(x, y, z) = x^2 + 4y^2 - z^2$, so a normal vector to the surface at $(x_0, y_0, z_0)$

is $\nabla F(x_0, y_0, z_0) = \langle 2x_0, 8y_0, -2z_0 \rangle$. A normal vector for the plane $2x + 2y + z = 5$ is $\langle 2, 2, 1 \rangle$. For the planes to be

parallel, we need the normal vectors to be parallel, so $\langle 2x_0, 8y_0, -2z_0 \rangle = k\,\langle 2, 2, 1 \rangle$, or $x_0 = k$, $y_0 = \frac{1}{4}k$, and $z_0 = -\frac{1}{2}k$.

But $x_0^2 + 4y_0^2 - z_0^2 = 4$ $\Rightarrow$ $k^2 + \frac{1}{4}k^2 - \frac{1}{4}k^2 = 4$ $\Rightarrow$ $k^2 = 4$ $\Rightarrow$ $k = \pm 2$. So there are two such points:

$\left(2, \frac{1}{2}, -1\right)$ and $\left(-2, -\frac{1}{2}, 1\right)$.

**33.** $f(x, y, z) = x^3 \sqrt{y^2 + z^2}$ $\Rightarrow$ $f_x(x, y, z) = 3x^2 \sqrt{y^2 + z^2}$, $f_y(x, y, z) = \dfrac{yx^3}{\sqrt{y^2 + z^2}}$, $f_z(x, y, z) = \dfrac{zx^3}{\sqrt{y^2 + z^2}}$,

so $f(2, 3, 4) = 8(5) = 40$, $f_x(2, 3, 4) = 3(4)\sqrt{25} = 60$, $f_y(2, 3, 4) = \frac{3(8)}{\sqrt{25}} = \frac{24}{5}$, and $f_z(2, 3, 4) = \frac{4(8)}{\sqrt{25}} = \frac{32}{5}$. Then the

linear approximation of $f$ at $(2, 3, 4)$ is

$$f(x, y, z) \approx f(2, 3, 4) + f_x(2, 3, 4)(x - 2) + f_y(2, 3, 4)(y - 3) + f_z(2, 3, 4)(z - 4)$$

$$= 40 + 60(x - 2) + \tfrac{24}{5}(y - 3) + \tfrac{32}{5}(z - 4) = 60x + \tfrac{24}{5}y + \tfrac{32}{5}z - 120$$

Then $(1.98)^3 \sqrt{(3.01)^2 + (3.97)^2} = f(1.98, 3.01, 3.97) \approx 60(1.98) + \tfrac{24}{5}(3.01) + \tfrac{32}{5}(3.97) - 120 = 38.656$.

**35.** $\dfrac{du}{dp} = \dfrac{\partial u}{\partial x}\dfrac{dx}{dp} + \dfrac{\partial u}{\partial y}\dfrac{dy}{dp} + \dfrac{\partial u}{\partial z}\dfrac{dz}{dp} = 2xy^3(1 + 6p) + 3x^2y^2(pe^p + e^p) + 4z^3(p\cos p + \sin p)$

**37.** By the Chain Rule, $\dfrac{\partial z}{\partial s} = \dfrac{\partial z}{\partial x}\dfrac{\partial x}{\partial s} + \dfrac{\partial z}{\partial y}\dfrac{\partial y}{\partial s}$. When $s = 1$ and $t = 2$, $x = g(1, 2) = 3$ and $y = h(1, 2) = 6$, so

$\dfrac{\partial z}{\partial s} = f_x(3, 6)g_s(1, 2) + f_y(3, 6)h_s(1, 2) = (7)(-1) + (8)(-5) = -47$. Similarly, $\dfrac{\partial z}{\partial t} = \dfrac{\partial z}{\partial x}\dfrac{\partial x}{\partial t} + \dfrac{\partial z}{\partial y}\dfrac{\partial y}{\partial t}$, so

$\dfrac{\partial z}{\partial t} = f_x(3, 6)g_t(1, 2) + f_y(3, 6)h_t(1, 2) = (7)(4) + (8)(10) = 108$.

**39.** $\dfrac{\partial z}{\partial x} = 2xf'(x^2 - y^2)$,   $\dfrac{\partial z}{\partial y} = 1 - 2yf'(x^2 - y^2)$   $\left[\text{where } f' = \dfrac{df}{d(x^2 - y^2)}\right]$. Then

$$y\,\frac{\partial z}{\partial x} + x\,\frac{\partial z}{\partial y} = 2xyf'(x^2 - y^2) + x - 2xyf'(x^2 - y^2) = x.$$

**41.** $\dfrac{\partial z}{\partial x} = \dfrac{\partial z}{\partial u}\,y + \dfrac{\partial z}{\partial v}\,\dfrac{-y}{x^2}$ and

$$\frac{\partial^2 z}{\partial x^2} = y\,\frac{\partial}{\partial x}\left(\frac{\partial z}{\partial u}\right) + \frac{2y}{x^3}\,\frac{\partial z}{\partial v} + \frac{-y}{x^2}\,\frac{\partial}{\partial x}\left(\frac{\partial z}{\partial v}\right) = \frac{2y}{x^3}\,\frac{\partial z}{\partial v} + y\left(\frac{\partial^2 z}{\partial u^2}\,y + \frac{\partial^2 z}{\partial v\,\partial u}\,\frac{-y}{x^2}\right) + \frac{-y}{x^2}\left(\frac{\partial^2 z}{\partial v^2}\,\frac{-y}{x^2} + \frac{\partial^2 z}{\partial u\,\partial v}\,y\right)$$

$$= \frac{2y}{x^3}\,\frac{\partial z}{\partial v} + y^2\,\frac{\partial^2 z}{\partial u^2} - \frac{2y^2}{x^2}\,\frac{\partial^2 z}{\partial u\,\partial v} + \frac{y^2}{x^4}\,\frac{\partial^2 z}{\partial v^2}$$

Also $\dfrac{\partial z}{\partial y} = x\,\dfrac{\partial z}{\partial u} + \dfrac{1}{x}\,\dfrac{\partial z}{\partial v}$ and

$$\frac{\partial^2 z}{\partial y^2} = x\,\frac{\partial}{\partial y}\left(\frac{\partial z}{\partial u}\right) + \frac{1}{x}\,\frac{\partial}{\partial y}\left(\frac{\partial z}{\partial v}\right) = x\left(\frac{\partial^2 z}{\partial u^2}\,x + \frac{\partial^2 z}{\partial v\,\partial u}\,\frac{1}{x}\right) + \frac{1}{x}\left(\frac{\partial^2 z}{\partial v^2}\,\frac{1}{x} + \frac{\partial^2 z}{\partial u\,\partial v}\,x\right) = x^2\,\frac{\partial^2 z}{\partial u^2} + 2\,\frac{\partial^2 z}{\partial u\,\partial v} + \frac{1}{x^2}\,\frac{\partial^2 z}{\partial v^2}$$

Thus

$$x^2\,\frac{\partial^2 z}{\partial x^2} - y^2\,\frac{\partial^2 z}{\partial y^2} = \frac{2y}{x}\,\frac{\partial z}{\partial v} + x^2 y^2\,\frac{\partial^2 z}{\partial u^2} - 2y^2\,\frac{\partial^2 z}{\partial u\,\partial v} + \frac{y^2}{x^2}\,\frac{\partial^2 z}{\partial v^2} - x^2 y^2\,\frac{\partial^2 z}{\partial u^2} - 2y^2\,\frac{\partial^2 z}{\partial u\,\partial v} - \frac{y^2}{x^2}\,\frac{\partial^2 z}{\partial v^2}$$

$$= \frac{2y}{x}\,\frac{\partial z}{\partial v} - 4y^2\,\frac{\partial^2 z}{\partial u\,\partial v} = 2v\,\frac{\partial z}{\partial v} - 4uv\,\frac{\partial^2 z}{\partial u\,\partial v}$$

since $y = xv = \dfrac{uv}{y}$ or $y^2 = uv$.

**43.** $f(x, y, z) = x^2 e^{yz^2}$   $\Rightarrow$   $\nabla f = \langle f_x, f_y, f_z \rangle = \left\langle 2xe^{yz^2}, x^2 e^{yz^2} \cdot z^2, x^2 e^{yz^2} \cdot 2yz \right\rangle = \left\langle 2xe^{yz^2}, x^2 z^2 e^{yz^2}, 2x^2 yze^{yz^2} \right\rangle$

**45.** $f(x, y) = x^2 e^{-y}$   $\Rightarrow$   $\nabla f = \langle 2xe^{-y}, -x^2 e^{-y} \rangle$, $\nabla f(-2, 0) = \langle -4, -4 \rangle$. The direction is given by $\langle 4, -3 \rangle$, so

$\mathbf{u} = \dfrac{1}{\sqrt{4^2 + (-3)^2}}\langle 4, -3 \rangle = \frac{1}{5}\langle 4, -3 \rangle$ and $D_{\mathbf{u}} f(-2, 0) = \nabla f(-2, 0) \cdot \mathbf{u} = \langle -4, -4 \rangle \cdot \frac{1}{5}\langle 4, -3 \rangle = \frac{1}{5}(-16 + 12) = -\frac{4}{5}$.

**47.** $\nabla f = \langle 2xy, x^2 + 1/(2\sqrt{y}) \rangle$, $|\nabla f(2, 1)| = |\langle 4, \frac{9}{2} \rangle|$. Thus the maximum rate of change of $f$ at $(2, 1)$ is $\frac{\sqrt{145}}{2}$ in the

direction $\langle 4, \frac{9}{2} \rangle$.

**49.** First we draw a line passing through Homestead and the eye of the hurricane. We can approximate the directional derivative at

Homestead in the direction of the eye of the hurricane by the average rate of change of wind speed between the points where

this line intersects the contour lines closest to Homestead. In the direction of the eye of the hurricane, the wind speed changes

from 45 to 50 knots. We estimate the distance between these two points to be approximately 8 miles, so the rate of change of

wind speed in the direction given is approximately $\frac{50 - 45}{8} = \frac{5}{8} = 0.625$ knot/mi.

**51.** $f(x, y) = x^2 - xy + y^2 + 9x - 6y + 10$   $\Rightarrow$   $f_x = 2x - y + 9$,

$f_y = -x + 2y - 6$, $f_{xx} = 2 = f_{yy}$, $f_{xy} = -1$. Then $f_x = 0$ and $f_y = 0$ imply

$y = 1$, $x = -4$. Thus the only critical point is $(-4, 1)$ and $f_{xx}(-4, 1) > 0$,

$D(-4, 1) = 3 > 0$, so $f(-4, 1) = -11$ is a local minimum.

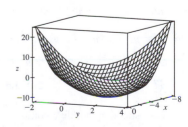

**53.** $f(x,y) = 3xy - x^2 y - xy^2 \Rightarrow f_x = 3y - 2xy - y^2, \ f_y = 3x - x^2 - 2xy,$

$f_{xx} = -2y, \ f_{yy} = -2x, \ f_{xy} = 3 - 2x - 2y.$ Then $f_x = 0$ implies

$y(3 - 2x - y) = 0$ so $y = 0$ or $y = 3 - 2x.$ Substituting into $f_y = 0$ implies

$x(3 - x) = 0$ or $3x(-1 + x) = 0.$ Hence the critical points are $(0,0), (3,0),$

$(0,3)$ and $(1,1).$ $D(0,0) = D(3,0) = D(0,3) = -9 < 0$ so $(0,0), (3,0),$ and

$(0,3)$ are saddle points. $D(1,1) = 3 > 0$ and $f_{xx}(1,1) = -2 < 0,$ so

$f(1,1) = 1$ is a local maximum.

**55.** First solve inside $D.$ Here $f_x = 4y^2 - 2xy^2 - y^3, \ f_y = 8xy - 2x^2 y - 3xy^2.$

Then $f_x = 0$ implies $y = 0$ or $y = 4 - 2x,$ but $y = 0$ isn't inside $D.$ Substituting

$y = 4 - 2x$ into $f_y = 0$ implies $x = 0, x = 2$ or $x = 1,$ but $x = 0$ isn't inside $D,$

and when $x = 2, y = 0$ but $(2,0)$ isn't inside $D.$ Thus the only critical point inside

$D$ is $(1,2)$ and $f(1,2) = 4.$ Secondly we consider the boundary of $D.$

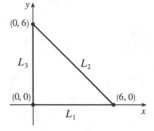

On $L_1$: $f(x,0) = 0$ and so $f = 0$ on $L_1.$ On $L_2$: $x = -y + 6$ and

$f(-y + 6, y) = y^2(6 - y)(-2) = -2(6y^2 - y^3)$ which has critical points

at $y = 0$ and $y = 4.$ Then $f(6,0) = 0$ while $f(2,4) = -64.$ On $L_3$: $f(0,y) = 0,$ so $f = 0$ on $L_3.$ Thus on $D$ the absolute

maximum of $f$ is $f(1,2) = 4$ while the absolute minimum is $f(2,4) = -64.$

**57.** $f(x,y) = x^3 - 3x + y^4 - 2y^2$

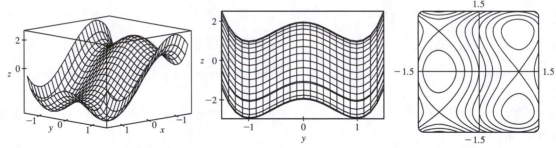

From the graphs, it appears that $f$ has a local maximum $f(-1,0) \approx 2,$ local minima $f(1, \pm 1) \approx -3,$ and saddle points at

$(-1, \pm 1)$ and $(1,0).$

To find the exact quantities, we calculate $f_x = 3x^2 - 3 = 0 \ \Leftrightarrow \ x = \pm 1$ and $f_y = 4y^3 - 4y = 0 \ \Leftrightarrow$

$y = 0, \pm 1,$ giving the critical points estimated above. Also $f_{xx} = 6x, \ f_{xy} = 0, \ f_{yy} = 12y^2 - 4,$ so using the Second

Derivatives Test, $D(-1,0) = 24 > 0$ and $f_{xx}(-1,0) = -6 < 0$ indicating a local maximum $f(-1,0) = 2;$

$D(1, \pm 1) = 48 > 0$ and $f_{xx}(1, \pm 1) = 6 > 0$ indicating local minima $f(1, \pm 1) = -3;$ and $D(-1, \pm 1) = -48$ and

$D(1,0) = -24,$ indicating saddle points.

**59.** $f(x,y) = x^2 y, \ g(x,y) = x^2 + y^2 = 1 \ \Rightarrow \ \nabla f = \langle 2xy, x^2 \rangle = \lambda \nabla g = \langle 2\lambda x, 2\lambda y \rangle.$ Then $2xy = 2\lambda x$ implies $x = 0$ or

$y = \lambda.$ If $x = 0$ then $x^2 + y^2 = 1$ gives $y = \pm 1$ and we have possible points $(0, \pm 1)$ where $f(0, \pm 1) = 0.$ If $y = \lambda$ then

$x^2 = 2\lambda y$ implies $x^2 = 2y^2$ and substitution into $x^2 + y^2 = 1$ gives $3y^2 = 1 \ \Rightarrow \ y = \pm \frac{1}{\sqrt{3}}$ and $x = \pm \sqrt{\frac{2}{3}}.$ The

corresponding possible points are $\left( \pm \sqrt{\frac{2}{3}}, \pm \frac{1}{\sqrt{3}} \right).$ The absolute maximum is $f\left( \pm \sqrt{\frac{2}{3}}, \frac{1}{\sqrt{3}} \right) = \frac{2}{3\sqrt{3}}$ while the absolute

minimum is $f\left( \pm \sqrt{\frac{2}{3}}, -\frac{1}{\sqrt{3}} \right) = -\frac{2}{3\sqrt{3}}.$

**61.** $f(x, y, z) = xyz$, $g(x, y, z) = x^2 + y^2 + z^2 = 3$. $\nabla f = \lambda \nabla g \Rightarrow \langle yz, xz, xy \rangle = \lambda \langle 2x, 2y, 2z \rangle$. If any of $x$, $y$, or $z$ is

zero, then $x = y = z = 0$ which contradicts $x^2 + y^2 + z^2 = 3$. Then $\lambda = \dfrac{yz}{2x} = \dfrac{xz}{2y} = \dfrac{xy}{2z} \Rightarrow 2y^2 z = 2x^2 z \Rightarrow$

$y^2 = x^2$, and similarly $2yz^2 = 2x^2 y \Rightarrow z^2 = x^2$. Substituting into the constraint equation gives $x^2 + x^2 + x^2 = 3 \Rightarrow$

$x^2 = 1 = y^2 = z^2$. Thus the possible points are $(1, 1, \pm 1)$, $(1, -1, \pm 1)$, $(-1, 1, \pm 1)$, $(-1, -1, \pm 1)$. The absolute maximum

is $f(1, 1, 1) = f(1, -1, -1) = f(-1, 1, -1) = f(-1, -1, 1) = 1$ and the absolute

minimum is $f(1, 1, -1) = f(1, -1, 1) = f(-1, 1, 1) = f(-1, -1, -1) = -1$.

**63.** $f(x, y, z) = x^2 + y^2 + z^2$, $g(x, y, z) = xy^2 z^3 = 2 \Rightarrow \nabla f = \langle 2x, 2y, 2z \rangle = \lambda \nabla g = \langle \lambda y^2 z^3, 2\lambda xyz^3, 3\lambda xy^2 z^2 \rangle$.

Since $xy^2 z^3 = 2$, $x \neq 0$, $y \neq 0$ and $z \neq 0$, so $2x = \lambda y^2 z^3$ **(1)**, $1 = \lambda xz^3$ **(2)**, $2 = 3\lambda xy^2 z$ **(3)**. Then **(2)** and **(3)** imply

$\dfrac{1}{xz^3} = \dfrac{2}{3xy^2 z}$ or $y^2 = \tfrac{2}{3} z^2$ so $y = \pm z \sqrt{\tfrac{2}{3}}$. Similarly **(1)** and **(3)** imply $\dfrac{2x}{y^2 z^3} = \dfrac{2}{3xy^2 z}$ or $3x^2 = z^2$ so $x = \pm \tfrac{1}{\sqrt{3}} z$. But

$xy^2 z^3 = 2$ so $x$ and $z$ must have the same sign, that is, $x = \tfrac{1}{\sqrt{3}} z$. Thus $g(x, y, z) = 2$ implies $\tfrac{1}{\sqrt{3}} z \left( \tfrac{2}{3} z^2 \right) z^3 = 2$ or

$z = \pm 3^{1/4}$ and the possible points are $(\pm 3^{-1/4}, 3^{-1/4}\sqrt{2}, \pm 3^{1/4})$, $(\pm 3^{-1/4}, -3^{-1/4}\sqrt{2}, \pm 3^{1/4})$. However at each of these

points $f$ takes on the same value, $2\sqrt{3}$. But $(2, 1, 1)$ also satisfies $g(x, y, z) = 2$ and $f(2, 1, 1) = 6 > 2\sqrt{3}$. Thus $f$ has an

absolute minimum value of $2\sqrt{3}$ and no absolute maximum subject to the constraint $xy^2 z^3 = 2$.

*Alternate solution:* $g(x, y, z) = xy^2 z^3 = 2$ implies $y^2 = \dfrac{2}{xz^3}$, so minimize $f(x, z) = x^2 + \dfrac{2}{xz^3} + z^2$. Then

$f_x = 2x - \dfrac{2}{x^2 z^3}$, $f_z = -\dfrac{6}{xz^4} + 2z$, $f_{xx} = 2 + \dfrac{4}{x^3 z^3}$, $f_{zz} = \dfrac{24}{xz^5} + 2$ and $f_{xz} = \dfrac{6}{x^2 z^4}$. Now $f_x = 0$ implies

$2x^3 z^3 - 2 = 0$ or $z = 1/x$. Substituting into $f_y = 0$ implies $-6x^3 + 2x^{-1} = 0$ or $x = \dfrac{1}{\sqrt[4]{3}}$, so the two critical points are

$\left( \pm \dfrac{1}{\sqrt[4]{3}}, \pm \sqrt[4]{3} \right)$. Then $D \left( \pm \dfrac{1}{\sqrt[4]{3}}, \pm \sqrt[4]{3} \right) = (2 + 4)\left( 2 + \tfrac{24}{3} \right) - \left( \dfrac{6}{\sqrt{3}} \right)^2 > 0$ and $f_{xx} \left( \pm \dfrac{1}{\sqrt[4]{3}}, \pm \sqrt[4]{3} \right) = 6 > 0$, so each point

is a minimum. Finally, $y^2 = \dfrac{2}{xz^3}$, so the four points closest to the origin are $\left( \pm \dfrac{1}{\sqrt[4]{3}}, \dfrac{\sqrt{2}}{\sqrt[4]{3}}, \pm \sqrt[4]{3} \right)$, $\left( \pm \dfrac{1}{\sqrt[4]{3}}, -\dfrac{\sqrt{2}}{\sqrt[4]{3}}, \pm \sqrt[4]{3} \right)$.

**65.**

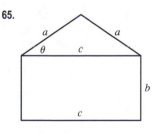

The area of the triangle is $\tfrac{1}{2} ca \sin \theta$ and the area of the rectangle is $bc$. Thus, the area of the whole object is $f(a, b, c) = \tfrac{1}{2} ca \sin \theta + bc$. The perimeter of the object is $g(a, b, c) = 2a + 2b + c = P$. To simplify $\sin \theta$ in terms of $a$, $b$, and $c$ notice that $a^2 \sin^2 \theta + \left( \tfrac{1}{2} c \right)^2 = a^2 \Rightarrow \sin \theta = \dfrac{1}{2a} \sqrt{4a^2 - c^2}$.

Thus $f(a, b, c) = \dfrac{c}{4} \sqrt{4a^2 - c^2} + bc$. (Instead of using $\theta$, we could just have

used the Pythagorean Theorem.) As a result, by Lagrange's method, we must find $a$, $b$, $c$, and $\lambda$ by solving $\nabla f = \lambda \nabla g$ which

gives the following equations: $ca(4a^2 - c^2)^{-1/2} = 2\lambda$ **(1)**, $c = 2\lambda$ **(2)**, $\tfrac{1}{4}(4a^2 - c^2)^{1/2} - \tfrac{1}{4} c^2 (4a^2 - c^2)^{-1/2} + b = \lambda$

**(3)**, and $2a + 2b + c = P$ **(4)**. From **(2)**, $\lambda = \tfrac{1}{2} c$ and so **(1)** produces $ca(4a^2 - c^2)^{-1/2} = c \Rightarrow (4a^2 - c^2)^{1/2} = a \Rightarrow$

$4a^2 - c^2 = a^2 \Rightarrow c = \sqrt{3}\, a$ **(5)**. Similarly, since $(4a^2 - c^2)^{1/2} = a$ and $\lambda = \tfrac{1}{2} c$, **(3)** gives $\dfrac{a}{4} - \dfrac{c^2}{4a} + b = \dfrac{c}{2}$, so from

**(5)**, $\dfrac{a}{4} - \dfrac{3a}{4} + b = \dfrac{\sqrt{3}\, a}{2} \Rightarrow -\dfrac{a}{2} - \dfrac{\sqrt{3}\, a}{2} = -b \Rightarrow b = \dfrac{a}{2} \left( 1 + \sqrt{3} \right)$ **(6)**. Substituting **(5)** and **(6)** into **(4)** we get:

$2a + a(1 + \sqrt{3}) + \sqrt{3}\, a = P \Rightarrow 3a + 2\sqrt{3}\, a = P \Rightarrow a = \dfrac{P}{3 + 2\sqrt{3}} = \dfrac{2\sqrt{3} - 3}{3} P$ and thus

$b = \dfrac{(2\sqrt{3} - 3)(1 + \sqrt{3})}{6} P = \dfrac{3 - \sqrt{3}}{6} P$ and $c = \left( 2 - \sqrt{3} \right) P$.

# ☐  FOCUS ON PROBLEM SOLVING

1. The areas of the smaller rectangles are $A_1 = xy$, $A_2 = (L - x)y$,

   $A_3 = (L - x)(W - y)$, $A_4 = x(W - y)$. For $0 \le x \le L$, $0 \le y \le W$, let

   $$f(x, y) = A_1^2 + A_2^2 + A_3^2 + A_4^2$$
   $$= x^2 y^2 + (L - x)^2 y^2 + (L - x)^2 (W - y)^2 + x^2 (W - y)^2$$
   $$= [x^2 + (L - x)^2][y^2 + (W - y)^2]$$

   Then we need to find the maximum and minimum values of $f(x, y)$. Here

   $$f_x(x, y) = [2x - 2(L - x)][y^2 + (W - y)^2] = 0 \quad \Rightarrow \quad 4x - 2L = 0 \text{ or } x = \tfrac{1}{2}L, \text{ and}$$

   $$f_y(x, y) = [x^2 + (L - x)^2][2y - 2(W - y)] = 0 \quad \Rightarrow \quad 4y - 2W = 0 \text{ or } y = W/2. \text{ Also}$$

   $f_{xx} = 4[y^2 + (W - y)^2]$, $f_{yy} = 4[x^2 + (L - x)^2]$, and $f_{xy} = (4x - 2L)(4y - 2W)$. Then

   $D = 16[y^2 + (W - y)^2][x^2 + (L - x)^2] - (4x - 2L)^2(4y - 2W)^2$. Thus when $x = \tfrac{1}{2}L$ and $y = \tfrac{1}{2}W$, $D > 0$ and

   $f_{xx} = 2W^2 > 0$. Thus a minimum of $f$ occurs at $\left(\tfrac{1}{2}L, \tfrac{1}{2}W\right)$ and this minimum value is $f\left(\tfrac{1}{2}L, \tfrac{1}{2}W\right) = \tfrac{1}{4}L^2 W^2$.

   There are no other critical points, so the maximum must occur on the boundary. Now along the width of the rectangle let

   $g(y) = f(0, y) = f(L, y) = L^2[y^2 + (W - y)^2]$, $0 \le y \le W$. Then $g'(y) = L^2[2y - 2(W - y)] = 0 \quad \Leftrightarrow \quad y = \tfrac{1}{2}W$.

   And $g\left(\tfrac{1}{2}\right) = \tfrac{1}{2}L^2 W^2$. Checking the endpoints, we get $g(0) = g(W) = L^2 W^2$. Along the length of the rectangle let

   $h(x) = f(x, 0) = f(x, W) = W^2[x^2 + (L - x)^2]$, $0 \le x \le L$. By symmetry $h'(x) = 0 \quad \Leftrightarrow \quad x = \tfrac{1}{2}L$ and

   $h\left(\tfrac{1}{2}L\right) = \tfrac{1}{2}L^2 W^2$. At the endpoints we have $h(0) = h(L) = L^2 W^2$. Therefore $L^2 W^2$ is the maximum value of $f$.

   This maximum value of $f$ occurs when the "cutting" lines correspond to sides of the rectangle.

3. (a) The area of a trapezoid is $\tfrac{1}{2}h(b_1 + b_2)$, where $h$ is the height (the distance between the two parallel sides) and $b_1$, $b_2$ are

   the lengths of the bases (the parallel sides). From the figure in the text, we see that $h = x \sin \theta$, $b_1 = w - 2x$, and

   $b_2 = w - 2x + 2x \cos \theta$. Therefore the cross-sectional area of the rain gutter is

   $$A(x, \theta) = \tfrac{1}{2}x \sin \theta \left[(w - 2x) + (w - 2x + 2x \cos \theta)\right] = (x \sin \theta)(w - 2x + x \cos \theta)$$
   $$= wx \sin \theta - 2x^2 \sin \theta + x^2 \sin \theta \cos \theta, \; 0 < x \le \tfrac{1}{2}w, 0 < \theta \le \tfrac{\pi}{2}$$

   We look for the critical points of $A$: $\partial A/\partial x = w \sin \theta - 4x \sin \theta + 2x \sin \theta \cos \theta$ and

   $\partial A/\partial \theta = wx \cos \theta - 2x^2 \cos \theta + x^2(\cos^2 \theta - \sin^2 \theta)$, so $\partial A/\partial x = 0 \quad \Leftrightarrow \quad \sin \theta \, (w - 4x + 2x \cos \theta) = 0 \quad \Leftrightarrow$

   $\cos \theta = \dfrac{4x - w}{2x} = 2 - \dfrac{w}{2x} \quad (0 < \theta \le \tfrac{\pi}{2} \quad \Rightarrow \quad \sin \theta > 0)$. If, in addition, $\partial A/\partial \theta = 0$, then

$$0 = wx\cos\theta - 2x^2\cos\theta + x^2(2\cos^2\theta - 1)$$

$$= wx\left(2 - \frac{w}{2x}\right) - 2x^2\left(2 - \frac{w}{2x}\right) + x^2\left[2\left(2 - \frac{w}{2x}\right)^2 - 1\right]$$

$$= 2wx - \tfrac{1}{2}w^2 - 4x^2 + wx + x^2\left[8 - \frac{4w}{x} + \frac{w^2}{2x^2} - 1\right] = -wx + 3x^2 = x(3x - w)$$

Since $x > 0$, we must have $x = \tfrac{1}{3}w$, in which case $\cos\theta = \tfrac{1}{2}$, so $\theta = \tfrac{\pi}{3}$, $\sin\theta = \tfrac{\sqrt{3}}{2}$, $k = \tfrac{\sqrt{3}}{6}w$, $b_1 = \tfrac{1}{3}w$, $b_2 = \tfrac{2}{3}w$,

and $A = \tfrac{\sqrt{3}}{12}w^2$. As in Example 11.7.6, we can argue from the physical nature of this problem that we have found a local

maximum of $A$. Now checking the boundary of $A$, let

$g(\theta) = A(w/2, \theta) = \tfrac{1}{2}w^2\sin\theta - \tfrac{1}{2}w^2\sin\theta + \tfrac{1}{4}w^2\sin\theta\cos\theta = \tfrac{1}{8}w^2\sin 2\theta$, $0 < \theta \le \tfrac{\pi}{2}$. Clearly $g$ is maximized when

$\sin 2\theta = 1$ in which case $A = \tfrac{1}{8}w^2$. Also along the line $\theta = \tfrac{\pi}{2}$, let $h(x) = A\left(x, \tfrac{\pi}{2}\right) = wx - 2x^2$, $0 < x < \tfrac{1}{2}w$  $\Rightarrow$

$h'(x) = w - 4x = 0$  $\Leftrightarrow$  $x = \tfrac{1}{4}w$, and $h\left(\tfrac{1}{4}w\right) = w\left(\tfrac{1}{4}w\right) - 2\left(\tfrac{1}{4}w\right)^2 = \tfrac{1}{8}w^2$. Since $\tfrac{1}{8}w^2 < \tfrac{\sqrt{3}}{12}w^2$, we conclude that

the local maximum found earlier was an absolute maximum.

(b) If the metal were bent into a semi-circular gutter of radius $r$, we would have $w = \pi r$ and $A = \tfrac{1}{2}\pi r^2 = \tfrac{1}{2}\pi\left(\frac{w}{\pi}\right)^2 = \frac{w^2}{2\pi}$.

Since $\frac{w^2}{2\pi} > \frac{\sqrt{3}\,w^2}{12}$, it *would* be better to bend the metal into a gutter with a semicircular cross-section.

**5.** Let $g(x, y) = xf\left(\frac{y}{x}\right)$. Then $g_x(x, y) = f\left(\frac{y}{x}\right) + xf'\left(\frac{y}{x}\right)\left(-\frac{y}{x^2}\right) = f\left(\frac{y}{x}\right) - \frac{y}{x}f'\left(\frac{y}{x}\right)$ and

$g_y(x, y) = xf'\left(\frac{y}{x}\right)\left(\frac{1}{x}\right) = f'\left(\frac{y}{x}\right)$. Thus the tangent plane at $(x_0, y_0, z_0)$ on the surface has equation

$$z - x_0 f\left(\frac{y_0}{x_0}\right) = \left[f\left(\frac{y_0}{x_0}\right) - y_0 x_0^{-1} f'\left(\frac{y_0}{x_0}\right)\right](x - x_0) + f'\left(\frac{y_0}{x_0}\right)(y - y_0)  \Rightarrow$$

$\left[f\left(\frac{y_0}{x_0}\right) - y_0 x_0^{-1} f'\left(\frac{y_0}{x_0}\right)\right]x + \left[f'\left(\frac{y_0}{x_0}\right)\right]y - z = 0$. But any plane whose equation is of the form $ax + by + cz = 0$

passes through the origin. Thus the origin is the common point of intersection.

**7.** (a) $x = r\cos\theta$, $y = r\sin\theta$, $z = z$.  Then $\dfrac{\partial u}{\partial r} = \dfrac{\partial u}{\partial x}\dfrac{\partial x}{\partial r} + \dfrac{\partial u}{\partial y}\dfrac{\partial y}{\partial r} + \dfrac{\partial u}{\partial z}\dfrac{\partial z}{\partial r} = \dfrac{\partial u}{\partial x}\cos\theta + \dfrac{\partial u}{\partial y}\sin\theta$ and

$$\frac{\partial^2 u}{\partial r^2} = \cos\theta\left[\frac{\partial^2 u}{\partial x^2}\frac{\partial x}{\partial r} + \frac{\partial^2 u}{\partial y\,\partial x}\frac{\partial y}{\partial r} + \frac{\partial^2 u}{\partial z\,\partial x}\frac{\partial z}{\partial r}\right] + \sin\theta\left[\frac{\partial^2 u}{\partial y^2}\frac{\partial y}{\partial r} + \frac{\partial^2 u}{\partial x\,\partial y}\frac{\partial x}{\partial r} + \frac{\partial^2 u}{\partial z\,\partial y}\frac{\partial z}{\partial r}\right]$$

$$= \frac{\partial^2 u}{\partial x^2}\cos^2\theta + \frac{\partial^2 u}{\partial y^2}\sin^2\theta + 2\frac{\partial^2 u}{\partial y\,\partial x}\cos\theta\sin\theta$$

Similarly $\dfrac{\partial u}{\partial\theta} = -\dfrac{\partial u}{\partial x}r\sin\theta + \dfrac{\partial u}{\partial y}r\cos\theta$ and

$$\frac{\partial^2 u}{\partial \theta^2} = \frac{\partial^2 u}{\partial x^2} r^2 \sin^2\theta + \frac{\partial^2 u}{\partial y^2} r^2 \cos^2\theta - 2\frac{\partial^2 u}{\partial y\,\partial x} r^2 \sin\theta \cos\theta - \frac{\partial u}{\partial x} r\cos\theta - \frac{\partial u}{\partial y} r\sin\theta. \text{ So}$$

$$\frac{\partial^2 u}{\partial r^2} + \frac{1}{r}\frac{\partial u}{\partial r} + \frac{1}{r^2}\frac{\partial^2 u}{\partial \theta^2} + \frac{\partial^2 u}{\partial z^2} = \frac{\partial^2 u}{\partial x^2}\cos^2\theta + \frac{\partial^2 u}{\partial y^2}\sin^2\theta + 2\frac{\partial^2 u}{\partial y\,\partial x}\cos\theta\sin\theta + \frac{\partial u}{\partial x}\frac{\cos\theta}{r} + \frac{\partial u}{\partial y}\frac{\sin\theta}{r}$$

$$+ \frac{\partial^2 u}{\partial x^2}\sin^2\theta + \frac{\partial^2 u}{\partial y^2}\cos^2\theta - 2\frac{\partial^2 u}{\partial y\,\partial x}\sin\theta\cos\theta$$

$$- \frac{\partial u}{\partial x}\frac{\cos\theta}{r} - \frac{\partial u}{\partial y}\frac{\sin\theta}{r} + \frac{\partial^2 u}{\partial z^2}$$

$$= \frac{\partial^2 u}{\partial x^2} + \frac{\partial^2 u}{\partial y^2} + \frac{\partial^2 u}{\partial z^2}$$

(b) $x = \rho\sin\phi\cos\theta$, $y = \rho\sin\phi\sin\theta$, $z = \rho\cos\phi$. Then

$$\frac{\partial u}{\partial \rho} = \frac{\partial u}{\partial x}\frac{\partial x}{\partial \rho} + \frac{\partial u}{\partial y}\frac{\partial y}{\partial \rho} + \frac{\partial u}{\partial z}\frac{\partial z}{\partial \rho} = \frac{\partial u}{\partial x}\sin\phi\cos\theta + \frac{\partial u}{\partial y}\sin\phi\sin\theta + \frac{\partial u}{\partial z}\cos\phi, \text{ and}$$

$$\frac{\partial^2 u}{\partial \rho^2} = \sin\phi\cos\theta\left[\frac{\partial^2 u}{\partial x^2}\frac{\partial x}{\partial \rho} + \frac{\partial^2 u}{\partial y\,\partial x}\frac{\partial y}{\partial \rho} + \frac{\partial^2 u}{\partial z\,\partial x}\frac{\partial z}{\partial \rho}\right]$$

$$+ \sin\phi\sin\theta\left[\frac{\partial^2 u}{\partial y^2}\frac{\partial y}{\partial \rho} + \frac{\partial^2 u}{\partial x\,\partial y}\frac{\partial x}{\partial \rho} + \frac{\partial^2 u}{\partial z\,\partial y}\frac{\partial z}{\partial \rho}\right]$$

$$+ \cos\phi\left[\frac{\partial^2 u}{\partial z^2}\frac{\partial z}{\partial \rho} + \frac{\partial^2 u}{\partial x\,\partial z}\frac{\partial x}{\partial \rho} + \frac{\partial^2 u}{\partial y\,\partial z}\frac{\partial y}{\partial \rho}\right]$$

$$= 2\frac{\partial^2 u}{\partial y\,\partial x}\sin^2\phi\sin\theta\cos\theta + 2\frac{\partial^2 u}{\partial z\,\partial x}\sin\phi\cos\phi\cos\theta + 2\frac{\partial^2 u}{\partial y\,\partial z}\sin\phi\cos\phi\sin\theta$$

$$+ \frac{\partial^2 u}{\partial x^2}\sin^2\phi\cos^2\theta + \frac{\partial^2 u}{\partial y^2}\sin^2\phi\sin^2\theta + \frac{\partial^2 u}{\partial z^2}\cos^2\phi$$

Similarly $\dfrac{\partial u}{\partial \phi} = \dfrac{\partial u}{\partial x}\rho\cos\phi\cos\theta + \dfrac{\partial u}{\partial y}\rho\cos\phi\sin\theta - \dfrac{\partial u}{\partial z}\rho\sin\phi$, and

$$\frac{\partial^2 u}{\partial \phi^2} = 2\frac{\partial^2 u}{\partial y\,\partial x}\rho^2\cos^2\phi\sin\theta\cos\theta - 2\frac{\partial^2 u}{\partial x\,\partial z}\rho^2\sin\phi\cos\phi\cos\theta$$

$$- 2\frac{\partial^2 u}{\partial y\,\partial z}\rho^2\sin\phi\cos\phi\sin\theta + \frac{\partial^2 u}{\partial x^2}\rho^2\cos^2\phi\cos^2\theta + \frac{\partial^2 u}{\partial y^2}\rho^2\cos^2\phi\sin^2\theta$$

$$+ \frac{\partial^2 u}{\partial z^2}\rho^2\sin^2\phi - \frac{\partial u}{\partial x}\rho\sin\phi\cos\theta - \frac{\partial u}{\partial y}\rho\sin\phi\sin\theta - \frac{\partial u}{\partial z}\rho\cos\phi$$

And $\dfrac{\partial u}{\partial \theta} = -\dfrac{\partial u}{\partial x}\rho\sin\phi\sin\theta + \dfrac{\partial u}{\partial y}\rho\sin\phi\cos\theta$, while

$$\frac{\partial^2 u}{\partial \theta^2} = -2\frac{\partial^2 u}{\partial y\,\partial x}\rho^2\sin^2\phi\cos\theta\sin\theta + \frac{\partial^2 u}{\partial x^2}\rho^2\sin^2\phi\sin^2\theta$$

$$+ \frac{\partial^2 u}{\partial y^2}\rho^2\sin^2\phi\cos^2\theta - \frac{\partial u}{\partial x}\rho\sin\phi\cos\theta - \frac{\partial u}{\partial y}\rho\sin\phi\sin\theta$$

[continued]

Therefore

$$\frac{\partial^2 u}{\partial \rho^2} + \frac{2}{\rho}\frac{\partial u}{\partial \rho} + \frac{\cot \phi}{\rho^2}\frac{\partial u}{\partial \phi} + \frac{1}{\rho^2}\frac{\partial^2 u}{\partial \phi^2} + \frac{1}{\rho^2 \sin^2 \phi}\frac{\partial^2 u}{\partial \theta^2}$$

$$= \frac{\partial^2 u}{\partial x^2}\left[(\sin^2 \phi \, \cos^2 \theta) + (\cos^2 \phi \, \cos^2 \theta) + \sin^2 \theta\right]$$

$$+ \frac{\partial^2 u}{\partial y^2}\left[(\sin^2 \phi \, \sin^2 \theta) + (\cos^2 \phi \, \sin^2 \theta) + \cos^2 \theta\right] + \frac{\partial^2 u}{\partial z^2}\left[\cos^2 \phi + \sin^2 \phi\right]$$

$$+ \frac{\partial u}{\partial x}\left[\frac{2\sin^2 \phi \, \cos \theta + \cos^2 \phi \, \cos \theta - \sin^2 \phi \, \cos \theta - \cos \theta}{\rho \sin \phi}\right]$$

$$+ \frac{\partial u}{\partial y}\left[\frac{2\sin^2 \phi \, \sin \theta + \cos^2 \phi \, \sin \theta - \sin^2 \phi \, \sin \theta - \sin \theta}{\rho \sin \phi}\right]$$

But $2\sin^2 \phi \cos \theta + \cos^2 \phi \cos \theta - \sin^2 \phi \cos \theta - \cos \theta = (\sin^2 \phi + \cos^2 \phi - 1)\cos \theta = 0$ and similarly the coefficient of

$\partial u/\partial y$ is 0. Also $\sin^2 \phi \cos^2 \theta + \cos^2 \phi \cos^2 \theta + \sin^2 \theta = \cos^2 \theta\,(\sin^2 \phi + \cos^2 \phi) + \sin^2 \theta = 1$, and similarly the

coefficient of $\partial^2 u/\partial y^2$ is 1. So Laplace's Equation in spherical coordinates is as stated.

**9.** Since we are minimizing the area of the ellipse, and the circle lies above the $x$-axis,

the ellipse will intersect the circle for only one value of $y$. This $y$-value must

satisfy both the equation of the circle and the equation of the ellipse. Now

$\dfrac{x^2}{a^2} + \dfrac{y^2}{b^2} = 1 \;\Rightarrow\; x^2 = \dfrac{a^2}{b^2}\left(b^2 - y^2\right)$. Substituting into the equation of the

circle gives $\dfrac{a^2}{b^2}\left(b^2 - y^2\right) + y^2 - 2y = 0 \;\Rightarrow\; \left(\dfrac{b^2 - a^2}{b^2}\right)y^2 - 2y + a^2 = 0$.

In order for there to be only one solution to this quadratic equation, the discriminant must be 0, so $4 - 4a^2\,\dfrac{b^2 - a^2}{b^2} = 0 \;\Rightarrow$

$b^2 - a^2 b^2 + a^4 = 0$. The area of the ellipse is $A(a, b) = \pi ab$, and we minimize this function subject to the constraint

$g(a, b) = b^2 - a^2 b^2 + a^4 = 0$.

Now $\nabla A = \lambda \nabla g \;\Leftrightarrow\; \pi b = \lambda(4a^3 - 2ab^2),\; \pi a = \lambda(2b - 2ba^2) \;\Rightarrow\; \lambda = \dfrac{\pi b}{2a(2a^2 - b^2)}$ **(1)**,

$\lambda = \dfrac{\pi a}{2b(1 - a^2)}$ **(2)**, $b^2 - a^2 b^2 + a^4 = 0$ **(3)**. Comparing **(1)** and **(2)** gives $\dfrac{\pi b}{2a(2a^2 - b^2)} = \dfrac{\pi a}{2b(1 - a^2)} \;\Rightarrow$

$2\pi b^2 = 4\pi a^4 \;\Leftrightarrow\; a^2 = \frac{1}{\sqrt{2}}\,b$. Substitute this into **(3)** to get $b = \frac{3}{\sqrt{2}} \;\Rightarrow\; a = \sqrt{\frac{3}{2}}$.

# 12 ☐ MULTIPLE INTEGRALS

## 12.1 Double Integrals over Rectangles

**1.** (a) The subrectangles are shown in the figure.

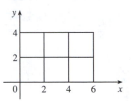

The surface is the graph of $f(x, y) = xy$ and $\Delta A = 4$, so we estimate

$$V \approx \sum_{i=1}^{3} \sum_{j=1}^{2} f(x_i, y_j) \, \Delta A$$

$$= f(2, 2) \, \Delta A + f(2, 4) \, \Delta A + f(4, 2) \, \Delta A + f(4, 4) \, \Delta A + f(6, 2) \, \Delta A + f(6, 4) \, \Delta A$$

$$= 4(4) + 8(4) + 8(4) + 16(4) + 12(4) + 24(4) = 288$$

(b) $V \approx \displaystyle\sum_{i=1}^{3} \sum_{j=1}^{2} f(\overline{x}_i, \overline{y}_j) \, \Delta A = f(1, 1) \, \Delta A + f(1, 3) \, \Delta A + f(3, 1) \, \Delta A + f(3, 3) \, \Delta A + f(5, 1) \, \Delta A + f(5, 3) \, \Delta A$

$$= 1(4) + 3(4) + 3(4) + 9(4) + 5(4) + 15(4) = 144$$

**3.** (a) The subrectangles are shown in the figure. Since $\Delta A = \pi^2/4$, we estimate

$$\iint_R \sin(x + y) \, dA \approx \sum_{i=1}^{2} \sum_{j=1}^{2} f\left(x_{ij}^*, y_{ij}^*\right) \Delta A$$

$$= f(0, 0) \, \Delta A + f\left(0, \tfrac{\pi}{2}\right) \Delta A + f\left(\tfrac{\pi}{2}, 0\right) \Delta A + f\left(\tfrac{\pi}{2}, \tfrac{\pi}{2}\right) \Delta A$$

$$= 0\left(\tfrac{\pi^2}{4}\right) + 1\left(\tfrac{\pi^2}{4}\right) + 1\left(\tfrac{\pi^2}{4}\right) + 0\left(\tfrac{\pi^2}{4}\right) = \tfrac{\pi^2}{2} \approx 4.935$$

(b) $\iint_R \sin(x + y) \, dA \approx \displaystyle\sum_{i=1}^{2} \sum_{j=1}^{2} f(\overline{x}_i, \overline{y}_j) \, \Delta A$

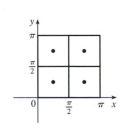

$$= f\left(\tfrac{\pi}{4}, \tfrac{\pi}{4}\right) \Delta A + f\left(\tfrac{\pi}{4}, \tfrac{3\pi}{4}\right) \Delta A + f\left(\tfrac{3\pi}{4}, \tfrac{\pi}{4}\right) \Delta A + f\left(\tfrac{3\pi}{4}, \tfrac{3\pi}{4}\right) \Delta A$$

$$= 1\left(\tfrac{\pi^2}{4}\right) + 0\left(\tfrac{\pi^2}{4}\right) + 0\left(\tfrac{\pi^2}{4}\right) + (-1)\left(\tfrac{\pi^2}{4}\right) = 0$$

**5.** (a) Each subrectangle and its midpoint are shown in the figure.

The area of each subrectangle is $\Delta A = 2$, so we evaluate $f$

at each midpoint and estimate

$$\iint_R f(x, y) \, dA \approx \sum_{i=1}^{2} \sum_{j=1}^{2} f(\overline{x}_i, \overline{y}_j) \, \Delta A$$

$$= f(1, 2.5) \, \Delta A + f(1, 3.5) \, \Delta A$$

$$+ f(3, 2.5) \, \Delta A + f(3, 3.5) \, \Delta A$$

$$= -2(2) + (-1)(2) + 2(2) + 3(2) = 4$$

(b) The subrectangles are shown in the figure.

In each subrectangle, the sample point closest to the origin

is the lower left corner, and the area of each subrectangle is $\Delta A = \frac{1}{2}$.

Thus we estimate

$$\iint_R f(x,y)\,dA \approx \sum_{i=1}^{4}\sum_{j=1}^{4} f\left(x_{ij}^*, y_{ij}^*\right)\Delta A$$

$$= f(0,2)\,\Delta A + f(0,2.5)\,\Delta A + f(0,3)\,\Delta A + f(0,3.5)\,\Delta A$$

$$+ f(1,2)\,\Delta A + f(1,2.5)\,\Delta A + f(1,3)\,\Delta A + f(1,3.5)\,\Delta A$$

$$+ f(2,2)\,\Delta A + f(2,2.5)\,\Delta A + f(2,3)\,\Delta A + f(2,3.5)\,\Delta A$$

$$+ f(3,2)\,\Delta A + f(3,2.5)\,\Delta A + f(3,3)\,\Delta A + f(3,3.5)\,\Delta A$$

$$= -3\left(\tfrac{1}{2}\right) + (-5)\left(\tfrac{1}{2}\right) + (-6)\left(\tfrac{1}{2}\right) + (-4)\left(\tfrac{1}{2}\right) + (-1)\left(\tfrac{1}{2}\right) + (-2)\left(\tfrac{1}{2}\right) + (-3)\left(\tfrac{1}{2}\right) + (-1)\left(\tfrac{1}{2}\right)$$

$$+ 1\left(\tfrac{1}{2}\right) + 0\left(\tfrac{1}{2}\right) + (-1)\left(\tfrac{1}{2}\right) + 1\left(\tfrac{1}{2}\right) + 2\left(\tfrac{1}{2}\right) + 2\left(\tfrac{1}{2}\right) + 1\left(\tfrac{1}{2}\right) + 3\left(\tfrac{1}{2}\right)$$

$$= -8$$

**7.** The values of $f(x,y) = \sqrt{52 - x^2 - y^2}$ get smaller as we move farther from the origin, so on any of the subrectangles in the problem, the function will have its largest value at the lower left corner of the subrectangle and its smallest value at the upper right corner, and any other value will lie between these two. So using these subrectangles we have $U < V < L$. (Note that this is true no matter how $R$ is divided into subrectangles.)

**9.** (a) With $m = n = 2$, we have $\Delta A = 4$. Using the contour map to estimate the value of $f$ at the center of each subrectangle, we have

$$\iint_R f(x,y)\,dA \approx \sum_{i=1}^{2}\sum_{j=1}^{2} f(\overline{x}_i, \overline{y}_j)\,\Delta A = \Delta A[f(1,1) + f(1,3) + f(3,1) + f(3,3)] \approx 4(27 + 4 + 14 + 17) = 248$$

(b) $f_{\text{ave}} = \frac{1}{A(R)}\iint_R f(x,y)\,dA \approx \frac{1}{16}(248) = 15.5$

**11.** $z = 3 > 0$, so we can interpret the integral as the volume of the solid $S$ that lies below the plane $z = 3$ and above the rectangle $[-2, 2] \times [1, 6]$. $S$ is a rectangular solid, thus $\iint_R 3\,dA = 4 \cdot 5 \cdot 3 = 60$.

**13.** $z = f(x,y) = 4 - 2y \geq 0$ for $0 \leq y \leq 1$. Thus the integral represents the volume of that part of the rectangular solid $[0, 1] \times [0, 1] \times [0, 4]$ which lies below the plane $z = 4 - 2y$. So

$$\iint_R (4 - 2y)\,dA = (1)(1)(2) + \tfrac{1}{2}(1)(1)(2) = 3$$

**15.** To calculate the estimates using a programmable calculator, we can use an algorithm similar to that of Exercise 5.1.7. In Maple, we can define the function

$f(x, y) = \sqrt{1 + xe^{-y}}$ (calling it f), load the student package, and then use the command

middlesum(middlesum(f,x=0..1,m),

y=0..1,m);

to get the estimate with $n = m^2$ squares of equal size. Mathematica has no special Riemann sum command, but we can define f and then use nested Sum commands to calculate the estimates.

| $n$ | estimate |
|---|---|
| 1 | 1.141606 |
| 4 | 1.143191 |
| 16 | 1.143535 |
| 64 | 1.143617 |
| 256 | 1.143637 |
| 1024 | 1.143642 |

**17.** If we divide $R$ into $mn$ subrectangles, $\iint_R k \, dA \approx \sum_{i=1}^{m} \sum_{j=1}^{n} f(x_{ij}^*, y_{ij}^*) \, \Delta A$ for any choice of sample points $(x_{ij}^*, y_{ij}^*)$.

But $f(x_{ij}^*, y_{ij}^*) = k$ always and $\sum_{i=1}^{m} \sum_{j=1}^{n} \Delta A = $ area of $R = (b - a)(d - c)$. Thus, no matter how we choose the sample

points, $\sum_{i=1}^{m} \sum_{j=1}^{n} f(x_{ij}^*, y_{ij}^*) \, \Delta A = k \sum_{i=1}^{m} \sum_{j=1}^{n} \Delta A = k(b-a)(d-c)$ and so

$\iint_R k \, dA = \lim_{m,n\to\infty} \sum_{i=1}^{m} \sum_{j=1}^{n} f(x_{ij}^*, y_{ij}^*) \, \Delta A = \lim_{m,n\to\infty} k \sum_{i=1}^{m} \sum_{j=1}^{n} \Delta A = \lim_{m,n\to\infty} k(b-a)(d-c) = k(b-a)(d-c)$.

## 12.2  Iterated Integrals

**1.** $\int_0^5 12x^2y^3 \, dx = \left[12 \dfrac{x^3}{3} y^3\right]_{x=0}^{x=5} = 4x^3y^3\big]_{x=0}^{x=5} = 4(5)^3 y^3 - 4(0)^3 y^3 = 500y^3$,

$\int_0^1 12x^2y^3 \, dy = \left[12x^2 \dfrac{y^4}{4}\right]_{y=0}^{y=1} = 3x^2y^4\big]_{y=0}^{y=1} = 3x^2(1)^4 - 3x^2(0)^4 = 3x^2$

**3.** $\int_1^3 \int_0^1 (1 + 4xy) \, dx \, dy = \int_1^3 \left[x + 2x^2y\right]_{x=0}^{x=1} dy = \int_1^3 (1 + 2y) \, dy = \left[y + y^2\right]_1^3 = (3 + 9) - (1 + 1) = 10$

**5.** $\int_0^2 \int_0^{\pi/2} x \sin y \, dy \, dx = \int_0^2 x \, dx \int_0^{\pi/2} \sin y \, dy$  [as in Example 5]  $= \left[\dfrac{x^2}{2}\right]_0^2 \left[-\cos y\right]_0^{\pi/2} = (2 - 0)(0 + 1) = 2$

**7.** $\int_0^2 \int_0^1 (2x + y)^8 \, dx \, dy = \int_0^2 \left[\dfrac{1}{2} \dfrac{(2x+y)^9}{9}\right]_{x=0}^{x=1} dy$  [substitute $u = 2x + y \Rightarrow dx = \frac{1}{2} \, du$]

$= \dfrac{1}{18} \int_0^2 \left[(2+y)^9 - (0+y)^9\right] dy = \dfrac{1}{18} \left[\dfrac{(2+y)^{10}}{10} - \dfrac{y^{10}}{10}\right]_0^2$

$= \frac{1}{180}\left[(4^{10} - 2^{10}) - (2^{10} - 0^{10})\right] = \frac{1{,}046{,}528}{180} = \frac{261{,}632}{45}$

**9.** $\int_1^4 \int_1^2 \left(\dfrac{x}{y} + \dfrac{y}{x}\right) dy \, dx = \int_1^4 \left[x \ln|y| + \dfrac{1}{x} \cdot \dfrac{1}{2} y^2\right]_{y=1}^{y=2} dx = \int_1^4 \left(x \ln 2 + \dfrac{3}{2x}\right) dx = \left[\frac{1}{2}x^2 \ln 2 + \frac{3}{2} \ln|x|\right]_1^4$

$= 8 \ln 2 + \frac{3}{2} \ln 4 - \frac{1}{2} \ln 2 = \frac{15}{2} \ln 2 + 3 \ln 4^{1/2} = \frac{21}{2} \ln 2$

**11.** $\int_0^1 \int_0^1 (u-v)^5 \, du \, dv = \int_0^1 \left[\frac{1}{6}(u-v)^6\right]_{u=0}^{u=1} dv = \frac{1}{6}\int_0^1 \left[(1-v)^6 - (0-v)^6\right] dv$

$\qquad = \frac{1}{6}\int_0^1 \left[(1-v)^6 - v^6\right] dv = \frac{1}{6}\left[-\frac{1}{7}(1-v)^7 - \frac{1}{7}v^7\right]_0^1$

$\qquad = -\frac{1}{42}\left[(0+1) - (1+0)\right] = 0$

**13.** $\int_0^2 \int_0^\pi r\sin^2\theta \, d\theta \, dr = \int_0^2 r \, dr \int_0^\pi \sin^2\theta \, d\theta$   [as in Example 5]   $= \int_0^2 r \, dr \int_0^\pi \frac{1}{2}(1-\cos 2\theta) \, d\theta$

$\qquad = \left[\frac{1}{2}r^2\right]_0^2 \cdot \frac{1}{2}\left[\theta - \frac{1}{2}\sin 2\theta\right]_0^\pi = (2-0)\cdot\frac{1}{2}\left[\left(\pi - \frac{1}{2}\sin 2\pi\right) - \left(0 - \frac{1}{2}\sin 0\right)\right]$

$\qquad = 2\cdot\frac{1}{2}[(\pi-0)-(0-0)] = \pi$

**15.** $\iint_R (6x^2y^3 - 5y^4)\, dA = \int_0^3 \int_0^1 (6x^2y^3 - 5y^4) \, dy \, dx = \int_0^3 \left[\frac{3}{2}x^2y^4 - y^5\right]_{y=0}^{y=1} dx = \int_0^3 \left(\frac{3}{2}x^2 - 1\right) dx$

$\qquad = \left[\frac{1}{2}x^3 - x\right]_0^3 = \frac{27}{2} - 3 = \frac{21}{2}$

**17.** $\iint_R \frac{xy^2}{x^2+1} \, dA = \int_0^1 \int_{-3}^3 \frac{xy^2}{x^2+1} \, dy \, dx = \int_0^1 \frac{x}{x^2+1} \, dx \int_{-3}^3 y^2 \, dy = \left[\frac{1}{2}\ln(x^2+1)\right]_0^1 \left[\frac{1}{3}y^3\right]_{-3}^3$

$\qquad = \frac{1}{2}(\ln 2 - \ln 1) \cdot \frac{1}{3}(27+27) = 9\ln 2$

**19.** $\int_0^{\pi/6} \int_0^{\pi/3} x\sin(x+y) \, dy \, dx$

$\qquad = \int_0^{\pi/6} \left[-x\cos(x+y)\right]_{y=0}^{y=\pi/3} dx = \int_0^{\pi/6} \left[x\cos x - x\cos\left(x+\frac{\pi}{3}\right)\right] dx$

$\qquad = x\left[\sin x - \sin\left(x+\frac{\pi}{3}\right)\right]_0^{\pi/6} - \int_0^{\pi/6}\left[\sin x - \sin\left(x+\frac{\pi}{3}\right)\right] dx$    [by integrating by parts separately for each term]

$\qquad = \frac{\pi}{6}\left[\frac{1}{2} - 1\right] - \left[-\cos x + \cos\left(x+\frac{\pi}{3}\right)\right]_0^{\pi/6} = -\frac{\pi}{12} - \left[-\frac{\sqrt{3}}{2} + 0 - \left(-1+\frac{1}{2}\right)\right] = \frac{\sqrt{3}-1}{2} - \frac{\pi}{12}$

**21.** $\iint_R xye^{x^2y} \, dA = \int_0^2 \int_0^1 xye^{x^2y} \, dx \, dy = \int_0^2 \left[\frac{1}{2}e^{x^2y}\right]_{x=0}^{x=1} dy = \frac{1}{2}\int_0^2 (e^y - 1)\, dy = \frac{1}{2}\left[e^y - y\right]_0^2$

$\qquad = \frac{1}{2}[(e^2 - 2) - (1-0)] = \frac{1}{2}(e^2 - 3)$

**23.** $z = f(x,y) = 4 - x - 2y \geq 0$ for $0 \leq x \leq 1$ and $0 \leq y \leq 1$. So the solid

is the region in the first octant which lies below the plane $z = 4 - x - 2y$

and above $[0,1]\times[0,1]$.

**25.** $V = \iint_R (12 - 3x - 2y)\, dA = \int_{-2}^3 \int_0^1 (12 - 3x - 2y) \, dx \, dy = \int_{-2}^3 \left[12x - \frac{3}{2}x^2 - 2xy\right]_{x=0}^{x=1} dy$

$\qquad = \int_{-2}^3 \left(\frac{21}{2} - 2y\right) dy = \left[\frac{21}{2}y - y^2\right]_{-2}^3 = \frac{95}{2}$

**27.** $V = \int_{-2}^2 \int_{-1}^1 \left(1 - \frac{1}{4}x^2 - \frac{1}{9}y^2\right) dx \, dy = 4\int_0^2 \int_0^1 \left(1 - \frac{1}{4}x^2 - \frac{1}{9}y^2\right) dx \, dy$

$\qquad = 4\int_0^2 \left[x - \frac{1}{12}x^3 - \frac{1}{9}y^2x\right]_{x=0}^{x=1} dy = 4\int_0^2 \left(\frac{11}{12} - \frac{1}{9}y^2\right) dy = 4\left[\frac{11}{12}y - \frac{1}{27}y^3\right]_0^2 = 4\cdot\frac{83}{54} = \frac{166}{27}$

**29.** Here we need the volume of the solid lying under the surface $z = x \sec^2 y$ and above the rectangle $R = [0, 2] \times [0, \pi/4]$ in

the $xy$-plane.

$$V = \int_0^2 \int_0^{\pi/4} x \sec^2 y \, dy \, dx = \int_0^2 x \, dx \int_0^{\pi/4} \sec^2 y \, dy = \left[\tfrac{1}{2}x^2\right]_0^2 \left[\tan y\right]_0^{\pi/4}$$
$$= (2 - 0)(\tan \tfrac{\pi}{4} - \tan 0) = 2(1 - 0) = 2$$

**31.** The solid lies below the surface $z = 2 + x^2 + (y - 2)^2$ and above the plane $z = 1$ for $-1 \le x \le 1, 0 \le y \le 4$. The volume

of the solid is the difference in volumes between the solid that lies under $z = 2 + x^2 + (y - 2)^2$ over the rectangle

$R = [-1, 1] \times [0, 4]$ and the solid that lies under $z = 1$ over $R$.

$$V = \int_0^4 \int_{-1}^1 [2 + x^2 + (y - 2)^2] \, dx \, dy - \int_0^4 \int_{-1}^1 (1) \, dx \, dy = \int_0^4 \left[2x + \tfrac{1}{3}x^3 + x(y - 2)^2\right]_{x=-1}^{x=1} \, dy - \int_{-1}^1 dx \int_0^4 dy$$
$$= \int_0^4 \left[(2 + \tfrac{1}{3} + (y - 2)^2) - (-2 - \tfrac{1}{3} - (y - 2)^2)\right] dy - [x]_{-1}^1 [y]_0^4$$
$$= \int_0^4 \left[\tfrac{14}{3} + 2(y - 2)^2\right] dy - [1 - (-1)][4 - 0] = \left[\tfrac{14}{3}y + \tfrac{2}{3}(y - 2)^3\right]_0^4 - (2)(4)$$
$$= \left[\left(\tfrac{56}{3} + \tfrac{16}{3}\right) - \left(0 - \tfrac{16}{3}\right)\right] - 8 = \tfrac{88}{3} - 8 = \tfrac{64}{3}$$

**33.** In Maple, we can calculate the integral by defining the integrand as `f`

and then using the command `int(int(f,x=0..1),y=0..1);`.

In Mathematica, we can use the command

`Integrate[f,{x,0,1},{y,0,1}]`

We find that $\iint_R x^5 y^3 e^{xy} \, dA = 21e - 57 \approx 0.0839$. We can use `plot3d`

(in Maple) or `Plot3D` (in Mathematica) to graph the function.

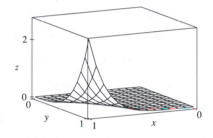

**35.** $R$ is the rectangle $[-1, 1] \times [0, 5]$. Thus, $A(R) = 2 \cdot 5 = 10$ and

$$f_{ave} = \frac{1}{A(R)} \iint_R f(x, y) \, dA = \tfrac{1}{10} \int_0^5 \int_{-1}^1 x^2 y \, dx \, dy = \tfrac{1}{10} \int_0^5 \left[\tfrac{1}{3}x^3 y\right]_{x=-1}^{x=1} \, dy = \tfrac{1}{10} \int_0^5 \tfrac{2}{3}y \, dy = \tfrac{1}{10} \left[\tfrac{1}{3}y^2\right]_0^5 = \tfrac{5}{6}.$$

**37.** $\displaystyle\iint_R \frac{xy}{1 + x^4} \, dA = \int_{-1}^1 \int_0^1 \frac{xy}{1 + x^4} \, dy \, dx = \int_{-1}^1 \frac{x}{1 + x^4} \, dx \int_0^1 y \, dy$ [by Equation 5] but $f(x) = \dfrac{x}{1 + x^4}$ is an odd

function so $\displaystyle\int_{-1}^1 f(x) \, dx = 0$ by (6) in Section 5.5. Thus $\displaystyle\iint_R \frac{xy}{1 + x^4} \, dA = 0 \cdot \int_0^1 y \, dy = 0.$

**39.** Let $f(x, y) = \dfrac{x - y}{(x + y)^3}$. Then a CAS gives $\int_0^1 \int_0^1 f(x, y) \, dy \, dx = \tfrac{1}{2}$ and $\int_0^1 \int_0^1 f(x, y) \, dx \, dy = -\tfrac{1}{2}$.

To explain the seeming violation of Fubini's Theorem, note that $f$ has an infinite discontinuity at $(0, 0)$ and thus does not

satisfy the conditions of Fubini's Theorem. In fact, both iterated integrals involve improper integrals which diverge at their

lower limits of integration.

## 12.3 Double Integrals over General Regions

**1.** $\int_0^4 \int_0^{\sqrt{y}} xy^2 \, dx \, dy = \int_0^4 \left[ \frac{1}{2} x^2 y^2 \right]_{x=0}^{x=\sqrt{y}} dy = \int_0^4 \frac{1}{2} y^2 [(\sqrt{y})^2 - 0^2] dy = \frac{1}{2} \int_0^4 y^3 \, dy = \frac{1}{2} \left[ \frac{1}{4} y^4 \right]_0^4 = \frac{1}{2}(64 - 0) = 32$

**3.** $\int_0^1 \int_{x^2}^{x} (1 + 2y) dy \, dx = \int_0^1 \left[ y + y^2 \right]_{y=x^2}^{y=x} dx = \int_0^1 \left[ x + x^2 - x^2 - (x^2)^2 \right] dx$

$\qquad = \int_0^1 (x - x^4) dx = \left[ \frac{1}{2} x^2 - \frac{1}{5} x^5 \right]_0^1 = \frac{1}{2} - \frac{1}{5} - 0 + 0 = \frac{3}{10}$

**5.** $\int_0^{\pi/2} \int_0^{\cos\theta} e^{\sin\theta} dr \, d\theta = \int_0^{\pi/2} \left[ r e^{\sin\theta} \right]_{r=0}^{r=\cos\theta} d\theta = \int_0^{\pi/2} (\cos\theta) e^{\sin\theta} d\theta = e^{\sin\theta} \Big]_0^{\pi/2} = e^{\sin(\pi/2)} - e^0 = e - 1$

**7.** $\iint_D y^2 \, dA = \int_{-1}^1 \int_{-y-2}^{y} y^2 \, dx \, dy = \int_{-1}^1 \left[ xy^2 \right]_{x=-y-2}^{x=y} dy = \int_{-1}^1 y^2 \left[ y - (-y - 2) \right] dy$

$\qquad = \int_{-1}^1 (2y^3 + 2y^2) dy = \left[ \frac{1}{2} y^4 + \frac{2}{3} y^3 \right]_{-1}^1 = \frac{1}{2} + \frac{2}{3} - \frac{1}{2} + \frac{2}{3} = \frac{4}{3}$

**9.** $\iint_D x \, dA = \int_0^\pi \int_0^{\sin x} x \, dy \, dx = \int_0^\pi \left[ xy \right]_{y=0}^{y=\sin x} dx = \int_0^\pi x \sin x \, dx \qquad \left[ \begin{array}{c} \text{integrate by parts} \\ \text{with } u = x, \, dv = \sin x \, dx \end{array} \right]$

$\qquad = \left[ -x \cos x + \sin x \right]_0^\pi = -\pi \cos\pi + \sin\pi + 0 - \sin 0 = \pi$

**11.** (a) At the right we sketch an example of a region $D$ that can be described as lying

between the graphs of two continuous functions of $x$ (a type I region) but not as

lying between graphs of two continuous functions of $y$ (a type II region). The

regions shown in Figures 6 and 8 in the text are additional examples.

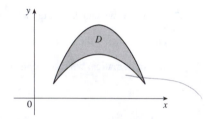

(b) Now we sketch an example of a region $D$ that can be described as lying between

the graphs of two continuous functions of $y$ but not as lying between graphs of two

continuous functions of $x$. The first region shown in Figure 7 is another example.

**13.**

As a type I region, $D$ lies between the lower boundary $y = 0$ and the upper

boundary $y = x$ for $0 \le x \le 1$, so $D = \{(x, y) \mid 0 \le x \le 1, 0 \le y \le x\}$. If we

describe $D$ as a type II region, $D$ lies between the left boundary $x = y$ and the

right boundary $x = 1$ for $0 \le y \le 1$, so $D = \{(x, y) \mid 0 \le y \le 1, y \le x \le 1\}$.

Thus $\iint_D x \, dA = \int_0^1 \int_0^x x \, dy \, dx = \int_0^1 \left[ xy \right]_{y=0}^{y=x} dx = \int_0^1 x^2 \, dx = \frac{1}{3} x^3 \Big]_0^1 = \frac{1}{3}(1 - 0) = \frac{1}{3}$ or

$\iint_D x \, dA = \int_0^1 \int_y^1 x \, dx \, dy = \int_0^1 \left[ \frac{1}{2} x^2 \right]_{x=y}^{x=1} dy = \frac{1}{2} \int_0^1 (1 - y^2) \, dy = \frac{1}{2} \left[ y - \frac{1}{3} y^3 \right]_0^1 = \frac{1}{2} \left[ (1 - \frac{1}{3}) - 0 \right] = \frac{1}{3}.$

**15.**

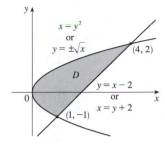

The curves $y = x - 2$ or $x = y + 2$ and $x = y^2$ intersect when $y + 2 = y^2$ $\Leftrightarrow$

$y^2 - y - 2 = 0$ $\Leftrightarrow$ $(y - 2)(y + 1) = 0$ $\Leftrightarrow$ $y = -1, y = 2$, so the points of

intersection are $(1, -1)$ and $(4, 2)$. If we describe $D$ as a type I region, the upper

boundary curve is $y = \sqrt{x}$ but the lower boundary curve consists of two parts,

$y = -\sqrt{x}$ for $0 \le x \le 1$ and $y = x - 2$ for $1 \le x \le 4$.

Thus $D = \{(x, y) \mid 0 \le x \le 1, \ -\sqrt{x} \le y \le \sqrt{x}\} \cup \{(x, y) \mid 1 \le x \le 4, x - 2 \le y \le \sqrt{x}\}$ and

$\iint_D y \, dA = \int_0^1 \int_{-\sqrt{x}}^{\sqrt{x}} y \, dy \, dx + \int_1^4 \int_{x-2}^{\sqrt{x}} y \, dy \, dx$. If we describe $D$ as a type II region, $D$ is enclosed by the left boundary

$x = y^2$ and the right boundary $x = y + 2$ for $-1 \le y \le 2$, so $D = \{(x, y) \mid -1 \le y \le 2, y^2 \le x \le y + 2\}$ and

$\iint_D y \, dA = \int_{-1}^2 \int_{y^2}^{y+2} y \, dx \, dy$. In either case, the resulting iterated integrals are not difficult to evaluate but the region $D$ is

more simply described as a type II region, giving one iterated integral rather than a sum of two, so we evaluate the latter

integral:

$$\iint_D y \, dA = \int_{-1}^2 \int_{y^2}^{y+2} y \, dx \, dy = \int_{-1}^2 \left[ xy \right]_{x=y^2}^{x=y+2} dy = \int_{-1}^2 (y + 2 - y^2) y \, dy = \int_{-1}^2 (y^2 + 2y - y^3) \, dy$$

$$= \left[ \tfrac{1}{3} y^3 + y^2 - \tfrac{1}{4} y^4 \right]_{-1}^2 = \left( \tfrac{8}{3} + 4 - 4 \right) - \left( -\tfrac{1}{3} + 1 - \tfrac{1}{4} \right) = \tfrac{9}{4}$$

**17.** $\int_0^1 \int_0^{x^2} x \cos y \, dy \, dx = \int_0^1 \left[ x \sin y \right]_{y=0}^{y=x^2} dx = \int_0^1 x \sin x^2 \, dx = -\tfrac{1}{2} \cos x^2 \Big]_0^1 = \tfrac{1}{2} (1 - \cos 1)$

**19.**

$$\int_1^2 \int_{2-y}^{2y-1} y^3 \, dx \, dy = \int_1^2 \left[ xy^3 \right]_{x=2-y}^{x=2y-1} dy = \int_1^2 \left[ (2y - 1) - (2 - y) \right] y^3 \, dy$$

$$= \int_1^2 (3y^4 - 3y^3) \, dy = \left[ \tfrac{3}{5} y^5 - \tfrac{3}{4} y^4 \right]_1^2$$

$$= \tfrac{96}{5} - 12 - \tfrac{3}{5} + \tfrac{3}{4} = \tfrac{147}{20}$$

**21.**

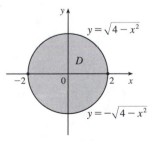

$$\int_{-2}^2 \int_{-\sqrt{4-x^2}}^{\sqrt{4-x^2}} (2x - y) \, dy \, dx$$

$$= \int_{-2}^2 \left[ 2xy - \tfrac{1}{2} y^2 \right]_{y=-\sqrt{4-x^2}}^{y=\sqrt{4-x^2}} dx$$

$$= \int_{-2}^2 \left[ 2x\sqrt{4-x^2} - \tfrac{1}{2}(4 - x^2) + 2x\sqrt{4-x^2} + \tfrac{1}{2}(4 - x^2) \right] dx$$

$$= \int_{-2}^2 4x\sqrt{4-x^2} \, dx = -\tfrac{4}{3}(4 - x^2)^{3/2} \Big]_{-2}^2 = 0$$

[Or, note that $4x\sqrt{4-x^2}$ is an odd function, so $\int_{-2}^2 4x\sqrt{4-x^2} \, dx = 0$.]

**23.**

$V = \int_0^1 \int_{x^4}^x (x + 2y)\, dy\, dx$

$= \int_0^1 \left[xy + y^2\right]_{y=x^4}^{y=x}\, dx = \int_0^1 (2x^2 - x^5 - x^8)\, dx$

$= \left[\frac{2}{3}x^3 - \frac{1}{6}x^6 - \frac{1}{9}x^9\right]_0^1 = \frac{2}{3} - \frac{1}{6} - \frac{1}{9} = \frac{7}{18}$

**25.**

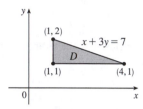

$V = \int_1^2 \int_1^{7-3y} xy\, dx\, dy = \int_1^2 \left[\frac{1}{2}x^2 y\right]_{x=1}^{x=7-3y}\, dy$

$= \frac{1}{2}\int_1^2 (48y - 42y^2 + 9y^3)\, dy$

$= \frac{1}{2}\left[24y^2 - 14y^3 + \frac{9}{4}y^4\right]_1^2 = \frac{31}{8}$

**27.**

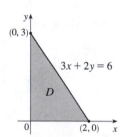

$V = \int_0^2 \int_0^{3-\frac{3}{2}x} (6 - 3x - 2y)\, dy\, dx$

$= \int_0^2 \left[6y - 3xy - y^2\right]_{y=0}^{y=3-\frac{3}{2}x}\, dx$

$= \int_0^2 \left[6(3 - \frac{3}{2}x) - 3x(3 - \frac{3}{2}x) - (3 - \frac{3}{2}x)^2\right]\, dx$

$= \int_0^2 \left(\frac{9}{4}x^2 - 9x + 9\right) dx = \left[\frac{3}{4}x^3 - \frac{9}{2}x^2 + 9x\right]_0^2 = 6 - 0 = 6$

**29.**

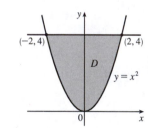

$V = \int_{-2}^2 \int_{x^2}^4 x^2\, dy\, dx$

$= \int_{-2}^2 x^2 \left[y\right]_{y=x^2}^{y=4}\, dx = \int_{-2}^2 (4x^2 - x^4)\, dx$

$= \left[\frac{4}{3}x^3 - \frac{1}{5}x^5\right]_{-2}^2 = \frac{32}{3} - \frac{32}{5} + \frac{32}{3} - \frac{32}{5} = \frac{128}{15}$

**31.**

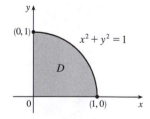

$V = \int_0^1 \int_0^{\sqrt{1-x^2}} y\, dy\, dx = \int_0^1 \left[\frac{y^2}{2}\right]_{y=0}^{y=\sqrt{1-x^2}}\, dx$

$= \int_0^1 \frac{1-x^2}{2}\, dx = \frac{1}{2}\left[x - \frac{1}{3}x^3\right]_0^1 = \frac{1}{3}$

**33.**

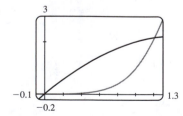

From the graph, it appears that the two curves intersect at $x = 0$ and at $x \approx 1.213$. Thus the desired integral is

$\iint_D x\, dA \approx \int_0^{1.213} \int_{x^4}^{3x - x^2} x\, dy\, dx = \int_0^{1.213} \left[xy\right]_{y=x^4}^{y=3x-x^2}\, dx$

$= \int_0^{1.213} (3x^2 - x^3 - x^5)\, dx = \left[x^3 - \frac{1}{4}x^4 - \frac{1}{6}x^6\right]_0^{1.213}$

$\approx 0.713$

**35.** The two bounding curves $y = 1 - x^2$ and $y = x^2 - 1$ intersect at $(\pm 1, 0)$ with $1 - x^2 \geq x^2 - 1$ on $[-1, 1]$. Within this

region, the plane $z = 2x + 2y + 10$ is above the plane $z = 2 - x - y$, so

$$V = \int_{-1}^{1} \int_{x^2-1}^{1-x^2} (2x + 2y + 10) \, dy \, dx - \int_{-1}^{1} \int_{x^2-1}^{1-x^2} (2 - x - y) \, dy \, dx$$

$$= \int_{-1}^{1} \int_{x^2-1}^{1-x^2} (2x + 2y + 10 - (2 - x - y)) \, dy \, dx$$

$$= \int_{-1}^{1} \int_{x^2-1}^{1-x^2} (3x + 3y + 8) \, dy \, dx = \int_{-1}^{1} \left[ 3xy + \tfrac{3}{2}y^2 + 8y \right]_{y=x^2-1}^{y=1-x^2} dx$$

$$= \int_{-1}^{1} \left[ 3x(1 - x^2) + \tfrac{3}{2}(1 - x^2)^2 + 8(1 - x^2) - 3x(x^2 - 1) - \tfrac{3}{2}(x^2 - 1)^2 - 8(x^2 - 1) \right] dx$$

$$= \int_{-1}^{1} (-6x^3 - 16x^2 + 6x + 16) \, dx = \left[ -\tfrac{3}{2}x^4 - \tfrac{16}{3}x^3 + 3x^2 + 16x \right]_{-1}^{1}$$

$$= -\tfrac{3}{2} - \tfrac{16}{3} + 3 + 16 + \tfrac{3}{2} - \tfrac{16}{3} - 3 + 16 = \tfrac{64}{3}$$

**37.** The solid lies below the plane $z = 1 - x - y$

or $x + y + z = 1$ and above the region

$D = \{(x, y) \mid 0 \leq x \leq 1, 0 \leq y \leq 1 - x\}$

in the $xy$-plane. The solid is a tetrahedron.

**39.** The two bounding curves $y = x^3 - x$ and $y = x^2 + x$ intersect at the origin and at $x = 2$, with $x^2 + x > x^3 - x$ on $(0, 2)$.

Using a CAS, we find that the volume is

$$V = \int_{0}^{2} \int_{x^3-x}^{x^2+x} z \, dy \, dx = \int_{0}^{2} \int_{x^3-x}^{x^2+x} (x^3 y^4 + xy^2) \, dy \, dx = \frac{13{,}984{,}735{,}616}{14{,}549{,}535}$$

**41.**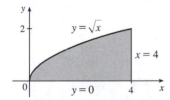

Because the region of integration is

$D = \{(x, y) \mid 0 \leq y \leq \sqrt{x}, 0 \leq x \leq 4\} = \{(x, y) \mid y^2 \leq x \leq 4, 0 \leq y \leq 2\}$

we have $\int_{0}^{4} \int_{0}^{\sqrt{x}} f(x, y) \, dy \, dx = \iint_{D} f(x, y) \, dA = \int_{0}^{2} \int_{y^2}^{4} f(x, y) \, dx \, dy$.

**43.**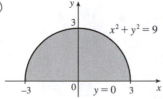

Because the region of integration is

$D = \left\{ (x, y) \mid -\sqrt{9 - y^2} \leq x \leq \sqrt{9 - y^2}, 0 \leq y \leq 3 \right\}$

$= \left\{ (x, y) \mid 0 \leq y \leq \sqrt{9 - x^2}, -3 \leq x \leq 3 \right\}$

we have

$$\int_{0}^{3} \int_{-\sqrt{9-y^2}}^{\sqrt{9-y^2}} f(x, y) \, dx \, dy = \iint_{D} f(x, y) \, dA$$

$$= \int_{-3}^{3} \int_{0}^{\sqrt{9-x^2}} f(x, y) \, dy \, dx$$

**45.**

Because the region of integration is

$$D = \{(x,y) \mid 0 \le y \le \ln x, 1 \le x \le 2\} = \{(x,y) \mid e^y \le x \le 2, 0 \le y \le \ln 2\}$$

we have

$$\int_1^2 \int_0^{\ln x} f(x,y)\, dy\, dx = \iint_D f(x,y)\, dA = \int_0^{\ln 2} \int_{e^y}^2 f(x,y)\, dx\, dy$$

**47.**
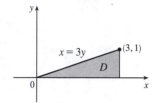

$$\int_0^1 \int_{3y}^3 e^{x^2}\, dx\, dy = \int_0^3 \int_0^{x/3} e^{x^2}\, dy\, dx = \int_0^3 \Big[e^{x^2} y\Big]_{y=0}^{y=x/3}\, dx$$

$$= \int_0^3 \left(\frac{x}{3}\right) e^{x^2}\, dx = \tfrac{1}{6} e^{x^2}\Big]_0^3 = \frac{e^9 - 1}{6}$$

**49.**
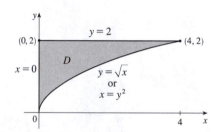

$$\int_0^4 \int_{\sqrt{x}}^2 \frac{1}{y^3 + 1}\, dy\, dx = \int_0^2 \int_0^{y^2} \frac{1}{y^3 + 1}\, dx\, dy$$

$$= \int_0^2 \frac{1}{y^3 + 1}\, \big[x\big]_{x=0}^{x=y^2}\, dy = \int_0^2 \frac{y^2}{y^3 + 1}\, dy$$

$$= \tfrac{1}{3} \ln\big|y^3 + 1\big|\,\Big]_0^2 = \tfrac{1}{3}(\ln 9 - \ln 1) = \tfrac{1}{3} \ln 9$$

**51.**

$$\int_0^1 \int_{\arcsin y}^{\pi/2} \cos x \sqrt{1 + \cos^2 x}\, dx\, dy$$

$$= \int_0^{\pi/2} \int_0^{\sin x} \cos x \sqrt{1 + \cos^2 x}\, dy\, dx$$

$$= \int_0^{\pi/2} \cos x \sqrt{1 + \cos^2 x}\, \big[y\big]_{y=0}^{y=\sin x}\, dx$$

$$= \int_0^{\pi/2} \cos x \sqrt{1 + \cos^2 x}\, \sin x\, dx \qquad \left[\begin{array}{l} \text{Let } u = \cos x,\, du = -\sin x\, dx, \\ dx = du/(-\sin x) \end{array}\right]$$

$$= \int_1^0 -u\sqrt{1 + u^2}\, du = -\tfrac{1}{3}\big(1 + u^2\big)^{3/2}\Big]_1^0$$

$$= \tfrac{1}{3}\big(\sqrt{8} - 1\big) = \tfrac{1}{3}\big(2\sqrt{2} - 1\big)$$

**53.** $D = \{(x,y) \mid 0 \le x \le 1,\, -x+1 \le y \le 1\} \cup \{(x,y) \mid -1 \le x \le 0,\, x+1 \le y \le 1\}$

$$\cup \{(x,y) \mid 0 \le x \le 1,\, -1 \le y \le x-1\} \cup \{(x,y) \mid -1 \le x \le 0,\, -1 \le y \le -x-1\}, \quad \text{all type I.}$$

$$\iint_D x^2\, dA = \int_0^1 \int_{1-x}^1 x^2\, dy\, dx + \int_{-1}^0 \int_{x+1}^1 x^2\, dy\, dx + \int_0^1 \int_{-1}^{x-1} x^2\, dy\, dx + \int_{-1}^0 \int_{-1}^{-x-1} x^2\, dy\, dx$$

$$= 4 \int_0^1 \int_{1-x}^1 x^2\, dy\, dx \qquad \text{[by symmetry of the regions and because } f(x,y) = x^2 \ge 0]$$

$$= 4 \int_0^1 x^3\, dx = 4\big[\tfrac{1}{4} x^4\big]_0^1 = 1$$

**55.** Here $Q = \{(x,y) \mid x^2 + y^2 \le \tfrac{1}{4}, x \ge 0, y \ge 0\}$, and $0 \le (x^2 + y^2)^2 \le \left(\tfrac{1}{4}\right)^2 \;\Rightarrow\; -\tfrac{1}{16} \le -(x^2 + y^2)^2 \le 0$ so

$e^{-1/16} \le e^{-(x^2 + y^2)^2} \le e^0 = 1$ since $e^t$ is an increasing function. We have $A(Q) = \tfrac{1}{4}\pi\left(\tfrac{1}{2}\right)^2 = \tfrac{\pi}{16}$, so by Property 11,

$e^{-1/16} A(Q) \le \iint_Q e^{-(x^2+y^2)^2} dA \le 1 \cdot A(Q) \Rightarrow \frac{\pi}{16} e^{-1/16} \le \iint_Q e^{-(x^2+y^2)^2} dA \le \frac{\pi}{16}$ or we can say

$0.1844 < \iint_Q e^{-(x^2+y^2)^2} dA < 0.1964.$ (We have rounded the lower bound down and the upper bound up to preserve the inequalities.)

**57.** The average value of a function $f$ of two variables defined on a rectangle $R$ was defined in Section 12.1 as $f_{ave} = \frac{1}{A(R)} \iint_R f(x,y)dA$. Extending this definition to general regions $D$, we have $f_{ave} = \frac{1}{A(D)} \iint_D f(x,y)dA$.

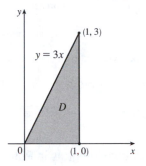

Here $D = \{(x,y) \mid 0 \le x \le 1, 0 \le y \le 3x\}$, so $A(D) = \frac{1}{2}(1)(3) = \frac{3}{2}$ and

$$f_{ave} = \frac{1}{A(D)} \iint_D f(x,y)dA = \frac{1}{3/2} \int_0^1 \int_0^{3x} xy \, dy \, dx$$

$$= \frac{2}{3} \int_0^1 \left[\frac{1}{2}xy^2\right]_{y=0}^{y=3x} dx = \frac{1}{3} \int_0^1 9x^3 \, dx = \frac{3}{4}x^4\Big]_0^1 = \frac{3}{4}$$

**59.** Since $m \le f(x,y) \le M$, $\iint_D m \, dA \le \iint_D f(x,y) \, dA \le \iint_D M \, dA$ by (8) $\Rightarrow$

$m \iint_D 1 \, dA \le \iint_D f(x,y) \, dA \le M \iint_D 1 \, dA$ by (7) $\Rightarrow$ $mA(D) \le \iint_D f(x,y) \, dA \le MA(D)$ by (10).

**61.**

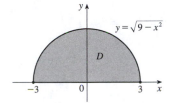

First we can write $\iint_D (x+2) \, dA = \iint_D x \, dA + \iint_D 2 \, dA$. But $f(x,y) = x$ is an odd function with respect to $x$ [that is, $f(-x,y) = -f(x,y)$] and $D$ is symmetric with respect to $x$. Consequently, the volume above $D$ and below the graph of $f$ is the same as the volume below $D$ and above the graph of $f$, so $\iint_D x \, dA = 0$. Also, $\iint_D 2 \, dA = 2 \cdot A(D) = 2 \cdot \frac{1}{2}\pi(3)^2 = 9\pi$ since $D$ is a half disk of radius 3. Thus $\iint_D (x+2) \, dA = 0 + 9\pi = 9\pi$.

**63.** We can write $\iint_D (2x + 3y) \, dA = \iint_D 2x \, dA + \iint_D 3y \, dA$. $\iint_D 2x \, dA$ represents the volume of the solid lying under the plane $z = 2x$ and above the rectangle $D$. This solid region is a triangular cylinder with length $b$ and whose cross-section is a triangle with width $a$ and height $2a$. (See the first figure.)

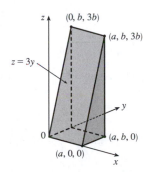

Thus its volume is $\frac{1}{2} \cdot a \cdot 2a \cdot b = a^2 b$. Similarly, $\iint_D 3y \, dA$ represents the volume of a triangular cylinder with length $a$, triangular cross-section with width $b$ and height $3b$, and volume $\frac{1}{2} \cdot b \cdot 3b \cdot a = \frac{3}{2}ab^2$. (See the second figure.) Thus

$$\iint_D (2x + 3y) \, dA = a^2 b + \frac{3}{2}ab^2$$

**65.** $\iint_D \left(ax^3 + by^3 + \sqrt{a^2 - x^2}\right) dA = \iint_D ax^3 \, dA + \iint_D by^3 \, dA + \iint_D \sqrt{a^2 - x^2} \, dA.$ Now $ax^3$ is odd with respect

to $x$ and $by^3$ is odd with respect to $y$, and the region of integration is symmetric with respect to both $x$ and $y$,

so $\iint_D ax^3 \, dA = \iint_D by^3 \, dA = 0.$

$\iint_D \sqrt{a^2 - x^2} \, dA$ represents the volume of the solid region under the

graph of $z = \sqrt{a^2 - x^2}$ and above the rectangle $D$, namely a half circular

cylinder with radius $a$ and length $2b$ (see the figure) whose volume is

$\frac{1}{2} \cdot \pi r^2 h = \frac{1}{2}\pi a^2 (2b) = \pi a^2 b.$ Thus

$\iint_D \left(ax^3 + by^3 + \sqrt{a^2 - x^2}\right) dA = 0 + 0 + \pi a^2 b = \pi a^2 b.$

## 12.4 Double Integrals in Polar Coordinates

**1.** The region $R$ is more easily described by polar coordinates: $R = \left\{(r, \theta) \mid 0 \le r \le 4, 0 \le \theta \le \frac{3\pi}{2}\right\}.$

Thus $\iint_R f(x, y) \, dA = \int_0^{3\pi/2} \int_0^4 f(r\cos\theta, r\sin\theta) \, r \, dr \, d\theta.$

**3.** The region $R$ is more easily described by rectangular coordinates: $R = \left\{(x, y) \mid -1 \le x \le 1, 0 \le y \le \frac{1}{2}x + \frac{1}{2}\right\}.$

Thus $\iint_R f(x, y) \, dA = \int_{-1}^1 \int_0^{(x+1)/2} f(x, y) \, dy \, dx.$

**5.** The integral $\int_\pi^{2\pi} \int_4^7 r \, dr \, d\theta$ represents the area of the region

$R = \left\{(r, \theta) \mid 4 \le r \le 7, \pi \le \theta \le 2\pi\right\},$ the lower half of a ring.

$\int_\pi^{2\pi} \int_4^7 r \, dr \, d\theta = \int_\pi^{2\pi} d\theta \int_4^7 r \, dr$

$\qquad = \left[\theta\right]_\pi^{2\pi} \left[\frac{1}{2}r^2\right]_4^7 = \pi \cdot \frac{1}{2}(49 - 16) = \frac{33\pi}{2}$

**7.** The disk $D$ can be described in polar coordinates as $D = \left\{(r, \theta) \mid 0 \le r \le 3, 0 \le \theta \le 2\pi\right\}.$ Then

$\iint_D xy \, dA = \int_0^{2\pi} \int_0^3 (r\cos\theta)(r\sin\theta) \, r \, dr \, d\theta = \int_0^{2\pi} \sin\theta\cos\theta \, d\theta \int_0^3 r^3 \, dr = \left[\frac{1}{2}\sin^2\theta\right]_0^{2\pi} \left[\frac{1}{4}r^4\right]_0^3 = 0.$

**9.** $\iint_R \cos(x^2 + y^2) \, dA = \int_0^\pi \int_0^3 \cos(r^2) \, r \, dr \, d\theta = \int_0^\pi d\theta \int_0^3 r\cos(r^2) \, dr$

$\qquad\qquad = \left[\theta\right]_0^\pi \left[\frac{1}{2}\sin(r^2)\right]_0^3 = \pi \cdot \frac{1}{2}(\sin 9 - \sin 0) = \frac{\pi}{2}\sin 9$

**11.** $\iint_D e^{-x^2 - y^2} \, dA = \int_{-\pi/2}^{\pi/2} \int_0^2 e^{-r^2} r \, dr \, d\theta = \int_{-\pi/2}^{\pi/2} d\theta \int_0^2 re^{-r^2} \, dr$

$\qquad\qquad = \left[\theta\right]_{-\pi/2}^{\pi/2} \left[-\frac{1}{2}e^{-r^2}\right]_0^2 = \pi\left(-\frac{1}{2}\right)(e^{-4} - e^0) = \frac{\pi}{2}(1 - e^{-4})$

**13.** $R$ is the region shown in the figure, and can be described

by $R = \left\{(r, \theta) \mid 0 \le \theta \le \pi/4, 1 \le r \le 2\right\}.$ Thus

$\iint_R \arctan(y/x) \, dA = \int_0^{\pi/4} \int_1^2 \arctan(\tan\theta) \, r \, dr \, d\theta$ since $y/x = \tan\theta.$

Also, $\arctan(\tan\theta) = \theta$ for $0 \le \theta \le \pi/4,$ so the integral becomes

$\int_0^{\pi/4} \int_1^2 \theta r \, dr \, d\theta = \int_0^{\pi/4} \theta \, d\theta \int_1^2 r \, dr = \left[\frac{1}{2}\theta^2\right]_0^{\pi/4} \left[\frac{1}{2}r^2\right]_1^2 = \frac{\pi^2}{32} \cdot \frac{3}{2} = \frac{3}{64}\pi^2.$

**15.** $V = \iint_{x^2+y^2 \le 4} \sqrt{x^2+y^2}\, dA = \int_0^{2\pi} \int_0^2 \sqrt{r^2}\, r\, dr\, d\theta = \int_0^{2\pi} d\theta \int_0^2 r^2\, dr = \big[\theta\big]_0^{2\pi} \big[\tfrac{1}{3}r^3\big]_0^2 = 2\pi\big(\tfrac{8}{3}\big) = \tfrac{16}{3}\pi$

**17.** The hyperboloid of two sheets $-x^2 - y^2 + z^2 = 1$ intersects the plane $z = 2$ when $-x^2 - y^2 + 4 = 1$ or $x^2 + y^2 = 3$. So the

solid region lies above the surface $z = \sqrt{1+x^2+y^2}$ and below the plane $z = 2$ for $x^2 + y^2 \le 3$, and its volume is

$$V = \iint\limits_{x^2+y^2 \le 3} \left(2 - \sqrt{1+x^2+y^2}\right) dA = \int_0^{2\pi} \int_0^{\sqrt{3}} \left(2 - \sqrt{1+r^2}\right) r\, dr\, d\theta$$

$$= \int_0^{2\pi} d\theta \int_0^{\sqrt{3}} \left(2r - r\sqrt{1+r^2}\right) dr = \big[\theta\big]_0^{2\pi} \big[r^2 - \tfrac{1}{3}(1+r^2)^{3/2}\big]_0^{\sqrt{3}}$$

$$= 2\pi\left(3 - \tfrac{8}{3} - 0 + \tfrac{1}{3}\right) = \tfrac{4}{3}\pi$$

**19.** By symmetry,

$$V = 2 \iint\limits_{x^2+y^2 \le a^2} \sqrt{a^2 - x^2 - y^2}\, dA = 2\int_0^{2\pi} \int_0^a \sqrt{a^2 - r^2}\, r\, dr\, d\theta = 2\int_0^{2\pi} d\theta \int_0^a r\sqrt{a^2 - r^2}\, dr$$

$$= 2\big[\theta\big]_0^{2\pi} \big[-\tfrac{1}{3}(a^2 - r^2)^{3/2}\big]_0^a = 2(2\pi)\big(0 + \tfrac{1}{3}a^3\big) = \tfrac{4\pi}{3}a^3$$

**21.** The cone $z = \sqrt{x^2+y^2}$ intersects the sphere $x^2 + y^2 + z^2 = 1$ when $x^2 + y^2 + \left(\sqrt{x^2+y^2}\right)^2 = 1$ or $x^2 + y^2 = \tfrac{1}{2}$. So

$$V = \iint\limits_{x^2+y^2 \le 1/2} \left(\sqrt{1-x^2-y^2} - \sqrt{x^2+y^2}\right) dA = \int_0^{2\pi} \int_0^{1/\sqrt{2}} \left(\sqrt{1-r^2} - r\right) r\, dr\, d\theta$$

$$= \int_0^{2\pi} d\theta \int_0^{1/\sqrt{2}} \left(r\sqrt{1-r^2} - r^2\right) dr = \big[\theta\big]_0^{2\pi} \big[-\tfrac{1}{3}(1-r^2)^{3/2} - \tfrac{1}{3}r^3\big]_0^{1/\sqrt{2}} = 2\pi\big(-\tfrac{1}{3}\big)\big(\tfrac{1}{\sqrt{2}} - 1\big) = \tfrac{\pi}{3}\big(2 - \sqrt{2}\big)$$

**23.** The given solid is the region inside the cylinder $x^2 + y^2 = 4$ between the surfaces $z = \sqrt{64 - 4x^2 - 4y^2}$

and $z = -\sqrt{64 - 4x^2 - 4y^2}$. So

$$V = \iint\limits_{x^2+y^2 \le 4} \left[\sqrt{64 - 4x^2 - 4y^2} - \left(-\sqrt{64 - 4x^2 - 4y^2}\right)\right] dA = \iint\limits_{x^2+y^2 \le 4} 2\sqrt{64 - 4x^2 - 4y^2}\, dA$$

$$= 4\int_0^{2\pi} \int_0^2 \sqrt{16 - r^2}\, r\, dr\, d\theta = 4\int_0^{2\pi} d\theta \int_0^2 r\sqrt{16 - r^2}\, dr = 4\big[\theta\big]_0^{2\pi} \big[-\tfrac{1}{3}(16 - r^2)^{3/2}\big]_0^2$$

$$= 8\pi\big(-\tfrac{1}{3}\big)(12^{3/2} - 16^{2/3}) = \tfrac{8\pi}{3}\big(64 - 24\sqrt{3}\big)$$

**25.** One loop is given by the region

$D = \{(r, \theta)\,|\,-\pi/6 \le \theta \le \pi/6,\, 0 \le r \le \cos 3\theta\,\}$, so the area is

$$\iint_D dA = \int_{-\pi/6}^{\pi/6} \int_0^{\cos 3\theta} r\, dr\, d\theta = \int_{-\pi/6}^{\pi/6} \left[\tfrac{1}{2}r^2\right]_{r=0}^{r=\cos 3\theta} d\theta$$

$$= \int_{-\pi/6}^{\pi/6} \tfrac{1}{2}\cos^2 3\theta\, d\theta = 2\int_0^{\pi/6} \tfrac{1}{2}\left(\frac{1+\cos 6\theta}{2}\right) d\theta$$

$$= \tfrac{1}{2}\left[\theta + \tfrac{1}{6}\sin 6\theta\right]_0^{\pi/6} = \frac{\pi}{12}$$

**27.**

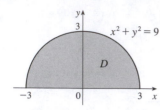

$$\int_{-3}^{3}\int_{0}^{\sqrt{9-x^2}}\sin(x^2+y^2)\,dy\,dx = \int_{0}^{\pi}\int_{0}^{3}\sin\left(r^2\right)r\,dr\,d\theta$$

$$= \int_{0}^{\pi}d\theta\int_{0}^{3}r\sin\left(r^2\right)dr = [\theta]_{0}^{\pi}\left[-\tfrac{1}{2}\cos\left(r^2\right)\right]_{0}^{3}$$

$$= \pi\left(-\tfrac{1}{2}\right)(\cos 9 - 1) = \tfrac{\pi}{2}(1-\cos 9)$$

**29.**

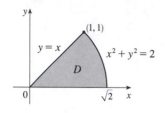

$$\int_{0}^{\pi/4}\int_{0}^{\sqrt{2}}(r\cos\theta + r\sin\theta)\,r\,dr\,d\theta = \int_{0}^{\pi/4}(\cos\theta+\sin\theta)\,d\theta\int_{0}^{\sqrt{2}}r^2\,dr$$

$$= [\sin\theta - \cos\theta]_{0}^{\pi/4}\left[\tfrac{1}{3}r^3\right]_{0}^{\sqrt{2}}$$

$$= \left[\tfrac{\sqrt{2}}{2} - \tfrac{\sqrt{2}}{2} - 0 + 1\right]\cdot\tfrac{1}{3}(2\sqrt{2}-0) = \tfrac{2\sqrt{2}}{3}$$

**31.** The surface of the water in the pool is a circular disk $D$ with radius 20 ft. If we place $D$ on coordinate axes with the origin at the center of $D$ and define $f(x,y)$ to be the depth of the water at $(x,y)$, then the volume of water in the pool is the volume of the solid that lies above $D = \{(x,y) \mid x^2 + y^2 \le 400\}$ and below the graph of $f(x,y)$. We can associate north with the positive $y$-direction, so we are given that the depth is constant in the $x$-direction and the depth increases linearly in the $y$-direction from $f(0,-20) = 2$ to $f(0,20) = 7$. The trace in the $yz$-plane is a line segment from $(0,-20,2)$ to $(0,20,7)$. The slope of this line is $\frac{7-2}{20-(-20)} = \frac{1}{8}$, so an equation of the line is $z - 7 = \frac{1}{8}(y-20) \Rightarrow z = \frac{1}{8}y + \frac{9}{2}$. Since $f(x,y)$ is independent of $x$, $f(x,y) = \frac{1}{8}y + \frac{9}{2}$. Thus the volume is given by $\iint_D f(x,y)\,dA$, which is most conveniently evaluated using polar coordinates. Then $D = \{(r,\theta) \mid 0 \le r \le 20, 0 \le \theta \le 2\pi\}$ and substituting $x = r\cos\theta$, $y = r\sin\theta$ the integral becomes

$$\int_{0}^{2\pi}\int_{0}^{20}\left(\tfrac{1}{8}r\sin\theta + \tfrac{9}{2}\right)r\,dr\,d\theta = \int_{0}^{2\pi}\left[\tfrac{1}{24}r^3\sin\theta + \tfrac{9}{4}r^2\right]_{r=0}^{r=20}d\theta = \int_{0}^{2\pi}\left(\tfrac{1000}{3}\sin\theta + 900\right)d\theta$$

$$= \left[-\tfrac{1000}{3}\cos\theta + 900\theta\right]_{0}^{2\pi} = 1800\pi$$

Thus the pool contains $1800\pi \approx 5655$ ft$^3$ of water.

**33.** As in Exercise 12.3.57, $f_{ave} = \frac{1}{A(D)}\iint_D f(x,y)\,dA$. Here $D = \{(r,\theta) \mid a \le r \le b, 0 \le \theta \le 2\pi\}$, so $A(D) = \pi b^2 - \pi a^2 = \pi(b^2 - a^2)$ and

$$f_{ave} = \frac{1}{A(D)}\iint_D \frac{1}{\sqrt{x^2+y^2}}\,dA = \frac{1}{\pi(b^2-a^2)}\int_{0}^{2\pi}\int_{a}^{b}\frac{1}{\sqrt{r^2}}r\,dr\,d\theta = \frac{1}{\pi(b^2-a^2)}\int_{0}^{2\pi}d\theta\int_{a}^{b}dr$$

$$= \frac{1}{\pi(b^2-a^2)}[\theta]_{0}^{2\pi}[r]_{a}^{b} = \frac{1}{\pi(b^2-a^2)}(2\pi)(b-a) = \frac{2(b-a)}{(b+a)(b-a)} = \frac{2}{a+b}$$

**35.**

$$\int_{1/\sqrt{2}}^{1}\int_{\sqrt{1-x^2}}^{x}xy\,dy\,dx + \int_{1}^{\sqrt{2}}\int_{0}^{x}xy\,dy\,dx + \int_{\sqrt{2}}^{2}\int_{0}^{\sqrt{4-x^2}}xy\,dy\,dx$$

$$= \int_{0}^{\pi/4}\int_{1}^{2}r^3\cos\theta\sin\theta\,dr\,d\theta = \int_{0}^{\pi/4}\left[\frac{r^4}{4}\cos\theta\sin\theta\right]_{r=1}^{r=2}d\theta$$

$$= \frac{15}{4}\int_{0}^{\pi/4}\sin\theta\cos\theta\,d\theta = \frac{15}{4}\left[\frac{\sin^2\theta}{2}\right]_{0}^{\pi/4} = \frac{15}{16}$$

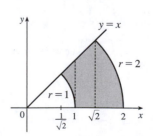

**37.** (a) We integrate by parts with $u = x$ and $dv = xe^{-x^2}\,dx$. Then $du = dx$ and $v = -\frac{1}{2}e^{-x^2}$, so

$$\int_0^\infty x^2 e^{-x^2}\,dx = \lim_{t\to\infty}\int_0^t x^2 e^{-x^2}\,dx = \lim_{t\to\infty}\left(\left[-\tfrac{1}{2}xe^{-x^2}\right]_0^t + \int_0^t \tfrac{1}{2}e^{-x^2}\,dx\right)$$

$$= \lim_{t\to\infty}\left(-\tfrac{1}{2}te^{-t^2}\right) + \tfrac{1}{2}\int_0^\infty e^{-x^2}\,dx = 0 + \tfrac{1}{2}\int_0^\infty e^{-x^2}\,dx \qquad \text{[by l'Hospital's Rule]}$$

$$= \tfrac{1}{4}\int_{-\infty}^\infty e^{-x^2}\,dx \qquad \text{[since } e^{-x^2} \text{ is an even function]}$$

$$= \tfrac{1}{4}\sqrt{\pi} \qquad \text{[by Exercise 36(c)]}$$

(b) Let $u = \sqrt{x}$. Then $u^2 = x \ \Rightarrow\ dx = 2u\,du \ \Rightarrow$

$$\int_0^\infty \sqrt{x}\,e^{-x}\,dx = \lim_{t\to\infty}\int_0^t \sqrt{x}\,e^{-x}\,dx = \lim_{t\to\infty}\int_0^{\sqrt{t}} ue^{-u^2}2u\,du = 2\int_0^\infty u^2 e^{-u^2}\,du = 2\left(\tfrac{1}{4}\sqrt{\pi}\right) \quad \text{[by part(a)]} = \tfrac{1}{2}\sqrt{\pi}.$$

## 12.5  Applications of Double Integrals

**1.** $Q = \iint_D \sigma(x,y)\,dA = \int_1^3 \int_0^2 (2xy + y^2)\,dy\,dx = \int_1^3 \left[xy^2 + \tfrac{1}{3}y^3\right]_{y=0}^{y=2}dx$

$= \int_1^3 \left(4x + \tfrac{8}{3}\right)dx = \left[2x^2 + \tfrac{8}{3}x\right]_1^3 = 16 + \tfrac{16}{3} = \tfrac{64}{3}$ C

**3.** $m = \iint_D \rho(x,y)\,dA = \int_0^2 \int_{-1}^1 xy^2\,dy\,dx = \int_0^2 x\,dx \int_{-1}^1 y^2\,dy = \left[\tfrac{1}{2}x^2\right]_0^2 \left[\tfrac{1}{3}y^3\right]_{-1}^1 = 2\cdot\tfrac{2}{3} = \tfrac{4}{3}$,

$\bar{x} = \tfrac{1}{m}\iint_D x\rho(x,y)\,dA = \tfrac{3}{4}\int_0^2 \int_{-1}^1 x^2 y^2\,dy\,dx = \tfrac{3}{4}\int_0^2 x^2\,dx \int_{-1}^1 y^2\,dy = \tfrac{3}{4}\left[\tfrac{1}{3}x^3\right]_0^2 \left[\tfrac{1}{3}y^3\right]_{-1}^1 = \tfrac{3}{4}\cdot\tfrac{8}{3}\cdot\tfrac{2}{3} = \tfrac{4}{3}$,

$\bar{y} = \tfrac{1}{m}\iint_D y\rho(x,y)\,dA = \tfrac{3}{4}\int_0^2 \int_{-1}^1 xy^3\,dy\,dx = \tfrac{3}{4}\int_0^2 x\,dx \int_{-1}^1 y^3\,dy = \tfrac{3}{4}\left[\tfrac{1}{2}x^2\right]_0^2 \left[\tfrac{1}{4}y^4\right]_{-1}^1 = \tfrac{3}{4}\cdot 2\cdot 0 = 0.$

Hence, $(\bar{x}, \bar{y}) = \left(\tfrac{4}{3}, 0\right)$.

**5.** $m = \int_0^2 \int_{x/2}^{3-x}(x+y)\,dy\,dx = \int_0^2 \left[xy + \tfrac{1}{2}y^2\right]_{y=x/2}^{y=3-x}dx = \int_0^2 \left[x\left(3 - \tfrac{3}{2}x\right) + \tfrac{1}{2}(3-x)^2 - \tfrac{1}{8}x^2\right]dx$

$= \int_0^2 \left(-\tfrac{9}{8}x^2 + \tfrac{9}{2}\right)dx = \left[-\tfrac{9}{8}\left(\tfrac{1}{3}x^3\right) + \tfrac{9}{2}x\right]_0^2 = 6,$

$M_y = \int_0^2 \int_{x/2}^{3-x}(x^2 + xy)\,dy\,dx = \int_0^2 \left[x^2 y + \tfrac{1}{2}xy^2\right]_{y=x/2}^{y=3-x}dx = \int_0^2 \left(\tfrac{9}{2}x - \tfrac{9}{8}x^3\right)dx = \tfrac{9}{2},$

$M_x = \int_0^2 \int_{x/2}^{3-y}(xy + y^2)\,dy\,dx = \int_0^2 \left[\tfrac{1}{2}xy^2 + \tfrac{1}{3}y^3\right]_{y=x/2}^{y=3-x}dx = \int_0^2 \left(9 - \tfrac{9}{2}x\right)dx = 9.$

Hence $m = 6$, $(\bar{x}, \bar{y}) = \left(\dfrac{M_y}{m}, \dfrac{M_x}{m}\right) = \left(\dfrac{3}{4}, \dfrac{3}{2}\right)$.

**7.** $m = \int_0^1 \int_0^{e^x} y\,dy\,dx = \int_0^1 \left[\tfrac{1}{2}y^2\right]_{y=0}^{y=e^x}dx = \tfrac{1}{2}\int_0^1 e^{2x}\,dx = \tfrac{1}{4}\left[e^{2x}\right]_0^1 = \tfrac{1}{4}(e^2 - 1),$

$M_y = \int_0^1 \int_0^{e^x} xy\,dy\,dx = \tfrac{1}{2}\int_0^1 xe^{2x}\,dx = \tfrac{1}{2}\left[\tfrac{1}{2}xe^{2x} - \tfrac{1}{4}e^{2x}\right]_0^1 = \tfrac{1}{8}(e^2 + 1),$

$M_x = \int_0^1 \int_0^{e^x} y^2\,dy\,dx = \int_0^1 \left[\tfrac{1}{3}y^3\right]_{y=0}^{y=e^x}dx = \tfrac{1}{3}\int_0^1 e^{3x}\,dx = \tfrac{1}{3}\left[\tfrac{1}{3}e^{3x}\right]_0^1 = \tfrac{1}{9}(e^3 - 1).$

Hence $m = \tfrac{1}{4}(e^2 - 1)$, $(\bar{x}, \bar{y}) = \left(\dfrac{\tfrac{1}{8}(e^2 + 1)}{\tfrac{1}{4}(e^2 - 1)}, \dfrac{\tfrac{1}{9}(e^3 - 1)}{\tfrac{1}{4}(e^2 - 1)}\right) = \left(\dfrac{e^2 + 1}{2(e^2 - 1)}, \dfrac{4(e^3 - 1)}{9(e^2 - 1)}\right).$

**9.** Note that $\sin(\pi x/L) \geq 0$ for $0 \leq x \leq L$.

$$m = \int_0^L \int_0^{\sin(\pi x/L)} y \, dy \, dx = \int_0^L \tfrac{1}{2} \sin^2(\pi x/L) \, dx = \tfrac{1}{2} \left[\tfrac{1}{2}x - \tfrac{L}{4\pi} \sin(2\pi x/L)\right]_0^L = \tfrac{1}{4}L,$$

$$M_y = \int_0^L \int_0^{\sin(\pi x/L)} x \cdot y \, dy \, dx = \tfrac{1}{2} \int_0^L x \sin^2(\pi x/L) \, dx \qquad \begin{bmatrix} \text{integrate by parts with} \\ u = x, \, dv = \sin^2(\pi x/L) \, dx \end{bmatrix}$$

$$= \tfrac{1}{2} \cdot x \left(\tfrac{1}{2}x - \tfrac{L}{4\pi} \sin(2\pi x/L)\right)\Big]_0^L - \tfrac{1}{2} \int_0^L \left[\tfrac{1}{2}x - \tfrac{L}{4\pi} \sin(2\pi x/L)\right] dx$$

$$= \tfrac{1}{4}L^2 - \tfrac{1}{2}\left[\tfrac{1}{4}x^2 + \tfrac{L^2}{4\pi^2} \cos(2\pi x/L)\right]_0^L = \tfrac{1}{4}L^2 - \tfrac{1}{2}\left(\tfrac{1}{4}L^2 + \tfrac{L^2}{4\pi^2} - \tfrac{L^2}{4\pi^2}\right) = \tfrac{1}{8}L^2,$$

$$M_x = \int_0^L \int_0^{\sin(\pi x/L)} y \cdot y \, dy \, dx = \int_0^L \tfrac{1}{3} \sin^3(\pi x/L) \, dx = \tfrac{1}{3} \int_0^L \left[1 - \cos^2(\pi x/L)\right] \sin(\pi x/L) \, dx$$

$$[\text{substitute } u = \cos(\pi x/L)] \quad \Rightarrow \quad du = -\tfrac{\pi}{L} \sin(\pi x/L)]$$

$$= \tfrac{1}{3}\left(-\tfrac{L}{\pi}\right)\left[\cos(\pi x/L) - \tfrac{1}{3}\cos^3(\pi x/L)\right]_0^L = -\tfrac{L}{3\pi}\left(-1 + \tfrac{1}{3} - 1 + \tfrac{1}{3}\right) = \tfrac{4}{9\pi}L.$$

Hence $m = \dfrac{L}{4}$, $(\bar{x}, \bar{y}) = \left(\dfrac{L^2/8}{L/4}, \dfrac{4L/(9\pi)}{L/4}\right) = \left(\dfrac{L}{2}, \dfrac{16}{9\pi}\right)$.

**11.** $\rho(x,y) = ky = kr \sin\theta$, $m = \int_0^{\pi/2} \int_0^1 kr^2 \sin\theta \, dr \, d\theta = \tfrac{1}{3}k \int_0^{\pi/2} \sin\theta \, d\theta = \tfrac{1}{3}k\left[-\cos\theta\right]_0^{\pi/2} = \tfrac{1}{3}k$,

$$M_y = \int_0^{\pi/2} \int_0^1 kr^3 \sin\theta \cos\theta \, dr \, d\theta = \tfrac{1}{4}k \int_0^{\pi/2} \sin\theta \cos\theta \, d\theta = \tfrac{1}{8}k\left[-\cos 2\theta\right]_0^{\pi/2} = \tfrac{1}{8}k,$$

$$M_x = \int_0^{\pi/2} \int_0^1 kr^3 \sin^2\theta \, dr \, d\theta = \tfrac{1}{4}k \int_0^{\pi/2} \sin^2\theta \, d\theta = \tfrac{1}{8}k\left[\theta + \sin 2\theta\right]_0^{\pi/2} = \tfrac{\pi}{16}k.$$

Hence $(\bar{x}, \bar{y}) = \left(\tfrac{3}{8}, \tfrac{3\pi}{16}\right)$.

**13.**

$$\rho(x,y) = k\sqrt{x^2 + y^2} = kr,$$

$$m = \iint_D \rho(x,y) \, dA = \int_0^\pi \int_1^2 kr \cdot r \, dr \, d\theta$$

$$= k \int_0^\pi d\theta \int_1^2 r^2 \, dr = k(\pi)\left[\tfrac{1}{3}r^3\right]_1^2 = \tfrac{7}{3}\pi k,$$

$$M_y = \iint_D x\rho(x,y) \, dA = \int_0^\pi \int_1^2 (r\cos\theta)(kr) \, r \, dr \, d\theta = k \int_0^\pi \cos\theta \, d\theta \int_1^2 r^3 \, dr$$

$$= k\left[\sin\theta\right]_0^\pi \left[\tfrac{1}{4}r^4\right]_1^2 = k(0)\left(\tfrac{15}{4}\right) = 0 \qquad \begin{array}{l} \text{[this is to be expected as the region and density} \\ \text{function are symmetric about the } y\text{-axis]} \end{array}$$

$$M_x = \iint_D y\rho(x,y) \, dA = \int_0^\pi \int_1^2 (r\sin\theta)(kr) \, r \, dr \, d\theta = k \int_0^\pi \sin\theta \, d\theta \int_1^2 r^3 \, dr$$

$$= k\left[-\cos\theta\right]_0^\pi \left[\tfrac{1}{4}r^4\right]_1^2 = k(1+1)\left(\tfrac{15}{4}\right) = \tfrac{15}{2}k.$$

Hence $(\bar{x}, \bar{y}) = \left(0, \dfrac{15k/2}{7\pi k/3}\right) = \left(0, \dfrac{45}{14\pi}\right)$.

**15.** Placing the vertex opposite the hypotenuse at $(0,0)$, $\rho(x,y) = k(x^2 + y^2)$. Then

$$m = \int_0^a \int_0^{a-x} k(x^2 + y^2) \, dy \, dx = k \int_0^a \left[ax^2 - x^3 + \tfrac{1}{3}(a-x)^3\right] dx = k\left[\tfrac{1}{3}ax^3 - \tfrac{1}{4}x^4 - \tfrac{1}{12}(a-x)^4\right]_0^a = \tfrac{1}{6}ka^4.$$

By symmetry,

$$M_y = M_x = \int_0^a \int_0^{a-x} ky(x^2 + y^2) \, dy \, dx = k \int_0^a \left[\tfrac{1}{2}(a-x)^2 x^2 + \tfrac{1}{4}(a-x)^4\right] dx$$

$$= k\left[\tfrac{1}{6}a^2 x^3 - \tfrac{1}{4}ax^4 + \tfrac{1}{10}x^5 - \tfrac{1}{20}(a-x)^5\right]_0^a = \tfrac{1}{15}ka^5$$

Hence $(\bar{x}, \bar{y}) = \left(\tfrac{2}{5}a, \tfrac{2}{5}a\right)$.

**17.** $I_x = \iint_D y^2 \rho(x,y) \, dA = \int_0^1 \int_0^{e^x} y^2 \cdot y \, dy \, dx = \int_0^1 \left[\frac{1}{4} y^4\right]_{y=0}^{y=e^x} dx = \frac{1}{4} \int_0^1 e^{4x} \, dx = \frac{1}{4} \left[\frac{1}{4} e^{4x}\right]_0^1 = \frac{1}{16}(e^4 - 1)$,

$I_y = \iint_D x^2 \rho(x,y) \, dA = \int_0^1 \int_0^{e^x} x^2 y \, dy \, dx = \int_0^1 x^2 \left[\frac{1}{2} y^2\right]_{y=0}^{y=e^x} dx = \frac{1}{2} \int_0^1 x^2 e^{2x} \, dx$

$\qquad = \frac{1}{2}\left[\left(\frac{1}{2} x^2 - \frac{1}{2} x + \frac{1}{4}\right) e^{2x}\right]_0^1$  [integrate by parts twice] $= \frac{1}{8}(e^2 - 1)$,

and $I_0 = I_x + I_y = \frac{1}{16}(e^4 - 1) + \frac{1}{8}(e^2 - 1) = \frac{1}{16}(e^4 + 2e^2 - 3)$.

**19.** As in Exercise 15, we place the vertex opposite the hypotenuse at $(0,0)$ and the equal sides along the positive axes.

$I_x = \int_0^a \int_0^{a-x} y^2 k(x^2 + y^2) \, dy \, dx = k \int_0^a \int_0^{a-x} (x^2 y^2 + y^4) \, dy \, dx = k \int_0^a \left[\frac{1}{3} x^2 y^3 + \frac{1}{5} y^5\right]_{y=0}^{y=a-x} dx$

$\qquad = k \int_0^a \left[\frac{1}{3} x^2 (a-x)^3 + \frac{1}{5}(a-x)^5\right] dx = k \left[\frac{1}{3}\left(\frac{1}{3} a^3 x^3 - \frac{3}{4} a^2 x^4 + \frac{3}{5} a x^5 - \frac{1}{6} x^6\right) - \frac{1}{30}(a-x)^6\right]_0^a = \frac{7}{180} k a^6$,

$I_y = \int_0^a \int_0^{a-x} x^2 k(x^2 + y^2) \, dy \, dx = k \int_0^a \int_0^{a-x} (x^4 + x^2 y^2) \, dy \, dx = k \int_0^a \left[x^4 y + \frac{1}{3} x^2 y^3\right]_{y=0}^{y=a-x} dx$

$\qquad = k \int_0^a \left[x^4 (a-x) + \frac{1}{3} x^2 (a-x)^3\right] dx = k \left[\frac{1}{5} a x^5 - \frac{1}{6} x^6 + \frac{1}{3}\left(\frac{1}{3} a^3 x^3 - \frac{3}{4} a^2 x^4 + \frac{3}{5} a x^5 - \frac{1}{6} x^6\right)\right]_0^a = \frac{7}{180} k a^6$,

and $I_0 = I_x + I_y = \frac{7}{90} k a^6$.

**21.** Using a CAS, we find $m = \iint_D \rho(x,y) \, dA = \int_0^\pi \int_0^{\sin x} xy \, dy \, dx = \dfrac{\pi^2}{8}$. Then

$\bar{x} = \frac{1}{m} \iint_D x \rho(x,y) \, dA = \frac{8}{\pi^2} \int_0^\pi \int_0^{\sin x} x^2 y \, dy \, dx = \frac{2\pi}{3} - \frac{1}{\pi}$ and

$\bar{y} = \frac{1}{m} \iint_D y \rho(x,y) \, dA = \frac{8}{\pi^2} \int_0^\pi \int_0^{\sin x} xy^2 \, dy \, dx = \frac{16}{9\pi}$, so $(\bar{x}, \bar{y}) = \left(\dfrac{2\pi}{3} - \dfrac{1}{\pi}, \dfrac{16}{9\pi}\right)$.

The moments of inertia are $I_x = \iint_D y^2 \rho(x,y) \, dA = \int_0^\pi \int_0^{\sin x} xy^3 \, dy \, dx = \dfrac{3\pi^2}{64}$,

$I_y = \iint_D x^2 \rho(x,y) \, dA = \int_0^\pi \int_0^{\sin x} x^3 y \, dy \, dx = \dfrac{\pi^2}{16}(\pi^2 - 3)$, and $I_0 = I_x + I_y = \dfrac{\pi^2}{64}(4\pi^2 - 9)$.

**23.** (a) $f(x,y)$ is a joint density function, so we know $\iint_{\mathbb{R}^2} f(x,y) \, dA = 1$. Since $f(x,y) = 0$ outside the

rectangle $[0,1] \times [0,2]$, we can say

$$\iint_{\mathbb{R}^2} f(x,y) \, dA = \int_{-\infty}^\infty \int_{-\infty}^\infty f(x,y) \, dy \, dx = \int_0^1 \int_0^2 Cx(1+y) \, dy \, dx$$
$$= C \int_0^1 x \left[y + \frac{1}{2} y^2\right]_{y=0}^{y=2} dx = C \int_0^1 4x \, dx = C\left[2x^2\right]_0^1 = 2C$$

Then $2C = 1 \;\Rightarrow\; C = \frac{1}{2}$.

(b) $P(X \le 1, Y \le 1) = \int_{-\infty}^1 \int_{-\infty}^1 f(x,y) \, dy \, dx = \int_0^1 \int_0^1 \frac{1}{2} x(1+y) \, dy \, dx$

$\qquad = \int_0^1 \frac{1}{2} x \left[y + \frac{1}{2} y^2\right]_{y=0}^{y=1} dx = \int_0^1 \frac{1}{2} x \left(\frac{3}{2}\right) dx = \frac{3}{4}\left[\frac{1}{2} x^2\right]_0^1 = \frac{3}{8}$ or $0.375$

(c) $P(X + Y \le 1) = P((X,Y) \in D)$ where $D$ is the triangular region shown in

the figure. Thus
$P(X + Y \le 1) = \iint_D f(x,y) \, dA = \int_0^1 \int_0^{1-x} \frac{1}{2} x(1+y) \, dy \, dx$

$\qquad = \int_0^1 \frac{1}{2} x \left[y + \frac{1}{2} y^2\right]_{y=0}^{y=1-x} dx = \int_0^1 \frac{1}{2} x \left(\frac{1}{2} x^2 - 2x + \frac{3}{2}\right) dx$

$\qquad = \frac{1}{4} \int_0^1 (x^3 - 4x^2 + 3x) \, dx = \frac{1}{4}\left[\frac{x^4}{4} - 4\frac{x^3}{3} + 3\frac{x^2}{2}\right]_0^1$

$\qquad = \frac{5}{48} \approx 0.1042$

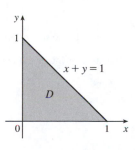

**25.** (a) $f(x, y) \geq 0$, so $f$ is a joint density function if $\iint_{\mathbb{R}^2} f(x, y) \, dA = 1$. Here, $f(x, y) = 0$ outside the first quadrant, so

$$\iint_{\mathbb{R}^2} f(x, y) \, dA = \int_0^\infty \int_0^\infty 0.1 e^{-(0.5x + 0.2y)} \, dy \, dx = 0.1 \int_0^\infty \int_0^\infty e^{-0.5x} e^{-0.2y} \, dy \, dx = 0.1 \int_0^\infty e^{-0.5x} \, dx \int_0^\infty e^{-0.2y} \, dy$$

$$= 0.1 \lim_{t \to \infty} \int_0^t e^{-0.5x} \, dx \lim_{t \to \infty} \int_0^t e^{-0.2y} \, dy = 0.1 \lim_{t \to \infty} \left[ -2e^{-0.5x} \right]_0^t \lim_{t \to \infty} \left[ -5e^{-0.2y} \right]_0^t$$

$$= 0.1 \lim_{t \to \infty} \left[ -2(e^{-0.5t} - 1) \right] \lim_{t \to \infty} \left[ -5(e^{-0.2t} - 1) \right] = (0.1) \cdot (-2)(0 - 1) \cdot (-5)(0 - 1) = 1$$

Thus $f(x, y)$ is a joint density function.

(b) (i) No restriction is placed on $X$, so

$$P(Y \geq 1) = \int_{-\infty}^\infty \int_1^\infty f(x, y) \, dy \, dx = \int_0^\infty \int_1^\infty 0.1 e^{-(0.5x + 0.2y)} \, dy \, dx$$

$$= 0.1 \int_0^\infty e^{-0.5x} \, dx \int_1^\infty e^{-0.2y} \, dy = 0.1 \lim_{t \to \infty} \int_0^t e^{-0.5x} \, dx \lim_{t \to \infty} \int_1^t e^{-0.2y} \, dy$$

$$= 0.1 \lim_{t \to \infty} \left[ -2e^{-0.5x} \right]_0^t \lim_{t \to \infty} \left[ -5e^{-0.2y} \right]_1^t = 0.1 \lim_{t \to \infty} \left[ -2(e^{-0.5t} - 1) \right] \lim_{t \to \infty} \left[ -5(e^{-0.2t} - e^{-0.2}) \right]$$

$$(0.1) \cdot (-2)(0 - 1) \cdot (-5)(0 - e^{-0.2}) = e^{-0.2} \approx 0.8187$$

(ii) $P(X \leq 2, Y \leq 4) = \int_{-\infty}^2 \int_{-\infty}^4 f(x, y) \, dy \, dx = \int_0^2 \int_0^4 0.1 e^{-(0.5x + 0.2y)} \, dy \, dx$

$$= 0.1 \int_0^2 e^{-0.5x} \, dx \int_0^4 e^{-0.2y} \, dy = 0.1 \left[ -2e^{-0.5x} \right]_0^2 \left[ -5e^{-0.2y} \right]_0^4$$

$$= (0.1) \cdot (-2)(e^{-1} - 1) \cdot (-5)(e^{-0.8} - 1)$$

$$= (e^{-1} - 1)(e^{-0.8} - 1) = 1 + e^{-1.8} - e^{-0.8} - e^{-1} \approx 0.3481$$

(c) The expected value of $X$ is given by

$$\mu_1 = \iint_{\mathbb{R}^2} x f(x, y) \, dA = \int_0^\infty \int_0^\infty x \left[ 0.1 e^{-(0.5x + 0.2y)} \right] dy \, dx$$

$$= 0.1 \int_0^\infty x e^{-0.5x} \, dx \int_0^\infty e^{-0.2y} \, dy = 0.1 \lim_{t \to \infty} \int_0^t x e^{-0.5x} \, dx \lim_{t \to \infty} \int_0^t e^{-0.2y} \, dy$$

To evaluate the first integral, we integrate by parts with $u = x$ and $dv = e^{-0.5x} \, dx$ (or we can use Formula 96 in the Table of Integrals): $\int x e^{-0.5x} \, dx = -2x e^{-0.5x} - \int -2e^{-0.5x} \, dx = -2x e^{-0.5x} - 4e^{-0.5x} = -2(x + 2)e^{-0.5x}$.

Thus

$$\mu_1 = 0.1 \lim_{t \to \infty} \left[ -2(x + 2)e^{-0.5x} \right]_0^t \lim_{t \to \infty} \left[ -5e^{-0.2y} \right]_0^t$$

$$= 0.1 \lim_{t \to \infty} (-2) \left[ (t + 2)e^{-0.5t} - 2 \right] \lim_{t \to \infty} (-5) \left[ e^{-0.2t} - 1 \right]$$

$$= 0.1(-2) \left( \lim_{t \to \infty} \frac{t + 2}{e^{0.5t}} - 2 \right) (-5)(-1) = 2 \qquad \text{[by l'Hospital's Rule]}$$

The expected value of $Y$ is given by

$$\mu_2 = \iint_{\mathbb{R}^2} y f(x, y) \, dA = \int_0^\infty \int_0^\infty y \left[ 0.1 e^{-(0.5 + 0.2y)} \right] dy \, dx$$

$$= 0.1 \int_0^\infty e^{-0.5x} \, dx \int_0^\infty y e^{-0.2y} \, dy = 0.1 \lim_{t \to \infty} \int_0^t e^{-0.5x} \, dx \lim_{t \to \infty} \int_0^t y e^{-0.2y} \, dy$$

To evaluate the second integral, we integrate by parts with $u = y$ and $dv = e^{-0.2y} \, dy$ (or again we can use Formula 96 in the Table of Integrals) which gives $\int y e^{-0.2y} \, dy = -5y e^{-0.2y} + \int 5e^{-0.2y} \, dy = -5(y + 5)e^{-0.2y}$. Then

$$\mu_2 = 0.1 \lim_{t \to \infty} \left[ -2e^{-0.5x} \right]_0^t \lim_{t \to \infty} \left[ -5(y + 5)e^{-0.2y} \right]_0^t$$

$$= 0.1 \lim_{t \to \infty} \left[ -2(e^{-0.5t} - 1) \right] \lim_{t \to \infty} \left( -5 \left[ (t + 5)e^{-0.2t} - 5 \right] \right)$$

$$= 0.1(-2)(-1) \cdot (-5) \left( \lim_{t \to \infty} \frac{t + 5}{e^{0.2t}} - 5 \right) = 5 \qquad \text{[by l'Hospital's Rule]}$$

**27.** (a) The random variables $X$ and $Y$ are normally distributed with $\mu_1 = 45$, $\mu_2 = 20$, $\sigma_1 = 0.5$, and $\sigma_2 = 0.1$.

The individual density functions for $X$ and $Y$, then, are $f_1(x) = \dfrac{1}{0.5\sqrt{2\pi}} e^{-(x-45)^2/0.5}$ and

$f_2(y) = \dfrac{1}{0.1\sqrt{2\pi}} e^{-(y-20)^2/0.02}$. Since $X$ and $Y$ are independent, the joint density function is the product

$f(x,y) = f_1(x)f_2(y) = \dfrac{1}{0.5\sqrt{2\pi}} e^{-(x-45)^2/0.5}\,\dfrac{1}{0.1\sqrt{2\pi}} e^{-(y-20)^2/0.02} = \dfrac{10}{\pi} e^{-2(x-45)^2-50(y-20)^2}$.

Then $P(40 \le X \le 50,\, 20 \le Y \le 25) = \int_{40}^{50} \int_{20}^{25} f(x,y)\,dy\,dx = \dfrac{10}{\pi} \int_{40}^{50} \int_{20}^{25} e^{-2(x-45)^2-50(y-20)^2}\,dy\,dx$.

Using a CAS or calculator to evaluate the integral, we get $P(40 \le X \le 50,\, 20 \le Y \le 25) \approx 0.500$.

(b) $P(4(X-45)^2 + 100(Y-20)^2 \le 2) = \iint_D \dfrac{10}{\pi} e^{-2(x-45)^2-50(y-20)^2}\,dA$, where $D$ is the region enclosed by the ellipse

$4(x-45)^2 + 100(y-20)^2 = 2$. Solving for $y$ gives $y = 20 \pm \tfrac{1}{10}\sqrt{2 - 4(x-45)^2}$, the upper and lower halves of the

ellipse, and these two halves meet where $y = 20$  [since the ellipse is centered at $(45, 20)$]  $\Rightarrow$  $4(x-45)^2 = 2$  $\Rightarrow$

$x = 45 \pm \tfrac{1}{\sqrt{2}}$. Thus

$$\iint_D \dfrac{10}{\pi} e^{-2(x-45)^2-50(y-20)^2}\,dA = \dfrac{10}{\pi} \int_{45-1/\sqrt{2}}^{45+1/\sqrt{2}} \int_{20-\frac{1}{10}\sqrt{2-4(x-45)^2}}^{20+\frac{1}{10}\sqrt{2-4(x-45)^2}} e^{-2(x-45)^2-50(y-20)^2}\,dy\,dx.$$

Using a CAS or calculator to evaluate the integral, we get $P(4(X-45)^2 + 100(Y-20)^2 \le 2) \approx 0.632$.

**29.** (a) If $f(P, A)$ is the probability that an individual at $A$ will be infected by an individual at $P$, and $k\,dA$ is the number of
infected individuals in an element of area $dA$, then $f(P, A)k\,dA$ is the number of infections that should result from
exposure of the individual at $A$ to infected people in the element of area $dA$. Integration over $D$ gives the number of
infections of the person at $A$ due to all the infected people in $D$. In rectangular coordinates (with the origin at the city's
center), the exposure of a person at $A$ is

$$E = \iint_D k f(P, A)\,dA = k \iint_D \tfrac{1}{20}\left[20 - d(P, A)\right]\,dA = k \iint_D \left[1 - \tfrac{1}{20}\sqrt{(x-x_0)^2 + (y-y_0)^2}\right]\,dA$$

(b) If $A = (0, 0)$, then

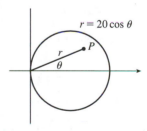

$$E = k \iint_D \left[1 - \tfrac{1}{20}\sqrt{x^2 + y^2}\right]\,dA$$

$$= k \int_0^{2\pi} \int_0^{10} \left(1 - \tfrac{1}{20}r\right) r\,dr\,d\theta = 2\pi k \left[\tfrac{1}{2}r^2 - \tfrac{1}{60}r^3\right]_0^{10}$$

$$= 2\pi k \left(50 - \tfrac{50}{3}\right) = \tfrac{200}{3}\pi k \approx 209k$$

$r = 20\cos\theta$

For $A$ at the edge of the city, it is convenient to use a polar coordinate system centered at $A$. Then the polar equation for
the circular boundary of the city becomes $r = 20\cos\theta$ instead of $r = 10$, and the distance from $A$ to a point $P$ in the city
is again $r$ (see the figure). So

$$E = k \int_{-\pi/2}^{\pi/2} \int_0^{20\cos\theta} \left(1 - \tfrac{1}{20}r\right) r\,dr\,d\theta = k \int_{-\pi/2}^{\pi/2} \left[\tfrac{1}{2}r^2 - \tfrac{1}{60}r^3\right]_{r=0}^{r=20\cos\theta}\,d\theta$$

$$= k \int_{-\pi/2}^{\pi/2} \left(200\cos^2\theta - \tfrac{400}{3}\cos^3\theta\right)\,d\theta = 200k \int_{-\pi/2}^{\pi/2} \left[\tfrac{1}{2} + \tfrac{1}{2}\cos 2\theta - \tfrac{2}{3}(1 - \sin^2\theta)\cos\theta\right]\,d\theta$$

$$= 200k \left[\tfrac{1}{2}\theta + \tfrac{1}{4}\sin 2\theta - \tfrac{2}{3}\sin\theta + \tfrac{2}{3}\cdot\tfrac{1}{3}\sin^3\theta\right]_{-\pi/2}^{\pi/2} = 200k \left[\tfrac{\pi}{4} + 0 - \tfrac{2}{3} + \tfrac{2}{9} + \tfrac{\pi}{4} + 0 - \tfrac{2}{3} + \tfrac{2}{9}\right]$$

$$= 200k \left(\tfrac{\pi}{2} - \tfrac{8}{9}\right) \approx 136k$$

Therefore the risk of infection is much lower at the edge of the city than in the middle, so it is better to live at the edge.

## 12.6 Surface Area

**1.** Here we can write $z = f(x, y) = \frac{1}{3} - \frac{1}{3}x - \frac{2}{3}y$ and $D$ is the disk $x^2 + y^2 \leq 3$, so by Formula 6 the area of the surface is

$$A(S) = \iint_D \sqrt{1 + \left(\frac{\partial z}{\partial x}\right)^2 + \left(\frac{\partial z}{\partial y}\right)^2} \, dA = \iint_D \sqrt{1 + \left(-\frac{1}{3}\right)^2 + \left(-\frac{2}{3}\right)^2} \, dA = \frac{\sqrt{14}}{3} \iint_D dA$$

$$= \frac{\sqrt{14}}{3} A(D) = \frac{\sqrt{14}}{3} \cdot \pi \left(\sqrt{3}\right)^2 = \sqrt{14}\,\pi$$

**3.** The surface $S$ is given by $z = f(x, y) = 6 - 3x - 2y$ which intersects the $xy$-plane in the line $3x + 2y = 6$, so $D$ is the triangular region given by $\{(x, y) \mid 0 \leq x \leq 2, 0 \leq y \leq 3 - \frac{3}{2}x\}$. By Formula 6, the surface area of $S$ is

$$A(S) = \iint_D \sqrt{1 + \left(\frac{\partial z}{\partial x}\right)^2 + \left(\frac{\partial z}{\partial y}\right)^2} \, dA$$

$$= \iint_D \sqrt{1 + (-3)^2 + (-2)^2} \, dA = \sqrt{14} \iint_D dA = \sqrt{14}\, A(D) = \sqrt{14} \left(\frac{1}{2} \cdot 2 \cdot 3\right) = 3\sqrt{14}.$$

**5.** $z = f(x, y) = \sqrt{x^2 + y^2} \ \Rightarrow\ \dfrac{\partial z}{\partial x} = \dfrac{1}{2} \left(x^2 + y^2\right)^{-1/2} \cdot 2x = \dfrac{x}{\sqrt{x^2 + y^2}}, \dfrac{\partial z}{\partial y} = \dfrac{y}{\sqrt{x^2 + y^2}}$, and

$$\sqrt{1 + \left(\frac{\partial z}{\partial x}\right)^2 + \left(\frac{\partial z}{\partial y}\right)^2} = \sqrt{1 + \frac{x^2}{x^2 + y^2} + \frac{y^2}{x^2 + y^2}} = \sqrt{1 + \frac{x^2 + y^2}{x^2 + y^2}} = \sqrt{2}$$

Here $D$ is given by $\{(x, y) \mid 0 \leq x \leq 1, x^2 \leq y \leq x\}$, so by Formula 6, the surface area of $S$ is

$$A(S) = \iint_D \sqrt{2} \, dA = \int_0^1 \int_{x^2}^x \sqrt{2} \, dy \, dx = \sqrt{2} \int_0^1 \left(x - x^2\right) dx = \sqrt{2} \left[\frac{1}{2}x^2 - \frac{1}{3}x^3\right]_0^1 = \sqrt{2} \left(\frac{1}{2} - \frac{1}{3}\right) = \frac{\sqrt{2}}{6}$$

**7.** $\mathbf{r}_u = \langle 2u, v, 0 \rangle$, $\mathbf{r}_v = \langle 0, u, v \rangle$, and $\mathbf{r}_u \times \mathbf{r}_v = \langle v^2, -2uv, 2u^2 \rangle$. Then

$$A(S) = \iint_D |\mathbf{r}_u \times \mathbf{r}_v| \, dA = \int_0^1 \int_0^2 \sqrt{v^4 + 4u^2v^2 + 4u^4} \, dv \, du = \int_0^1 \int_0^2 \sqrt{(v^2 + 2u^2)^2} \, dv \, du$$

$$= \int_0^1 \int_0^2 (v^2 + 2u^2) \, dv \, du = \int_0^1 \left[\frac{1}{3}v^3 + 2u^2v\right]_{v=0}^{v=2} du = \int_0^1 \left(\frac{8}{3} + 4u^2\right) du = \left[\frac{8}{3}u + \frac{4}{3}u^3\right]_0^1 = 4$$

**9.** $z = f(x, y) = xy$ with $0 \leq x^2 + y^2 \leq 1$, so $f_x = y$, $f_y = x \ \Rightarrow$

$$A(S) = \iint_D \sqrt{1 + y^2 + x^2} \, dA = \int_0^{2\pi} \int_0^1 \sqrt{r^2 + 1}\, r \, dr \, d\theta = \int_0^{2\pi} \left[\frac{1}{3}\left(r^2 + 1\right)^{3/2}\right]_{r=0}^{r=1} d\theta$$

$$= \int_0^{2\pi} \frac{1}{3} \left(2\sqrt{2} - 1\right) d\theta = \frac{2\pi}{3} \left(2\sqrt{2} - 1\right)$$

**11.** The cylinder encloses separate portions of the sphere in the upper and lower halves. The top half of the sphere is $z = f(x, y) = \sqrt{b^2 - x^2 - y^2}$ and $D$ is given by $\{(x, y) \mid x^2 + y^2 \leq a^2\}$. By Formula 6, the surface area of the upper enclosed portion is

$$A = \iint_D \sqrt{1 + \left(\frac{-x}{\sqrt{b^2 - x^2 - y^2}}\right)^2 + \left(\frac{-y}{\sqrt{b^2 - x^2 - y^2}}\right)^2} \, dA = \iint_D \sqrt{1 + \frac{x^2 + y^2}{b^2 - x^2 - y^2}} \, dA$$

$$= \iint_D \sqrt{\frac{b^2}{b^2 - x^2 - y^2}} \, dA = \int_0^{2\pi} \int_0^a \frac{b}{\sqrt{b^2 - r^2}}\, r \, dr \, d\theta = b \int_0^{2\pi} d\theta \int_0^a \frac{r}{\sqrt{b^2 - r^2}} \, dr$$

$$= b \, [\theta]_0^{2\pi} \left[-\sqrt{b^2 - r^2}\right]_0^a = 2\pi b\left(-\sqrt{b^2 - a^2} + \sqrt{b^2 - 0}\right) = 2\pi b\left(b - \sqrt{b^2 - a^2}\right)$$

The lower portion of the sphere enclosed by the cylinder has identical shape, so the total area is $2A = 4\pi b\left(b - \sqrt{b^2 - a^2}\right)$.

**13.** $z = f(x, y) = e^{-x^2 - y^2}$, $f_x = -2xe^{-x^2 - y^2}$, $f_y = -2ye^{-x^2 - y^2}$. Then

$$A(S) = \iint_{x^2 + y^2 \le 4} \sqrt{1 + (-2xe^{-x^2 - y^2})^2 + (-2ye^{-x^2 - y^2})^2} \, dA = \iint_{x^2 + y^2 \le 4} \sqrt{1 + 4(x^2 + y^2)e^{-2(x^2 + y^2)}} \, dA.$$

Converting to polar coordinates we have

$$A(S) = \int_0^{2\pi} \int_0^2 \sqrt{1 + 4r^2 e^{-2r^2}} \, r \, dr \, d\theta = \int_0^{2\pi} d\theta \int_0^2 r \sqrt{1 + 4r^2 e^{-2r^2}} \, dr$$

$$= 2\pi \int_0^2 r \sqrt{1 + 4r^2 e^{-2r^2}} \, dr \approx 13.9783 \text{ using a calculator.}$$

**15.** (a) $A(S) = \iint_D \sqrt{1 + \left(\dfrac{\partial z}{\partial x}\right)^2 + \left(\dfrac{\partial z}{\partial y}\right)^2} \, dA = \int_0^6 \int_0^4 \sqrt{1 + \dfrac{4x^2 + 4y^2}{(1 + x^2 + y^2)^4}} \, dy \, dx.$

Using the Midpoint Rule with $f(x, y) = \sqrt{1 + \dfrac{4x^2 + 4y^2}{(1 + x^2 + y^2)^4}}$, $m = 3$, $n = 2$ we have

$$A(S) \approx \sum_{i=1}^3 \sum_{j=1}^2 f(\overline{x}_i, \overline{y}_j) \, \Delta A = 4 \left[ f(1, 1) + f(1, 3) + f(3, 1) + f(3, 3) + f(5, 1) + f(5, 3) \right] \approx 24.2055$$

(b) Using a CAS we have $A(S) = \int_0^6 \int_0^4 \sqrt{1 + \dfrac{4x^2 + 4y^2}{(1 + x^2 + y^2)^4}} \, dy \, dx \approx 24.2476$. This agrees with the estimate in part (a)

to the first decimal place.

**17.** $\mathbf{r}(u, v) = \langle \cos^3 u \cos^3 v, \sin^3 u \cos^3 v, \sin^3 v \rangle$, so $\mathbf{r}_u = \langle -3 \cos^2 u \sin u \cos^3 v, 3 \sin^2 u \cos u \cos^3 v, 0 \rangle$,

$\mathbf{r}_v = \langle -3 \cos^3 u \cos^2 v \sin v, -3 \sin^3 u \cos^2 v \sin v, 3 \sin^2 v \cos v \rangle$, and

$\mathbf{r}_u \times \mathbf{r}_v = \langle 9 \cos u \sin^2 u \cos^4 v \sin^2 v, 9 \cos^2 u \sin u \cos^4 v \sin^2 v, 9 \cos^2 u \sin^2 u \cos^5 v \sin v \rangle$. Then

$$|\mathbf{r}_u \times \mathbf{r}_v| = 9 \sqrt{\cos^2 u \sin^4 u \cos^8 v \sin^4 v + \cos^4 u \sin^2 u \cos^8 v \sin^4 v + \cos^4 u \sin^4 u \cos^{10} v \sin^2 v}$$

$$= 9 \sqrt{\cos^2 u \sin^2 u \cos^8 v \sin^2 v (\sin^2 v + \cos^2 u \sin^2 u \cos^2 v)}$$

$$= 9 \cos^4 v \, |\cos u \sin u \sin v| \sqrt{\sin^2 v + \cos^2 u \sin^2 u \cos^2 v}$$

Using a CAS, we have $A(S) = \int_0^\pi \int_0^{2\pi} 9 \cos^4 v \, |\cos u \sin u \sin v| \sqrt{\sin^2 v + \cos^2 u \sin^2 u \cos^2 v} \, dv \, du \approx 4.4506$.

**19.** $z = 1 + 2x + 3y + 4y^2$, so

$$A(S) = \iint_D \sqrt{1 + \left(\dfrac{\partial z}{\partial x}\right)^2 + \left(\dfrac{\partial z}{\partial y}\right)^2} \, dA = \int_1^4 \int_0^1 \sqrt{1 + 4 + (3 + 8y)^2} \, dy \, dx = \int_1^4 \int_0^1 \sqrt{14 + 48y + 64y^2} \, dy \, dx.$$

Using a CAS, we have

$\int_1^4 \int_0^1 \sqrt{14 + 48y + 64y^2} \, dy \, dx = \dfrac{45}{8} \sqrt{14} + \dfrac{15}{16} \ln\left(11\sqrt{5} + 3\sqrt{14}\sqrt{5}\right) - \dfrac{15}{16} \ln\left(3\sqrt{5} + \sqrt{14}\sqrt{5}\right)$

or $\dfrac{45}{8} \sqrt{14} + \dfrac{15}{16} \ln \dfrac{11\sqrt{5} + 3\sqrt{70}}{3\sqrt{5} + \sqrt{70}}$.

**21.** (a) $x = a \sin u \cos v$, $y = b \sin u \sin v$, $z = c \cos u$ $\Rightarrow$

$$\dfrac{x^2}{a^2} + \dfrac{y^2}{b^2} + \dfrac{z^2}{c^2} = (\sin u \cos v)^2 + (\sin u \sin v)^2 + (\cos u)^2$$

$$= \sin^2 u + \cos^2 u = 1$$

and since the ranges of $u$ and $v$ are sufficient to generate the entire graph, the parametric equations represent an ellipsoid.

(b)

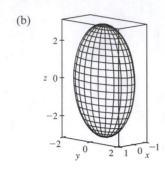

(c) From the parametric equations (with $a = 1$, $b = 2$, and $c = 3$),

we calculate $\mathbf{r}_u = \cos u \cos v \, \mathbf{i} + 2 \cos u \sin v \, \mathbf{j} - 3 \sin u \, \mathbf{k}$ and

$\mathbf{r}_v = -\sin u \sin v \, \mathbf{i} + 2 \sin u \cos v \, \mathbf{j}$. So

$\mathbf{r}_u \times \mathbf{r}_v = 6 \sin^2 u \cos v \, \mathbf{i} + 3 \sin^2 u \sin v \, \mathbf{j} + 2 \sin u \cos u \, \mathbf{k}$,

and the surface area is given by

$$A(S) = \int_0^{2\pi} \int_0^{\pi} |\mathbf{r}_u \times \mathbf{r}_v| \, du \, dv$$

$$= \int_0^{2\pi} \int_0^{\pi} \sqrt{36 \sin^4 u \cos^2 v + 9 \sin^4 u \sin^2 v + 4 \cos^2 u \sin^2 u} \, du \, dv$$

**23.** If we revolve the curve $y = f(x)$, $a \le x \le b$ about the $x$-axis, where $f(x) \ge 0$, then from Equations 10.5.3 we know we can

parametrize the surface using $x = x$, $y = f(x) \cos \theta$, and $z = f(x) \sin \theta$, where $a \le x \le b$ and $0 \le \theta \le 2\pi$. Thus we can

say the surface is represented by $\mathbf{r}(x, \theta) = x \, \mathbf{i} + f(x) \cos \theta \, \mathbf{j} + f(x) \sin \theta \, \mathbf{k}$, with $a \le x \le b$ and $0 \le \theta \le 2\pi$. Then by (4),

the surface area is given by $A(S) = \iint_D |\mathbf{r}_x \times \mathbf{r}_\theta| \, dA$ where $D$ is the rectangular parameter region $[a, b] \times [0, 2\pi]$. Here,

$\mathbf{r}_x(x, \theta) = \mathbf{i} + f'(x) \cos \theta \, \mathbf{j} + f'(x) \sin \theta \, \mathbf{k}$ and $\mathbf{r}_\theta(x) = -f(x) \sin \theta \, \mathbf{j} + f(x) \cos \theta \, \mathbf{k}$. So

$$\mathbf{r}_x \times \mathbf{r}_\theta = \begin{vmatrix} \mathbf{i} & \mathbf{j} & \mathbf{k} \\ 1 & f'(x) \cos \theta & f'(x) \sin \theta \\ 0 & -f(x) \sin \theta & f(x) \cos \theta \end{vmatrix} = \left[ f(x) f'(x) \cos^2 \theta + f(x) f'(x) \sin^2 \theta \right] \mathbf{i} - f(x) \cos \theta \, \mathbf{j} - f(x) \sin \theta \, \mathbf{k}$$

$$= f(x) f'(x) \mathbf{i} - f(x) \cos \theta \, \mathbf{j} - f(x) \sin \theta \, \mathbf{k} \text{ and}$$

$$|\mathbf{r}_x \times \mathbf{r}_\theta| = \sqrt{[f(x) f'(x)]^2 + [f(x)]^2 \cos^2 \theta + [f(x)]^2 \sin^2 \theta}$$

$$= \sqrt{[f(x)]^2 \left( [f'(x)]^2 + 1 \right)} = f(x) \sqrt{1 + [f'(x)]^2} \text{ [since } f(x) \ge 0 \text{]. Thus}$$

$$A(S) = \iint_D |\mathbf{r}_x \times \mathbf{r}_\theta| \, dA = \int_a^b \int_0^{2\pi} f(x) \sqrt{1 + [f'(x)]^2} \, d\theta \, dx$$

$$= \int_a^b f(x) \sqrt{1 + [f'(x)]^2} \, [\theta]_0^{2\pi} \, dx = 2\pi \int_a^b f(x) \sqrt{1 + [f'(x)]^2} \, dx$$

**25.** $y = f(x) = \sqrt{1 + 4x} \quad \Rightarrow \quad f'(x) = \frac{1}{2}(1 + 4x)^{-1/2} \cdot 4 = 2/\sqrt{1 + 4x}$. Then by Formula 7,

$$A = 2\pi \int_1^5 f(x) \sqrt{1 + [f'(x)]^2} \, dx = 2\pi \int_1^5 \sqrt{1 + 4x} \sqrt{1 + \left( 2/\sqrt{1 + 4x} \right)^2} \, dx$$

$$= 2\pi \int_1^5 \sqrt{1 + 4x} \sqrt{\frac{5 + 4x}{1 + 4x}} \, dx = 2\pi \int_1^5 \sqrt{5 + 4x} \, dx = 2\pi \left[ \frac{1}{6} (5 + 4x)^{3/2} \right]_1^5$$

$$= \frac{\pi}{3} (25^{3/2} - 9^{3/2}) = \frac{\pi}{3} (125 - 27) = \frac{98}{3} \pi$$

**27.** To find the region $D$: $z = x^2 + y^2$ implies $z + z^2 = 4z$ or $z^2 - 3z = 0$. Thus $z = 0$ or $z = 3$ are the planes where the

surfaces intersect. But $x^2 + y^2 + z^2 = 4z$ implies $x^2 + y^2 + (z - 2)^2 = 4$, so $z = 3$ intersects the upper hemisphere.

Thus $(z-2)^2 = 4 - x^2 - y^2$ or $z = 2 + \sqrt{4 - x^2 - y^2}$. Therefore $D$ is the region inside the circle $x^2 + y^2 + (3-2)^2 = 4$, that is, $D = \{(x,y) \mid x^2 + y^2 \le 3\}$.

$$A(S) = \iint_D \sqrt{1 + [(-x)(4-x^2-y^2)^{-1/2}]^2 + [(-y)(4-x^2-y^2)^{-1/2}]^2}\, dA$$

$$= \int_0^{2\pi} \int_0^{\sqrt{3}} \sqrt{1 + \frac{r^2}{4-r^2}}\, r\, dr\, d\theta = \int_0^{2\pi} \int_0^{\sqrt{3}} \frac{2r\, dr}{\sqrt{4-r^2}}\, d\theta = \int_0^{2\pi} \left[-2(4-r^2)^{1/2}\right]_{r=0}^{r=\sqrt{3}} d\theta$$

$$= \int_0^{2\pi} (-2+4)\, d\theta = 2\theta\Big]_0^{2\pi} = 4\pi$$

## 12.7　Triple Integrals

**1.** $\iiint_B xyz^2\, dV = \int_0^1 \int_0^3 \int_{-1}^2 xyz^2\, dy\, dz\, dx = \int_0^1 \int_0^3 \left[\frac{1}{2}xy^2z^2\right]_{y=-1}^{y=2} dz\, dx = \int_0^1 \int_0^3 \frac{3}{2}xz^2\, dz\, dx$

$$= \int_0^1 \left[\frac{1}{2}xz^3\right]_{z=0}^{z=3} dx = \int_0^1 \frac{27}{2}x\, dx = \frac{27}{4}x^2\Big]_0^1 = \frac{27}{4}$$

**3.** $\int_0^1 \int_0^z \int_0^{x+z} 6xz\, dy\, dx\, dz = \int_0^1 \int_0^z \left[6xyz\right]_{y=0}^{y=x+z} dx\, dz = \int_0^1 \int_0^z 6xz(x+z)\, dx\, dz$

$$= \int_0^1 \left[2x^3z + 3x^2z^2\right]_{x=0}^{x=z} dz = \int_0^1 (2z^4 + 3z^4)\, dz = \int_0^1 5z^4\, dz = z^5\Big]_0^1 = 1$$

**5.** $\int_0^3 \int_0^1 \int_0^{\sqrt{1-z^2}} ze^y\, dx\, dz\, dy = \int_0^3 \int_0^1 \left[xze^y\right]_{x=0}^{x=\sqrt{1-z^2}} dz\, dy = \int_0^3 \int_0^1 ze^y\sqrt{1-z^2}\, dz\, dy$

$$= \int_0^3 \left[-\frac{1}{3}(1-z^2)^{3/2}e^y\right]_{z=0}^{z=1} dy = \int_0^3 \frac{1}{3}e^y\, dy = \frac{1}{3}e^y\Big]_0^3 = \frac{1}{3}(e^3 - 1)$

**7.** $\int_0^{\pi/2} \int_0^y \int_0^x \cos(x+y+z)dz\, dx\, dy = \int_0^{\pi/2} \int_0^y \left[\sin(x+y+z)\right]_{z=0}^{z=x} dx\, dy$

$$= \int_0^{\pi/2} \int_0^y [\sin(2x+y) - \sin(x+y)]\, dx\, dy$$

$$= \int_0^{\pi/2} \left[-\frac{1}{2}\cos(2x+y) + \cos(x+y)\right]_{x=0}^{x=y} dy$$

$$= \int_0^{\pi/2} \left[-\frac{1}{2}\cos 3y + \cos 2y + \frac{1}{2}\cos y - \cos y\right] dy$$

$$= \left[-\frac{1}{6}\sin 3y + \frac{1}{2}\sin 2y - \frac{1}{2}\sin y\right]_0^{\pi/2} = \frac{1}{6} - \frac{1}{2} = -\frac{1}{3}$$

**9.** $\iiint_E 2x\, dV = \int_0^2 \int_0^{\sqrt{4-y^2}} \int_0^y 2x\, dz\, dx\, dy = \int_0^2 \int_0^{\sqrt{4-y^2}} \left[2xz\right]_{z=0}^{z=y} dx\, dy = \int_0^2 \int_0^{\sqrt{4-y^2}} 2xy\, dx\, dy$

$$= \int_0^2 \left[x^2 y\right]_{x=0}^{x=\sqrt{4-y^2}} dy = \int_0^2 (4-y^2)y\, dy = \left[2y^2 - \frac{1}{4}y^4\right]_0^2 = 4$$

**11.** Here $E = \{(x,y,z) \mid 0 \le x \le 1, 0 \le y \le \sqrt{x}, 0 \le z \le 1 + x + y\}$, so

$$\iiint_E 6xy\, dV = \int_0^1 \int_0^{\sqrt{x}} \int_0^{1+x+y} 6xy\, dz\, dy\, dx = \int_0^1 \int_0^{\sqrt{x}} \left[6xyz\right]_{z=0}^{z=1+x+y} dy\, dx = \int_0^1 \int_0^{\sqrt{x}} 6xy(1+x+y)\, dy\, dx$$

$$= \int_0^1 \left[3xy^2 + 3x^2y^2 + 2xy^3\right]_{y=0}^{y=\sqrt{x}} dx = \int_0^1 (3x^2 + 3x^3 + 2x^{5/2})\, dx = \left[x^3 + \frac{3}{4}x^4 + \frac{4}{7}x^{7/2}\right]_0^1 = \frac{65}{28}$$

**13.**

$E$ is the region below the parabolic cylinder $z = 1 - y^2$ and above the square $[-1, 1] \times [-1, 1]$ in the $xy$-plane.

$$\iiint_E x^2 e^y \, dV = \int_{-1}^{1} \int_{-1}^{1} \int_{0}^{1-y^2} x^2 e^y \, dz \, dy \, dx$$

$$= \int_{-1}^{1} \int_{-1}^{1} x^2 e^y (1 - y^2) \, dy \, dx$$

$$= \int_{-1}^{1} x^2 \, dx \int_{-1}^{1} (e^y - y^2 e^y) \, dy$$

$$= \left[ \tfrac{1}{3} x^3 \right]_{-1}^{1} \left[ e^y - (y^2 - 2y + 2)e^y \right]_{-1}^{1} \qquad \begin{bmatrix} \text{integrate by} \\ \text{parts twice} \end{bmatrix}$$

$$= \tfrac{1}{3}(2)[e - e - e^{-1} + 5e^{-1}] = \tfrac{8}{3e}$$

**15.**

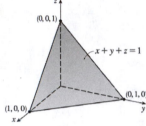

Here $T = \{(x, y, z) \mid 0 \le x \le 1, 0 \le y \le 1 - x, 0 \le z \le 1 - x - y\}$, so

$$\iiint_T x^2 \, dV = \int_0^1 \int_0^{1-x} \int_0^{1-x-y} x^2 \, dz \, dy \, dx = \int_0^1 \int_0^{1-x} x^2(1 - x - y) \, dy \, dx$$

$$= \int_0^1 \int_0^{1-x} (x^2 - x^3 - x^2 y) \, dy \, dx = \int_0^1 \left[ x^2 y - x^3 y - \tfrac{1}{2} x^2 y^2 \right]_{y=0}^{y=1-x} dx$$

$$= \int_0^1 \left[ x^2(1 - x) - x^3(1 - x) - \tfrac{1}{2} x^2 (1 - x)^2 \right] dx$$

$$= \int_0^1 \left( \tfrac{1}{2} x^4 - x^3 + \tfrac{1}{2} x^2 \right) dx = \left[ \tfrac{1}{10} x^5 - \tfrac{1}{4} x^4 + \tfrac{1}{6} x^3 \right]_0^1$$

$$= \tfrac{1}{10} - \tfrac{1}{4} + \tfrac{1}{6} = \tfrac{1}{60}$$

**17.**

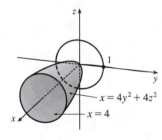

The projection $E$ on the $yz$-plane is the disk $y^2 + z^2 \le 1$. Using polar coordinates $y = r \cos \theta$ and $z = r \sin \theta$, we get

$$\iiint_E x \, dV = \iint_D \left[ \int_{4y^2+4z^2}^{4} x \, dx \right] dA = \tfrac{1}{2} \iint_D \left[ 4^2 - (4y^2 + 4z^2)^2 \right] dA$$

$$= 8 \int_0^{2\pi} \int_0^1 (1 - r^4) r \, dr \, d\theta = 8 \int_0^{2\pi} d\theta \int_0^1 (r - r^5) \, dr$$

$$= 8(2\pi) \left[ \tfrac{1}{2} r^2 - \tfrac{1}{6} r^6 \right]_0^1 = \tfrac{16\pi}{3}$$

**19.** The plane $2x + y + z = 4$ intersects the $xy$-plane when

$2x + y + 0 = 4 \implies y = 4 - 2x$, so

$E = \{(x, y, z) \mid 0 \le x \le 2, 0 \le y \le 4 - 2x, 0 \le z \le 4 - 2x - y\}$ and

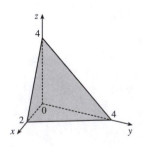

$$V = \int_0^2 \int_0^{4-2x} \int_0^{4-2x-y} dz \, dy \, dx = \int_0^2 \int_0^{4-2x} (4 - 2x - y) \, dy \, dx$$

$$= \int_0^2 \left[ 4y - 2xy - \tfrac{1}{2} y^2 \right]_{y=0}^{y=4-2x} dx$$

$$= \int_0^2 \left[ 4(4 - 2x) - 2x(4 - 2x) - \tfrac{1}{2}(4 - 2x)^2 \right] dx$$

$$= \int_0^2 (2x^2 - 8x + 8) \, dx = \left[ \tfrac{2}{3} x^3 - 4x^2 + 8x \right]_0^2 = \tfrac{16}{3}$$

**21.** $V = \int_{-3}^{3} \int_{-\sqrt{9-x^2}}^{\sqrt{9-x^2}} \int_{1}^{5-y} dz \, dy \, dx = \int_{-3}^{3} \int_{-\sqrt{9-x^2}}^{\sqrt{9-x^2}} (5 - y - 1) \, dy \, dx = \int_{-3}^{3} \left[ 4y - \tfrac{1}{2}y^2 \right]_{y=-\sqrt{9-x^2}}^{y=\sqrt{9-x^2}} dx$

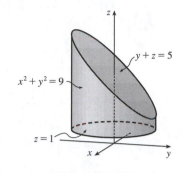

$= \int_{-3}^{3} 8\sqrt{9 - x^2} \, dx = 8 \left[ \tfrac{x}{2}\sqrt{9 - x^2} + \tfrac{9}{2}\sin^{-1}\left(\tfrac{x}{3}\right) \right]_{-3}^{3}$   $\begin{bmatrix} \text{using trigonometric substitution or} \\ \text{Formula 30 in the Table of Integrals} \end{bmatrix}$

$= 8\left[ \tfrac{9}{2}\sin^{-1}(1) - \tfrac{9}{2}\sin^{-1}(-1) \right] = 36\left(\tfrac{\pi}{2} - \left(-\tfrac{\pi}{2}\right)\right) = 36\pi$

Alternatively, use polar coordinates to evaluate the double integral:

$\int_{-3}^{3} \int_{-\sqrt{9-x^2}}^{\sqrt{9-x^2}} (4 - y) \, dy \, dx = \int_{0}^{2\pi} \int_{0}^{3} (4 - r\sin\theta)\, r \, dr \, d\theta$

$= \int_{0}^{2\pi} \left[ 2r^2 - \tfrac{1}{3}r^3 \sin\theta \right]_{r=0}^{r=3} d\theta$

$= \int_{0}^{2\pi} (18 - 9\sin\theta) \, d\theta$

$= 18\theta + 9\cos\theta \Big]_{0}^{2\pi} = 36\pi$

**23.** (a) The wedge can be described as the region

$D = \left\{ (x, y, z) \mid y^2 + z^2 \le 1, 0 \le x \le 1, 0 \le y \le x \right\}$

$= \left\{ (x, y, z) \mid 0 \le x \le 1, 0 \le y \le x, 0 \le z \le \sqrt{1 - y^2} \right\}$

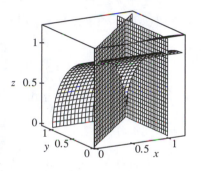

So the integral expressing the volume of the wedge is

$\iiint_D dV = \int_0^1 \int_0^x \int_0^{\sqrt{1-y^2}} dz \, dy \, dx.$

(b) A CAS gives $\int_0^1 \int_0^x \int_0^{\sqrt{1-y^2}} dz \, dy \, dx = \tfrac{\pi}{4} - \tfrac{1}{3}.$

(Or use Formulas 30 and 87 from the Table of Integrals.)

**25.** Here $f(x, y, z) = \dfrac{1}{\ln(1 + x + y + z)}$ and $\Delta V = 2 \cdot 4 \cdot 2 = 16$, so the Midpoint Rule gives

$\iiint_B f(x, y, z) \, dV \approx \sum_{i=1}^{l} \sum_{j=1}^{m} \sum_{k=1}^{n} f(\overline{x}_i, \overline{y}_j, \overline{z}_k) \, \Delta V$

$= 16[f(1, 2, 1) + f(1, 2, 3) + f(1, 6, 1) + f(1, 6, 3)$

$+ f(3, 2, 1) + f(3, 2, 3) + f(3, 6, 1) + f(3, 6, 3)]$

$= 16\left[ \tfrac{1}{\ln 5} + \tfrac{1}{\ln 7} + \tfrac{1}{\ln 9} + \tfrac{1}{\ln 11} + \tfrac{1}{\ln 7} + \tfrac{1}{\ln 9} + \tfrac{1}{\ln 11} + \tfrac{1}{\ln 13} \right] \approx 60.533$

**27.** $E = \{(x, y, z) \mid 0 \le x \le 1, 0 \le z \le 1 - x, 0 \le y \le 2 - 2z\}$,

the solid bounded by the three coordinate planes and the planes

$z = 1 - x, \; y = 2 - 2z.$

**29.**

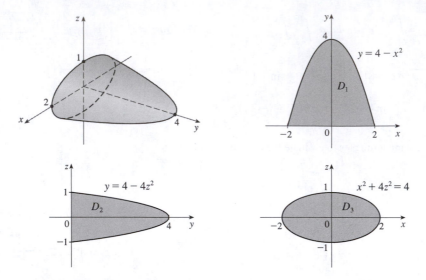

If $D_1$, $D_2$, $D_3$ are the projections of $E$ on the $xy$-, $yz$-, and $xz$-planes, then

$$D_1 = \left\{(x,y) \mid -2 \le x \le 2, 0 \le y \le 4 - x^2\right\} = \left\{(x,y) \mid 0 \le y \le 4, -\sqrt{4-y} \le x \le \sqrt{4-y}\right\}$$

$$D_2 = \left\{(y,z) \mid 0 \le y \le 4, -\tfrac{1}{2}\sqrt{4-y} \le z \le \tfrac{1}{2}\sqrt{4-y}\right\} = \left\{(y,z) \mid -1 \le z \le 1, 0 \le y \le 4 - 4z^2\right\}$$

$$D_3 = \left\{(x,z) \mid x^2 + 4z^2 \le 4\right\}$$

Therefore

$$E = \left\{(x,y,z) \mid -2 \le x \le 2, 0 \le y \le 4 - x^2, -\tfrac{1}{2}\sqrt{4-x^2-y} \le z \le \tfrac{1}{2}\sqrt{4-x^2-y}\right\}$$

$$= \left\{(x,y,z) \mid 0 \le y \le 4, -\sqrt{4-y} \le x \le \sqrt{4-y}, -\tfrac{1}{2}\sqrt{4-x^2-y} \le z \le \tfrac{1}{2}\sqrt{4-x^2-y}\right\}$$

$$= \left\{(x,y,z) \mid -1 \le z \le 1, 0 \le y \le 4 - 4z^2, -\sqrt{4-y-4z^2} \le x \le \sqrt{4-y-4z^2}\right\}$$

$$= \left\{(x,y,z) \mid 0 \le y \le 4, -\tfrac{1}{2}\sqrt{4-y} \le z \le \tfrac{1}{2}\sqrt{4-y}, -\sqrt{4-y-4z^2} \le x \le \sqrt{4-y-4z^2}\right\}$$

$$= \left\{(x,y,z) \mid -2 \le x \le 2, -\tfrac{1}{2}\sqrt{4-x^2} \le z \le \tfrac{1}{2}\sqrt{4-x^2}, 0 \le y \le 4 - x^2 - 4z^2\right\}$$

$$= \left\{(x,y,z) \mid -1 \le z \le 1, -\sqrt{4-4z^2} \le x \le \sqrt{4-4z^2}, 0 \le y \le 4 - x^2 - 4z^2\right\}$$

Then

$$\iiint_E f(x,y,z)\,dV = \int_{-2}^{2} \int_{0}^{4-x^2} \int_{-\sqrt{4-x^2-y}/2}^{\sqrt{4-x^2-y}/2} f(x,y,z)\,dz\,dy\,dx = \int_{0}^{4} \int_{-\sqrt{4-y}}^{\sqrt{4-y}} \int_{-\sqrt{4-x^2-y}/2}^{\sqrt{4-x^2-y}/2} f(x,y,z)\,dz\,dx\,dy$$

$$= \int_{-1}^{1} \int_{0}^{4-4z^2} \int_{-\sqrt{4-y-4z^2}}^{\sqrt{4-y-4z^2}} f(x,y,z)\,dx\,dy\,dz = \int_{0}^{4} \int_{-\sqrt{4-y}/2}^{\sqrt{4-y}/2} \int_{-\sqrt{4-y-4z^2}}^{\sqrt{4-y-4z^2}} f(x,y,z)\,dx\,dz\,dy$$

$$= \int_{-2}^{2} \int_{-\sqrt{4-x^2}/2}^{\sqrt{4-x^2}/2} \int_{0}^{4-x^2-4z^2} f(x,y,z)\,dy\,dz\,dx = \int_{-1}^{1} \int_{-\sqrt{4-4z^2}}^{\sqrt{4-4z^2}} \int_{0}^{4-x^2-4z^2} f(x,y,z)\,dy\,dx\,dz$$

**31.**

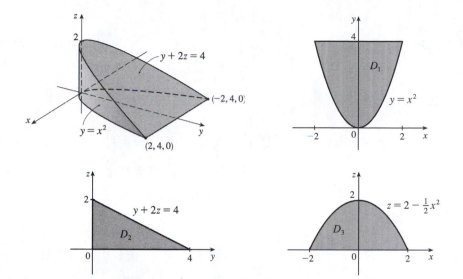

If $D_1$, $D_2$, and $D_3$ are the projections of $E$ on the $xy$-, $yz$-, and $xz$-planes, then

$$D_1 = \left\{ (x, y) \mid -2 \le x \le 2, x^2 \le y \le 4 \right\} = \left\{ (x, y) \mid 0 \le y \le 4, -\sqrt{y} \le x \le \sqrt{y} \right\},$$

$$D_2 = \left\{ (y, z) \mid 0 \le y \le 4, 0 \le z \le 2 - \tfrac{1}{2} y \right\} = \left\{ (y, z) \mid 0 \le z \le 2, 0 \le y \le 4 - 2z \right\}, \text{ and}$$

$$D_3 = \left\{ (x, z) \mid -2 \le x \le 2, 0 \le z \le 2 - \tfrac{1}{2} x^2 \right\} = \left\{ (x, z) \mid 0 \le z \le 2, -\sqrt{4 - 2z} \le x \le \sqrt{4 - 2z} \right\}$$

Therefore

$$E = \left\{ (x, y, z) \mid -2 \le x \le 2, x^2 \le y \le 4, 0 \le z \le 2 - \tfrac{1}{2} y \right\}$$

$$= \left\{ (x, y, z) \mid 0 \le y \le 4, -\sqrt{y} \le x \le \sqrt{y}, 0 \le z \le 2 - \tfrac{1}{2} y \right\}$$

$$= \left\{ (x, y, z) \mid 0 \le y \le 4, 0 \le z \le 2 - \tfrac{1}{2} y, -\sqrt{y} \le x \le \sqrt{y} \right\}$$

$$= \left\{ (x, y, z) \mid 0 \le z \le 2, 0 \le y \le 4 - 2z, -\sqrt{y} \le x \le \sqrt{y} \right\}$$

$$= \left\{ (x, y, z) \mid -2 \le x \le 2, 0 \le z \le 2 - \tfrac{1}{2} x^2, x^2 \le y \le 4 - 2z \right\}$$

$$= \left\{ (x, y, z) \mid 0 \le z \le 2, -\sqrt{4 - 2z} \le x \le \sqrt{4 - 2z}, x^2 \le y \le 4 - 2z \right\}$$

Then

$$\iiint_E f(x, y, z)\, dV = \int_{-2}^{2} \int_{x^2}^{4} \int_{0}^{2 - y/2} f(x, y, z)\, dz\, dy\, dx = \int_{0}^{4} \int_{-\sqrt{y}}^{\sqrt{y}} \int_{0}^{2 - y/2} f(x, y, z)\, dz\, dx\, dy$$

$$= \int_{0}^{4} \int_{0}^{2 - y/2} \int_{-\sqrt{y}}^{\sqrt{y}} f(x, y, z)\, dx\, dz\, dy = \int_{0}^{2} \int_{0}^{4 - 2z} \int_{-\sqrt{y}}^{\sqrt{y}} f(x, y, z)\, dx\, dy\, dz$$

$$= \int_{-2}^{2} \int_{0}^{2 - x^2/2} \int_{x^2}^{4 - 2z} f(x, y, z)\, dy\, dz\, dx = \int_{0}^{2} \int_{-\sqrt{4 - 2z}}^{\sqrt{4 - 2z}} \int_{x^2}^{4 - 2z} f(x, y, z)\, dy\, dx\, dz$$

The diagrams show the projections of $E$ on the $xy$-, $yz$-, and $xz$-planes.

Therefore

$$\int_0^1 \int_{\sqrt{x}}^1 \int_0^{1-y} f(x,y,z)\,dz\,dy\,dx = \int_0^1 \int_0^{y^2} \int_0^{1-y} f(x,y,z)\,dz\,dx\,dy = \int_0^1 \int_0^{1-z} \int_0^{y^2} f(x,y,z)\,dx\,dy\,dz$$

$$= \int_0^1 \int_0^{1-y} \int_0^{y^2} f(x,y,z)\,dx\,dz\,dy = \int_0^1 \int_0^{1-\sqrt{x}} \int_{\sqrt{x}}^{1-z} f(x,y,z)\,dy\,dz\,dx$$

$$= \int_0^1 \int_0^{(1-z)^2} \int_{\sqrt{x}}^{1-z} f(x,y,z)\,dy\,dx\,dz$$

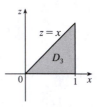

$\int_0^1 \int_y^1 \int_0^y f(x,y,z)\,dz\,dx\,dy = \iiint_E f(x,y,z)\,dV$ where $E = \{(x,y,z) \mid 0 \le z \le y, y \le x \le 1, 0 \le y \le 1\}$.

If $D_1$, $D_2$, and $D_3$ are the projections of $E$ on the $xy$-, $yz$- and $xz$-planes then

$$D_1 = \{(x,y) \mid 0 \le y \le 1, y \le x \le 1\} = \{(x,y) \mid 0 \le x \le 1, 0 \le y \le x\},$$

$$D_2 = \{(y,z) \mid 0 \le y \le 1, 0 \le z \le y\} = \{(y,z) \mid 0 \le z \le 1, z \le y \le 1\},\text{ and}$$

$$D_3 = \{(x,z) \mid 0 \le x \le 1, 0 \le z \le x\} = \{(x,z) \mid 0 \le z \le 1, z \le x \le 1\}.$$

Thus we also have

$$E = \{(x,y,z) \mid 0 \le x \le 1, 0 \le y \le x, 0 \le z \le y\} = \{(x,y,z) \mid 0 \le y \le 1, 0 \le z \le y, y \le x \le 1\}$$

$$= \{(x,y,z) \mid 0 \le z \le 1, z \le y \le 1, y \le x \le 1\} = \{(x,y,z) \mid 0 \le x \le 1, 0 \le z \le x, z \le y \le x\}$$

$$= \{(x,y,z) \mid 0 \le z \le 1, z \le x \le 1, z \le y \le x\}.$$

Then

$$\int_0^1 \int_y^1 \int_0^y f(x,y,z)\,dz\,dx\,dy = \int_0^1 \int_0^x \int_0^y f(x,y,z)\,dz\,dy\,dx = \int_0^1 \int_0^y \int_y^1 f(x,y,z)\,dx\,dz\,dy$$

$$= \int_0^1 \int_z^1 \int_y^1 f(x,y,z)\,dx\,dy\,dz = \int_0^1 \int_0^x \int_z^x f(x,y,z)\,dy\,dz\,dx$$

$$= \int_0^1 \int_z^1 \int_z^x f(x,y,z)\,dy\,dx\,dz$$

37) $m = \iiint_E \rho(x,y,z)\,dV = \int_0^1 \int_0^{\sqrt{x}} \int_0^{1+x+y} 2\,dz\,dy\,dx = \int_0^1 \int_0^{\sqrt{x}} 2(1+x+y)\,dy\,dx$

$= \int_0^1 \left[2y + 2xy + y^2\right]_{y=0}^{y=\sqrt{x}}\,dx = \int_0^1 \left(2\sqrt{x} + 2x^{3/2} + x\right)\,dx = \left[\frac{4}{3}x^{3/2} + \frac{4}{5}x^{5/2} + \frac{1}{2}x^2\right]_0^1 = \frac{79}{30}$

$M_{yz} = \iiint_E x\rho(x,y,z)\,dV = \int_0^1 \int_0^{\sqrt{x}} \int_0^{1+x+y} 2x\,dz\,dy\,dx = \int_0^1 \int_0^{\sqrt{x}} 2x(1+x+y)\,dy\,dx$

$= \int_0^1 \left[2xy + 2x^2 y + xy^2\right]_{y=0}^{y=\sqrt{x}}\,dx = \int_0^1 (2x^{3/2} + 2x^{5/2} + x^2)\,dx = \left[\frac{4}{5}x^{5/2} + \frac{4}{7}x^{7/2} + \frac{1}{3}x^3\right]_0^1 = \frac{179}{105}$

$M_{xz} = \iiint_E y\rho(x,y,z)\,dV = \int_0^1 \int_0^{\sqrt{x}} \int_0^{1+x+y} 2y\,dz\,dy\,dx = \int_0^1 \int_0^{\sqrt{x}} 2y(1+x+y)\,dy\,dx$

$= \int_0^1 \left[y^2 + xy^2 + \frac{2}{3}y^3\right]_{y=0}^{y=\sqrt{x}}\,dx = \int_0^1 \left(x + x^2 + \frac{2}{3}x^{3/2}\right)\,dx = \left[\frac{1}{2}x^2 + \frac{1}{3}x^3 + \frac{4}{15}x^{5/2}\right]_0^1 = \frac{11}{10}$

$M_{xy} = \iiint_E z\rho(x,y,z)\,dV = \int_0^1 \int_0^{\sqrt{x}} \int_0^{1+x+y} 2z\,dz\,dy\,dx = \int_0^1 \int_0^{\sqrt{x}} \left[z^2\right]_{z=0}^{z=1+x+y}\,dy\,dx = \int_0^1 \int_0^{\sqrt{x}} (1+x+y)^2\,dy\,dx$

$= \int_0^1 \int_0^{\sqrt{x}} (1 + 2x + 2y + 2xy + x^2 + y^2)\,dy\,dx = \int_0^1 \left[y + 2xy + y^2 + xy^2 + x^2 y + \frac{1}{3}y^3\right]_{y=0}^{y=\sqrt{x}}\,dx$

$= \int_0^1 \left(\sqrt{x} + \frac{7}{3}x^{3/2} + x + x^2 + x^{5/2}\right)\,dx = \left[\frac{2}{3}x^{3/2} + \frac{14}{15}x^{5/2} + \frac{1}{2}x^2 + \frac{1}{3}x^3 + \frac{2}{7}x^{7/2}\right]_0^1 = \frac{571}{210}$

Thus the mass is $\frac{79}{30}$ and the center of mass is $(\overline{x}, \overline{y}, \overline{z}) = \left(\dfrac{M_{yz}}{m}, \dfrac{M_{xz}}{m}, \dfrac{M_{xy}}{m}\right) = \left(\dfrac{358}{553}, \dfrac{33}{79}, \dfrac{571}{553}\right)$.

**39.** $m = \int_0^a \int_0^a \int_0^a (x^2 + y^2 + z^2)\,dx\,dy\,dz = \int_0^a \int_0^a \left[\frac{1}{3}x^3 + xy^2 + xz^2\right]_{x=0}^{x=a}\,dy\,dz = \int_0^a \int_0^a \left(\frac{1}{3}a^3 + ay^2 + az^2\right)\,dy\,dz$

$= \int_0^a \left[\frac{1}{3}a^3 y + \frac{1}{3}ay^3 + ayz^2\right]_{y=0}^{y=a}\,dz = \int_0^a \left(\frac{2}{3}a^4 + a^2 z^2\right)\,dz = \left[\frac{2}{3}a^4 z + \frac{1}{3}a^2 z^3\right]_0^a = \frac{2}{3}a^5 + \frac{1}{3}a^5 = a^5$

$M_{yz} = \int_0^a \int_0^a \int_0^a \left[x^3 + x(y^2 + z^2)\right]\,dx\,dy\,dz = \int_0^a \int_0^a \left[\frac{1}{4}a^4 + \frac{1}{2}a^2(y^2 + z^2)\right]\,dy\,dz$

$= \int_0^a \left(\frac{1}{4}a^5 + \frac{1}{6}a^5 + \frac{1}{2}a^3 z^2\right)\,dz = \frac{1}{4}a^6 + \frac{1}{3}a^6 = \frac{7}{12}a^6 = M_{xz} = M_{xy}$ by symmetry of $E$ and $\rho(x,y,z)$.

Hence $(\overline{x}, \overline{y}, \overline{z}) = \left(\frac{7}{12}a, \frac{7}{12}a, \frac{7}{12}a\right)$.

**41.** $I_x = \int_0^L \int_0^L \int_0^L k(y^2 + z^2)\,dz\,dy\,dx = k\int_0^L \int_0^L \left(Ly^2 + \frac{1}{3}L^3\right)\,dy\,dx = k\int_0^L \frac{2}{3}L^4\,dx = \frac{2}{3}kL^5$.

By symmetry, $I_x = I_y = I_z = \frac{2}{3}kL^5$.

**43.** $I_z = \iiint_E (x^2 + y^2)\,\rho(x,y,z)\,dV = \iint_{x^2+y^2 \le a^2} \left[\int_0^h k(x^2 + y^2)\,dz\right]dA = \iint_{x^2+y^2 \le a^2} k(x^2 + y^2)h\,dA$

$= kh\int_0^{2\pi}\int_0^a (r^2)\,r\,dr\,d\theta = kh\int_0^{2\pi} d\theta \int_0^a r^3\,dr = kh(2\pi)\left[\frac{1}{4}r^4\right]_0^a = 2\pi kh \cdot \frac{1}{4}a^4 = \frac{1}{2}\pi kha^4$

**45.** (a) $m = \int_{-3}^3 \int_{-\sqrt{9-x^2}}^{\sqrt{9-x^2}} \int_1^{5-y} \sqrt{x^2 + y^2}\,dz\,dy\,dx$

(b) $(\overline{x}, \overline{y}, \overline{z}) = \left(\dfrac{M_{yz}}{m}, \dfrac{M_{xz}}{m}, \dfrac{M_{xy}}{m}\right)$ where

$M_{yz} = \int_{-3}^3 \int_{-\sqrt{9-x^2}}^{\sqrt{9-x^2}} \int_1^{5-y} x\sqrt{x^2 + y^2}\,dz\,dy\,dx$, $M_{xz} = \int_{-3}^3 \int_{-\sqrt{9-x^2}}^{\sqrt{9-x^2}} \int_1^{5-y} y\sqrt{x^2 + y^2}\,dz\,dy\,dx$, and

$M_{xy} = \int_{-3}^3 \int_{-\sqrt{9-x^2}}^{\sqrt{9-x^2}} \int_1^{5-y} z\sqrt{x^2 + y^2}\,dz\,dy\,dx$.

(c) $I_z = \int_{-3}^3 \int_{-\sqrt{9-x^2}}^{\sqrt{9-x^2}} \int_1^{5-y} (x^2 + y^2)\sqrt{x^2 + y^2}\,dz\,dy\,dx = \int_{-3}^3 \int_{-\sqrt{9-x^2}}^{\sqrt{9-x^2}} \int_1^{5-y} (x^2 + y^2)^{3/2}\,dz\,dy\,dx$

**47.** (a) $m = \int_0^1 \int_0^{\sqrt{1-x^2}} \int_0^y (1+x+y+z)\,dz\,dy\,dx = \frac{3\pi}{32} + \frac{11}{24}$

(b) $(\bar{x}, \bar{y}, \bar{z}) = \left( m^{-1} \int_0^1 \int_0^{\sqrt{1-x^2}} \int_0^y x(1+x+y+z)\,dz\,dy\,dx, \right.$

$$m^{-1} \int_0^1 \int_0^{\sqrt{1-x^2}} \int_0^y y(1+x+y+z)\,dz\,dy\,dx,$$

$$\left. m^{-1} \int_0^1 \int_0^{\sqrt{1-x^2}} \int_0^y z(1+x+y+z)\,dz\,dy\,dx \right)$$

$$= \left( \frac{28}{9\pi + 44}, \frac{30\pi + 128}{45\pi + 220}, \frac{45\pi + 208}{135\pi + 660} \right)$$

(c) $I_z = \int_0^1 \int_0^{\sqrt{1-x^2}} \int_0^y (x^2 + y^2)(1+x+y+z)\,dz\,dy\,dx = \dfrac{68 + 15\pi}{240}$

**49.** (a) $f(x, y, z)$ is a joint density function, so we know $\iiint_{\mathbb{R}^3} f(x, y, z)\,dV = 1$. Here we have

$$\iiint_{\mathbb{R}^3} f(x, y, z)\,dV = \int_{-\infty}^{\infty} \int_{-\infty}^{\infty} \int_{-\infty}^{\infty} f(x, y, z)\,dz\,dy\,dx = \int_0^2 \int_0^2 \int_0^2 Cxyz\,dz\,dy\,dx$$

$$= C \int_0^2 x\,dx \int_0^2 y\,dy \int_0^2 z\,dz = C \left[ \tfrac{1}{2}x^2 \right]_0^2 \left[ \tfrac{1}{2}y^2 \right]_0^2 \left[ \tfrac{1}{2}z^2 \right]_0^2 = 8C$$

Then we must have $8C = 1 \;\Rightarrow\; C = \tfrac{1}{8}$.

(b) $P(X \leq 1, Y \leq 1, Z \leq 1) = \int_{-\infty}^1 \int_{-\infty}^1 \int_{-\infty}^1 f(x, y, z)\,dz\,dy\,dx = \int_0^1 \int_0^1 \int_0^1 \tfrac{1}{8}xyz\,dz\,dy\,dx$

$$= \tfrac{1}{8} \int_0^1 x\,dx \int_0^1 y\,dy \int_0^1 z\,dz = \tfrac{1}{8} \left[ \tfrac{1}{2}x^2 \right]_0^1 \left[ \tfrac{1}{2}y^2 \right]_0^1 \left[ \tfrac{1}{2}z^2 \right]_0^1 = \tfrac{1}{8} \left( \tfrac{1}{2} \right)^3 = \tfrac{1}{64}$$

(c) $P(X + Y + Z \leq 1) = P((X, Y, Z) \in E)$ where $E$ is the solid region in the first octant bounded by the coordinate planes

and the plane $x + y + z = 1$. The plane $x + y + z = 1$ meets the $xy$-plane in the line $x + y = 1$, so we have

$$P(X + Y + Z \leq 1) = \iiint_E f(x, y, z)\,dV = \int_0^1 \int_0^{1-x} \int_0^{1-x-y} \tfrac{1}{8}xyz\,dz\,dy\,dx$$

$$= \tfrac{1}{8} \int_0^1 \int_0^{1-x} xy \left[ \tfrac{1}{2}z^2 \right]_{z=0}^{z=1-x-y}\,dy\,dx = \tfrac{1}{16} \int_0^1 \int_0^{1-x} xy(1-x-y)^2\,dy\,dx$$

$$= \tfrac{1}{16} \int_0^1 \int_0^{1-x} [(x^3 - 2x^2 + x)y + (2x^2 - 2x)y^2 + xy^3]\,dy\,dx$$

$$= \tfrac{1}{16} \int_0^1 \left[ (x^3 - 2x^2 + x)\tfrac{1}{2}y^2 + (2x^2 - 2x)\tfrac{1}{3}y^3 + x\left( \tfrac{1}{4}y^4 \right) \right]_{y=0}^{y=1-x}\,dx$$

$$= \tfrac{1}{192} \int_0^1 (x - 4x^2 + 6x^3 - 4x^4 + x^5)\,dx = \tfrac{1}{192} \left( \tfrac{1}{30} \right) = \tfrac{1}{5760}$$

**51.** $V(E) = L^3 \;\Rightarrow\; f_{\text{ave}} = \dfrac{1}{L^3} \int_0^L \int_0^L \int_0^L xyz\,dx\,dy\,dz = \dfrac{1}{L^3} \int_0^L x\,dx \int_0^L y\,dy \int_0^L z\,dz$

$$= \dfrac{1}{L^3} \left[ \dfrac{x^2}{2} \right]_0^L \left[ \dfrac{y^2}{2} \right]_0^L \left[ \dfrac{z^2}{2} \right]_0^L = \dfrac{1}{L^3} \dfrac{L^2}{2} \dfrac{L^2}{2} \dfrac{L^2}{2} = \dfrac{L^3}{8}$$

**53.** (a) The triple integral will attain its maximum when the integrand $1 - x^2 - 2y^2 - 3z^2$ is positive in the region $E$ and negative

everywhere else. For if $E$ contains some region $F$ where the integrand is negative, the integral could be increased by

excluding $F$ from $E$, and if $E$ fails to contain some part $G$ of the region where the integrand is positive, the integral could be increased by including $G$ in $E$. So we require that $x^2 + 2y^2 + 3z^2 \le 1$. This describes the region bounded by the ellipsoid $x^2 + 2y^2 + 3z^2 = 1$.

(b) The maximum value of $\iiint_E (1 - x^2 - 2y^2 - 3z^2)\, dV$ occurs when $E$ is the solid region bounded by the ellipsoid $x^2 + 2y^2 + 3z^2 = 1$. The projection of $E$ on the $xy$-plane is the planar region bounded by the ellipse $x^2 + 2y^2 = 1$, so

$$E = \left\{ (x, y, z) \mid -1 \le x \le 1, -\sqrt{\tfrac{1}{2}(1 - x^2)} \le y \le \sqrt{\tfrac{1}{2}(1 - x^2)}, -\sqrt{\tfrac{1}{3}(1 - x^2 - 2y^2)} \le z \le \sqrt{\tfrac{1}{3}(1 - x^2 - 2y^2)} \right\}$$

and

$$\iiint_E (1 - x^2 - 2y^2 - 3z^2)\, dV = \int_{-1}^{1} \int_{-\sqrt{\frac{1}{2}(1-x^2)}}^{\sqrt{\frac{1}{2}(1-x^2)}} \int_{-\sqrt{\frac{1}{3}(1-x^2-2y^2)}}^{\sqrt{\frac{1}{3}(1-x^2-2y^2)}} (1 - x^2 - 2y^2 - 3z^2)\, dz\, dy\, dx = \frac{4\sqrt{6}}{45}\pi$$

using a CAS.

## 12.8    Triple Integrals in Cylindrical and Spherical Coordinates

**1.**

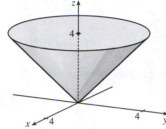

The region of integration is given in cylindrical coordinates by

$E = \{(r, \theta, z) \mid 0 \le \theta \le 2\pi, 0 \le r \le 4, r \le z \le 4\}$. This represents the solid region bounded below by the cone $z = r$ and above by the horizontal plane $z = 4$.

$$\int_0^4 \int_0^{2\pi} \int_r^4 r\, dz\, d\theta\, dr = \int_0^4 \int_0^{2\pi} [rz]_{z=r}^{z=4}\, d\theta\, dr = \int_0^4 \int_0^{2\pi} r(4 - r)\, d\theta\, dr$$
$$= \int_0^4 (4r - r^2)\, dr \int_0^{2\pi} d\theta = \left[2r^2 - \tfrac{1}{3}r^3\right]_0^4 \left[\theta\right]_0^{2\pi}$$
$$= \left(32 - \tfrac{64}{3}\right)(2\pi) = \frac{64\pi}{3}$$

**3.**

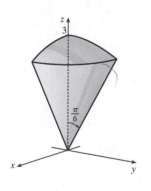

The region of integration is given in spherical coordinates by

$E = \{(\rho, \theta, \phi) \mid 0 \le \rho \le 3, 0 \le \theta \le \pi/2, 0 \le \phi \le \pi/6\}$. This represents the solid region in the first octant bounded above by the sphere $\rho = 3$ and below by the cone $\phi = \pi/6$.

$$\int_0^{\pi/6} \int_0^{\pi/2} \int_0^3 \rho^2 \sin\phi\, d\rho\, d\theta\, d\phi = \int_0^{\pi/6} \sin\phi\, d\phi \int_0^{\pi/2} d\theta \int_0^3 \rho^2\, d\rho$$
$$= \left[-\cos\phi\right]_0^{\pi/6} \left[\theta\right]_0^{\pi/2} \left[\tfrac{1}{3}\rho^3\right]_0^3$$
$$= \left(1 - \frac{\sqrt{3}}{2}\right)\left(\frac{\pi}{2}\right)(9) = \frac{9\pi}{4}\left(2 - \sqrt{3}\right)$$

**5.** The solid $E$ is most conveniently described if we use cylindrical coordinates:

$E = \{(r, \theta, z) \mid 0 \le \theta \le \tfrac{\pi}{2}, 0 \le r \le 3, 0 \le z \le 2\}$. Then

$\iiint_E f(x, y, z)\, dV = \int_0^{\pi/2} \int_0^3 \int_0^2 f(r\cos\theta, r\sin\theta, z)\, r\, dz\, dr\, d\theta$.

**7.** In cylindrical coordinates, $E$ is given by $\{(r, \theta, z) \mid 0 \le \theta \le 2\pi, 0 \le r \le 4, -5 \le z \le 4\}$. So

$$\iiint_E \sqrt{x^2 + y^2}\, dV = \int_0^{2\pi} \int_0^4 \int_{-5}^4 \sqrt{r^2}\, r\, dz\, dr\, d\theta = \int_0^{2\pi} d\theta \int_0^4 r^2\, dr \int_{-5}^4 dz$$
$$= \left[\theta\right]_0^{2\pi} \left[\tfrac{1}{3}r^3\right]_0^4 \left[z\right]_{-5}^4 = (2\pi)\left(\tfrac{64}{3}\right)(9) = 384\pi$$

9. In cylindrical coordinates $E$ is bounded by the paraboloid $z = 1 + r^2$, the cylinder $r^2 = 5$ or $r = \sqrt{5}$, and the $xy$-plane, so $E$ is given by $\{(r, \theta, z) \mid 0 \le \theta \le 2\pi, 0 \le r \le \sqrt{5}, 0 \le z \le 1 + r^2\}$. Thus

$$\iiint_E e^z \, dV = \int_0^{2\pi} \int_0^{\sqrt{5}} \int_0^{1+r^2} e^z \, r \, dz \, dr \, d\theta = \int_0^{2\pi} \int_0^{\sqrt{5}} r \left[e^z\right]_{z=0}^{z=1+r^2} dr \, d\theta = \int_0^{2\pi} \int_0^{\sqrt{5}} r(e^{1+r^2} - 1) \, dr \, d\theta$$

$$= \int_0^{2\pi} d\theta \int_0^{\sqrt{5}} \left(re^{1+r^2} - r\right) dr = 2\pi \left[\tfrac{1}{2}e^{1+r^2} - \tfrac{1}{2}r^2\right]_0^{\sqrt{5}} = \pi(e^6 - e - 5)$$

11. In cylindrical coordinates, $E$ is bounded by the cylinder $r = 1$, the plane $z = 0$, and the cone $z = 2r$. So $E = \{(r, \theta, z) \mid 0 \le \theta \le 2\pi, 0 \le r \le 1, 0 \le z \le 2r\}$ and

$$\iiint_E x^2 \, dV = \int_0^{2\pi} \int_0^1 \int_0^{2r} r^2 \cos^2 \theta \, r \, dz \, dr \, d\theta = \int_0^{2\pi} \int_0^1 \left[r^3 \cos^2 \theta \, z\right]_{z=0}^{z=2r} dr \, d\theta = \int_0^{2\pi} \int_0^1 2r^4 \cos^2 \theta \, dr \, d\theta$$

$$= \int_0^{2\pi} \left[\tfrac{2}{5} r^5 \cos^2 \theta\right]_{r=0}^{r=1} d\theta = \tfrac{2}{5} \int_0^{2\pi} \cos^2 \theta \, d\theta = \tfrac{2}{5} \int_0^{2\pi} \tfrac{1}{2}(1 + \cos 2\theta) \, d\theta = \tfrac{1}{5}\left[\theta + \tfrac{1}{2} \sin 2\theta\right]_0^{2\pi} = \tfrac{2\pi}{5}$$

13. (a) The paraboloids intersect when $x^2 + y^2 = 36 - 3x^2 - 3y^2 \Rightarrow x^2 + y^2 = 9$, so the region of integration is $D = \{(x, y) \mid x^2 + y^2 \le 9\}$. Then, in cylindrical coordinates,

$E = \{(r, \theta, z) \mid r^2 \le z \le 36 - 3r^2, 0 \le r \le 3, 0 \le \theta \le 2\pi\}$ and

$$V = \int_0^{2\pi} \int_0^3 \int_{r^2}^{36 - 3r^2} r \, dz \, dr \, d\theta = \int_0^{2\pi} \int_0^3 \left(36r - 4r^3\right) dr \, d\theta = \int_0^{2\pi} \left[18r^2 - r^4\right]_{r=0}^{r=3} d\theta = \int_0^{2\pi} 81 \, d\theta = 162\pi.$$

(b) For constant density $K$, $m = KV = 162\pi K$ from part (a). Since the region is homogeneous and symmetric, $M_{yz} = M_{xz} = 0$ and

$$M_{xy} = \int_0^{2\pi} \int_0^3 \int_{r^2}^{36 - 3r^2} (zK) \, r \, dz \, dr \, d\theta = K \int_0^{2\pi} \int_0^3 r\left[\tfrac{1}{2}z^2\right]_{z=r^2}^{z=36-3r^2} dr \, d\theta$$

$$= \tfrac{K}{2} \int_0^{2\pi} \int_0^3 r((36 - 3r^2)^2 - r^4) \, dr \, d\theta = \tfrac{K}{2} \int_0^{2\pi} d\theta \int_0^3 (8r^5 - 216r^3 + 1296r) \, dr$$

$$= \tfrac{K}{2} (2\pi) \left[\tfrac{8}{6}r^6 - \tfrac{216}{4}r^4 + \tfrac{1296}{2}r^2\right]_0^3 = \pi K(2430) = 2430\pi K$$

Thus $(\overline{x}, \overline{y}, \overline{z}) = \left(\dfrac{M_{yz}}{m}, \dfrac{M_{xz}}{m}, \dfrac{M_{xy}}{m}\right) = \left(0, 0, \dfrac{2430\pi K}{162\pi K}\right) = (0, 0, 15).$

15. The paraboloid $z = 4x^2 + 4y^2$ intersects the plane $z = a$ when $a = 4x^2 + 4y^2$ or $x^2 + y^2 = \tfrac{1}{4}a$. So, in cylindrical coordinates, $E = \{(r, \theta, z) \mid 0 \le r \le \tfrac{1}{2}\sqrt{a}, 0 \le \theta \le 2\pi, 4r^2 \le z \le a\}$. Thus

$$m = \int_0^{2\pi} \int_0^{\sqrt{a}/2} \int_{4r^2}^a Kr \, dz \, dr \, d\theta = K \int_0^{2\pi} \int_0^{\sqrt{a}/2} (ar - 4r^3) \, dr \, d\theta$$

$$= K \int_0^{2\pi} \left[\tfrac{1}{2}ar^2 - r^4\right]_{r=0}^{r=\sqrt{a}/2} d\theta = K \int_0^{2\pi} \tfrac{1}{16}a^2 \, d\theta = \tfrac{1}{8}a^2 \pi K$$

Since the region is homogeneous and symmetric, $M_{yz} = M_{xz} = 0$ and

$$M_{xy} = \int_0^{2\pi} \int_0^{\sqrt{a}/2} \int_{4r^2}^a Krz \, dz \, dr \, d\theta = K \int_0^{2\pi} \int_0^{\sqrt{a}/2} \left(\tfrac{1}{2}a^2 r - 8r^5\right) dr \, d\theta$$

$$= K \int_0^{2\pi} \left[\tfrac{1}{4}a^2 r^2 - \tfrac{4}{3}r^6\right]_{r=0}^{r=\sqrt{a}/2} d\theta = K \int_0^{2\pi} \tfrac{1}{24}a^3 \, d\theta = \tfrac{1}{12}a^3 \pi K$$

Hence $(\overline{x}, \overline{y}, \overline{z}) = \left(0, 0, \tfrac{2}{3}a\right)$.

**17.** In spherical coordinates, $B$ is represented by $\{(\rho, \theta, \phi) \mid 0 \leq \rho \leq 5, 0 \leq \theta \leq 2\pi, 0 \leq \phi \leq \pi\}$. Thus

$$\iiint_B (x^2 + y^2 + z^2)^2 \, dV = \int_0^\pi \int_0^{2\pi} \int_0^5 (\rho^2)^2 \rho^2 \sin\phi \, d\rho \, d\theta \, d\phi = \int_0^\pi \sin\phi \, d\phi \int_0^{2\pi} d\theta \int_0^5 \rho^6 \, d\rho$$

$$= \left[-\cos\phi\right]_0^\pi \left[\theta\right]_0^{2\pi} \left[\tfrac{1}{7}\rho^7\right]_0^5 = (2)(2\pi)\left(\tfrac{78,125}{7}\right)$$

$$= \tfrac{312,500}{7}\pi \approx 140{,}249.7$$

**19.** In spherical coordinates, $E$ is represented by $\left\{(\rho, \theta, \phi) \mid 1 \leq \rho \leq 2, 0 \leq \theta \leq \tfrac{\pi}{2}, 0 \leq \phi \leq \tfrac{\pi}{2}\right\}$. Thus

$$\iiint_E z \, dV = \int_0^{\pi/2} \int_0^{\pi/2} \int_1^2 (\rho\cos\phi)\,\rho^2 \sin\phi \, d\rho \, d\theta \, d\phi = \int_0^{\pi/2} \cos\phi\sin\phi \, d\phi \int_0^{\pi/2} d\theta \int_1^2 \rho^3 \, d\rho$$

$$= \left[\tfrac{1}{2}\sin^2\phi\right]_0^{\pi/2} \left[\theta\right]_0^{\pi/2} \left[\tfrac{1}{4}\rho^4\right]_1^2 = \left(\tfrac{1}{2}\right)\left(\tfrac{\pi}{2}\right)\left(\tfrac{15}{4}\right) = \tfrac{15\pi}{16}$$

**21.** $\iiint_E x^2 \, dV = \int_0^\pi \int_0^\pi \int_3^4 (\rho\sin\phi\cos\theta)^2 \, \rho^2 \sin\phi \, d\rho \, d\phi \, d\theta = \int_0^\pi \cos^2\theta \, d\theta \int_0^\pi \sin^3\phi \, d\phi \int_3^4 \rho^4 \, d\rho$

$$= \left[\tfrac{1}{2}\theta + \tfrac{1}{4}\sin 2\theta\right]_0^\pi \left[-\tfrac{1}{3}(2 + \sin^2\phi)\cos\phi\right]_0^\pi \left[\tfrac{1}{5}\rho^5\right]_3^4 = \left(\tfrac{\pi}{2}\right)\left(\tfrac{2}{3} + \tfrac{2}{3}\right)\tfrac{1}{5}(4^5 - 3^5) = \tfrac{1562}{15}\pi$$

**23.** The solid region is given by $E = \left\{(\rho, \theta, \phi) \mid 0 \leq \rho \leq a, 0 \leq \theta \leq 2\pi, \tfrac{\pi}{6} \leq \phi \leq \tfrac{\pi}{3}\right\}$ and its volume is

$$V = \iiint_E dV = \int_{\pi/6}^{\pi/3} \int_0^{2\pi} \int_0^a \rho^2 \sin\phi \, d\rho \, d\theta \, d\phi = \int_{\pi/6}^{\pi/3} \sin\phi \, d\phi \int_0^{2\pi} d\theta \int_0^a \rho^2 \, d\rho$$

$$= \left[-\cos\phi\right]_{\pi/6}^{\pi/3} \left[\theta\right]_0^{2\pi} \left[\tfrac{1}{3}\rho^3\right]_0^a = \left(-\tfrac{1}{2} + \tfrac{\sqrt{3}}{2}\right)(2\pi)\left(\tfrac{1}{3}a^3\right) = \tfrac{\sqrt{3}-1}{3}\pi a^3$$

**25.** (a) Since $\rho = 4\cos\phi$ implies $\rho^2 = 4\rho\cos\phi$, the equation is that of a sphere of radius 2 with center at $(0, 0, 2)$. Thus

$$V = \int_0^{2\pi} \int_0^{\pi/3} \int_0^{4\cos\phi} \rho^2 \sin\phi \, d\rho \, d\phi \, d\theta = \int_0^{2\pi} \int_0^{\pi/3} \left[\tfrac{1}{3}\rho^3\right]_{\rho=0}^{\rho=4\cos\phi} \sin\phi \, d\phi \, d\theta = \int_0^{2\pi} \int_0^{\pi/3} \left(\tfrac{64}{3}\cos^3\phi\right)\sin\phi \, d\phi \, d\theta$$

$$= \int_0^{2\pi} \left[-\tfrac{16}{3}\cos^4\phi\right]_{\phi=0}^{\phi=\pi/3} d\theta = \int_0^{2\pi} -\tfrac{16}{3}\left(\tfrac{1}{16} - 1\right) d\theta = 5\theta\Big]_0^{2\pi} = 10\pi$$

(b) By the symmetry of the problem $M_{yz} = M_{xz} = 0$. Then

$$M_{xy} = \int_0^{2\pi} \int_0^{\pi/3} \int_0^{4\cos\phi} \rho^3 \cos\phi\sin\phi \, d\rho \, d\phi \, d\theta = \int_0^{2\pi} \int_0^{\pi/3} \cos\phi\sin\phi\left(64\cos^4\phi\right) d\phi \, d\theta$$

$$= \int_0^{2\pi} 64\left[-\tfrac{1}{6}\cos^6\phi\right]_{\phi=0}^{\phi=\pi/3} d\theta = \int_0^{2\pi} \tfrac{21}{2} \, d\theta = 21\pi$$

Hence $(\overline{x}, \overline{y}, \overline{z}) = (0, 0, 2.1)$.

**27.** By the symmetry of the region, $M_{xy} = 0$ and $M_{yz} = 0$. Assuming constant density $K$,

$$m = \iiint_E K \, dV = K \int_0^\pi \int_0^\pi \int_3^4 \rho^2 \sin\phi \, d\rho \, d\phi \, d\theta = K \int_0^\pi d\theta \int_0^\pi \sin\phi \, d\phi \int_3^4 \rho^2 \, d\rho$$

$$= K\pi\left[-\cos\phi\right]_0^\pi \left[\tfrac{1}{3}\rho^3\right]_3^4 = 2K\pi \cdot \tfrac{37}{3} = \tfrac{74}{3}\pi K$$

and $M_{xz} = \iiint_E y K \, dV = K \int_0^\pi \int_0^\pi \int_3^4 (\rho\sin\phi\sin\theta)\,\rho^2 \sin\phi \, d\rho \, d\phi \, d\theta = K \int_0^\pi \sin\theta \, d\theta \int_0^\pi \sin^2\phi \, d\phi \int_3^4 \rho^3 \, d\rho$

$$= K \left[-\cos\theta\right]_0^\pi \left[\tfrac{1}{2}\phi - \tfrac{1}{4}\sin 2\phi\right]_0^\pi \left[\tfrac{1}{4}\rho^4\right]_3^4 = K(2)\left(\tfrac{\pi}{2}\right)\tfrac{1}{4}(256 - 81) = \tfrac{175}{4}\pi K$$

Thus the centroid is $(\overline{x}, \overline{y}, \overline{z}) = \left(\dfrac{M_{yz}}{m}, \dfrac{M_{xz}}{m}, \dfrac{M_{xy}}{m}\right) = \left(0, \dfrac{175\pi K/4}{74\pi K/3}, 0\right) = \left(0, \tfrac{525}{296}, 0\right)$.

**29.** (a) The density function is $\rho(x, y, z) = K$, a constant, and by the symmetry of the problem $M_{xz} = M_{yz} = 0$. Then

$M_{xy} = \int_0^{2\pi} \int_0^{\pi/2} \int_0^a K\rho^3 \sin\phi \cos\phi \, d\rho \, d\phi \, d\theta = \frac{1}{2}\pi K a^4 \int_0^{\pi/2} \sin\phi \cos\phi \, d\phi = \frac{1}{8}\pi K a^4$. But the mass is $K$(volume of

the hemisphere) $= \frac{2}{3}\pi K a^3$, so the centroid is $\left(0, 0, \frac{3}{8}a\right)$.

(b) Place the center of the base at $(0, 0, 0)$; the density function is $\rho(x, y, z) = K$. By symmetry, the moments of inertia about

any two such diameters will be equal, so we just need to find $I_x$:

$$I_x = \int_0^{2\pi} \int_0^{\pi/2} \int_0^a (K\rho^2 \sin\phi) \, \rho^2 \left(\sin^2\phi \sin^2\theta + \cos^2\phi\right) d\rho \, d\phi \, d\theta$$

$$= K \int_0^{2\pi} \int_0^{\pi/2} (\sin^3\phi \sin^2\theta + \sin\phi \cos^2\phi)\left(\frac{1}{5}a^5\right) d\phi \, d\theta$$

$$= \frac{1}{5}Ka^5 \int_0^{2\pi} \left[\sin^2\theta \left(-\cos\phi + \frac{1}{3}\cos^3\phi\right) + \left(-\frac{1}{3}\cos^3\phi\right)\right]_{\phi=0}^{\phi=\pi/2} d\theta = \frac{1}{5}Ka^5 \int_0^{2\pi} \left[\frac{2}{3}\sin^2\theta + \frac{1}{3}\right] d\theta$$

$$= \frac{1}{5}Ka^5 \left[\frac{2}{3}\left(\frac{1}{2}\theta - \frac{1}{4}\sin 2\theta\right) + \frac{1}{3}\theta\right]_0^{2\pi} = \frac{1}{5}Ka^5 \left[\frac{2}{3}(\pi - 0) + \frac{1}{3}(2\pi - 0)\right] = \frac{4}{15}Ka^5\pi$$

**31.** In spherical coordinates $z = \sqrt{x^2 + y^2}$ becomes $\cos\phi = \sin\phi$ or $\phi = \frac{\pi}{4}$. Then

$V = \int_0^{2\pi} \int_0^{\pi/4} \int_0^1 \rho^2 \sin\phi \, d\rho \, d\phi \, d\theta = \int_0^{2\pi} d\theta \int_0^{\pi/4} \sin\phi \, d\phi \int_0^1 \rho^2 \, d\rho = 2\pi \left(-\frac{\sqrt{2}}{2} + 1\right)\left(\frac{1}{3}\right) = \frac{1}{3}\pi(2 - \sqrt{2})$,

$M_{xy} = \int_0^{2\pi} \int_0^{\pi/4} \int_0^1 \rho^3 \sin\phi \cos\phi \, d\rho \, d\phi \, d\theta = 2\pi\left[-\frac{1}{4}\cos 2\phi\right]_0^{\pi/4} \left(\frac{1}{4}\right) = \frac{\pi}{8}$ and by symmetry $M_{yz} = M_{xz} = 0$.

Hence $(\overline{x}, \overline{y}, \overline{z}) = \left(0, 0, \dfrac{3}{8(2 - \sqrt{2})}\right)$.

**33.** In cylindrical coordinates the paraboloid is given by $z = r^2$ and the plane by $z = 2r\sin\theta$ and they intersect in the circle

$r = 2\sin\theta$. Then $\iiint_E z \, dV = \int_0^\pi \int_0^{2\sin\theta} \int_{r^2}^{2r\sin\theta} rz \, dz \, dr \, d\theta = \frac{5\pi}{6}$  [using a CAS].

**35.** The region of integration is the region above the cone $z = \sqrt{x^2 + y^2}$, or $z = r$, and below the plane $z = 2$. Also, we have

$-2 \leq y \leq 2$ with $-\sqrt{4 - y^2} \leq x \leq \sqrt{4 - y^2}$ which describes a circle of radius 2 in the $xy$-plane centered at $(0, 0)$. Thus,

$$\int_{-2}^2 \int_{-\sqrt{4-y^2}}^{\sqrt{4-y^2}} \int_{\sqrt{x^2+y^2}}^2 xz \, dz \, dx \, dy = \int_0^{2\pi} \int_0^2 \int_r^2 (r\cos\theta) \, zr \, dz \, dr \, d\theta = \int_0^{2\pi} \int_0^2 \int_r^2 r^2 (\cos\theta) \, z \, dz \, dr \, d\theta$$

$$= \int_0^{2\pi} \int_0^2 r^2 (\cos\theta) \left[\frac{1}{2}z^2\right]_{z=r}^{z=2} dr \, d\theta = \frac{1}{2}\int_0^{2\pi} \int_0^2 r^2 (\cos\theta) \left(4 - r^2\right) dr \, d\theta$$

$$= \frac{1}{2}\int_0^{2\pi} \cos\theta \, d\theta \int_0^2 \left(4r^2 - r^4\right) dr = \frac{1}{2}\left[\sin\theta\right]_0^{2\pi} \left[\frac{4}{3}r^3 - \frac{1}{5}r^5\right]_0^2 = 0$$

**37.** The region $E$ of integration is the region above the cone $z = \sqrt{x^2 + y^2}$ and below the sphere $x^2 + y^2 + z^2 = 2$ in the first

octant. Because $E$ is in the first octant we have $0 \leq \theta \leq \frac{\pi}{2}$. The cone has equation $\phi = \frac{\pi}{4}$ (as in Example 4), so $0 \leq \phi \leq \frac{\pi}{4}$,

and $0 \leq \rho \leq \sqrt{2}$. So the integral becomes

$\int_0^{\pi/4} \int_0^{\pi/2} \int_0^{\sqrt{2}} (\rho\sin\phi\cos\theta)(\rho\sin\phi\sin\theta)\,\rho^2 \sin\phi \, d\rho \, d\theta \, d\phi$

$$= \int_0^{\pi/4} \sin^3\phi \, d\phi \int_0^{\pi/2} \sin\theta\cos\theta \, d\theta \int_0^{\sqrt{2}} \rho^4 \, d\rho = \left(\int_0^{\pi/4}(1 - \cos^2\phi)\sin\phi \, d\phi\right)\left[\frac{1}{2}\sin^2\theta\right]_0^{\pi/2} \left[\frac{1}{5}\rho^5\right]_0^{\sqrt{2}}$$

$$= \left[\frac{1}{3}\cos^3\phi - \cos\phi\right]_0^{\pi/4} \cdot \frac{1}{2} \cdot \frac{1}{5}\left(\sqrt{2}\right)^5 = \left[\frac{\sqrt{2}}{12} - \frac{\sqrt{2}}{2} - \left(\frac{1}{3} - 1\right)\right] \cdot \frac{2\sqrt{2}}{5} = \frac{4\sqrt{2}-5}{15}$$

**39.** If $E$ is the solid enclosed by the surface $\rho = 1 + \frac{1}{5}\sin 6\theta \sin 5\phi$, it can be described in spherical coordinates as

$E = \left\{(\rho, \theta, \phi) \mid 0 \leq \rho \leq 1 + \frac{1}{5}\sin 6\theta \sin 5\phi, 0 \leq \theta \leq 2\pi, 0 \leq \phi \leq \pi\right\}$. Its volume is given by

$V(E) = \iiint_E dV = \int_0^\pi \int_0^{2\pi} \int_0^{1+(\sin 6\theta \sin 5\phi)/5} \rho^2 \sin\phi \, d\rho \, d\theta \, d\phi = \frac{136\pi}{99}$  [using a CAS].

**41. (a)** The mountain comprises a solid conical region $C$. The work done in lifting a small volume of material $\Delta V$ with density

$g(P)$ to a height $h(P)$ above sea level is $h(P)g(P)\,\Delta V$. Summing over the whole mountain we get

$W = \iiint_C h(P)g(P)\, dV$.

**(b)** Here $C$ is a solid right circular cone with radius $R = 62{,}000$ ft, height $H = 12{,}400$ ft,

and density $g(P) = 200$ lb/ft$^3$ at all points $P$ in $C$. We use cylindrical coordinates:

$W = \int_0^{2\pi} \int_0^H \int_0^{R(1-z/H)} z \cdot 200r\, dr\, dz\, d\theta = 2\pi \int_0^H 200z \left[\frac{1}{2}r^2\right]_{r=0}^{r=R(1-z/H)} dz$

$= 400\pi \int_0^H z \frac{R^2}{2}\left(1 - \frac{z}{H}\right)^2 dz = 200\pi R^2 \int_0^H \left(z - \frac{2z^2}{H} + \frac{z^3}{H^2}\right) dz$

$= 200\pi R^2 \left[\frac{z^2}{2} - \frac{2z^3}{3H} + \frac{z^4}{4H^2}\right]_0^H = 200\pi R^2\left(\frac{H^2}{2} - \frac{2H^2}{3} + \frac{H^2}{4}\right)$

$= \frac{50}{3}\pi R^2 H^2 = \frac{50}{3}\pi(62{,}000)^2(12{,}400)^2 \approx 3.1 \times 10^{19}$ ft-lb

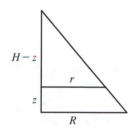

$\dfrac{r}{R} = \dfrac{H-z}{H} = 1 - \dfrac{z}{H}$

## 12.9   Change of Variables in Multiple Integrals

**1.** $x = 5u - v, \; y = u + 3v$.

The Jacobian is $\dfrac{\partial(x,y)}{\partial(u,v)} = \begin{vmatrix} \partial x/\partial u & \partial x/\partial v \\ \partial y/\partial u & \partial y/\partial v \end{vmatrix} = \begin{vmatrix} 5 & -1 \\ 1 & 3 \end{vmatrix} = 5(3) - (-1)(1) = 16.$

**3.** $x = e^{-r}\sin\theta, \; y = e^r \cos\theta$.

$\dfrac{\partial(x,y)}{\partial(r,\theta)} = \begin{vmatrix} \partial x/\partial r & \partial x/\partial \theta \\ \partial y/\partial r & \partial y/\partial \theta \end{vmatrix} = \begin{vmatrix} -e^{-r}\sin\theta & e^{-r}\cos\theta \\ e^r \cos\theta & -e^r \sin\theta \end{vmatrix} = e^{-r}e^r \sin^2\theta - e^{-r}e^r \cos^2\theta = \sin^2\theta - \cos^2\theta \text{ or } -\cos 2\theta$

**5.** $x = u/v, \; y = v/w, \; z = w/u$.

$\dfrac{\partial(x,y,z)}{\partial(u,v,w)} = \begin{vmatrix} \partial x/\partial u & \partial x/\partial v & \partial x/\partial w \\ \partial y/\partial u & \partial y/\partial v & \partial y/\partial w \\ \partial z/\partial u & \partial z/\partial v & \partial z/\partial w \end{vmatrix} = \begin{vmatrix} 1/v & -u/v^2 & 0 \\ 0 & 1/w & -v/w^2 \\ -w/u^2 & 0 & 1/u \end{vmatrix}$

$= \dfrac{1}{v}\begin{vmatrix} 1/w & -v/w^2 \\ 0 & 1/u \end{vmatrix} - \left(-\dfrac{u}{v^2}\right)\begin{vmatrix} 0 & -v/w^2 \\ -w/u^2 & 1/u \end{vmatrix} + 0\begin{vmatrix} 0 & 1/w \\ -w/u^2 & 0 \end{vmatrix}$

$= \dfrac{1}{v}\left(\dfrac{1}{uw} - 0\right) + \dfrac{u}{v^2}\left(0 - \dfrac{v}{u^2 w}\right) + 0 = \dfrac{1}{uvw} - \dfrac{1}{uvw} = 0$

**7.** The transformation maps the boundary of $S$ to the boundary of the image $R$, so we first look at side $S_1$ in the $uv$-plane. $S_1$ is

described by $v = 0, 0 \le u \le 3$, so $x = 2u + 3v = 2u$ and $y = u - v = u$. Eliminating $u$, we have $x = 2y, 0 \le x \le 6$. $S_2$ is

the line segment $u = 3, 0 \le v \le 2$, so $x = 6 + 3v$ and $y = 3 - v$. Then $v = 3 - y \Rightarrow x = 6 + 3(3 - y) = 15 - 3y$,

$6 \le x \le 12$. $S_3$ is the line segment $v = 2, 0 \le u \le 3$, so $x = 2u + 6$ and $y = u - 2$, giving $u = y + 2 \Rightarrow x = 2y + 10$,

$6 \le x \le 12$. Finally, $S_4$ is the segment $u = 0, 0 \le v \le 2$, so $x = 3v$ and $y = -v \Rightarrow x = -3y, 0 \le x \le 6$. The image of

set $S$ is the region $R$ shown in the $xy$-plane, a parallelogram bounded by these four segments.

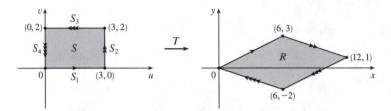

**9.** $S_1$ is the line segment $u = v, 0 \le u \le 1$, so $y = v = u$ and $x = u^2 = y^2$. Since $0 \le u \le 1$, the image is the portion of the

parabola $x = y^2, 0 \le y \le 1$. $S_2$ is the segment $v = 1, 0 \le u \le 1$, thus $y = v = 1$ and $x = u^2$, so $0 \le x \le 1$. The image is

the line segment $y = 1, 0 \le x \le 1$. $S_3$ is the segment $u = 0, 0 \le v \le 1$, so $x = u^2 = 0$ and $y = v \Rightarrow 0 \le y \le 1$. The

image is the segment $x = 0, 0 \le y \le 1$. Thus, the image of $S$ is the region $R$ in the first quadrant bounded by the parabola

$x = y^2$, the $y$-axis, and the line $y = 1$.

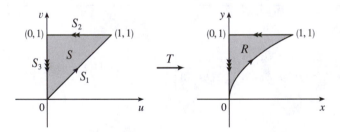

**11.** $R$ is a parallelogram enclosed by the parallel lines $y = 2x - 1, y = 2x + 1$ and the parallel lines $y = 1 - x, y = 3 - x$. The

first pair of equations can be written as $y - 2x = -1, y - 2x = 1$. If we let $u = y - 2x$ then these lines are mapped to the

vertical lines $u = -1, u = 1$ in the $uv$-plane. Similarly, the second pair of equations can be written as $x + y = 1, x + y = 3$,

and setting $v = x + y$ maps these lines to the horizontal lines $v = 1, v = 3$ in the $uv$-plane. Boundary curves are mapped to

boundary curves under a transformation, so here the equations $u = y - 2x, v = x + y$ define a transformation $T^{-1}$ that

maps $R$ in the $xy$-plane to the square $S$ enclosed by the lines $u = -1, u = 1, v = 1, v = 3$ in the $uv$-plane. To find the

transformation $T$ that maps $S$ to $R$ we solve $u = y - 2x, v = x + y$ for $x, y$: Subtracting the first equation from the second

gives $v - u = 3x \Rightarrow x = \frac{1}{3}(v - u)$ and adding twice the second equation to the first gives $u + 2v = 3y \Rightarrow$

$y = \frac{1}{3}(u + 2v)$. Thus one possible transformation $T$ (there are many) is given by $x = \frac{1}{3}(v - u), y = \frac{1}{3}(u + 2v)$.

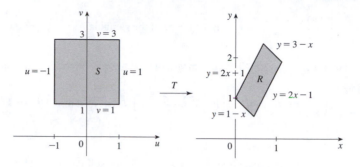

**13.** $R$ is a portion of an annular region (see the figure) that is easily described in polar coordinates as

$R = \{(r,\theta) \mid 1 \le r \le \sqrt{2}, 0 \le \theta \le \pi/2\}$. If we converted a double integral over $R$ to polar coordinates the resulting region

of integration is a rectangle (in the $r\theta$-plane), so we can create a transformation $T$ here by letting $u$ play the role of $r$ and $v$ the

role of $\theta$. Thus $T$ is defined by $x = u\cos v$, $y = u\sin v$ and $T$ maps the rectangle $S = \{(u,v) \mid 1 \le u \le \sqrt{2}, 0 \le v \le \pi/2\}$

in the $uv$-plane to $R$ in the $xy$-plane.

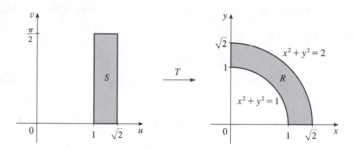

**15.** $\dfrac{\partial(x,y)}{\partial(u,v)} = \begin{vmatrix} 2 & 1 \\ 1 & 2 \end{vmatrix} = 3$ and $x - 3y = (2u + v) - 3(u + 2v) = -u - 5v$. To find the region $S$ in the $uv$-plane that

corresponds to $R$ we first find the corresponding boundary under the given transformation. The line through $(0,0)$ and $(2,1)$ is

$y = \frac{1}{2}x$ which is the image of $u + 2v = \frac{1}{2}(2u + v) \;\Rightarrow\; v = 0$; the line through $(2,1)$ and $(1,2)$ is $x + y = 3$ which is the

image of $(2u + v) + (u + 2v) = 3 \;\Rightarrow\; u + v = 1$; the line through $(0,0)$ and $(1,2)$ is $y = 2x$ which is the image of

$u + 2v = 2(2u + v) \;\Rightarrow\; u = 0$. Thus $S$ is the triangle $0 \le v \le 1 - u, 0 \le u \le 1$ in the $uv$-plane and

$$\iint_R (x - 3y)\, dA = \int_0^1 \int_0^{1-u} (-u - 5v)\, |3|\, dv\, du = -3\int_0^1 \left[ uv + \tfrac{5}{2}v^2 \right]_{v=0}^{v=1-u} du$$

$$= -3\int_0^1 \left( u - u^2 + \tfrac{5}{2}(1 - u)^2 \right) du = -3\left[ \tfrac{1}{2}u^2 - \tfrac{1}{3}u^3 - \tfrac{5}{6}(1 - u)^3 \right]_0^1 = -3\left( \tfrac{1}{2} - \tfrac{1}{3} + \tfrac{5}{6} \right) = -3$$

**17.** $\dfrac{\partial(x,y)}{\partial(u,v)} = \begin{vmatrix} 2 & 0 \\ 0 & 3 \end{vmatrix} = 6$, $x^2 = 4u^2$ and the planar ellipse $9x^2 + 4y^2 \le 36$ is the image of the disk $u^2 + v^2 \le 1$. Thus

$$\iint_R x^2\, dA = \iint_{u^2+v^2\le 1} (4u^2)(6)\, du\, dv = \int_0^{2\pi} \int_0^1 (24r^2 \cos^2 \theta)\, r\, dr\, d\theta = 24 \int_0^{2\pi} \cos^2 \theta\, d\theta \int_0^1 r^3\, dr$$

$$= 24\left[ \tfrac{1}{2}x + \tfrac{1}{4}\sin 2x \right]_0^{2\pi} \left[ \tfrac{1}{4}r^4 \right]_0^1 = 24(\pi)\left( \tfrac{1}{4} \right) = 6\pi$$

**19.** $\dfrac{\partial(x, y)}{\partial(u, v)} = \begin{vmatrix} 1/v & -u/v^2 \\ 0 & 1 \end{vmatrix} = \dfrac{1}{v}$, $xy = u$, $y = x$ is the image of the parabola $v^2 = u$, $y = 3x$ is the image of the parabola

$v^2 = 3u$, and the hyperbolas $xy = 1$, $xy = 3$ are the images of the lines $u = 1$ and $u = 3$ respectively. Thus

$$\iint_R xy\, dA = \int_1^3 \int_{\sqrt{u}}^{\sqrt{3u}} u\left(\frac{1}{v}\right) dv\, du = \int_1^3 u\left(\ln\sqrt{3u} - \ln\sqrt{u}\right) du = \int_1^3 u \ln\sqrt{3}\, du = 4\ln\sqrt{3} = 2\ln 3.$$

**21.** (a) $\dfrac{\partial(x, y, z)}{\partial(u, v, w)} = \begin{vmatrix} a & 0 & 0 \\ 0 & b & 0 \\ 0 & 0 & c \end{vmatrix} = abc$ and since $u = \dfrac{x}{a}$, $v = \dfrac{y}{b}$, $w = \dfrac{z}{c}$ the solid enclosed by the ellipsoid is the image of the

ball $u^2 + v^2 + w^2 \leq 1$. So

$$\iiint_E dV = \iiint_{u^2+v^2+w^2 \leq 1} abc\, du\, dv\, dw = (abc)(\text{volume of the ball}) = \tfrac{4}{3}\pi abc$$

(b) If we approximate the surface of the earth by the ellipsoid $\dfrac{x^2}{6378^2} + \dfrac{y^2}{6378^2} + \dfrac{z^2}{6356^2} = 1$, then we can estimate

the volume of the earth by finding the volume of the solid $E$ enclosed by the ellipsoid. From part (a), this is

$\iiint_E dV = \tfrac{4}{3}\pi(6378)(6378)(6356) \approx 1.083 \times 10^{12} \text{ km}^3$.

**23.** Letting $u = x - 2y$ and $v = 3x - y$, we have $x = \tfrac{1}{5}(2v - u)$ and $y = \tfrac{1}{5}(v - 3u)$. Then $\dfrac{\partial(x, y)}{\partial(u, v)} = \begin{vmatrix} -1/5 & 2/5 \\ -3/5 & 1/5 \end{vmatrix} = \dfrac{1}{5}$

and $R$ is the image of the rectangle enclosed by the lines $u = 0$, $u = 4$, $v = 1$, and $v = 8$. Thus

$$\iint_R \frac{x - 2y}{3x - y}\, dA = \int_0^4 \int_1^8 \frac{u}{v}\left|\frac{1}{5}\right| dv\, du = \frac{1}{5}\int_0^4 u\, du \int_1^8 \frac{1}{v}\, dv = \frac{1}{5}\left[\tfrac{1}{2}u^2\right]_0^4 \left[\ln|v|\right]_1^8 = \tfrac{8}{5}\ln 8.$$

**25.** Letting $u = y - x$, $v = y + x$, we have $y = \tfrac{1}{2}(u + v)$, $x = \tfrac{1}{2}(v - u)$. Then $\dfrac{\partial(x, y)}{\partial(u, v)} = \begin{vmatrix} -1/2 & 1/2 \\ 1/2 & 1/2 \end{vmatrix} = -\dfrac{1}{2}$ and $R$ is the

image of the trapezoidal region with vertices $(-1, 1)$, $(-2, 2)$, $(2, 2)$, and $(1, 1)$. Thus

$$\iint_R \cos\frac{y - x}{y + x}\, dA = \int_1^2 \int_{-v}^{v} \cos\frac{u}{v}\left|-\frac{1}{2}\right| du\, dv = \frac{1}{2}\int_1^2 \left[v\sin\frac{u}{v}\right]_{u=-v}^{u=v} dv = \frac{1}{2}\int_1^2 2v\sin(1)\, dv = \tfrac{3}{2}\sin 1$$

**27.** Let $u = x + y$ and $v = -x + y$. Then $u + v = 2y \Rightarrow y = \tfrac{1}{2}(u + v)$ and $u - v = 2x \Rightarrow x = \tfrac{1}{2}(u - v)$.

$\dfrac{\partial(x, y)}{\partial(u, v)} = \begin{vmatrix} 1/2 & -1/2 \\ 1/2 & 1/2 \end{vmatrix} = \dfrac{1}{2}$. Now $|u| = |x + y| \leq |x| + |y| \leq 1 \Rightarrow -1 \leq u \leq 1$, and

$|v| = |-x + y| \leq |x| + |y| \leq 1 \Rightarrow -1 \leq v \leq 1$. $R$ is the image of the square

region with vertices $(1, 1)$, $(1, -1)$, $(-1, -1)$, and $(-1, 1)$.

So $\iint_R e^{x+y}\, dA = \tfrac{1}{2}\int_{-1}^{1}\int_{-1}^{1} e^u\, du\, dv = \tfrac{1}{2}\left[e^u\right]_{-1}^{1}\left[v\right]_{-1}^{1} = e - e^{-1}$.

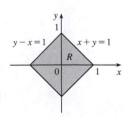

# 12   Review

## CONCEPT CHECK

**1.** (a) A double Riemann sum of $f$ is $\sum_{i=1}^{m} \sum_{j=1}^{n} f(x_{ij}^{*}, y_{ij}^{*}) \, \Delta A$, where $\Delta A$ is the area of each subrectangle and $(x_{ij}^{*}, y_{ij}^{*})$ is a

sample point in each subrectangle. If $f(x, y) \geq 0$, this sum represents an approximation to the volume of the solid that lies above the rectangle $R$ and below the graph of $f$.

(b) $\iint_R f(x, y) \, dA = \lim_{m, n \to \infty} \sum_{i=1}^{m} \sum_{j=1}^{n} f(x_{ij}^{*}, y_{ij}^{*}) \, \Delta A$

(c) If $f(x, y) \geq 0$, $\iint_R f(x, y) \, dA$ represents the volume of the solid that lies above the rectangle $R$ and below the surface $z = f(x, y)$. If $f$ takes on both positive and negative values, $\iint_R f(x, y) \, dA$ is the difference of the volume above $R$ but below the surface $z = f(x, y)$ and the volume below $R$ but above the surface $z = f(x, y)$.

(d) We usually evaluate $\iint_R f(x, y) \, dA$ as an iterated integral according to Fubini's Theorem (see Theorem 12.2.4).

(e) The Midpoint Rule for Double Integrals says that we approximate the double integral $\iint_R f(x, y) \, dA$ by the double

Riemann sum $\sum_{i=1}^{m} \sum_{j=1}^{n} f(\overline{x}_i, \overline{y}_j) \, \Delta A$ where the sample points $(\overline{x}_i, \overline{y}_j)$ are the centers of the subrectangles.

(f) $f_{\text{ave}} = \dfrac{1}{A(R)} \iint_R f(x, y) \, dA$ where $A(R)$ is the area of $R$.

**2.** (a) See (1) and (2) and the accompanying discussion in Section 12.3.

(b) See (3) and the accompanying discussion in Section 12.3.

(c) See (5) and the preceding discussion in Section 12.3.

(d) See (6)–(11) in Section 12.3.

**3.** We may want to change from rectangular to polar coordinates in a double integral if the region $R$ of integration is more easily

described in polar coordinates. To accomplish this, we use $\iint_R f(x, y) \, dA = \int_{\alpha}^{\beta} \int_{a}^{b} f(r \cos \theta, r \sin \theta) \, r \, dr \, d\theta$ where $R$ is given by $0 \leq a \leq r \leq b, \alpha \leq \theta \leq \beta$.

**4.** (a) $m = \iint_D \rho(x, y) \, dA$

(b) $M_x = \iint_D y\rho(x, y) \, dA,\ M_y = \iint_D x\rho(x, y) \, dA$

(c) The center of mass is $(\overline{x}, \overline{y})$ where $\overline{x} = \dfrac{M_y}{m}$ and $\overline{y} = \dfrac{M_x}{m}$.

(d) $I_x = \iint_D y^2 \rho(x, y) \, dA,\ I_y = \iint_D x^2 \rho(x, y) \, dA,\ I_0 = \iint_D (x^2 + y^2)\rho(x, y) \, dA$

**5.** (a) $P(a \leq X \leq b, c \leq Y \leq d) = \int_{a}^{b} \int_{c}^{d} f(x, y) \, dy \, dx$

(b) $f(x, y) \geq 0$ and $\iint_{\mathbb{R}^2} f(x, y) \, dA = 1$.

(c) The expected value of $X$ is $\mu_1 = \iint_{\mathbb{R}^2} x f(x, y) \, dA$; the expected value of $Y$ is $\mu_2 = \iint_{\mathbb{R}^2} y f(x, y) \, dA$.

**6.** (a) $A(S) = \iint_D |\mathbf{r}_u \times \mathbf{r}_v| \, dA$

(b) $A(S) = \iint_D \sqrt{1 + \left(\dfrac{\partial z}{\partial x}\right)^2 + \left(\dfrac{\partial z}{\partial y}\right)^2} \, dA$

(c) $A(S) = 2\pi \int_a^b f(x)\sqrt{1 + [f'(x)]^2} \, dx$

**7.** (a) $\iiint_B f(x, y, z) \, dV = \lim\limits_{l, m, n \to \infty} \sum\limits_{i=1}^{l} \sum\limits_{j=1}^{m} \sum\limits_{k=1}^{n} f\left(x_{ijk}^*, y_{ijk}^*, z_{ijk}^*\right) \Delta V$

(b) We usually evaluate $\iiint_B f(x, y, z) \, dV$ as an iterated integral according to Fubini's Theorem for Triple Integrals

(see Theorem 12.7.4).

(c) See the paragraph following Example 12.7.1.

(d) See (5) and (6) and the accompanying discussion in Section 12.7.

(e) See (10) and the accompanying discussion in Section 12.7.

(f) See (11) and the preceding discussion in Section 12.7.

**8.** (a) $m = \iiint_E \rho(x, y, z) \, dV$

(b) $M_{yz} = \iiint_E x\rho(x, y, z) \, dV, \ M_{xz} = \iiint_E y\rho(x, y, z) \, dV, \ M_{xy} = \iiint_E z\rho(x, y, z) \, dV.$

(c) The center of mass is $(\overline{x}, \overline{y}, \overline{z})$ where $\overline{x} = \dfrac{M_{yz}}{m}, \overline{y} = \dfrac{M_{xz}}{m}$, and $\overline{z} = \dfrac{M_{xy}}{m}.$

(d) $I_x = \iiint_E (y^2 + z^2)\rho(x, y, z) \, dV, \ I_y = \iiint_E (x^2 + z^2)\rho(x, y, z) \, dV, \ I_z = \iiint_E (x^2 + y^2)\rho(x, y, z) \, dV.$

**9.** (a) See Formula 12.8.2 and the accompanying discussion.

(b) See Formula 12.8.4 and the accompanying discussion.

(c) We may want to change from rectangular to cylindrical or spherical coordinates in a triple integral if the region $E$ of

integration is more easily described in cylindrical or spherical coordinates or if the triple integral is easier to evaluate using

cylindrical or spherical coordinates.

**10.** (a) $\dfrac{\partial (x, y)}{\partial (u, v)} = \begin{vmatrix} \partial x/\partial u & \partial x/\partial v \\ \partial y/\partial u & \partial y/\partial v \end{vmatrix} = \dfrac{\partial x}{\partial u}\dfrac{\partial y}{\partial v} - \dfrac{\partial x}{\partial v}\dfrac{\partial y}{\partial u}$

(b) See (9) and the accompanying discussion in Section 12.9.

(c) See (13) and the accompanying discussion in Section 12.9.

## TRUE-FALSE QUIZ

**1.** This is true by Fubini's Theorem.

**3.** True. See the discussion following Example 4 on page 842.

**5.** True. By Equation 12.2.5 we can write $\int_0^1 \int_0^1 f(x)\,f(y)\,dy\,dx = \int_0^1 f(x)\,dx \int_0^1 f(y)\,dy$. But $\int_0^1 f(y)\,dy = \int_0^1 f(x)\,dx$ so

this becomes $\int_0^1 f(x)\,dx \int_0^1 f(x)\,dx = \left[ \int_0^1 f(x)\,dx \right]^2$.

**7.** True:  $\iint_D \sqrt{4 - x^2 - y^2}\,dA$ = the volume under the surface $x^2 + y^2 + z^2 = 4$ and above the $xy$-plane

$\qquad = \frac{1}{2}$ (the volume of the sphere $x^2 + y^2 + z^2 = 4$) $= \frac{1}{2} \cdot \frac{4}{3}\pi(2)^3 = \frac{16}{3}\pi$

**9.** The volume enclosed by the cone $z = \sqrt{x^2 + y^2}$ and the plane $z = 2$ is, in cylindrical coordinates,

$V = \int_0^{2\pi} \int_0^2 \int_r^2 r\,dz\,dr\,d\theta \neq \int_0^{2\pi} \int_0^2 \int_r^2 dz\,dr\,d\theta$, so the assertion is false.

## EXERCISES

**1.** As shown in the contour map, we divide $R$ into 9 equally sized subsquares, each with area $\Delta A = 1$. Then we approximate

$\iint_R f(x,y)\,dA$ by a Riemann sum with $m = n = 3$ and the sample points the upper right corners of each square, so

$$\iint_R f(x,y)\,dA \approx \sum_{i=1}^{3} \sum_{j=1}^{3} f(x_i, y_j)\,\Delta A$$

$$= \Delta A \left[ f(1,1) + f(1,2) + f(1,3) + f(2,1) + f(2,2) + f(2,3) + f(3,1) + f(3,2) + f(3,3) \right]$$

Using the contour lines to estimate the function values, we have

$$\iint_R f(x,y)\,dA \approx 1[2.7 + 4.7 + 8.0 + 4.7 + 6.7 + 10.0 + 6.7 + 8.6 + 11.9] \approx 64.0$$

**3.** $\int_1^2 \int_0^2 (y + 2xe^y)\,dx\,dy = \int_1^2 \left[ xy + x^2 e^y \right]_{x=0}^{x=2} dy = \int_1^2 (2y + 4e^y)\,dy = \left[ y^2 + 4e^y \right]_1^2$

$\qquad = 4 + 4e^2 - 1 - 4e = 4e^2 - 4e + 3$

**5.** $\int_0^1 \int_0^x \cos(x^2)\,dy\,dx = \int_0^1 \left[ \cos(x^2)y \right]_{y=0}^{y=x} dx = \int_0^1 x\cos(x^2)\,dx = \frac{1}{2}\sin(x^2) \Big]_0^1 = \frac{1}{2}\sin 1$

**7.** $\int_0^\pi \int_0^1 \int_0^{\sqrt{1-y^2}} y \sin x\,dz\,dy\,dx = \int_0^\pi \int_0^1 \left[ (y \sin x)z \right]_{z=0}^{z=\sqrt{1-y^2}} dy\,dx = \int_0^\pi \int_0^1 y\sqrt{1-y^2}\sin x\,dy\,dx$

$\qquad = \int_0^\pi \left[ -\frac{1}{3}(1-y^2)^{3/2}\sin x \right]_{y=0}^{y=1} dx = \int_0^\pi \frac{1}{3}\sin x\,dx = -\frac{1}{3}\cos x \Big]_0^\pi = \frac{2}{3}$

**9.** The region $R$ is more easily described by polar coordinates: $R = \{(r, \theta) \mid 2 \leq r \leq 4, 0 \leq \theta \leq \pi\}$. Thus

$\iint_R f(x,y)\,dA = \int_0^\pi \int_2^4 f(r\cos\theta, r\sin\theta)\,r\,dr\,d\theta$.

**11.**

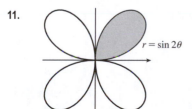

$r = \sin 2\theta$

The region whose area is given by $\int_0^{\pi/2} \int_0^{\sin 2\theta} r\,dr\,d\theta$ is

$\{(r,\theta) \mid 0 \leq \theta \leq \frac{\pi}{2}, 0 \leq r \leq \sin 2\theta\}$, which is the region contained in the

loop in the first quadrant of the four-leaved rose $r = \sin 2\theta$.

**13.**

$$\int_0^1 \int_x^1 \cos(y^2)\, dy\, dx = \int_0^1 \int_0^y \cos(y^2)\, dx\, dy$$

$$= \int_0^1 \cos(y^2)\big[x\big]_{x=0}^{x=y}\, dy = \int_0^1 y\cos(y^2)\, dy$$

$$= \big[\tfrac{1}{2}\sin(y^2)\big]_0^1 = \tfrac{1}{2}\sin 1$$

**15.** $\iint_R y e^{xy}\, dA = \int_0^3 \int_0^2 y e^{xy}\, dx\, dy = \int_0^3 \big[e^{xy}\big]_{x=0}^{x=2}\, dy = \int_0^3 (e^{2y}-1)\, dy = \big[\tfrac{1}{2}e^{2y}-y\big]_0^3 = \tfrac{1}{2}e^6 - 3 - \tfrac{1}{2} = \tfrac{1}{2}e^6 - \tfrac{7}{2}$

**17.**

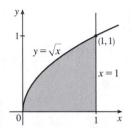

$$\iint_D \frac{y}{1+x^2}\, dA = \int_0^1 \int_0^{\sqrt{x}} \frac{y}{1+x^2}\, dy\, dx = \int_0^1 \frac{1}{1+x^2}\big[\tfrac{1}{2}y^2\big]_{y=0}^{y=\sqrt{x}}\, dx$$

$$= \tfrac{1}{2}\int_0^1 \frac{x}{1+x^2}\, dx = \big[\tfrac{1}{4}\ln(1+x^2)\big]_0^1 = \tfrac{1}{4}\ln 2$$

**19.**

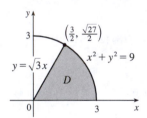

$$\iint_D y\, dA = \int_0^2 \int_{y^2}^{8-y^2} y\, dx\, dy$$

$$= \int_0^2 y\big[x\big]_{x=y^2}^{x=8-y^2}\, dy = \int_0^2 y(8 - y^2 - y^2)\, dy$$

$$= \int_0^2 (8y - 2y^3)\, dy = \big[4y^2 - \tfrac{1}{2}y^4\big]_0^2 = 8$$

**21.**

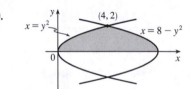

$$\iint_D (x^2+y^2)^{3/2}\, dA = \int_0^{\pi/3} \int_0^3 (r^2)^{3/2} r\, dr\, d\theta$$

$$= \int_0^{\pi/3} d\theta \int_0^3 r^4\, dr = \big[\theta\big]_0^{\pi/3}\big[\tfrac{1}{5}r^5\big]_0^3$$

$$= \frac{\pi}{3}\frac{3^5}{5} = \frac{81\pi}{5}$$

**23.** $\iiint_E xy\, dV = \int_0^3 \int_0^x \int_0^{x+y} xy\, dz\, dy\, dx = \int_0^3 \int_0^x xy\big[z\big]_{z=0}^{z=x+y}\, dy\, dx = \int_0^3 \int_0^x xy(x+y)\, dy\, dx$

$$= \int_0^3 \int_0^x (x^2 y + xy^2)\, dy\, dx = \int_0^3 \big[\tfrac{1}{2}x^2 y^2 + \tfrac{1}{3}xy^3\big]_{y=0}^{y=x}\, dx = \int_0^3 \big(\tfrac{1}{2}x^4 + \tfrac{1}{3}x^4\big)\, dx$$

$$= \tfrac{5}{6}\int_0^3 x^4\, dx = \big[\tfrac{1}{6}x^5\big]_0^3 = \tfrac{81}{2} = 40.5$$

**25.** $\iiint_E y^2 z^2\, dV = \int_{-1}^1 \int_{-\sqrt{1-y^2}}^{\sqrt{1-y^2}} \int_0^{1-y^2-z^2} y^2 z^2\, dx\, dz\, dy = \int_{-1}^1 \int_{-\sqrt{1-y^2}}^{\sqrt{1-y^2}} y^2 z^2 (1 - y^2 - z^2)\, dz\, dy$

$$= \int_0^{2\pi} \int_0^1 (r^2\cos^2\theta)(r^2\sin^2\theta)(1-r^2)\, r\, dr\, d\theta = \int_0^{2\pi} \int_0^1 \tfrac{1}{4}\sin^2 2\theta\,(r^5 - r^7)\, dr\, d\theta$$

$$= \int_0^{2\pi} \tfrac{1}{8}(1 - \cos 4\theta)\big[\tfrac{1}{6}r^6 - \tfrac{1}{8}r^8\big]_{r=0}^{r=1}\, d\theta = \tfrac{1}{192}\big[\theta - \tfrac{1}{4}\sin 4\theta\big]_0^{2\pi} = \tfrac{2\pi}{192} = \tfrac{\pi}{96}$$

**27.** $\iiint_E yz\, dV = \int_{-2}^2 \int_0^{\sqrt{4-x^2}} \int_0^y yz\, dz\, dy\, dx = \int_{-2}^2 \int_0^{\sqrt{4-x^2}} \tfrac{1}{2}y^3\, dy\, dx = \int_0^{\pi} \int_0^2 \tfrac{1}{2}r^3(\sin^3\theta)\, r\, dr\, d\theta$

$$= \tfrac{16}{5}\int_0^{\pi} \sin^3\theta\, d\theta = \tfrac{16}{5}\big[-\cos\theta + \tfrac{1}{3}\cos^3\theta\big]_0^{\pi} = \tfrac{64}{15}$$

**29.** $V = \int_0^2 \int_1^4 (x^2 + 4y^2)\, dy\, dx = \int_0^2 \big[x^2 y + \tfrac{4}{3}y^3\big]_{y=1}^{y=4}\, dx = \int_0^2 (3x^2 + 84)\, dx = 176$

**31.**

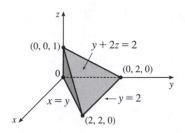

$V = \int_0^2 \int_0^y \int_0^{(2-y)/2} dz\,dx\,dy = \int_0^2 \int_0^y \left(1 - \tfrac{1}{2}y\right) dx\,dy$

$= \int_0^2 \left(y - \tfrac{1}{2}y^2\right) dy = \tfrac{2}{3}$

**33.** Using the wedge above the plane $z = 0$ and below the plane $z = mx$ and noting that we have the same volume for $m < 0$ as for $m > 0$ (so use $m > 0$), we have

$V = 2\int_0^{a/3} \int_0^{\sqrt{a^2-9y^2}} mx\,dx\,dy = 2\int_0^{a/3} \tfrac{1}{2}m(a^2 - 9y^2)\,dy = m\left[a^2 y - 3y^3\right]_0^{a/3} = m\left(\tfrac{1}{3}a^3 - \tfrac{1}{9}a^3\right) = \tfrac{2}{9}ma^3.$

**35.** (a) $m = \int_0^1 \int_0^{1-y^2} y\,dx\,dy = \int_0^1 (y - y^3)\,dy = \tfrac{1}{2} - \tfrac{1}{4} = \tfrac{1}{4}$

(b) $M_y = \int_0^1 \int_0^{1-y^2} xy\,dx\,dy = \int_0^1 \tfrac{1}{2}y(1-y^2)^2\,dy = -\tfrac{1}{12}(1-y^2)^3\Big]_0^1 = \tfrac{1}{12},$

$M_x = \int_0^1 \int_0^{1-y^2} y^2\,dx\,dy = \int_0^1 (y^2 - y^4)\,dy = \tfrac{2}{15}.$ Hence $(\overline{x}, \overline{y}) = \left(\tfrac{1}{3}, \tfrac{8}{15}\right).$

(c) $I_x = \int_0^1 \int_0^{1-y^2} y^3\,dx\,dy = \int_0^1 (y^3 - y^5)\,dy = \tfrac{1}{12},$

$I_y = \int_0^1 \int_0^{1-y^2} yx^2\,dx\,dy = \int_0^1 \tfrac{1}{3}y(1-y^2)^3\,dy = -\tfrac{1}{24}(1-y^2)^4\Big]_0^1 = \tfrac{1}{24},\ I_0 = I_x + I_y = \tfrac{1}{8}$

**37.** (a) The equation of the cone with the suggested orientation is $(h - z) = \tfrac{h}{a}\sqrt{x^2 + y^2}$, $0 \le z \le h$. Then $V = \tfrac{1}{3}\pi a^2 h$ is the volume of one frustum of a cone; by symmetry $M_{yz} = M_{xz} = 0$; and

$M_{xy} = \iint\limits_{x^2+y^2 \le a^2} \int_0^{h-(h/a)\sqrt{x^2+y^2}} z\,dz\,dA = \int_0^{2\pi} \int_0^a \int_0^{(h/a)(a-r)} rz\,dz\,dr\,d\theta = \pi \int_0^a r\frac{h^2}{a^2}(a-r)^2\,dr$

$= \frac{\pi h^2}{a^2} \int_0^a (a^2 r - 2ar^2 + r^3)\,dr = \frac{\pi h^2}{a^2}\left(\frac{a^4}{2} - \frac{2a^4}{3} + \frac{a^4}{4}\right) = \frac{\pi h^2 a^2}{12}$

Hence the centroid is $(\overline{x}, \overline{y}, \overline{z}) = \left(0, 0, \tfrac{1}{4}h\right).$

(b) $I_z = \int_0^{2\pi} \int_0^a \int_0^{(h/a)(a-r)} r^3\,dz\,dr\,d\theta = 2\pi \int_0^a \frac{h}{a}(ar^3 - r^4)\,dr = \frac{2\pi h}{a}\left(\frac{a^5}{4} - \frac{a^5}{5}\right) = \frac{\pi a^4 h}{10}$

**39.** Let $D$ represent the given triangle; then $D$ can be described as the area enclosed by the $x$- and $y$-axes and the line $y = 2 - 2x$, or equivalently $D = \{(x, y) \mid 0 \le x \le 1, 0 \le y \le 2 - 2x\}$. We want to find the surface area of the part of the graph of $z = x^2 + y$ that lies over $D$, so using Equation 12.6.6 we have

$A(S) = \iint\limits_D \sqrt{1 + \left(\frac{\partial z}{\partial x}\right)^2 + \left(\frac{\partial z}{\partial y}\right)^2}\,dA = \iint\limits_D \sqrt{1 + (2x)^2 + (1)^2}\,dA = \int_0^1 \int_0^{2-2x} \sqrt{2 + 4x^2}\,dy\,dx$

$= \int_0^1 \sqrt{2 + 4x^2}\,\big[y\big]_{y=0}^{y=2-2x}\,dx = \int_0^1 (2 - 2x)\sqrt{2 + 4x^2}\,dx = \int_0^1 2\sqrt{2 + 4x^2}\,dx - \int_0^1 2x\sqrt{2 + 4x^2}\,dx$

Using Formula 21 in the Table of Integrals with $a = \sqrt{2}$, $u = 2x$, and $du = 2\,dx$, we have

$\int 2\sqrt{2 + 4x^2}\,dx = x\sqrt{2 + 4x^2} + \ln\!\left(2x + \sqrt{2 + 4x^2}\,\right).$ If we substitute $u = 2 + 4x^2$ in the second integral, then

$du = 8x\,dx$ and $\int 2x\sqrt{2+4x^2}\,dx = \frac{1}{4}\int \sqrt{u}\,du = \frac{1}{4}\cdot\frac{2}{3}u^{3/2} = \frac{1}{6}(2+4x^2)^{3/2}$. Thus

$$A(S) = \left[x\sqrt{2+4x^2} + \ln\left(2x+\sqrt{2+4x^2}\right) - \frac{1}{6}(2+4x^2)^{3/2}\right]_0^1$$

$$= \sqrt{6} + \ln\left(2+\sqrt{6}\right) - \frac{1}{6}(6)^{3/2} - \ln\sqrt{2} + \frac{\sqrt{2}}{3} = \ln\frac{2+\sqrt{6}}{\sqrt{2}} + \frac{\sqrt{2}}{3}$$

$$= \ln\left(\sqrt{2}+\sqrt{3}\right) + \frac{\sqrt{2}}{3} \approx 1.6176$$

**41.**

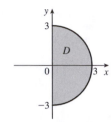

$$\int_0^3 \int_{-\sqrt{9-x^2}}^{\sqrt{9-x^2}} (x^3 + xy^2)\,dy\,dx = \int_0^3 \int_{-\sqrt{9-x^2}}^{\sqrt{9-x^2}} x(x^2 + y^2)\,dy\,dx$$

$$= \int_{-\pi/2}^{\pi/2} \int_0^3 (r\cos\theta)(r^2)\,r\,dr\,d\theta$$

$$= \int_{-\pi/2}^{\pi/2} \cos\theta\,d\theta \int_0^3 r^4\,dr$$

$$= \left[\sin\theta\right]_{-\pi/2}^{\pi/2} \left[\frac{1}{5}r^5\right]_0^3 = 2\cdot\frac{1}{5}(243) = \frac{486}{5} = 97.2$$

**43.** From the graph, it appears that $1 - x^2 = e^x$ at $x \approx -0.71$ and at $x = 0$, with $1 - x^2 > e^x$ on $(-0.71, 0)$. So the desired integral is

$$\iint_D y^2\,dA \approx \int_{-0.71}^0 \int_{e^x}^{1-x^2} y^2\,dy\,dx$$

$$= \frac{1}{3}\int_{-0.71}^0 [(1-x^2)^3 - e^{3x}]\,dx$$

$$= \frac{1}{3}\left[x - x^3 + \frac{3}{5}x^5 - \frac{1}{7}x^7 - \frac{1}{3}e^{3x}\right]_{-0.71}^0 \approx 0.0512$$

**45.** (a) $f(x,y)$ is a joint density function, so we know that $\iint_{\mathbb{R}^2} f(x,y)\,dA = 1$. Since $f(x,y) = 0$ outside the rectangle $[0,3] \times [0,2]$, we can say

$$\iint_{\mathbb{R}^2} f(x,y)\,dA = \int_{-\infty}^{\infty} \int_{-\infty}^{\infty} f(x,y)\,dy\,dx = \int_0^3 \int_0^2 C(x+y)\,dy\,dx$$

$$= C\int_0^3 \left[xy + \frac{1}{2}y^2\right]_{y=0}^{y=2}\,dx = C\int_0^3 (2x+2)\,dx = C\left[x^2 + 2x\right]_0^3 = 15C$$

Then $15C = 1 \Rightarrow C = \frac{1}{15}$.

(b) $P(X \le 2, Y \ge 1) = \int_{-\infty}^2 \int_1^\infty f(x,y)\,dy\,dx = \int_0^2 \int_1^2 \frac{1}{15}(x,y)\,dy\,dx = \frac{1}{15}\int_0^2 \left[xy + \frac{1}{2}y^2\right]_{y=1}^{y=2}\,dx$

$= \frac{1}{15}\int_0^2 \left(x + \frac{3}{2}\right)\,dx = \frac{1}{15}\left[\frac{1}{2}x^2 + \frac{3}{2}x\right]_0^2 = \frac{1}{3}$

(c) $P(X + Y \le 1) = P((X,Y) \in D)$ where $D$ is the triangular region shown in the figure. Thus

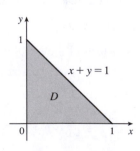

$$P(X+Y \le 1) = \iint_D f(x,y)\,dA = \int_0^1 \int_0^{1-x} \frac{1}{15}(x+y)\,dy\,dx$$

$$= \frac{1}{15}\int_0^1 \left[xy + \frac{1}{2}y^2\right]_{y=0}^{y=1-x}\,dx$$

$$= \frac{1}{15}\int_0^1 \left[x(1-x) + \frac{1}{2}(1-x)^2\right]\,dx$$

$$= \frac{1}{30}\int_0^1 (1-x^2)\,dx = \frac{1}{30}\left[x - \frac{1}{3}x^3\right]_0^1 = \frac{1}{45}$$

**47.**

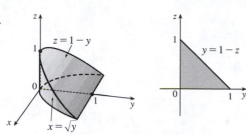

$$\int_{-1}^{1} \int_{x^2}^{1} \int_{0}^{1-y} f(x, y, z)\, dz\, dy\, dx = \int_{0}^{1} \int_{0}^{1-z} \int_{-\sqrt{y}}^{\sqrt{y}} f(x, y, z)\, dx\, dy\, dz$$

**49.** Since $u = x - y$ and $v = x + y$, $x = \frac{1}{2}(u + v)$ and $y = \frac{1}{2}(v - u)$.

Thus $\dfrac{\partial(x, y)}{\partial(u, v)} = \begin{vmatrix} 1/2 & 1/2 \\ -1/2 & 1/2 \end{vmatrix} = \dfrac{1}{2}$ and $\displaystyle\iint_{R} \dfrac{x - y}{x + y}\, dA = \int_{2}^{4} \int_{-2}^{0} \dfrac{u}{v}\left(\dfrac{1}{2}\right) du\, dv = -\int_{2}^{4} \dfrac{dv}{v} = -\ln 2.$

**51.** Let $u = y - x$ and $v = y + x$ so $x = y - u = (v - x) - u \;\Rightarrow\; x = \frac{1}{2}(v - u)$ and $y = v - \frac{1}{2}(v - u) = \frac{1}{2}(v + u)$.

$\left| \dfrac{\partial(x, y)}{\partial(u, v)} \right| = \left| \dfrac{\partial x}{\partial u}\dfrac{\partial y}{\partial v} - \dfrac{\partial x}{\partial v}\dfrac{\partial y}{\partial u} \right| = \left| -\frac{1}{2}\left(\frac{1}{2}\right) - \frac{1}{2}\left(\frac{1}{2}\right) \right| = \left| -\frac{1}{2} \right| = \frac{1}{2}$. $R$ is the image under this transformation of the square

with vertices $(u, v) = (0, 0)$, $(-2, 0)$, $(0, 2)$, and $(-2, 2)$. So

$$\iint_{R} xy\, dA = \int_{0}^{2} \int_{-2}^{0} \dfrac{v^2 - u^2}{4}\left(\dfrac{1}{2}\right) du\, dv = \tfrac{1}{8}\int_{0}^{2}\left[v^2 u - \tfrac{1}{3}u^3\right]_{u=-2}^{u=0} dv = \tfrac{1}{8}\int_{0}^{2}\left(2v^2 - \tfrac{8}{3}\right) dv = \tfrac{1}{8}\left[\tfrac{2}{3}v^3 - \tfrac{8}{3}v\right]_{0}^{2} = 0$$

This result could have been anticipated by symmetry, since the integrand is an odd function of $y$ and $R$ is symmetric about

the $x$-axis.

# □ FOCUS ON PROBLEM SOLVING

**1.** Let $R = \bigcup_{i=1}^{5} R_i$, where

$$R_i = \{(x, y) \mid x + y \geq i + 2, x + y < i + 3, 1 \leq x \leq 3, 2 \leq y \leq 5\}.$$

$$\iint_R [\![x + y]\!]\, dA = \sum_{i=1}^{5} \iint_{R_i} [\![x + y]\!]\, dA = \sum_{i=1}^{5} [\![x + y]\!] \iint_{R_i} dA, \text{ since}$$

$[\![x + y]\!] = \text{constant} = i + 2$ for $(x, y) \in R_i$. Therefore

$$\iint_R [\![x + y]\!]\, dA = \sum_{i=1}^{5} (i + 2)\,[A(R_i)]$$

$$= 3A(R_1) + 4A(R_2) + 5A(R_3) + 6A(R_4) + 7A(R_5)$$

$$= 3\left(\tfrac{1}{2}\right) + 4\left(\tfrac{3}{2}\right) + 5(2) + 6\left(\tfrac{3}{2}\right) + 7\left(\tfrac{1}{2}\right) = 30$$

**3.**
$$f_{\text{ave}} = \frac{1}{b - a} \int_a^b f(x)\, dx = \frac{1}{1 - 0} \int_0^1 \left[ \int_x^1 \cos(t^2)\, dt \right] dx$$

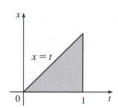

$$= \int_0^1 \int_x^1 \cos(t^2)\, dt\, dx = \int_0^1 \int_0^t \cos(t^2)\, dx\, dt \quad \text{[changing the order of integration]}$$

$$= \int_0^1 t \cos(t^2)\, dt = \tfrac{1}{2} \sin(t^2)\Big]_0^1 = \tfrac{1}{2} \sin 1$$

**5.** Since $|xy| < 1$, except at $(1, 1)$, the formula for the sum of a geometric series gives $\dfrac{1}{1 - xy} = \sum_{n=0}^{\infty} (xy)^n$, so

$$\int_0^1 \int_0^1 \frac{1}{1 - xy}\, dx\, dy = \int_0^1 \int_0^1 \sum_{n=0}^{\infty} (xy)^n\, dx\, dy = \sum_{n=0}^{\infty} \int_0^1 \int_0^1 (xy)^n\, dx\, dy = \sum_{n=0}^{\infty} \left[ \int_0^1 x^n\, dx \right]\left[ \int_0^1 y^n\, dy \right]$$

$$= \sum_{n=0}^{\infty} \frac{1}{n+1} \cdot \frac{1}{n+1} = \sum_{n=0}^{\infty} \frac{1}{(n+1)^2} = \frac{1}{1^2} + \frac{1}{2^2} + \frac{1}{3^2} + \cdots = \sum_{n=1}^{\infty} \frac{1}{n^2}$$

**7. (a)** Since $|xyz| < 1$ except at $(1, 1, 1)$, the formula for the sum of a geometric series gives $\dfrac{1}{1 - xyz} = \sum_{n=0}^{\infty} (xyz)^n$, so

$$\int_0^1 \int_0^1 \int_0^1 \frac{1}{1 - xyz}\, dx\, dy\, dz = \int_0^1 \int_0^1 \int_0^1 \sum_{n=0}^{\infty} (xyz)^n\, dx\, dy\, dz = \sum_{n=0}^{\infty} \int_0^1 \int_0^1 \int_0^1 (xyz)^n\, dx\, dy\, dz$$

$$= \sum_{n=0}^{\infty} \left[ \int_0^1 x^n\, dx \right]\left[ \int_0^1 y^n\, dy \right]\left[ \int_0^1 z^n\, dz \right] = \sum_{n=0}^{\infty} \frac{1}{n+1} \cdot \frac{1}{n+1} \cdot \frac{1}{n+1}$$

$$= \sum_{n=0}^{\infty} \frac{1}{(n+1)^3} = \frac{1}{1^3} + \frac{1}{2^3} + \frac{1}{3^3} + \cdots = \sum_{n=1}^{\infty} \frac{1}{n^3}$$

**(b)** Since $|-xyz| < 1$, except at $(1, 1, 1)$, the formula for the sum of a geometric series gives $\dfrac{1}{1 + xyz} = \sum_{n=0}^{\infty} (-xyz)^n$, so

$$\int_0^1 \int_0^1 \int_0^1 \frac{1}{1 + xyz}\, dx\, dy\, dz = \int_0^1 \int_0^1 \int_0^1 \sum_{n=0}^{\infty} (-xyz)^n\, dx\, dy\, dz = \sum_{n=0}^{\infty} \int_0^1 \int_0^1 \int_0^1 (-xyz)^n\, dx\, dy\, dz$$

$$= \sum_{n=0}^{\infty} (-1)^n \left[ \int_0^1 x^n\, dx \right]\left[ \int_0^1 y^n\, dy \right]\left[ \int_0^1 z^n\, dz \right] = \sum_{n=0}^{\infty} (-1)^n \frac{1}{n+1} \cdot \frac{1}{n+1} \cdot \frac{1}{n+1}$$

$$= \sum_{n=0}^{\infty} \frac{(-1)^n}{(n+1)^3} = \frac{1}{1^3} - \frac{1}{2^3} + \frac{1}{3^3} - \cdots = \sum_{n=0}^{\infty} \frac{(-1)^{n-1}}{n^3}$$

To evaluate this sum, we first write out a few terms: $s = 1 - \dfrac{1}{2^3} + \dfrac{1}{3^3} - \dfrac{1}{4^3} + \dfrac{1}{5^3} - \dfrac{1}{6^3} \approx 0.8998$. Notice that

$a_7 = \dfrac{1}{7^3} < 0.003$. By the Alternating Series Estimation Theorem from Section 8.4, we have $|s - s_6| \leq a_7 < 0.003$. This

error of $0.003$ will not affect the second decimal place, so we have $s \approx 0.90$.

**9.** $\int_0^x \int_0^y \int_0^z f(t)\, dt\, dz\, dy = \iiint_E f(t)\, dV$, where

$E = \{(t, z, y) \mid 0 \leq t \leq z, 0 \leq z \leq y, 0 \leq y \leq x\}$.

If we let $D$ be the projection of $E$ on the $yt$-plane then

$D = \{(y, t) \mid 0 \leq t \leq x, t \leq y \leq x\}$. And we see from the diagram

that $E = \{(t, z, y) \mid t \leq z \leq y, t \leq y \leq x, 0 \leq t \leq x\}$. So

$$\int_0^x \int_0^y \int_0^z f(t)\, dt\, dz\, dy = \int_0^x \int_t^x \int_t^y f(t)\, dz\, dy\, dt = \int_0^x \left[ \int_t^x (y - t)\, f(t)\, dy \right] dt$$

$$= \int_0^x \left[ \left( \tfrac{1}{2} y^2 - ty \right) f(t) \right]_{y=t}^{y=x} dt = \int_0^x \left[ \tfrac{1}{2} x^2 - tx - \tfrac{1}{2} t^2 + t^2 \right] f(t)\, dt$$

$$= \int_0^x \left[ \tfrac{1}{2} x^2 - tx + \tfrac{1}{2} t^2 \right] f(t)\, dt = \int_0^x \left( \tfrac{1}{2} x^2 - 2tx + t^2 \right) f(t)\, dt$$

$$= \tfrac{1}{2} \int_0^x (x - t)^2\, f(t)\, dt$$

**11.** The volume is $V = \iiint_R dV$ where $R$ is the solid region given. From Exercise 12.9.21(a), the transformation $x = au$,

$y = bv$, $z = cw$ maps the unit ball $u^2 + v^2 + w^2 \leq 1$ to the solid ellipsoid

$\dfrac{x^2}{a^2} + \dfrac{y^2}{b^2} + \dfrac{z^2}{c^2} \leq 1$ with $\dfrac{\partial(x, y, z)}{\partial(u, v, w)} = abc$. The same transformation maps the

plane $u + v + w = 1$ to $\dfrac{x}{a} + \dfrac{y}{b} + \dfrac{z}{c} = 1$. Thus the region $R$ in $xyz$-space

corresponds to the region $S$ in $uvw$-space consisting of the smaller piece of the

unit ball cut off by the plane $u + v + w = 1$, a "cap of a sphere" (see the figure).

We will need to compute the volume of $S$, but first consider the general case

where a horizontal plane slices the upper portion of a sphere of radius $r$ to produce

a cap of height $h$. We use spherical coordinates. From the figure, a line at angle $\phi$

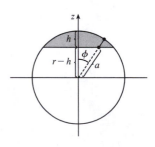

from the $z$-axis intersects the plane when $\cos \phi = (r - h)/a \;\; \Rightarrow$

$a = (r - h)/\cos \phi$, and the line passes through the outer rim of the cap when

$a = r \;\; \Rightarrow \;\; \cos \phi = (r - h)/r \;\; \Rightarrow \;\; \phi = \cos^{-1}((r - h)/r)$. Thus the cap

is described by $\{(\rho, \theta, \phi) \mid (r - h)/\cos \phi \leq \rho \leq r, 0 \leq \theta \leq 2\pi, 0 \leq \phi \leq \cos^{-1}((r - h)/r)\}$ and its volume is

$$V = \int_0^{2\pi} \int_0^{\cos^{-1}((r-h)/r)} \int_{(r-h)/\cos\phi}^{r} \rho^2 \sin\phi \, d\rho \, d\phi \, d\theta$$

$$= \int_0^{2\pi} \int_0^{\cos^{-1}((r-h)/r)} \left[ \tfrac{1}{3}\rho^3 \sin\phi \right]_{\rho=(r-h)/\cos\phi}^{\rho=r} d\phi \, d\theta$$

$$= \frac{1}{3} \int_0^{2\pi} \int_0^{\cos^{-1}((r-h)/r)} \left[ r^3 \sin\phi - \frac{(r-h)^3}{\cos^3\phi} \sin\phi \right] d\phi \, d\theta$$

$$= \tfrac{1}{3} \int_0^{2\pi} \left[ -r^3 \cos\phi - \tfrac{1}{2}(r-h)^3 \cos^{-2}\phi \right]_{\phi=0}^{\phi=\cos^{-1}((r-h)/r)} d\theta$$

$$= \frac{1}{3} \int_0^{2\pi} \left[ -r^3 \left( \frac{r-h}{r} \right) - \frac{1}{2}(r-h)^3 \left( \frac{r-h}{r} \right)^{-2} + r^3 + \frac{1}{2}(r-h)^3 \right] d\theta$$

$$= \tfrac{1}{3} \int_0^{2\pi} \left( \tfrac{3}{2}rh^2 - \tfrac{1}{2}h^3 \right) d\theta = \tfrac{1}{3} \left( \tfrac{3}{2}rh^2 - \tfrac{1}{2}h^3 \right)(2\pi) = \pi h^2 \left( r - \tfrac{1}{3}h \right)$$

(This volume can also be computed by treating the cap as a solid of revolution and using the single variable disk method; see Exercise 6.2.33.)

To determine the height $h$ of the cap cut from the unit ball by the plane $u + v + w = 1$, note that the line $u = v = w$ passes through the origin with direction vector $\langle 1, 1, 1 \rangle$ which is perpendicular to the plane. Therefore this line coincides with a radius of the sphere that passes through the center of the cap and $h$ is measured along this line. The line intersects the plane at $\left( \tfrac{1}{3}, \tfrac{1}{3}, \tfrac{1}{3} \right)$ and the sphere at $\left( \tfrac{1}{\sqrt{3}}, \tfrac{1}{\sqrt{3}}, \tfrac{1}{\sqrt{3}} \right)$. (See the figure.)

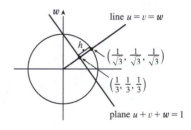

The distance between these points is $h = \sqrt{3 \left( \tfrac{1}{\sqrt{3}} - \tfrac{1}{3} \right)^2} = \sqrt{3} \left( \tfrac{1}{\sqrt{3}} - \tfrac{1}{3} \right) = 1 - \tfrac{1}{\sqrt{3}}$. Thus the volume of $R$ is

$$V = \iiint_R dV = \iiint_S \left| \frac{\partial(x, y, z)}{\partial(u, v, w)} \right| dV = abc \iiint_S dV = abc \, V(S)$$

$$= abc \cdot \pi h^2 \left( r - \tfrac{1}{3}h \right) = abc \cdot \pi \left( 1 - \tfrac{1}{\sqrt{3}} \right)^2 \left[ 1 - \tfrac{1}{3} \left( 1 - \tfrac{1}{\sqrt{3}} \right) \right]$$

$$= abc\pi \left( \tfrac{4}{3} - \tfrac{2}{\sqrt{3}} \right) \left( \tfrac{2}{3} + \tfrac{1}{3\sqrt{3}} \right) = abc\pi \left( \tfrac{2}{3} - \tfrac{8}{9\sqrt{3}} \right) \approx 0.482abc$$

# 13 □ VECTOR CALCULUS

## 13.1 Vector Fields

**1.** $\mathbf{F}(x, y) = 0.3\,\mathbf{i} - 0.4\,\mathbf{j}$

All vectors in this field are identical, with length 0.5 and parallel to $\langle 3, -4 \rangle$.

**3.** $\mathbf{F}(x, y) = y\,\mathbf{i} + \frac{1}{2}\,\mathbf{j}$

The length of the vector $y\,\mathbf{i} + \frac{1}{2}\,\mathbf{j}$ is $\sqrt{y^2 + \frac{1}{4}}$. Vectors are tangent to parabolas opening about the $x$-axis.

**5.** $\mathbf{F}(x, y) = \dfrac{y\,\mathbf{i} + x\,\mathbf{j}}{\sqrt{x^2 + y^2}}$

The length of the vector $\dfrac{y\,\mathbf{i} + x\,\mathbf{j}}{\sqrt{x^2 + y^2}}$ is 1.

**7.** $\mathbf{F}(x, y, z) = \mathbf{k}$

All vectors in this field are parallel to the $z$-axis and have length 1.

**9.** $\mathbf{F}(x, y, z) = x\,\mathbf{k}$

At each point $(x, y, z)$, $\mathbf{F}(x, y, z)$ is a vector of length $|x|$. For $x > 0$, all point in the direction of the positive $z$-axis, while for $x < 0$, all are in the direction of the negative $z$-axis. In each plane $x = k$, all the vectors are identical.

**11.** $\mathbf{F}(x, y) = \langle y, x \rangle$ corresponds to graph II. In the first quadrant all the vectors have positive $x$- and $y$-components, in the second quadrant all vectors have positive $x$-components and negative $y$-components, in the third quadrant all vectors have negative $x$- and $y$-components, and in the fourth quadrant all vectors have negative $x$-components and positive $y$-components. In addition, the vectors get shorter as we approach the origin.

**13.** $\mathbf{F}(x, y) = \langle x - 2, x + 1 \rangle$ corresponds to graph I since the vectors are independent of $y$ (vectors along vertical lines are identical) and, as we move to the right, both the $x$- and the $y$-components get larger.

**15.** $\mathbf{F}(x, y, z) = \mathbf{i} + 2\,\mathbf{j} + 3\,\mathbf{k}$ corresponds to graph IV, since all vectors have identical length and direction.

**17.** $\mathbf{F}(x, y, z) = x\,\mathbf{i} + y\,\mathbf{j} + 3\,\mathbf{k}$ corresponds to graph III; the projection of each vector onto the $xy$-plane is $x\,\mathbf{i} + y\,\mathbf{j}$, which points away from the origin, and the vectors point generally upward because their $z$-components are all 3.

**19.**

The vector field seems to have very short vectors near the line $y = 2x$.

For $\mathbf{F}(x, y) = \langle 0, 0 \rangle$ we must have $y^2 - 2xy = 0$ and $3xy - 6x^2 = 0$.

The first equation holds if $y = 0$ or $y = 2x$, and the second holds if $x = 0$ or $y = 2x$. So both equations hold [and thus $\mathbf{F}(x, y) = \mathbf{0}$] along the line $y = 2x$.

**21.** $f(x, y) = xe^{xy}$ $\Rightarrow$

$$\nabla f(x, y) = f_x(x, y)\,\mathbf{i} + f_y(x, y)\,\mathbf{j} = (xe^{xy} \cdot y + e^{xy})\,\mathbf{i} + (xe^{xy} \cdot x)\,\mathbf{j} = (xy + 1)e^{xy}\,\mathbf{i} + x^2 e^{xy}\,\mathbf{j}$$

**23.** $\nabla f(x, y, z) = f_x(x, y, z)\,\mathbf{i} + f_y(x, y, z)\,\mathbf{j} + f_z(x, y, z)\,\mathbf{k} = \dfrac{x}{\sqrt{x^2 + y^2 + z^2}}\,\mathbf{i} + \dfrac{y}{\sqrt{x^2 + y^2 + z^2}}\,\mathbf{j} + \dfrac{z}{\sqrt{x^2 + y^2 + z^2}}\,\mathbf{k}$

**25.** $f(x, y) = x^2 - y$ $\Rightarrow$ $\nabla f(x, y) = 2x\,\mathbf{i} - \mathbf{j}$.

The length of $\nabla f(x, y)$ is $\sqrt{4x^2 + 1}$. When $x \neq 0$, the vectors point away from the $y$-axis in a slightly downward direction with length that increases as the distance from the $y$-axis increases.

**27.** We graph $\nabla f$ along with a contour map of $f$.

The graph shows that the gradient vectors are perpendicular to the level curves. Also, the gradient vectors point in the direction in which $f$ is increasing and are longer where the level curves are closer together.

**29.** $f(x, y) = x^2 + y^2 \;\Rightarrow\; \nabla f(x, y) = 2x\,\mathbf{i} + 2y\,\mathbf{j}$. Thus, each vector $\nabla f(x, y)$ has the same direction and twice the length of the position vector of the point $(x, y)$, so the vectors all point directly away from the origin and their lengths increase as we move away from the origin. Hence, $\nabla f$ is graph III.

**31.** $f(x, y) = (x + y)^2 \;\Rightarrow\; \nabla f(x, y) = 2(x + y)\,\mathbf{i} + 2(x + y)\,\mathbf{j}$. The $x$- and $y$-components of each vector are equal, so all vectors are parallel to the line $y = x$. The vectors are $\mathbf{0}$ along the line $y = -x$ and their length increases as the distance from this line increases. Thus, $\nabla f$ is graph II.

**33.** At $t = 3$ the particle is at $(2, 1)$ so its velocity is $\mathbf{V}(2, 1) = \langle 4, 3\rangle$. After 0.01 units of time, the particle's change in location should be approximately $0.01\,\mathbf{V}(2, 1) = 0.01\,\langle 4, 3\rangle = \langle 0.04, 0.03\rangle$, so the particle should be approximately at the point $(2.04, 1.03)$.

**35.** (a) We sketch the vector field $\mathbf{F}(x, y) = x\,\mathbf{i} - y\,\mathbf{j}$ along with several approximate flow lines. The flow lines appear to be hyperbolas with shape similar to the graph of $y = \pm 1/x$, so we might guess that the flow lines have equations $y = C/x$.

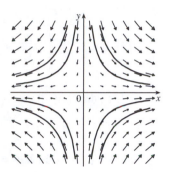

(b) If $x = x(t)$ and $y = y(t)$ are parametric equations of a flow line, then the velocity vector of the flow line at the point $(x, y)$ is $x'(t)\,\mathbf{i} + y'(t)\,\mathbf{j}$. Since the velocity vectors coincide with the vectors in the vector field, we have

$x'(t)\,\mathbf{i} + y'(t)\,\mathbf{j} = x\,\mathbf{i} - y\,\mathbf{j} \;\Rightarrow\; dx/dt = x,\; dy/dt = -y$. To solve these differential equations, we know

$dx/dt = x \;\Rightarrow\; dx/x = dt \;\Rightarrow\; \ln|x| = t + C \;\Rightarrow\; x = \pm e^{t+C} = Ae^t$ for some constant $A$, and

$dy/dt = -y \;\Rightarrow\; dy/y = -dt \;\Rightarrow\; \ln|y| = -t + K \;\Rightarrow\; y = \pm e^{-t+K} = Be^{-t}$ for some constant $B$. Therefore

$xy = Ae^t Be^{-t} = AB = $ constant. If the flow line passes through $(1, 1)$ then $(1)(1) = $ constant $= 1 \;\Rightarrow\; xy = 1 \;\Rightarrow\; y = 1/x,\; x > 0$.

## 13.2  Line Integrals

**1.** $x = t^3$ and $y = t$, $0 \le t \le 2$, so by Formula 3

$$\int_C y^3\,ds = \int_0^2 t^3 \sqrt{\left(\frac{dx}{dt}\right)^2 + \left(\frac{dy}{dt}\right)^2}\,dt = \int_0^2 t^3 \sqrt{(3t^2)^2 + (1)^2}\,dt = \int_0^2 t^3 \sqrt{9t^4 + 1}\,dt$$

$$= \tfrac{1}{36} \cdot \tfrac{2}{3}\left(9t^4 + 1\right)^{3/2}\Big]_0^2 = \tfrac{1}{54}(145^{3/2} - 1) \text{ or } \tfrac{1}{54}\left(145\sqrt{145} - 1\right)$$

**3.** Parametric equations for $C$ are $x = 4\cos t,\;\; y = 4\sin t,\;\; -\tfrac{\pi}{2} \le t \le \tfrac{\pi}{2}$. Then

$\int_C xy^4\,ds = \int_{-\pi/2}^{\pi/2}(4\cos t)(4\sin t)^4 \sqrt{(-4\sin t)^2 + (4\cos t)^2}\,dt = \int_{-\pi/2}^{\pi/2} 4^5 \cos t\,\sin^4 t\,\sqrt{16(\sin^2 t + \cos^2 t)}\,dt$

$= 4^5 \int_{-\pi/2}^{\pi/2}(\sin^4 t\cos t)(4)\,dt = (4)^6\left[\tfrac{1}{5}\sin^5 t\right]_{-\pi/2}^{\pi/2} = \tfrac{2 \cdot 4^6}{5} = 1638.4$

**5.** If we choose $x$ as the parameter, parametric equations for $C$ are $x = x$, $y = \sqrt{x}$ for $1 \le x \le 4$ and

$$\int_C \left( x^2 y^3 - \sqrt{x} \right) dy = \int_1^4 \left[ x^2 \cdot (\sqrt{x})^3 - \sqrt{x} \right] \frac{1}{2\sqrt{x}} \, dx = \tfrac{1}{2} \int_1^4 \left( x^3 - 1 \right) dx$$

$$= \tfrac{1}{2} \left[ \tfrac{1}{4} x^4 - x \right]_1^4 = \tfrac{1}{2} \left( 64 - 4 - \tfrac{1}{4} + 1 \right) = \tfrac{243}{8}$$

**7.**

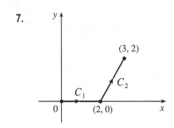

$C = C_1 + C_2$

On $C_1$: $x = x, y = 0 \;\Rightarrow\; dy = 0 \, dx, \; 0 \le x \le 2$.

On $C_2$: $x = x, y = 2x - 4 \;\Rightarrow\; dy = 2 \, dx, \; 2 \le x \le 3$.

Then

$$\int_C xy \, dx + (x - y) \, dy = \int_{C_1} xy \, dx + (x - y) \, dy + \int_{C_2} xy \, dx + (x - y) \, dy$$

$$= \int_0^2 (0 + 0) \, dx + \int_2^3 \left[ (2x^2 - 4x) + (-x + 4)(2) \right] dx$$

$$= \int_2^3 \left( 2x^2 - 6x + 8 \right) dx = \tfrac{17}{3}$$

**9.** $x = 2 \sin t$, $y = t$, $z = -2 \cos t$, $0 \le t \le \pi$. Then by Formula 9,

$$\int_C xyz \, ds = \int_0^\pi (2 \sin t)(t)(-2 \cos t) \sqrt{\left( \tfrac{dx}{dt} \right)^2 + \left( \tfrac{dy}{dt} \right)^2 + \left( \tfrac{dz}{dt} \right)^2} \, dt$$

$$= \int_0^\pi -4t \sin t \cos t \sqrt{(2 \cos t)^2 + (1)^2 + (2 \sin t)^2} \, dt = \int_0^\pi -2t \sin 2t \sqrt{4(\cos^2 t + \sin^2 t) + 1} \, dt$$

$$= -2\sqrt{5} \int_0^\pi t \sin 2t \, dt = -2\sqrt{5} \left[ -\tfrac{1}{2} t \cos 2t + \tfrac{1}{4} \sin 2t \right]_0^\pi \qquad \left[ \begin{array}{l} \text{integrate by parts with} \\ u = t, \, dv = \sin 2t \, dt \end{array} \right]$$

$$= -2\sqrt{5} \left( -\tfrac{\pi}{2} - 0 \right) = \sqrt{5}\,\pi$$

**11.** Parametric equations for $C$ are $x = t$, $y = 2t$, $z = 3t$, $0 \le t \le 1$. Then

$$\int_C xe^{yz} \, ds = \int_0^1 te^{(2t)(3t)} \sqrt{1^2 + 2^2 + 3^2} \, dt = \sqrt{14} \int_0^1 te^{6t^2} \, dt = \sqrt{14} \left[ \tfrac{1}{12} e^{6t^2} \right]_0^1 = \tfrac{\sqrt{14}}{12} (e^6 - 1).$$

**13.** $\int_C x^2 y \sqrt{z} \, dz = \int_0^1 (t^3)^2 (t) \sqrt{t^2} \cdot 2t \, dt = \int_0^1 2t^9 \, dt = \tfrac{1}{5} t^{10} \big]_0^1 = \tfrac{1}{5}$

**15.**

On $C_1$: $x = 1 + t \;\Rightarrow\; dx = dt, \, y = 3t \;\Rightarrow$

$\qquad dy = 3 \, dt, \; z = 1 \;\Rightarrow\; dz = 0 \, dt, \, 0 \le t \le 1$.

On $C_2$: $x = 2 \;\Rightarrow\; dx = 0 \, dt, \, y = 3 + 2t \;\Rightarrow$

$\qquad dy = 2 \, dt, \; z = 1 + t \;\Rightarrow\; dz = dt, \, 0 \le t \le 1$.

Then

$$\int_C (x + yz) \, dx + 2x \, dy + xyz \, dz$$

$$= \int_{C_1} (x + yz) \, dx + 2x \, dy + xyz \, dz + \int_{C_2} (x + yz) \, dx + 2x \, dy + xyz \, dz$$

$$= \int_0^1 (1 + t + (3t)(1)) \, dt + 2(1 + t) \cdot 3 \, dt + (1 + t)(3t)(1) \cdot 0 \, dt$$

$$\qquad + \int_0^1 (2 + (3 + 2t)(1 + t)) \cdot 0 \, dt + 2(2) \cdot 2 \, dt + (2)(3 + 2t)(1 + t) \, dt$$

$$= \int_0^1 (10t + 7) \, dt + \int_0^1 (4t^2 + 10t + 14) \, dt = \left[ 5t^2 + 7t \right]_0^1 + \left[ \tfrac{4}{3} t^3 + 5t^2 + 14t \right]_0^1 = 12 + \tfrac{61}{3} = \tfrac{97}{3}$$

**17.** (a) Along the line $x = -3$, the vectors of $\mathbf{F}$ have positive $y$-components, so since the path goes upward, the integrand $\mathbf{F} \cdot \mathbf{T}$ is

always positive. Therefore $\int_{C_1} \mathbf{F} \cdot d\mathbf{r} = \int_{C_1} \mathbf{F} \cdot \mathbf{T} \, ds$ is positive.

(b) All of the (nonzero) field vectors along the circle with radius 3 are pointed in the clockwise direction, that is, opposite the

direction to the path. So $\mathbf{F} \cdot \mathbf{T}$ is negative, and therefore $\int_{C_2} \mathbf{F} \cdot d\mathbf{r} = \int_{C_2} \mathbf{F} \cdot \mathbf{T} \, ds$ is negative.

**19.** $\mathbf{r}(t) = 11t^4 \, \mathbf{i} + t^3 \, \mathbf{j}$, so $\mathbf{F}(\mathbf{r}(t)) = (11t^4)(t^3) \, \mathbf{i} + 3(t^3)^2 \, \mathbf{j} = 11t^7 \, \mathbf{i} + 3t^6 \, \mathbf{j}$ and $\mathbf{r}'(t) = 44t^3 \, \mathbf{i} + 3t^2 \, \mathbf{j}$. Then

$\int_C \mathbf{F} \cdot d\mathbf{r} = \int_0^1 \mathbf{F}(\mathbf{r}(t)) \cdot \mathbf{r}'(t) \, dt = \int_0^1 (11t^7 \cdot 44t^3 + 3t^6 \cdot 3t^2) \, dt = \int_0^1 (484t^{10} + 9t^8) \, dt = \left[ 44t^{11} + t^9 \right]_0^1 = 45.$

**21.** $\int_C \mathbf{F} \cdot d\mathbf{r} = \int_0^1 \left\langle \sin t^3, \cos(-t^2), t^4 \right\rangle \cdot \left\langle 3t^2, -2t, 1 \right\rangle dt$

$= \int_0^1 (3t^2 \sin t^3 - 2t \cos t^2 + t^4) \, dt = \left[ -\cos t^3 - \sin t^2 + \tfrac{1}{5} t^5 \right]_0^1 = \tfrac{6}{5} - \cos 1 - \sin 1$

**23.** $\mathbf{F}(\mathbf{r}(t)) = (e^t)\left( e^{-t^2} \right) \mathbf{i} + \sin \left( e^{-t^2} \right) \mathbf{j} = e^{t-t^2} \, \mathbf{i} + \sin \left( e^{-t^2} \right) \mathbf{j}, \mathbf{r}'(t) = e^t \, \mathbf{i} - 2te^{-t^2} \, \mathbf{j}$. Then

$$\int_C \mathbf{F} \cdot d\mathbf{r} = \int_1^2 \mathbf{F}(\mathbf{r}(t)) \cdot \mathbf{r}'(t) \, dt = \int_1^2 \left[ e^{t-t^2} e^t + \sin \left( e^{-t^2} \right) \cdot \left( -2te^{-t^2} \right) \right] dt$$

$$= \int_1^2 \left[ e^{2t-t^2} - 2te^{-t^2} \sin \left( e^{-t^2} \right) \right] dt \approx 1.9633$$

**25.** $x = t^2, \ y = t^3, \ z = t^4$ so by Formula 9,

$$\int_C x \sin(y+z) \, ds = \int_0^5 (t^2) \sin(t^3 + t^4) \sqrt{(2t)^2 + (3t^2)^2 + (4t^3)^2} \, dt$$

$$= \int_0^5 t^2 \sin(t^3 + t^4) \sqrt{4t^2 + 9t^4 + 16t^6} \, dt \approx 15.0074$$

**27.** We graph $\mathbf{F}(x, y) = (x - y) \, \mathbf{i} + xy \, \mathbf{j}$ and the curve $C$. We see that most of the vectors starting on $C$ point in roughly the same

direction as $C$, so for these portions of $C$ the tangential component $\mathbf{F} \cdot \mathbf{T}$ is positive. Although some vectors in the third

quadrant which start on $C$ point in roughly the opposite direction, and hence give negative tangential components, it seems

reasonable that the effect of these portions of $C$ is outweighed by the positive tangential components. Thus, we would expect

$\int_C \mathbf{F} \cdot d\mathbf{r} = \int_C \mathbf{F} \cdot \mathbf{T} \, ds$ to be positive.

To verify, we evaluate $\int_C \mathbf{F} \cdot d\mathbf{r}$. The curve $C$ can be represented by $\mathbf{r}(t) = 2\cos t \, \mathbf{i} + 2\sin t \, \mathbf{j}, \ 0 \le t \le \tfrac{3\pi}{2}$,

so $\mathbf{F}(\mathbf{r}(t)) = (2\cos t - 2\sin t) \, \mathbf{i} + 4\cos t \sin t \, \mathbf{j}$ and $\mathbf{r}'(t) = -2\sin t \, \mathbf{i} + 2\cos t \, \mathbf{j}$. Then

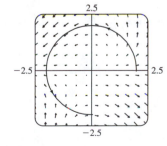

$\int_C \mathbf{F} \cdot d\mathbf{r} = \int_0^{3\pi/2} \mathbf{F}(\mathbf{r}(t)) \cdot \mathbf{r}'(t) \, dt$

$= \int_0^{3\pi/2} [-2\sin t (2\cos t - 2\sin t) + 2\cos t (4\cos t \sin t)] \, dt$

$= 4 \int_0^{3\pi/2} (\sin^2 t - \sin t \cos t + 2\sin t \cos^2 t) \, dt$

$= 3\pi + \tfrac{2}{3} \qquad$ [using a CAS]

**29.** (a) $\int_C \mathbf{F} \cdot d\mathbf{r} = \int_0^1 \left\langle e^{t^2-1}, t^5 \right\rangle \cdot \left\langle 2t, 3t^2 \right\rangle dt = \int_0^1 \left( 2te^{t^2-1} + 3t^7 \right) dt = \left[ e^{t^2-1} + \tfrac{3}{8} t^8 \right]_0^1 = \tfrac{11}{8} - 1/e$

(b) $\mathbf{r}(0) = \mathbf{0}$,  $\mathbf{F}(\mathbf{r}(0)) = \langle e^{-1}, 0 \rangle$;

$\mathbf{r}\left(\frac{1}{\sqrt{2}}\right) = \left\langle \frac{1}{2}, \frac{1}{2\sqrt{2}} \right\rangle$,  $\mathbf{F}\left(\mathbf{r}\left(\frac{1}{\sqrt{2}}\right)\right) = \left\langle e^{-1/2}, \frac{1}{4\sqrt{2}} \right\rangle$;

$\mathbf{r}(1) = \langle 1, 1 \rangle$,  $\mathbf{F}(\mathbf{r}(1)) = \langle 1, 1 \rangle$.

In order to generate the graph with Maple, we use the `line` command in the `plottools` package to define each of the vectors. For example,

$$\texttt{v1:=line([0,0],[exp(-1),0]):}$$

generates the vector from the vector field at the point $(0, 0)$ (but without an arrowhead) and gives it the name `v1`. To show everything on the same screen, we use the `display` command. In Mathematica, we use `ListPlot` (with the `PlotJoined - > True` option) to generate the vectors, and then `Show` to show everything on the same screen.

**31.** $x = e^{-t} \cos 4t$,  $y = e^{-t} \sin 4t$,  $z = e^{-t}$,  $0 \le t \le 2\pi$.

Then $\dfrac{dx}{dt} = e^{-t}(-\sin 4t)(4) - e^{-t} \cos 4t = -e^{-t}(4 \sin 4t + \cos 4t)$,

$\dfrac{dy}{dt} = e^{-t}(\cos 4t)(4) - e^{-t} \sin 4t = -e^{-t}(-4 \cos 4t + \sin 4t)$, and $\dfrac{dz}{dt} = -e^{-t}$, so

$$\sqrt{\left(\frac{dx}{dt}\right)^2 + \left(\frac{dy}{dt}\right)^2 + \left(\frac{dz}{dt}\right)^2} = \sqrt{(-e^{-t})^2[(4 \sin 4t + \cos 4t)^2 + (-4 \cos 4t + \sin 4t)^2 + 1]}$$

$$= e^{-t} \sqrt{16(\sin^2 4t + \cos^2 4t) + \sin^2 4t + \cos^2 4t + 1} = 3\sqrt{2}\, e^{-t}$$

Therefore $\int_C x^3 y^2 z \, ds = \int_0^{2\pi} (e^{-t} \cos 4t)^3 (e^{-t} \sin 4t)^2 (e^{-t}) \left(3\sqrt{2}\, e^{-t}\right) dt$

$$= \int_0^{2\pi} 3\sqrt{2}\, e^{-7t} \cos^3 4t \sin^2 4t \, dt = \tfrac{172,704}{5,632,705} \sqrt{2}\left(1 - e^{-14\pi}\right)$$

**33.** We use the parametrization $x = 2 \cos t$, $y = 2 \sin t$, $-\frac{\pi}{2} \le t \le \frac{\pi}{2}$. Then

$ds = \sqrt{\left(\frac{dx}{dt}\right)^2 + \left(\frac{dy}{dt}\right)^2}\, dt = \sqrt{(-2 \sin t)^2 + (2 \cos t)^2}\, dt = 2\, dt$, so $m = \int_C k \, ds = 2k \int_{-\pi/2}^{\pi/2} dt = 2k(\pi)$,

$\overline{x} = \frac{1}{2\pi k} \int_C xk \, ds = \frac{1}{2\pi} \int_{-\pi/2}^{\pi/2} (2 \cos t) 2 \, dt = \frac{1}{2\pi} \left[4 \sin t\right]_{-\pi/2}^{\pi/2} = \frac{4}{\pi}$, $\overline{y} = \frac{1}{2\pi k} \int_C yk \, ds = \frac{1}{2\pi} \int_{-\pi/2}^{\pi/2} (2 \sin t) 2 \, dt = 0$.

Hence $(\overline{x}, \overline{y}) = \left(\frac{4}{\pi}, 0\right)$.

**35.** (a) $\overline{x} = \dfrac{1}{m} \int_C x\rho(x, y, z) \, ds$, $\overline{y} = \dfrac{1}{m} \int_C y\rho(x, y, z) \, ds$, $\overline{z} = \dfrac{1}{m} \int_C z\rho(x, y, z) \, ds$ where $m = \int_C \rho(x, y, z) \, ds$.

(b) $m = \int_C k \, ds = k \int_0^{2\pi} \sqrt{4 \sin^2 t + 4 \cos^2 t + 9} \, dt = k\sqrt{13} \int_0^{2\pi} dt = 2\pi k \sqrt{13}$,

$$\overline{x} = \frac{1}{2\pi k \sqrt{13}} \int_0^{2\pi} 2k\sqrt{13} \sin t \, dt = 0, \overline{y} = \frac{1}{2\pi k \sqrt{13}} \int_0^{2\pi} 2k\sqrt{13} \cos t \, dt = 0,$$

$$\overline{z} = \frac{1}{2\pi k \sqrt{13}} \int_0^{2\pi} \left(k\sqrt{13}\right)(3t) \, dt = \frac{3}{2\pi}\left(2\pi^2\right) = 3\pi. \text{ Hence } (\overline{x}, \overline{y}, \overline{z}) = (0, 0, 3\pi).$$

**37.** From Example 3, $\rho(x, y) = k(1 - y)$,   $x = \cos t$,   $y = \sin t$, and $ds = dt$, $0 \le t \le \pi$   $\Rightarrow$

$$I_x = \int_C y^2 \rho(x, y)\, ds = \int_0^\pi \sin^2 t\, [k(1 - \sin t)]\, dt = k \int_0^\pi (\sin^2 t - \sin^3 t)\, dt$$

$$= \tfrac{1}{2} k \int_0^\pi (1 - \cos 2t)\, dt - k \int_0^\pi (1 - \cos^2 t) \sin t\, dt \qquad \left[ \begin{array}{l} \text{Let } u = \cos t,\, du = -\sin t\, dt \\ \text{in the second integral} \end{array} \right]$$

$$= k \left[ \tfrac{\pi}{2} + \int_1^{-1} (1 - u^2)\, du \right] = k \left( \tfrac{\pi}{2} - \tfrac{4}{3} \right)$$

$$I_y = \int_C x^2 \rho(x, y)\, ds = k \int_0^\pi \cos^2 t\, (1 - \sin t)\, dt = \tfrac{k}{2} \int_0^\pi (1 + \cos 2t)\, dt - k \int_0^\pi \cos^2 t \sin t\, dt$$

$$= k \left( \tfrac{\pi}{2} - \tfrac{2}{3} \right), \text{ using the same substitution as above.}$$

**39.** $W = \int_C \mathbf{F} \cdot d\mathbf{r} = \int_0^{2\pi} \langle t - \sin t, 3 - \cos t \rangle \cdot \langle 1 - \cos t, \sin t \rangle\, dt$

$$= \int_0^{2\pi} (t - t \cos t - \sin t + \sin t \cos t + 3 \sin t - \sin t \cos t)\, dt$$

$$= \int_0^{2\pi} (t - t \cos t + 2 \sin t)\, dt = \left[ \tfrac{1}{2} t^2 - (t \sin t + \cos t) - 2 \cos t \right]_0^{2\pi} \qquad \left[ \begin{array}{l} \text{integrate by parts} \\ \text{in the second term} \end{array} \right]$$

$$= 2\pi^2$$

**41.** $\mathbf{r}(t) = \langle 1 + 2t, 4t, 2t \rangle$,   $0 \le t \le 1$,

$$W = \int_C \mathbf{F} \cdot d\mathbf{r} = \int_0^1 \langle 6t, 1 + 4t, 1 + 6t \rangle \cdot \langle 2, 4, 2 \rangle\, dt = \int_0^1 (12t + 4(1 + 4t) + 2(1 + 6t))\, dt$$

$$= \int_0^1 (40t + 6)\, dt = \left[ 20t^2 + 6t \right]_0^1 = 26$$

**43.** Let $\mathbf{F} = 185\,\mathbf{k}$. To parametrize the staircase, let $x = 20 \cos t$,   $y = 20 \sin t$,   $z = \tfrac{90}{6\pi} t = \tfrac{15}{\pi} t$,   $0 \le t \le 6\pi$   $\Rightarrow$

$$W = \int_C \mathbf{F} \cdot d\mathbf{r} = \int_0^{6\pi} \langle 0, 0, 185 \rangle \cdot \langle -20 \sin t, 20 \cos t, \tfrac{15}{\pi} \rangle\, dt = (185) \tfrac{15}{\pi} \int_0^{6\pi} dt = (185)(90) \approx 1.67 \times 10^4 \text{ ft-lb}$$

**45.** (a) $\mathbf{r}(t) = \langle \cos t, \sin t \rangle$,   $0 \le t \le 2\pi$, and let $\mathbf{F} = \langle a, b \rangle$. Then

$$W = \int_C \mathbf{F} \cdot d\mathbf{r} = \int_0^{2\pi} \langle a, b \rangle \cdot \langle -\sin t, \cos t \rangle\, dt = \int_0^{2\pi} (-a \sin t + b \cos t)\, dt = \left[ a \cos t + b \sin t \right]_0^{2\pi}$$

$$= a + 0 - a + 0 = 0$$

(b) Yes. $\mathbf{F}(x, y) = k\,\mathbf{x} = \langle kx, ky \rangle$ and

$$W = \int_C \mathbf{F} \cdot d\mathbf{r} = \int_0^{2\pi} \langle k \cos t, k \sin t \rangle \cdot \langle -\sin t, \cos t \rangle\, dt = \int_0^{2\pi} (-k \sin t \cos t + k \sin t \cos t)\, dt = \int_0^{2\pi} 0\, dt = 0.$$

**47.** The work done in moving the object is $\int_C \mathbf{F} \cdot d\mathbf{r} = \int_C \mathbf{F} \cdot \mathbf{T}\, ds$. We can approximate this integral by dividing $C$ into

7 segments of equal length $\Delta s = 2$ and approximating $\mathbf{F} \cdot \mathbf{T}$, that is, the tangential component of force, at a point $(x_i^*, y_i^*)$ on

each segment. Since $C$ is composed of straight line segments, $\mathbf{F} \cdot \mathbf{T}$ is the scalar projection of each force vector onto $C$.

If we choose $(x_i^*, y_i^*)$ to be the point on the segment closest to the origin, then the work done is

$$\int_C \mathbf{F} \cdot \mathbf{T}\, ds \approx \sum_{i=1}^{7} [\mathbf{F}(x_i^*, y_i^*) \cdot \mathbf{T}(x_i^*, y_i^*)]\, \Delta s = [2 + 2 + 2 + 2 + 1 + 1 + 1](2) = 22. \text{ Thus, we estimate the work done to}$$

be approximately 22 J.

## 13.3 The Fundamental Theorem for Line Integrals

**1.** $C$ appears to be a smooth curve, and since $\nabla f$ is continuous, we know $f$ is differentiable. Then Theorem 2 says that the value of $\int_C \nabla f \cdot d\mathbf{r}$ is simply the difference of the values of $f$ at the terminal and initial points of $C$. From the graph, this is $50 - 10 = 40$.

**3.** $\partial(2x - 3y)/\partial y = -3 = \partial(-3x + 4y - 8)/\partial x$ and the domain of $\mathbf{F}$ is $\mathbb{R}^2$ which is open and simply-connected, so by Theorem 6 $\mathbf{F}$ is conservative. Thus, there exists a function $f$ such that $\nabla f = \mathbf{F}$, that is, $f_x(x, y) = 2x - 3y$ and $f_y(x, y) = -3x + 4y - 8$. But $f_x(x, y) = 2x - 3y$ implies $f(x, y) = x^2 - 3xy + g(y)$ and differentiating both sides of this equation with respect to $y$ gives $f_y(x, y) = -3x + g'(y)$. Thus $-3x + 4y - 8 = -3x + g'(y)$ so $g'(y) = 4y - 8$ and $g(y) = 2y^2 - 8y + K$ where $K$ is a constant. Hence $f(x, y) = x^2 - 3xy + 2y^2 - 8y + K$ is a potential function for $\mathbf{F}$.

**5.** $\partial(e^x \cos y)/\partial y = -e^x \sin y$, $\partial(e^x \sin y)/\partial x = e^x \sin y$. Since these are not equal, $\mathbf{F}$ is not conservative.

**7.** $\partial(ye^x + \sin y)/\partial y = e^x + \cos y = \partial(e^x + x\cos y)/\partial x$ and the domain of $\mathbf{F}$ is $\mathbb{R}^2$. Hence $\mathbf{F}$ is conservative so there exists a function $f$ such that $\nabla f = \mathbf{F}$. Then $f_x(x, y) = ye^x + \sin y$ implies $f(x, y) = ye^x + x\sin y + g(y)$ and $f_y(x, y) = e^x + x\cos y + g'(y)$. But $f_y(x, y) = e^x + x\cos y$ so $g(y) = K$ and $f(x, y) = ye^x + x\sin y + K$ is a potential function for $\mathbf{F}$.

**9.** $\partial(\ln y + 2xy^3)/\partial y = 1/y + 6xy^2 = \partial(3x^2y^2 + x/y)/\partial x$ and the domain of $\mathbf{F}$ is $\{(x, y) \mid y > 0\}$ which is open and simply connected. Hence $\mathbf{F}$ is conservative so there exists a function $f$ such that $\nabla f = \mathbf{F}$. Then $f_x(x, y) = \ln y + 2xy^3$ implies $f(x, y) = x\ln y + x^2y^3 + g(y)$ and $f_y(x, y) = x/y + 3x^2y^2 + g'(y)$. But $f_y(x, y) = 3x^2y^2 + x/y$ so $g'(y) = 0 \Rightarrow g(y) = K$ and $f(x, y) = x\ln y + x^2y^3 + K$ is a potential function for $\mathbf{F}$.

**11.** (a) $\mathbf{F}$ has continuous first-order partial derivatives and $\dfrac{\partial}{\partial y} 2xy = 2x = \dfrac{\partial}{\partial x}(x^2)$ on $\mathbb{R}^2$, which is open and simply-connected.

Thus, $\mathbf{F}$ is conservative by Theorem 6. Then we know that the line integral of $\mathbf{F}$ is independent of path; in particular, the value of $\int_C \mathbf{F} \cdot d\mathbf{r}$ depends only on the endpoints of $C$. Since all three curves have the same initial and terminal points, $\int_C \mathbf{F} \cdot d\mathbf{r}$ will have the same value for each curve.

(b) We first find a potential function $f$, so that $\nabla f = \mathbf{F}$. We know $f_x(x, y) = 2xy$ and $f_y(x, y) = x^2$. Integrating $f_x(x, y)$ with respect to $x$, we have $f(x, y) = x^2y + g(y)$. Differentiating both sides with respect to $y$ gives $f_y(x, y) = x^2 + g'(y)$, so we must have $x^2 + g'(y) = x^2 \Rightarrow g'(y) = 0 \Rightarrow g(y) = K$, a constant. Thus $f(x, y) = x^2y + K$. All three curves start at $(1, 2)$ and end at $(3, 2)$, so by Theorem 2, $\int_C \mathbf{F} \cdot d\mathbf{r} = f(3, 2) - f(1, 2) = 18 - 2 = 16$ for each curve.

**13.** (a) $f_x(x, y) = xy^2$ implies $f(x, y) = \frac{1}{2}x^2y^2 + g(y)$ and $f_y(x, y) = x^2y + g'(y)$. But $f_y(x, y) = x^2y$ so $g'(y) = 0 \Rightarrow g(y) = K$, a constant. We can take $K = 0$, so $f(x, y) = \frac{1}{2}x^2y^2$.

(b) The initial point of $C$ is $\mathbf{r}(0) = (0, 1)$ and the terminal point is $\mathbf{r}(1) = (2, 1)$, so $\int_C \mathbf{F} \cdot d\mathbf{r} = f(2, 1) - f(0, 1) = 2 - 0 = 2$.

**15.** (a) $f_x(x, y, z) = yz$ implies $f(x, y, z) = xyz + g(y, z)$ and so $f_y(x, y, z) = xz + g_y(y, z)$. But $f_y(x, y, z) = xz$ so

$g_y(y, z) = 0 \Rightarrow g(y, z) = h(z)$. Thus $f(x, y, z) = xyz + h(z)$ and $f_z(x, y, z) = xy + h'(z)$. But

$f_z(x, y, z) = xy + 2z$, so $h'(z) = 2z \Rightarrow h(z) = z^2 + K$. Hence $f(x, y, z) = xyz + z^2$ (taking $K = 0$).

(b) $\int_C \mathbf{F} \cdot d\mathbf{r} = f(4, 6, 3) - f(1, 0, -2) = 81 - 4 = 77$.

**17.** (a) $f_x(x, y, z) = y^2 \cos z$ implies $f(x, y, z) = xy^2 \cos z + g(y, z)$ and so $f_y(x, y, z) = 2xy \cos z + g_y(y, z)$. But

$f_y(x, y, z) = 2xy \cos z$ so $g_y(y, z) = 0 \Rightarrow g(y, z) = h(z)$. Thus $f(x, y, z) = xy^2 \cos z + h(z)$ and

$f_z(x, y, z) = -xy^2 \sin z + h'(z)$. But $f_z(x, y, z) = -xy^2 \sin z$, so $h'(z) = 0 \Rightarrow h(z) = K$. Hence

$f(x, y, z) = xy^2 \cos z$ (taking $K = 0$).

(b) $\mathbf{r}(0) = \langle 0, 0, 0 \rangle$, $\mathbf{r}(\pi) = \langle \pi^2, 0, \pi \rangle$ so $\int_C \mathbf{F} \cdot d\mathbf{r} = f(\pi^2, 0, \pi) - f(0, 0, 0) = 0 - 0 = 0$.

**19.** Here $\mathbf{F}(x, y) = \tan y\, \mathbf{i} + x \sec^2 y\, \mathbf{j}$. Then $f(x, y) = x \tan y$ is a potential function for $\mathbf{F}$, that is, $\nabla f = \mathbf{F}$ so

$\mathbf{F}$ is conservative and thus its line integral is independent of path. Hence

$\int_C \tan y\, dx + x \sec^2 y\, dy = \int_C \mathbf{F} \cdot d\mathbf{r} = f\left(2, \frac{\pi}{4}\right) - f(1, 0) = 2 \tan \frac{\pi}{4} - \tan 0 = 2$.

**21.** If $\mathbf{F}$ is conservative, then $\int_C \mathbf{F} \cdot d\mathbf{r}$ is independent of path. This means that the work done along all piecewise-smooth curves

that have the described initial and terminal points is the same. Your reply: It doesn't matter which curve is chosen.

**23.** $\mathbf{F}(x, y) = 2y^{3/2}\, \mathbf{i} + 3x \sqrt{y}\, \mathbf{j}$, $W = \int_C \mathbf{F} \cdot d\mathbf{r}$. Since $\partial(2y^{3/2})/\partial y = 3\sqrt{y} = \partial(3x\sqrt{y})/\partial x$, there exists a function $f$

such that $\nabla f = \mathbf{F}$. In fact, $f_x(x, y) = 2y^{3/2} \Rightarrow f(x, y) = 2xy^{3/2} + g(y) \Rightarrow f_y(x, y) = 3xy^{1/2} + g'(y)$. But

$f_y(x, y) = 3x \sqrt{y}$ so $g'(y) = 0$ or $g(y) = K$. We can take $K = 0 \Rightarrow f(x, y) = 2xy^{3/2}$. Thus

$W = \int_C \mathbf{F} \cdot d\mathbf{r} = f(2, 4) - f(1, 1) = 2(2)(8) - 2(1) = 30$.

**25.** We know that if the vector field (call it $\mathbf{F}$) is conservative, then around any closed path $C$, $\int_C \mathbf{F} \cdot d\mathbf{r} = 0$. But take $C$ to be a

circle centered at the origin, oriented counterclockwise. All of the field vectors that start on $C$ are roughly in the direction of

motion along $C$, so the integral around $C$ will be positive. Therefore the field is not conservative.

**27.**

From the graph, it appears that $\mathbf{F}$ is conservative, since around all closed

paths, the number and size of the field vectors pointing in directions similar

to that of the path seem to be roughly the same as the number and size of the

vectors pointing in the opposite direction. To check, we calculate

$$\frac{\partial}{\partial y}(\sin y) = \cos y = \frac{\partial}{\partial x}(1 + x \cos y).$$ Thus $\mathbf{F}$ is conservative, by

Theorem 6.

**29.** Since $\mathbf{F}$ is conservative, there exists a function $f$ such that $\mathbf{F} = \nabla f$, that is, $P = f_x$, $Q = f_y$, and $R = f_z$. Since $P$,

$Q$, and $R$ have continuous first order partial derivatives, Clairaut's Theorem says that $\partial P/\partial y = f_{xy} = f_{yx} = \partial Q/\partial x$,

$\partial P/\partial z = f_{xz} = f_{zx} = \partial R/\partial x$, and $\partial Q/\partial z = f_{yz} = f_{zy} = \partial R/\partial y$.

**31.** $D = \{(x, y) \mid 0 < y < 3\}$ consists of those points between, but not

on, the horizontal lines $y = 0$ and $y = 3$.

(a) Since $D$ does not include any of its boundary points, it is open. More

formally, at any point in $D$ there is a disk centered at that point that

lies entirely in $D$.

(b) Any two points chosen in $D$ can always be joined by a path that lies

entirely in $D$, so $D$ is connected. ($D$ consists of just one "piece.")

(c) $D$ is connected and it has no holes, so it's simply-connected. (Every simple closed curve in $D$ encloses only points that are

in $D$.)

**33.** $D = \{(x, y) \mid 1 \le x^2 + y^2 \le 4, \ y \ge 0\}$ is the semiannular region

in the upper half-plane between circles centered at the origin of radii

1 and 2 (including all boundary points).

(a) $D$ includes boundary points, so it is not open. [Note that at any

boundary point, $(1, 0)$ for instance, any disk centered there cannot lie

entirely in $D$.]

(b) The region consists of one piece, so it's connected.

(c) $D$ is connected and has no holes, so it's simply-connected.

**35.** (a) $P = -\dfrac{y}{x^2 + y^2}, \ \dfrac{\partial P}{\partial y} = \dfrac{y^2 - x^2}{(x^2 + y^2)^2}$ and $Q = \dfrac{x}{x^2 + y^2}, \ \dfrac{\partial Q}{\partial x} = \dfrac{y^2 - x^2}{(x^2 + y^2)^2}$. Thus $\dfrac{\partial P}{\partial y} = \dfrac{\partial Q}{\partial x}$.

(b) $C_1$: $x = \cos t, \ y = \sin t, 0 \le t \le \pi, C_2$: $\ x = \cos t, \ y = \sin t, t = 2\pi$ to $t = \pi$. Then

$$\int_{C_1} \mathbf{F} \cdot d\mathbf{r} = \int_0^\pi \frac{(-\sin t)(-\sin t) + (\cos t)(\cos t)}{\cos^2 t + \sin^2 t} \, dt = \int_0^\pi dt = \pi \text{ and } \int_{C_2} \mathbf{F} \cdot d\mathbf{r} = \int_{2\pi}^\pi dt = -\pi$$

Since these aren't equal, the line integral of $\mathbf{F}$ isn't independent of path. (Or notice that $\int_{C_3} \mathbf{F} \cdot d\mathbf{r} = \int_0^{2\pi} dt = 2\pi$ where

$C_3$ is the circle $x^2 + y^2 = 1$, and apply the contrapositive of Theorem 3.) This doesn't contradict Theorem 6, since the

domain of $\mathbf{F}$, which is $\mathbb{R}^2$ except the origin, isn't simply-connected.

## 13.4 Green's Theorem

**1.** (a) Parametric equations for $C$ are $x = 2\cos t, \ y = 2\sin t, \ 0 \le t \le 2\pi$. Then

$$\oint_C (x - y) \, dx + (x + y) \, dy = \int_0^{2\pi} [(2\cos t - 2\sin t)(-2\sin t) + (2\cos t + 2\sin t)(2\cos t)] \, dt$$

$$= \int_0^{2\pi} (4\sin^2 t + 4\cos^2 t) \, dt = \int_0^{2\pi} 4 \, dt = 4t\Big]_0^{2\pi} = 8\pi$$

(b) Note that $C$ as given in part (a) is a positively oriented, smooth, simple closed curve. Then by Green's Theorem,

$$\oint_C (x - y) \, dx + (x + y) \, dy = \iint_D \left[ \frac{\partial}{\partial x} (x + y) - \frac{\partial}{\partial y} (x - y) \right] dA = \iint_D [1 - (-1)] \, dA = 2 \iint_D dA$$

$$= 2A(D) = 2\pi(2)^2 = 8\pi$$

**3. (a)**

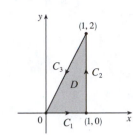

$C_1$: $x = t$ $\Rightarrow$ $dx = dt$, $y = 0$ $\Rightarrow$ $dy = 0\, dt$, $0 \le t \le 1$.

$C_2$: $x = 1$ $\Rightarrow$ $dx = 0\, dt$, $y = t$ $\Rightarrow$ $dy = dt$, $0 \le t \le 2$.

$C_3$: $x = 1 - t$ $\Rightarrow$ $dx = -dt$, $y = 2 - 2t$ $\Rightarrow$ $dy = -2\, dt$, $0 \le t \le 1$.

Thus $\displaystyle\oint_C xy\, dx + x^2 y^3\, dy = \oint_{C_1 + C_2 + C_3} xy\, dx + x^2 y^3\, dy$

$$= \int_0^1 0\, dt + \int_0^2 t^3\, dt + \int_0^1 \left[-(1-t)(2-2t) - 2(1-t)^2(2-2t)^3\right] dt$$

$$= 0 + \left[\tfrac{1}{4}t^4\right]_0^2 + \left[\tfrac{2}{3}(1-t)^3 + \tfrac{8}{3}(1-t)^6\right]_0^1 = 4 - \tfrac{10}{3} = \tfrac{2}{3}$$

**(b)** $\displaystyle\oint_C xy\, dx + x^2 y^3\, dy = \iint_D \left[\frac{\partial}{\partial x}(x^2 y^3) - \frac{\partial}{\partial y}(xy)\right] dA = \int_0^1 \int_0^{2x} (2xy^3 - x)\, dy\, dx$

$$= \int_0^1 \left[\tfrac{1}{2}xy^4 - xy\right]_{y=0}^{y=2x} dx = \int_0^1 (8x^5 - 2x^2)\, dx = \tfrac{4}{3} - \tfrac{2}{3} = \tfrac{2}{3}$$

**5.**

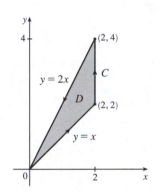

The region $D$ enclosed by $C$ is given by $\{(x,y) \mid 0 \le x \le 2, x \le y \le 2x\}$, so

$$\int_C xy^2\, dx + 2x^2 y\, dy = \iint_D \left[\frac{\partial}{\partial x}(2x^2 y) - \frac{\partial}{\partial y}(xy^2)\right] dA$$

$$= \int_0^2 \int_x^{2x} (4xy - 2xy)\, dy\, dx$$

$$= \int_0^2 \left[xy^2\right]_{y=x}^{y=2x} dx$$

$$= \int_0^2 3x^3\, dx = \tfrac{3}{4}x^4\Big]_0^2 = 12$$

**7.** $\displaystyle\int_C \left(y + e^{\sqrt{x}}\right) dx + (2x + \cos y^2)\, dy = \iint_D \left[\frac{\partial}{\partial x}(2x + \cos y^2) - \frac{\partial}{\partial y}\left(y + e^{\sqrt{x}}\right)\right] dA$

$$= \int_0^1 \int_{y^2}^{\sqrt{y}} (2 - 1)\, dx\, dy = \int_0^1 (y^{1/2} - y^2)\, dy = \tfrac{1}{3}$$

**9.** $\displaystyle\int_C y^3\, dx - x^3\, dy = \iint_D \left[\frac{\partial}{\partial x}(-x^3) - \frac{\partial}{\partial y}(y^3)\right] dA = \iint_D (-3x^2 - 3y^2)\, dA = \int_0^{2\pi} \int_0^2 (-3r^2)\, r\, dr\, d\theta$

$$= -3 \int_0^{2\pi} d\theta \int_0^2 r^3\, dr = -3(2\pi)(4) = -24\pi$$

**11.** $\mathbf{F}(x,y) = \left\langle \sqrt{x} + y^3, x^2 + \sqrt{y}\right\rangle$ and the region $D$ enclosed by $C$ is given by $\{(x,y) \mid 0 \le x \le \pi, 0 \le y \le \sin x\}$.

$C$ is traversed clockwise, so $-C$ gives the positive orientation.

$$\int_C \mathbf{F} \cdot d\mathbf{r} = -\int_{-C} \left(\sqrt{x} + y^3\right) dx + (x^2 + \sqrt{y})\, dy = -\iint_D \left[\frac{\partial}{\partial x}(x^2 + \sqrt{y}) - \frac{\partial}{\partial y}\left(\sqrt{x} + y^3\right)\right] dA$$

$$= -\int_0^\pi \int_0^{\sin x} (2x - 3y^2)\, dy\, dx = -\int_0^\pi \left[2xy - y^3\right]_{y=0}^{y=\sin x} dx$$

$$= -\int_0^\pi (2x \sin x - \sin^3 x)\, dx = -\int_0^\pi (2x \sin x - (1 - \cos^2 x)\sin x)\, dx$$

$$= -\left[2 \sin x - 2x \cos x + \cos x - \tfrac{1}{3}\cos^3 x\right]_0^\pi \quad \text{[integrate by parts in the first term]}$$

$$= -\left(2\pi - 2 + \tfrac{2}{3}\right) = \tfrac{4}{3} - 2\pi$$

**13.** $\mathbf{F}(x, y) = \langle e^x + x^2 y, e^y - xy^2 \rangle$ and the region $D$ enclosed by $C$ is the disk $x^2 + y^2 \le 25$.

C is traversed clockwise, so $-C$ gives the positive orientation.

$$\int_C \mathbf{F} \cdot d\mathbf{r} = -\int_{-C} (e^x + x^2 y)\, dx + (e^y - xy^2)\, dy = -\iint_D \left[ \frac{\partial}{\partial x}(e^y - xy^2) - \frac{\partial}{\partial y}(e^x + x^2 y) \right] dA$$

$$= -\iint_D (-y^2 - x^2)\, dA = \iint_D (x^2 + y^2)\, dA = \int_0^{2\pi} \int_0^5 (r^2)\, r\, dr\, d\theta$$

$$= \int_0^{2\pi} d\theta \int_0^5 r^3\, dr = 2\pi \left[ \tfrac{1}{4} r^4 \right]_0^5 = \frac{625}{2} \pi$$

**15.** Here $C = C_1 + C_2$ where

$C_1$ can be parametrized as $x = t$, $y = 1$, $-1 \le t \le 1$, and

$C_2$ is given by $x = -t$, $y = 2 - t^2$, $-1 \le t \le 1$.

Then the line integral is

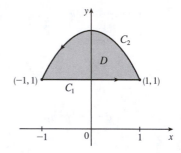

$$\oint_{C_1 + C_2} y^2 e^x\, dx + x^2 e^y\, dy = \int_{-1}^1 [1 \cdot e^t + t^2 e \cdot 0]\, dt$$

$$+ \int_{-1}^1 [(2 - t^2)^2 e^{-t}(-1) + (-t)^2 e^{2 - t^2}(-2t)]\, dt$$

$$= \int_{-1}^1 [e^t - (2 - t^2)^2 e^{-t} - 2t^3 e^{2 - t^2}]\, dt = -8e + 48e^{-1}$$

according to a CAS. The double integral is

$$\iint_D \left( \frac{\partial Q}{\partial x} - \frac{\partial P}{\partial y} \right) dA = \int_{-1}^1 \int_1^{2 - x^2} (2xe^y - 2ye^x)\, dy\, dx = -8e + 48e^{-1},$$ verifying Green's Theorem in this case.

**17.** By Green's Theorem, $W = \int_C \mathbf{F} \cdot d\mathbf{r} = \int_C x(x + y)\, dx + xy^2\, dy = \iint_D (y^2 - x)\, dA$ where $C$ is the path described in the question and $D$ is the triangle bounded by $C$. So

$$W = \int_0^1 \int_0^{1-x} (y^2 - x)\, dy\, dx = \int_0^1 \left[ \tfrac{1}{3} y^3 - xy \right]_{y=0}^{y=1-x} dx = \int_0^1 \left( \tfrac{1}{3}(1 - x)^3 - x(1 - x) \right) dx$$

$$= \left[ -\tfrac{1}{12}(1 - x)^4 - \tfrac{1}{2} x^2 + \tfrac{1}{3} x^3 \right]_0^1 = \left( -\tfrac{1}{2} + \tfrac{1}{3} \right) - \left( -\tfrac{1}{12} \right) = -\tfrac{1}{12}$$

**19.** Let $C_1$ be the arch of the cycloid from $(0, 0)$ to $(2\pi, 0)$, which corresponds to $0 \le t \le 2\pi$, and let $C_2$ be the segment from $(2\pi, 0)$ to $(0, 0)$, so $C_2$ is given by $x = 2\pi - t$, $y = 0$, $0 \le t \le 2\pi$. Then $C = C_1 \cup C_2$ is traversed clockwise, so $-C$ is oriented positively. Thus $-C$ encloses the area under one arch of the cycloid and from (5) we have

$$A = -\oint_{-C} y\, dx = \int_{C_1} y\, dx + \int_{C_2} y\, dx = \int_0^{2\pi} (1 - \cos t)(1 - \cos t)\, dt + \int_0^{2\pi} 0\, (-dt)$$

$$= \int_0^{2\pi} (1 - 2\cos t + \cos^2 t)\, dt + 0 = \left[ t - 2\sin t + \tfrac{1}{2} t + \tfrac{1}{4} \sin 2t \right]_0^{2\pi} = 3\pi$$

**21.** (a) Using Equation 13.2.8, we write parametric equations of the line segment as $x = (1 - t)x_1 + tx_2$, $y = (1 - t)y_1 + ty_2$, $0 \le t \le 1$. Then $dx = (x_2 - x_1)\, dt$ and $dy = (y_2 - y_1)\, dt$, so

$$\int_C x\, dy - y\, dx = \int_0^1 [(1 - t)x_1 + tx_2](y_2 - y_1)\, dt + [(1 - t)y_1 + ty_2](x_2 - x_1)\, dt$$

$$= \int_0^1 (x_1(y_2 - y_1) - y_1(x_2 - x_1) + t[(y_2 - y_1)(x_2 - x_1) - (x_2 - x_1)(y_2 - y_1)])\, dt$$

$$= \int_0^1 (x_1 y_2 - x_2 y_1)\, dt = x_1 y_2 - x_2 y_1$$

(b) We apply Green's Theorem to the path $C = C_1 \cup C_2 \cup \cdots \cup C_n$, where $C_i$ is the line segment that joins $(x_i, y_i)$ to $(x_{i+1}, y_{i+1})$ for $i = 1, 2, \ldots, n - 1$, and $C_n$ is the line segment that joins $(x_n, y_n)$ to $(x_1, y_1)$. From (5), $\tfrac{1}{2} \int_C x\, dy - y\, dx = \iint_D dA$, where $D$ is the polygon bounded by $C$. Therefore

$$\text{area of polygon} = A(D) = \iint_D dA = \tfrac{1}{2} \int_C x\, dy - y\, dx$$

$$= \tfrac{1}{2}\left( \int_{C_1} x\, dy - y\, dx + \int_{C_2} x\, dy - y\, dx + \cdots + \int_{C_{n-1}} x\, dy - y\, dx + \int_{C_n} x\, dy - y\, dx \right)$$

To evaluate these integrals we use the formula from (a) to get

$$A(D) = \tfrac{1}{2}[(x_1 y_2 - x_2 y_1) + (x_2 y_3 - x_3 y_2) + \cdots + (x_{n-1} y_n - x_n y_{n-1}) + (x_n y_1 - x_1 y_n)].$$

(c) $A = \tfrac{1}{2}[(0 \cdot 1 - 2 \cdot 0) + (2 \cdot 3 - 1 \cdot 1) + (1 \cdot 2 - 0 \cdot 3) + (0 \cdot 1 - (-1) \cdot 2) + (-1 \cdot 0 - 0 \cdot 1)]$

$\qquad = \tfrac{1}{2}(0 + 5 + 2 + 2) = \tfrac{9}{2}$

**23.** We orient the quarter-circular region as shown in the figure.

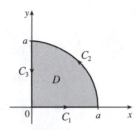

$A = \tfrac{1}{4}\pi a^2$ so $\overline{x} = \dfrac{1}{\pi a^2/2} \oint_C x^2\, dy$ and $\overline{y} = -\dfrac{1}{\pi a^2/2} \oint_C y^2\, dx$.

Here $C = C_1 + C_2 + C_3$ where $C_1$: $x = t$, $y = 0$, $0 \le t \le a$;

$C_2$: $x = a\cos t$, $y = a\sin t$, $0 \le t \le \tfrac{\pi}{2}$; and

$C_3$: $x = 0$, $y = a - t$, $0 \le t \le a$. Then

$\oint_C x^2\, dy = \int_{C_1} x^2\, dy + \int_{C_2} x^2\, dy + \int_{C_3} x^2\, dy = \int_0^a 0\, dt + \int_0^{\pi/2}(a\cos t)^2(a\cos t)\, dt + \int_0^a 0\, dt$

$\qquad = \int_0^{\pi/2} a^3 \cos^3 t\, dt = a^3 \int_0^{\pi/2}(1 - \sin^2 t)\cos t\, dt = a^3\left[\sin t - \tfrac{1}{3}\sin^3 t\right]_0^{\pi/2} = \tfrac{2}{3}a^3$

so $\overline{x} = \dfrac{1}{\pi a^2/2} \oint_C x^2\, dy = \dfrac{4a}{3\pi}$.

$\oint_C y^2\, dx = \int_{C_1} y^2\, dx + \int_{C_2} y^2\, dx + \int_{C_3} y^2\, dx = \int_0^a 0\, dt + \int_0^{\pi/2}(a\sin t)^2(-a\sin t)\, dt + \int_0^a 0\, dt$

$\qquad = \int_0^{\pi/2}(-a^3 \sin^3 t)\, dt = -a^3 \int_0^{\pi/2}(1 - \cos^2 t)\sin t\, dt = -a^3\left[\tfrac{1}{3}\cos^3 t - \cos t\right]_0^{\pi/2} = -\tfrac{2}{3}a^3$,

so $\overline{y} = -\dfrac{1}{\pi a^2/2} \oint_C y^2\, dx = \dfrac{4a}{3\pi}$. Thus $(\overline{x}, \overline{y}) = \left( \dfrac{4a}{3\pi}, \dfrac{4a}{3\pi} \right)$.

**25.** By Green's Theorem, $-\tfrac{1}{3}\rho \oint_C y^3\, dx = -\tfrac{1}{3}\rho \iint_D (-3y^2)\, dA = \iint_D y^2 \rho\, dA = I_x$ and

$\tfrac{1}{3}\rho \oint_C x^3\, dy = \tfrac{1}{3}\rho \iint_D (3x^2)\, dA = \iint_D x^2 \rho\, dA = I_y$.

**27.** As in Example 5, let $C'$ be a counterclockwise-oriented circle with center the origin and radius $a$, where $a$ is chosen to

be small enough so that $C'$ lies inside $C$, and $D$ the region bounded by $C$ and $C'$. Here

$P = \dfrac{2xy}{(x^2 + y^2)^2} \quad \Rightarrow \quad \dfrac{\partial P}{\partial y} = \dfrac{2x(x^2 + y^2)^2 - 2xy \cdot 2(x^2 + y^2) \cdot 2y}{(x^2 + y^2)^4} = \dfrac{2x^3 - 6xy^2}{(x^2 + y^2)^3}$ and

$Q = \dfrac{y^2 - x^2}{(x^2 + y^2)^2} \quad \Rightarrow \quad \dfrac{\partial Q}{\partial x} = \dfrac{-2x(x^2 + y^2)^2 - (y^2 - x^2) \cdot 2(x^2 + y^2) \cdot 2x}{(x^2 + y^2)^4} = \dfrac{2x^3 - 6xy^2}{(x^2 + y^2)^3}$. Thus, as in the example,

$$\int_C P\, dx + Q\, dy + \int_{-C'} P\, dx + Q\, dy = \iint_D \left( \dfrac{\partial Q}{\partial x} - \dfrac{\partial P}{\partial y} \right) dA = \iint_D 0\, dA = 0$$

and $\int_C \mathbf{F} \cdot d\mathbf{r} = \int_{C'} \mathbf{F} \cdot d\mathbf{r}$. We parametrize $C'$ as $\mathbf{r}(t) = a\cos t\, \mathbf{i} + a\sin t\, \mathbf{j}$, $0 \le t \le 2\pi$. Then

$$\int_C \mathbf{F} \cdot d\mathbf{r} = \int_{C'} \mathbf{F} \cdot d\mathbf{r} = \int_0^{2\pi} \dfrac{2(a\cos t)(a\sin t)\, \mathbf{i} + (a^2 \sin^2 t - a^2 \cos^2 t)\, \mathbf{j}}{(a^2 \cos^2 t + a^2 \sin^2 t)^2} \cdot \left( -a\sin t\, \mathbf{i} + a\cos t\, \mathbf{j} \right) dt$$

$$= \dfrac{1}{a} \int_0^{2\pi} \left( -\cos t \sin^2 t - \cos^3 t \right) dt = \dfrac{1}{a} \int_0^{2\pi} \left( -\cos t \sin^2 t - \cos t (1 - \sin^2 t) \right) dt$$

$$= -\dfrac{1}{a} \int_0^{2\pi} \cos t\, dt = -\dfrac{1}{a} \sin t \Big]_0^{2\pi} = 0$$

**29.** Since $C$ is a simple closed path which doesn't pass through or enclose the origin, there exists an open region that doesn't contain the origin but does contain $D$. Thus $P = -y/(x^2 + y^2)$ and $Q = x/(x^2 + y^2)$ have continuous partial derivatives on this open region containing $D$ and we can apply Green's Theorem. But by Exercise 13.3.35(a), $\partial P/\partial y = \partial Q/\partial x$, so $\oint_C \mathbf{F} \cdot d\mathbf{r} = \iint_D 0 \, dA = 0$.

**31.** Using the first part of (5), we have that $\iint_R dx \, dy = A(R) = \int_{\partial R} x \, dy$. But $x = g(u, v)$, and $dy = \dfrac{\partial h}{\partial u} du + \dfrac{\partial h}{\partial v} dv$, and we orient $\partial S$ by taking the positive direction to be that which corresponds, under the mapping, to the positive direction along $\partial R$, so

$$\int_{\partial R} x \, dy = \int_{\partial S} g(u, v) \left( \frac{\partial h}{\partial u} du + \frac{\partial h}{\partial v} dv \right) = \int_{\partial S} g(u, v) \frac{\partial h}{\partial u} du + g(u, v) \frac{\partial h}{\partial v} dv$$

$$= \pm \iint_S \left[ \frac{\partial}{\partial u} \left( g(u, v) \frac{\partial h}{\partial v} \right) - \frac{\partial}{\partial v} \left( g(u, v) \frac{\partial h}{\partial u} \right) \right] dA \quad \text{[using Green's Theorem in the } uv\text{-plane]}$$

$$= \pm \iint_S \left( \frac{\partial g}{\partial u} \frac{\partial h}{\partial v} + g(u, v) \frac{\partial^2 h}{\partial u \, \partial v} - \frac{\partial g}{\partial v} \frac{\partial h}{\partial u} - g(u, v) \frac{\partial^2 h}{\partial v \, \partial u} \right) dA \quad \text{[using the Chain Rule]}$$

$$= \pm \iint_S \left( \frac{\partial x}{\partial u} \frac{\partial y}{\partial v} - \frac{\partial x}{\partial v} \frac{\partial y}{\partial u} \right) dA \quad \text{[by the equality of mixed partials]} \quad = \pm \iint_S \frac{\partial(x,y)}{\partial(u,v)} du \, dv$$

The sign is chosen to be positive if the orientation that we gave to $\partial S$ corresponds to the usual positive orientation, and it is negative otherwise. In either case, since $A(R)$ is positive, the sign chosen must be the same as the sign of $\dfrac{\partial(x, y)}{\partial(u, v)}$.

Therefore $A(R) = \displaystyle\iint_R dx \, dy = \iint_S \left| \frac{\partial(x, y)}{\partial(u, v)} \right| du \, dv$.

## 13.5 Curl and Divergence

**1. (a)** $\operatorname{curl} \mathbf{F} = \nabla \times \mathbf{F} = \begin{vmatrix} \mathbf{i} & \mathbf{j} & \mathbf{k} \\ \partial/\partial x & \partial/\partial y & \partial/\partial z \\ xyz & 0 & -x^2 y \end{vmatrix} = (-x^2 - 0) \mathbf{i} - (-2xy - xy) \mathbf{j} + (0 - xz) \mathbf{k}$

$= -x^2 \mathbf{i} + 3xy \mathbf{j} - xz \mathbf{k}$

**(b)** $\operatorname{div} \mathbf{F} = \nabla \cdot \mathbf{F} = \dfrac{\partial}{\partial x} (xyz) + \dfrac{\partial}{\partial y} (0) + \dfrac{\partial}{\partial z} (-x^2 y) = yz + 0 + 0 = yz$

**3. (a)** $\operatorname{curl} \mathbf{F} = \nabla \times \mathbf{F} = \begin{vmatrix} \mathbf{i} & \mathbf{j} & \mathbf{k} \\ \partial/\partial x & \partial/\partial y & \partial/\partial z \\ xye^z & 0 & yze^x \end{vmatrix} = (ze^x - 0) \mathbf{i} - (yze^x - xye^z) \mathbf{j} + (0 - xe^z) \mathbf{k}$

$= ze^x \mathbf{i} + (xye^z - yze^x) \mathbf{j} - xe^z \mathbf{k}$

**(b)** $\operatorname{div} \mathbf{F} = \nabla \cdot \mathbf{F} = \dfrac{\partial}{\partial x} (xye^z) + \dfrac{\partial}{\partial y} (0) + \dfrac{\partial}{\partial z} (yze^x) = ye^z + 0 + ye^x = y(e^z + e^x)$

**5. (a)** $\operatorname{curl} \mathbf{F} = \nabla \times \mathbf{F} = \begin{vmatrix} \mathbf{i} & \mathbf{j} & \mathbf{k} \\ \partial/\partial x & \partial/\partial y & \partial/\partial z \\ \dfrac{x}{\sqrt{x^2 + y^2 + z^2}} & \dfrac{y}{\sqrt{x^2 + y^2 + z^2}} & \dfrac{z}{\sqrt{x^2 + y^2 + z^2}} \end{vmatrix}$

$= \dfrac{1}{(x^2 + y^2 + z^2)^{3/2}} [(-yz + yz) \mathbf{i} - (-xz + xz) \mathbf{j} + (-xy + xy) \mathbf{k}] = \mathbf{0}$

(b) $\operatorname{div}\mathbf{F} = \nabla\cdot\mathbf{F} = \dfrac{\partial}{\partial x}\left(\dfrac{x}{\sqrt{x^2+y^2+z^2}}\right) + \dfrac{\partial}{\partial y}\left(\dfrac{y}{\sqrt{x^2+y^2+z^2}}\right) + \dfrac{\partial}{\partial z}\left(\dfrac{z}{\sqrt{x^2+y^2+z^2}}\right)$

$= \dfrac{x^2+y^2+z^2-x^2}{(x^2+y^2+z^2)^{3/2}} + \dfrac{x^2+y^2+z^2-y^2}{(x^2+y^2+z^2)^{3/2}} + \dfrac{x^2+y^2+z^2-z^2}{(x^2+y^2+z^2)^{3/2}} = \dfrac{2x^2+2y^2+2z^2}{(x^2+y^2+z^2)^{3/2}} = \dfrac{2}{\sqrt{x^2+y^2+z^2}}$

**7.** (a) $\operatorname{curl}\mathbf{F} = \nabla\times\mathbf{F} = \begin{vmatrix} \mathbf{i} & \mathbf{j} & \mathbf{k} \\ \partial/\partial x & \partial/\partial y & \partial/\partial z \\ \ln x & \ln(xy) & \ln(xyz) \end{vmatrix} = \left(\dfrac{xz}{xyz}-0\right)\mathbf{i} - \left(\dfrac{yz}{xyz}-0\right)\mathbf{j} + \left(\dfrac{y}{xy}-0\right)\mathbf{k} = \left\langle\dfrac{1}{y},-\dfrac{1}{x},\dfrac{1}{x}\right\rangle$

(b) $\operatorname{div}\mathbf{F} = \nabla\cdot\mathbf{F} = \dfrac{\partial}{\partial x}(\ln x) + \dfrac{\partial}{\partial y}(\ln(xy)) + \dfrac{\partial}{\partial z}(\ln(xyz)) = \dfrac{1}{x} + \dfrac{x}{xy} + \dfrac{xy}{xyz} = \dfrac{1}{x} + \dfrac{1}{y} + \dfrac{1}{z}$

**9.** If the vector field is $\mathbf{F} = P\,\mathbf{i} + Q\,\mathbf{j} + R\,\mathbf{k}$, then we know $R = 0$. In addition, the $x$-component of each vector of $\mathbf{F}$ is 0, so

$P = 0$, hence $\dfrac{\partial P}{\partial x} = \dfrac{\partial P}{\partial y} = \dfrac{\partial P}{\partial z} = \dfrac{\partial R}{\partial x} = \dfrac{\partial R}{\partial y} = \dfrac{\partial R}{\partial z} = 0$. $Q$ decreases as $y$ increases, so $\dfrac{\partial Q}{\partial y} < 0$, but $Q$ doesn't change

in the $x$- or $z$-directions, so $\dfrac{\partial Q}{\partial x} = \dfrac{\partial Q}{\partial z} = 0$.

(a) $\operatorname{div}\mathbf{F} = \dfrac{\partial P}{\partial x} + \dfrac{\partial Q}{\partial y} + \dfrac{\partial R}{\partial z} = 0 + \dfrac{\partial Q}{\partial y} + 0 < 0$

(b) $\operatorname{curl}\mathbf{F} = \left(\dfrac{\partial R}{\partial y} - \dfrac{\partial Q}{\partial z}\right)\mathbf{i} + \left(\dfrac{\partial P}{\partial z} - \dfrac{\partial R}{\partial x}\right)\mathbf{j} + \left(\dfrac{\partial Q}{\partial x} - \dfrac{\partial P}{\partial y}\right)\mathbf{k} = (0-0)\,\mathbf{i} + (0-0)\,\mathbf{j} + (0-0)\,\mathbf{k} = \mathbf{0}$

**11.** If the vector field is $\mathbf{F} = P\,\mathbf{i} + Q\,\mathbf{j} + R\,\mathbf{k}$, then we know $R = 0$. In addition, the $y$-component of each vector of $\mathbf{F}$ is 0, so

$Q = 0$, hence $\dfrac{\partial Q}{\partial x} = \dfrac{\partial Q}{\partial y} = \dfrac{\partial Q}{\partial z} = \dfrac{\partial R}{\partial x} = \dfrac{\partial R}{\partial y} = \dfrac{\partial R}{\partial z} = 0$. $P$ increases as $y$ increases, so $\dfrac{\partial P}{\partial y} > 0$, but $P$ doesn't change in

the $x$- or $z$-directions, so $\dfrac{\partial P}{\partial x} = \dfrac{\partial P}{\partial z} = 0$.

(a) $\operatorname{div}\mathbf{F} = \dfrac{\partial P}{\partial x} + \dfrac{\partial Q}{\partial y} + \dfrac{\partial R}{\partial z} = 0 + 0 + 0 = 0$

(b) $\operatorname{curl}\mathbf{F} = \left(\dfrac{\partial R}{\partial y} - \dfrac{\partial Q}{\partial z}\right)\mathbf{i} + \left(\dfrac{\partial P}{\partial z} - \dfrac{\partial R}{\partial x}\right)\mathbf{j} + \left(\dfrac{\partial Q}{\partial x} - \dfrac{\partial P}{\partial y}\right)\mathbf{k} = (0-0)\,\mathbf{i} + (0-0)\,\mathbf{j} + \left(0 - \dfrac{\partial P}{\partial y}\right)\mathbf{k} = -\dfrac{\partial P}{\partial y}\mathbf{k}$

Since $\dfrac{\partial P}{\partial y} > 0$, $-\dfrac{\partial P}{\partial y}\mathbf{k}$ is a vector pointing in the negative $z$-direction.

**13.** $\operatorname{curl}\mathbf{F} = \nabla\times\mathbf{F} = \begin{vmatrix} \mathbf{i} & \mathbf{j} & \mathbf{k} \\ \partial/\partial x & \partial/\partial y & \partial/\partial z \\ y^2z^3 & 2xyz^3 & 3xy^2z^2 \end{vmatrix} = (6xyz^2 - 6xyz^2)\,\mathbf{i} - (3y^2z^2 - 3y^2z^2)\,\mathbf{j} + (2yz^3 - 2yz^3)\,\mathbf{k} = \mathbf{0}$

and $\mathbf{F}$ is defined on all of $\mathbb{R}^3$ with component functions which have continuous partial derivatives, so by Theorem 4,

$\mathbf{F}$ is conservative. Thus, there exists a function $f$ such that $\mathbf{F} = \nabla f$. Then $f_x(x, y, z) = y^2z^3$ implies

$f(x, y, z) = xy^2z^3 + g(y, z)$ and $f_y(x, y, z) = 2xyz^3 + g_y(y, z)$. But $f_y(x, y, z) = 2xyz^3$, so $g(y, z) = h(z)$ and

$f(x, y, z) = xy^2z^3 + h(z)$. Thus $f_z(x, y, z) = 3xy^2z^2 + h'(z)$ but $f_z(x, y, z) = 3xy^2z^2$ so $h(z) = K$, a constant.

Hence a potential function for $\mathbf{F}$ is $f(x, y, z) = xy^2z^3 + K$.

**15.** $\text{curl } \mathbf{F} = \nabla \times \mathbf{F} = \begin{vmatrix} \mathbf{i} & \mathbf{j} & \mathbf{k} \\ \partial/\partial x & \partial/\partial y & \partial/\partial z \\ 2xy & x^2 + 2yz & y^2 \end{vmatrix} = (2y - 2y)\,\mathbf{i} - (0 - 0)\,\mathbf{j} + (2x - 2x)\,\mathbf{k} = \mathbf{0}$, $\mathbf{F}$ is defined on all of $\mathbb{R}^3$,

and the partial derivatives of the component functions are continuous, so $\mathbf{F}$ is conservative. Thus there exists a function $f$

such that $\nabla f = \mathbf{F}$. Then $f_x(x, y, z) = 2xy$ implies $f(x, y, z) = x^2 y + g(y, z)$ and $f_y(x, y, z) = x^2 + g_y(y, z)$. But

$f_y(x, y, z) = x^2 + 2yz$, so $g(y, z) = y^2 z + h(z)$ and $f(x, y, z) = x^2 y + y^2 z + h(z)$. Thus $f_z(x, y, z) = y^2 + h'(z)$ but

$f_z(x, y, z) = y^2$ so $h(z) = K$ and $f(x, y, z) = x^2 y + y^2 z + K$.

**17.** $\text{curl } \mathbf{F} = \nabla \times \mathbf{F} = \begin{vmatrix} \mathbf{i} & \mathbf{j} & \mathbf{k} \\ \partial/\partial x & \partial/\partial y & \partial/\partial z \\ ye^{-x} & e^{-x} & 2z \end{vmatrix} = (0 - 0)\,\mathbf{i} - (0 - 0)\,\mathbf{j} + (-e^{-x} - e^{-x})\,\mathbf{k} = -2e^{-x}\,\mathbf{k} \neq \mathbf{0}$,

so $\mathbf{F}$ is not conservative.

**19.** No. Assume there is such a $\mathbf{G}$. Then $\text{div}(\text{curl } \mathbf{G}) = \dfrac{\partial}{\partial x}(x \sin y) + \dfrac{\partial}{\partial y}(\cos y) + \dfrac{\partial}{\partial z}(z - xy) = \sin y - \sin y + 1 \neq 0$,

which contradicts Theorem 11.

**21.** $\text{curl } \mathbf{F} = \begin{vmatrix} \mathbf{i} & \mathbf{j} & \mathbf{k} \\ \partial/\partial x & \partial/\partial y & \partial/\partial z \\ f(x) & g(y) & h(z) \end{vmatrix} = (0 - 0)\,\mathbf{i} + (0 - 0)\,\mathbf{j} + (0 - 0)\,\mathbf{k} = \mathbf{0}$. Hence $\mathbf{F} = f(x)\,\mathbf{i} + g(y)\,\mathbf{j} + h(z)\,\mathbf{k}$

is irrotational.

For Exercises 23–29, let $\mathbf{F}(x, y, z) = P_1\,\mathbf{i} + Q_1\,\mathbf{j} + R_1\,\mathbf{k}$ and $\mathbf{G}(x, y, z) = P_2\,\mathbf{i} + Q_2\,\mathbf{j} + R_2\,\mathbf{k}$.

**23.** $\text{div}(\mathbf{F} + \mathbf{G}) = \text{div}\langle P_1 + P_2, Q_1 + Q_2, R_1 + R_2 \rangle = \dfrac{\partial(P_1 + P_2)}{\partial x} + \dfrac{\partial(Q_1 + Q_2)}{\partial y} + \dfrac{\partial(R_1 + R_2)}{\partial z}$

$= \dfrac{\partial P_1}{\partial x} + \dfrac{\partial P_2}{\partial x} + \dfrac{\partial Q_1}{\partial y} + \dfrac{\partial Q_2}{\partial y} + \dfrac{\partial R_1}{\partial z} + \dfrac{\partial R_2}{\partial z} = \left( \dfrac{\partial P_1}{\partial x} + \dfrac{\partial Q_1}{\partial y} + \dfrac{\partial R_1}{\partial z} \right) + \left( \dfrac{\partial P_2}{\partial x} + \dfrac{\partial Q_2}{\partial y} + \dfrac{\partial R_2}{\partial z} \right)$

$= \text{div}\langle P_1, Q_1, R_1 \rangle + \text{div}\langle P_2, Q_2, R_2 \rangle = \text{div } \mathbf{F} + \text{div } \mathbf{G}$

**25.** $\text{div}(f\mathbf{F}) = \text{div}(f\,\langle P_1, Q_1, R_1 \rangle) = \text{div}\langle fP_1, fQ_1, fR_1 \rangle = \dfrac{\partial(fP_1)}{\partial x} + \dfrac{\partial(fQ_1)}{\partial y} + \dfrac{\partial(fR_1)}{\partial z}$

$= \left( f \dfrac{\partial P_1}{\partial x} + P_1 \dfrac{\partial f}{\partial x} \right) + \left( f \dfrac{\partial Q_1}{\partial y} + Q_1 \dfrac{\partial f}{\partial y} \right) + \left( f \dfrac{\partial R_1}{\partial z} + R_1 \dfrac{\partial f}{\partial z} \right)$

$= f \left( \dfrac{\partial P_1}{\partial x} + \dfrac{\partial Q_1}{\partial y} + \dfrac{\partial R_1}{\partial z} \right) + \langle P_1, Q_1, R_1 \rangle \cdot \left\langle \dfrac{\partial f}{\partial x}, \dfrac{\partial f}{\partial y}, \dfrac{\partial f}{\partial z} \right\rangle = f \, \text{div } \mathbf{F} + \mathbf{F} \cdot \nabla f$

**27.** $\text{div}(\mathbf{F} \times \mathbf{G}) = \nabla \cdot (\mathbf{F} \times \mathbf{G}) = \begin{vmatrix} \partial/\partial x & \partial/\partial y & \partial/\partial z \\ P_1 & Q_1 & R_1 \\ P_2 & Q_2 & R_2 \end{vmatrix} = \dfrac{\partial}{\partial x}\begin{vmatrix} Q_1 & R_1 \\ Q_2 & R_2 \end{vmatrix} - \dfrac{\partial}{\partial y}\begin{vmatrix} P_1 & R_1 \\ P_2 & R_2 \end{vmatrix} + \dfrac{\partial}{\partial z}\begin{vmatrix} P_1 & Q_1 \\ P_2 & Q_2 \end{vmatrix}$

$$= \left[ Q_1 \frac{\partial R_2}{\partial x} + R_2 \frac{\partial Q_1}{\partial x} - Q_2 \frac{\partial R_1}{\partial x} - R_1 \frac{\partial Q_2}{\partial x} \right] - \left[ P_1 \frac{\partial R_2}{\partial y} + R_2 \frac{\partial P_1}{\partial y} - P_2 \frac{\partial R_1}{\partial y} - R_1 \frac{\partial P_2}{\partial y} \right]$$

$$+ \left[ P_1 \frac{\partial Q_2}{\partial z} + Q_2 \frac{\partial P_1}{\partial z} - P_2 \frac{\partial Q_1}{\partial z} - Q_1 \frac{\partial P_2}{\partial z} \right]$$

$$= \left[ P_2 \left( \frac{\partial R_1}{\partial y} - \frac{\partial Q_1}{\partial z} \right) + Q_2 \left( \frac{\partial P_1}{\partial z} - \frac{\partial R_1}{\partial x} \right) + R_2 \left( \frac{\partial Q_1}{\partial x} - \frac{\partial P_1}{\partial y} \right) \right]$$

$$- \left[ P_1 \left( \frac{\partial R_2}{\partial y} - \frac{\partial Q_2}{\partial z} \right) + Q_1 \left( \frac{\partial P_2}{\partial z} - \frac{\partial R_2}{\partial x} \right) + R_1 \left( \frac{\partial Q_2}{\partial x} - \frac{\partial P_2}{\partial y} \right) \right]$$

$$= \mathbf{G} \cdot \text{curl}\,\mathbf{F} - \mathbf{F} \cdot \text{curl}\,\mathbf{G}$$

**29.** $\text{curl}(\text{curl}\,\mathbf{F}) = \nabla \times (\nabla \times \mathbf{F}) = \begin{vmatrix} \mathbf{i} & \mathbf{j} & \mathbf{k} \\ \partial/\partial x & \partial/\partial y & \partial/\partial z \\ \partial R_1/\partial y - \partial Q_1/\partial z & \partial P_1/\partial z - \partial R_1/\partial x & \partial Q_1/\partial x - \partial P_1/\partial y \end{vmatrix}$

$$= \left( \frac{\partial^2 Q_1}{\partial y \partial x} - \frac{\partial^2 P_1}{\partial y^2} - \frac{\partial^2 P_1}{\partial z^2} + \frac{\partial^2 R_1}{\partial z \partial x} \right) \mathbf{i} + \left( \frac{\partial^2 R_1}{\partial z \partial y} - \frac{\partial^2 Q_1}{\partial z^2} - \frac{\partial^2 Q_1}{\partial x^2} + \frac{\partial^2 P_1}{\partial x \partial y} \right) \mathbf{j}$$

$$+ \left( \frac{\partial^2 P_1}{\partial x \partial z} - \frac{\partial^2 R_1}{\partial x^2} - \frac{\partial^2 R_1}{\partial y^2} + \frac{\partial^2 Q_1}{\partial y \partial z} \right) \mathbf{k}$$

Now let's consider $\text{grad}(\text{div}\,\mathbf{F}) - \nabla^2 \mathbf{F}$ and compare with the above.

(Note that $\nabla^2 \mathbf{F}$ is defined on page 945.)

$$\text{grad}(\text{div}\,\mathbf{F}) - \nabla^2 \mathbf{F} = \left[ \left( \frac{\partial^2 P_1}{\partial x^2} + \frac{\partial^2 Q_1}{\partial x \partial y} + \frac{\partial^2 R_1}{\partial x \partial z} \right) \mathbf{i} + \left( \frac{\partial^2 P_1}{\partial y \partial x} + \frac{\partial^2 Q_1}{\partial y^2} + \frac{\partial^2 R_1}{\partial y \partial z} \right) \mathbf{j} + \left( \frac{\partial^2 P_1}{\partial z \partial x} + \frac{\partial^2 Q_1}{\partial z \partial y} + \frac{\partial^2 R_1}{\partial z^2} \right) \mathbf{k} \right]$$

$$- \left[ \left( \frac{\partial^2 P_1}{\partial x^2} + \frac{\partial^2 P_1}{\partial y^2} + \frac{\partial^2 P_1}{\partial z^2} \right) \mathbf{i} + \left( \frac{\partial^2 Q_1}{\partial x^2} + \frac{\partial^2 Q_1}{\partial y^2} + \frac{\partial^2 Q_1}{\partial z^2} \right) \mathbf{j} \right.$$

$$\left. + \left( \frac{\partial^2 R_1}{\partial x^2} + \frac{\partial^2 R_1}{\partial y^2} + \frac{\partial^2 R_1}{\partial z^2} \right) \mathbf{k} \right]$$

$$= \left( \frac{\partial^2 Q_1}{\partial x \partial y} + \frac{\partial^2 R_1}{\partial x \partial z} - \frac{\partial^2 P_1}{\partial y^2} - \frac{\partial^2 P_1}{\partial z^2} \right) \mathbf{i} + \left( \frac{\partial^2 P_1}{\partial y \partial x} + \frac{\partial^2 R_1}{\partial y \partial z} - \frac{\partial^2 Q_1}{\partial x^2} - \frac{\partial^2 Q_1}{\partial z^2} \right) \mathbf{j}$$

$$+ \left( \frac{\partial^2 P_1}{\partial z \partial x} + \frac{\partial^2 Q_1}{\partial z \partial y} - \frac{\partial^2 R_1}{\partial x^2} - \frac{\partial^2 R_2}{\partial y^2} \right) \mathbf{k}$$

Then applying Clairaut's Theorem to reverse the order of differentiation in the second partial derivatives as needed and comparing, we have $\text{curl}\,\text{curl}\,\mathbf{F} = \text{grad}\,\text{div}\,\mathbf{F} - \nabla^2 \mathbf{F}$ as desired.

**31. (a)** $\nabla r = \nabla\sqrt{x^2 + y^2 + z^2} = \dfrac{x}{\sqrt{x^2 + y^2 + z^2}}\,\mathbf{i} + \dfrac{y}{\sqrt{x^2 + y^2 + z^2}}\,\mathbf{j} + \dfrac{z}{\sqrt{x^2 + y^2 + z^2}}\,\mathbf{k} = \dfrac{x\,\mathbf{i} + y\,\mathbf{j} + z\,\mathbf{k}}{\sqrt{x^2 + y^2 + z^2}} = \dfrac{\mathbf{r}}{r}$

**(b)** $\nabla \times \mathbf{r} = \begin{vmatrix} \mathbf{i} & \mathbf{j} & \mathbf{k} \\ \dfrac{\partial}{\partial x} & \dfrac{\partial}{\partial y} & \dfrac{\partial}{\partial z} \\ x & y & z \end{vmatrix} = \left[\dfrac{\partial}{\partial y}(z) - \dfrac{\partial}{\partial z}(y)\right]\mathbf{i} + \left[\dfrac{\partial}{\partial z}(x) - \dfrac{\partial}{\partial x}(z)\right]\mathbf{j} + \left[\dfrac{\partial}{\partial x}(y) - \dfrac{\partial}{\partial y}(x)\right]\mathbf{k} = \mathbf{0}$

**(c)** $\nabla\left(\dfrac{1}{r}\right) = \nabla\left(\dfrac{1}{\sqrt{x^2 + y^2 + z^2}}\right)$

$= \dfrac{-\dfrac{1}{2\sqrt{x^2 + y^2 + z^2}}(2x)}{x^2 + y^2 + z^2}\,\mathbf{i} - \dfrac{\dfrac{1}{2\sqrt{x^2 + y^2 + z^2}}(2y)}{x^2 + y^2 + z^2}\,\mathbf{j} - \dfrac{\dfrac{1}{2\sqrt{x^2 + y^2 + z^2}}(2z)}{x^2 + y^2 + z^2}\,\mathbf{k}$

$= -\dfrac{x\,\mathbf{i} + y\,\mathbf{j} + z\,\mathbf{k}}{(x^2 + y^2 + z^2)^{3/2}} = -\dfrac{\mathbf{r}}{r^3}$

**(d)** $\nabla \ln r = \nabla \ln(x^2 + y^2 + z^2)^{1/2} = \tfrac{1}{2}\nabla \ln(x^2 + y^2 + z^2)$

$= \dfrac{x}{x^2 + y^2 + z^2}\,\mathbf{i} + \dfrac{y}{x^2 + y^2 + z^2}\,\mathbf{j} + \dfrac{z}{x^2 + y^2 + z^2}\,\mathbf{k} = \dfrac{x\,\mathbf{i} + y\,\mathbf{j} + z\,\mathbf{k}}{x^2 + y^2 + z^2} = \dfrac{\mathbf{r}}{r^2}$

**33.** By (13), $\oint_C f(\nabla g) \cdot \mathbf{n}\,ds = \iint_D \operatorname{div}(f\nabla g)\,dA = \iint_D [f\operatorname{div}(\nabla g) + \nabla g \cdot \nabla f]\,dA$ by Exercise 25. But $\operatorname{div}(\nabla g) = \nabla^2 g$.

Hence $\iint_D f\nabla^2 g\,dA = \oint_C f(\nabla g) \cdot \mathbf{n}\,ds - \iint_D \nabla g \cdot \nabla f\,dA$.

**35.** Let $f(x, y) = 1$. Then $\nabla f = \mathbf{0}$ and Green's first identity (see Exercise 33) says

$\iint_D \nabla^2 g\,dA = \oint_C (\nabla g) \cdot \mathbf{n}\,ds - \iint_D \mathbf{0} \cdot \nabla g\,dA \quad \Rightarrow \quad \iint_D \nabla^2 g\,dA = \oint_C \nabla g \cdot \mathbf{n}\,ds$. But $g$ is harmonic on $D$, so

$\nabla^2 g = 0 \quad \Rightarrow \quad \oint_C \nabla g \cdot \mathbf{n}\,ds = 0$ and $\oint_C D_{\mathbf{n}} g\,ds = \oint_C (\nabla g \cdot \mathbf{n})\,ds = 0$.

**37. (a)** We know that $\omega = v/d$, and from the diagram $\sin\theta = d/r \quad \Rightarrow \quad v = d\omega = (\sin\theta)r\omega = |\mathbf{w} \times \mathbf{r}|$. But $\mathbf{v}$ is perpendicular to both $\mathbf{w}$ and $\mathbf{r}$, so that $\mathbf{v} = \mathbf{w} \times \mathbf{r}$.

**(b)** From (a), $\mathbf{v} = \mathbf{w} \times \mathbf{r} = \begin{vmatrix} \mathbf{i} & \mathbf{j} & \mathbf{k} \\ 0 & 0 & \omega \\ x & y & z \end{vmatrix} = (0 \cdot z - \omega y)\,\mathbf{i} + (\omega x - 0 \cdot z)\,\mathbf{j} + (0 \cdot y - x \cdot 0)\,\mathbf{k} = -\omega y\,\mathbf{i} + \omega x\,\mathbf{j}$

**(c)** $\operatorname{curl}\mathbf{v} = \nabla \times \mathbf{v} = \begin{vmatrix} \mathbf{i} & \mathbf{j} & \mathbf{k} \\ \partial/\partial x & \partial/\partial y & \partial/\partial z \\ -\omega y & \omega x & 0 \end{vmatrix}$

$= \left[\dfrac{\partial}{\partial y}(0) - \dfrac{\partial}{\partial z}(\omega x)\right]\mathbf{i} + \left[\dfrac{\partial}{\partial z}(-\omega y) - \dfrac{\partial}{\partial x}(0)\right]\mathbf{j} + \left[\dfrac{\partial}{\partial x}(\omega x) - \dfrac{\partial}{\partial y}(-\omega y)\right]\mathbf{k}$

$= [\omega - (-\omega)]\,\mathbf{k} = 2\omega\,\mathbf{k} = 2\mathbf{w}$

**39.** For any continuous function $f$ on $\mathbb{R}^3$, define a vector field $\mathbf{G}(x, y, z) = \langle g(x, y, z), 0, 0\rangle$ where $g(x, y, z) = \int_0^x f(t, y, z)\,dt$.

Then $\operatorname{div}\mathbf{G} = \dfrac{\partial}{\partial x}(g(x, y, z)) + \dfrac{\partial}{\partial y}(0) + \dfrac{\partial}{\partial z}(0) = \dfrac{\partial}{\partial x}\int_0^x f(t, y, z)\,dt = f(x, y, z)$ by the Fundamental Theorem of

Calculus. Thus every continuous function $f$ on $\mathbb{R}^3$ is the divergence of some vector field.

## 13.6   Surface Integrals

**1.** The faces of the box in the planes $x = 0$ and $x = 2$ have surface area 24 and centers $(0, 2, 3)$, $(2, 2, 3)$. The faces in $y = 0$ and $y = 4$ have surface area 12 and centers $(1, 0, 3)$, $(1, 4, 3)$, and the faces in $z = 0$ and $z = 6$ have area 8 and centers $(1, 2, 0)$, $(1, 2, 6)$. For each face we take the point $P_{ij}^*$ to be the center of the face and $f(x, y, z) = e^{-0.1(x+y+z)}$, so by Definition 1,

$$\iint_S f(x, y, z)\, dS \approx [f(0, 2, 3)](24) + [f(2, 2, 3)](24) + [f(1, 0, 3)](12)$$

$$+ [f(1, 4, 3)](12) + [f(1, 2, 0)](8) + [f(1, 2, 6)](8)$$

$$= 24(e^{-0.5} + e^{-0.7}) + 12(e^{-0.4} + e^{-0.8}) + 8(e^{-0.3} + e^{-0.9}) \approx 49.09$$

**3.** We can use the $xz$- and $yz$-planes to divide $H$ into four patches of equal size, each with surface area equal to $\frac{1}{8}$ the surface area of a sphere with radius $\sqrt{50}$, so $\Delta S = \frac{1}{8}(4)\pi\left(\sqrt{50}\right)^2 = 25\pi$. Then $(\pm 3, \pm 4, 5)$ are sample points in the four patches, and using a Riemann sum as in Definition 1, we have

$$\iint_H f(x, y, z)\, dS \approx f(3, 4, 5)\,\Delta S + f(3, -4, 5)\,\Delta S + f(-3, 4, 5)\,\Delta S + f(-3, -4, 5)\,\Delta S$$

$$= (7 + 8 + 9 + 12)(25\pi) = 900\pi \approx 2827$$

**5.** $\mathbf{r}(u, v) = (u + v)\,\mathbf{i} + (u - v)\,\mathbf{j} + (1 + 2u + v)\,\mathbf{k}$, $0 \le u \le 2$, $0 \le v \le 1$ and

$\mathbf{r}_u \times \mathbf{r}_v = (\mathbf{i} + \mathbf{j} + 2\,\mathbf{k}) \times (\mathbf{i} - \mathbf{j} + \mathbf{k}) = 3\,\mathbf{i} + \mathbf{j} - 2\,\mathbf{k} \quad \Rightarrow \quad |\mathbf{r}_u \times \mathbf{r}_v| = \sqrt{3^2 + 1^2 + (-2)^2} = \sqrt{14}$. Then by Formula 2,

$$\iint_S (x + y + z)\, dS = \iint_D (u + v + u - v + 1 + 2u + v)\,|\mathbf{r}_u \times \mathbf{r}_v|\, dA = \int_0^1 \int_0^2 (4u + v + 1) \cdot \sqrt{14}\, du\, dv$$

$$= \sqrt{14} \int_0^1 \left[2u^2 + uv + u\right]_{u=0}^{u=2} dv = \sqrt{14} \int_0^1 (2v + 10)\, dv = \sqrt{14}\left[v^2 + 10v\right]_0^1 = 11\sqrt{14}$$

**7.** $\mathbf{r}(u, v) = \langle u\cos v, u\sin v, v\rangle$, $0 \le u \le 1$, $0 \le v \le \pi$ and

$\mathbf{r}_u \times \mathbf{r}_v = \langle \cos v, \sin v, 0\rangle \times \langle -u\sin v, u\cos v, 1\rangle = \langle \sin v, -\cos v, u\rangle \quad \Rightarrow$

$|\mathbf{r}_u \times \mathbf{r}_v| = \sqrt{\sin^2 v + \cos^2 v + u^2} = \sqrt{u^2 + 1}$. Then

$$\iint_S y\, dS = \iint_D (u\sin v)\,|\mathbf{r}_u \times \mathbf{r}_v|\, dA = \int_0^1 \int_0^\pi (u\sin v) \cdot \sqrt{u^2 + 1}\, dv\, du = \int_0^1 u\sqrt{u^2 + 1}\, du \int_0^\pi \sin v\, dv$$

$$= \left[\frac{1}{3}(u^2 + 1)^{3/2}\right]_0^1 [-\cos v]_0^\pi = \frac{1}{3}(2^{3/2} - 1) \cdot 2 = \frac{2}{3}(2\sqrt{2} - 1)$$

**9.** $z = 1 + 2x + 3y$ so $\dfrac{\partial z}{\partial x} = 2$ and $\dfrac{\partial z}{\partial y} = 3$. Then by Formula 4,

$$\iint_S x^2 yz\, dS = \iint_D x^2 yz \sqrt{\left(\frac{\partial z}{\partial x}\right)^2 + \left(\frac{\partial z}{\partial y}\right)^2 + 1}\, dA = \int_0^3 \int_0^2 x^2 y(1 + 2x + 3y)\sqrt{4 + 9 + 1}\, dy\, dx$$

$$= \sqrt{14} \int_0^3 \int_0^2 (x^2 y + 2x^3 y + 3x^2 y^2)\, dy\, dx = \sqrt{14} \int_0^3 \left[\frac{1}{2}x^2 y^2 + x^3 y^2 + x^2 y^3\right]_{y=0}^{y=2} dx$$

$$= \sqrt{14} \int_0^3 (10x^2 + 4x^3)\, dx = \sqrt{14}\left[\frac{10}{3}x^3 + x^4\right]_0^3 = 171\sqrt{14}$$

**11.** $S$ is the part of the plane $z = 1 - x - y$ over the region $D = \{(x, y)\mid 0 \le x \le 1, 0 \le y \le 1 - x\}$. Thus

$$\iint_S yz\, dS = \iint_D y(1 - x - y)\sqrt{(-1)^2 + (-1)^2 + 1}\, dA = \sqrt{3}\int_0^1 \int_0^{1-x}(y - xy - y^2)\, dy\, dx$$

$$= \sqrt{3}\int_0^1 \left[\frac{1}{2}y^2 - \frac{1}{2}xy^2 - \frac{1}{3}y^3\right]_{y=0}^{y=1-x} dx = \sqrt{3}\int_0^1 \frac{1}{6}(1 - x)^3\, dx = -\frac{\sqrt{3}}{24}(1 - x)^4\Big|_0^1 = \frac{\sqrt{3}}{24}$$

**13.** $S$ is the portion of the cone $z^2 = x^2 + y^2$ for $1 \le z \le 3$, or equivalently, $S$ is the part of the surface $z = \sqrt{x^2 + y^2}$ over the

region $D = \{(x, y) \mid 1 \le x^2 + y^2 \le 9\}$. Thus

$$\iint_S x^2 z^2 \, dS = \iint_D x^2(x^2 + y^2) \sqrt{\left(\frac{x}{\sqrt{x^2+y^2}}\right)^2 + \left(\frac{y}{\sqrt{x^2+y^2}}\right)^2 + 1} \, dA$$

$$= \iint_D x^2(x^2 + y^2) \sqrt{\frac{x^2+y^2}{x^2+y^2} + 1} \, dA = \iint_D \sqrt{2}\, x^2(x^2 + y^2) \, dA = \sqrt{2} \int_0^{2\pi} \int_1^3 (r\cos\theta)^2 (r^2)\, r \, dr \, d\theta$$

$$= \sqrt{2} \int_0^{2\pi} \cos^2\theta \, d\theta \int_1^3 r^5 \, dr = \sqrt{2} \left[\tfrac{1}{2}\theta + \tfrac{1}{4}\sin 2\theta\right]_0^{2\pi} \left[\tfrac{1}{6}r^6\right]_1^3 = \sqrt{2}\,(\pi) \cdot \tfrac{1}{6}(3^6 - 1) = \frac{364\sqrt{2}}{3}\pi$$

**15.** Using $x$ and $z$ as parameters, we have $\mathbf{r}(x, z) = x\,\mathbf{i} + (x^2 + z^2)\,\mathbf{j} + z\,\mathbf{k}$, $x^2 + z^2 \le 4$. Then

$\mathbf{r}_x \times \mathbf{r}_z = (\mathbf{i} + 2x\,\mathbf{j}) \times (2z\,\mathbf{j} + \mathbf{k}) = 2x\,\mathbf{i} - \mathbf{j} + 2z\,\mathbf{k}$ and $|\mathbf{r}_x \times \mathbf{r}_z| = \sqrt{4x^2 + 1 + 4z^2} = \sqrt{1 + 4(x^2 + z^2)}$. Thus

$$\iint_S y \, dS = \iint_{x^2+z^2\le 4} (x^2 + z^2)\sqrt{1 + 4(x^2 + z^2)} \, dA = \int_0^{2\pi}\int_0^2 r^2\sqrt{1 + 4r^2}\, r \, dr \, d\theta = \int_0^{2\pi} d\theta \int_0^2 r^2\sqrt{1 + 4r^2}\, r \, dr$$

$$= 2\pi \int_0^2 r^2\sqrt{1 + 4r^2}\, r \, dr \qquad \left[\text{let } u = 1 + 4r^2 \;\Rightarrow\; r^2 = \tfrac{1}{4}(u - 1) \text{ and } \tfrac{1}{8}du = r\,dr\right]$$

$$= 2\pi \int_1^{17} \tfrac{1}{4}(u - 1)\sqrt{u} \cdot \tfrac{1}{8}du = \tfrac{1}{16}\pi \int_1^{17}(u^{3/2} - u^{1/2})\, du$$

$$= \tfrac{1}{16}\pi\left[\tfrac{2}{5}u^{5/2} - \tfrac{2}{3}u^{3/2}\right]_1^{17} = \tfrac{1}{16}\pi\left[\tfrac{2}{5}(17)^{5/2} - \tfrac{2}{3}(17)^{3/2} - \tfrac{2}{5} + \tfrac{2}{3}\right] = \frac{\pi}{60}\left(391\sqrt{17} + 1\right)$$

**17.** Using spherical coordinates and Example 12.6.1 we have $\mathbf{r}(\phi, \theta) = 2\sin\phi\cos\theta\,\mathbf{i} + 2\sin\phi\sin\theta\,\mathbf{j} + 2\cos\phi\,\mathbf{k}$ and

$|\mathbf{r}_\phi \times \mathbf{r}_\theta| = 4\sin\phi$. Then $\iint_S (x^2 z + y^2 z)\, dS = \int_0^{2\pi}\int_0^{\pi/2}(4\sin^2\phi)(2\cos\phi)(4\sin\phi)\, d\phi\, d\theta = 16\pi\sin^4\phi\big]_0^{\pi/2} = 16\pi$.

**19.** $S$ is given by $\mathbf{r}(u, v) = u\,\mathbf{i} + \cos v\,\mathbf{j} + \sin v\,\mathbf{k}$, $0 \le u \le 3$, $0 \le v \le \pi/2$. Then

$\mathbf{r}_u \times \mathbf{r}_v = \mathbf{i} \times (-\sin v\,\mathbf{j} + \cos v\,\mathbf{k}) = -\cos v\,\mathbf{j} - \sin v\,\mathbf{k}$ and $|\mathbf{r}_u \times \mathbf{r}_v| = \sqrt{\cos^2 v + \sin^2 v} = 1$, so

$$\iint_S (z + x^2 y)\, dS = \int_0^{\pi/2}\int_0^3 (\sin v + u^2 \cos v)(1)\, du\, dv = \int_0^{\pi/2}(3\sin v + 9\cos v)\, dv$$

$$= [-3\cos v + 9\sin v]_0^{\pi/2} = 0 + 9 + 3 - 0 = 12$$

**21.** $\mathbf{F}(x, y, z) = xy\,\mathbf{i} + yz\,\mathbf{j} + zx\,\mathbf{k}$, $z = g(x, y) = 4 - x^2 - y^2$, and $D$ is the square $[0, 1] \times [0, 1]$, so by Equation 10

$$\iint_S \mathbf{F} \cdot d\mathbf{S} = \iint_D [-xy(-2x) - yz(-2y) + zx]\, dA = \int_0^1\int_0^1 [2x^2 y + 2y^2(4 - x^2 - y^2) + x(4 - x^2 - y^2)]\, dy\, dx$$

$$= \int_0^1 \left(\tfrac{1}{3}x^2 + \tfrac{11}{3}x - x^3 + \tfrac{34}{15}\right) dx = \frac{713}{180}$$

**23.** $\mathbf{F}(x, y, z) = xze^y\,\mathbf{i} - xze^y\,\mathbf{j} + z\,\mathbf{k}$, $z = g(x, y) = 1 - x - y$, and $D = \{(x, y) \mid 0 \le x \le 1, 0 \le y \le 1 - x\}$. Since $S$ has

downward orientation, we have

$$\iint_S \mathbf{F} \cdot d\mathbf{S} = -\iint_D [-xze^y(-1) - (-xze^y)(-1) + z]\, dA = -\int_0^1\int_0^{1-x}(1 - x - y)\, dy\, dx$$

$$= -\int_0^1 \left(\tfrac{1}{2}x^2 - x + \tfrac{1}{2}\right) dx = -\tfrac{1}{6}$$

**25.** $\mathbf{F}(x, y, z) = x\,\mathbf{i} - z\,\mathbf{j} + y\,\mathbf{k}$, $z = g(x, y) = \sqrt{4 - x^2 - y^2}$ and $D$ is the quarter disk

$\{(x, y) \mid 0 \le x \le 2, 0 \le y \le \sqrt{4 - x^2}\}$. $S$ has downward orientation, so by Formula 10,

$$\iint_S \mathbf{F} \cdot d\mathbf{S} = -\iint_D \left[ -x \cdot \tfrac{1}{2}(4 - x^2 - y^2)^{-1/2}(-2x) - (-z) \cdot \tfrac{1}{2}(4 - x^2 - y^2)^{-1/2}(-2y) + y \right] dA$$

$$= -\iint_D \left( \frac{x^2}{\sqrt{4 - x^2 - y^2}} - \sqrt{4 - x^2 - y^2} \cdot \frac{y}{\sqrt{4 - x^2 - y^2}} + y \right) dA$$

$$= -\iint_D x^2(4 - (x^2 + y^2))^{-1/2} \, dA = -\int_0^{\pi/2} \int_0^2 (r\cos\theta)^2 (4 - r^2)^{-1/2} \, r \, dr \, d\theta$$

$$= -\int_0^{\pi/2} \cos^2\theta \, d\theta \int_0^2 r^3 (4 - r^2)^{-1/2} \, dr \qquad [\text{let } u = 4 - r^2 \Rightarrow r^2 = 4 - u \text{ and } -\tfrac{1}{2} \, du = r \, dr]$$

$$= -\int_0^{\pi/2} \left( \tfrac{1}{2} + \tfrac{1}{2}\cos 2\theta \right) d\theta \int_4^0 -\tfrac{1}{2}(4 - u)(u)^{-1/2} \, du$$

$$= -\left[ \tfrac{1}{2}\theta + \tfrac{1}{4}\sin 2\theta \right]_0^{\pi/2} \left( -\tfrac{1}{2} \right) \left[ 8\sqrt{u} - \tfrac{2}{3}u^{3/2} \right]_4^0 = -\tfrac{\pi}{4} \left( -\tfrac{1}{2} \right) \left( -16 + \tfrac{16}{3} \right) = -\tfrac{4}{3}\pi$$

**27.** Let $S_1$ be the paraboloid $y = x^2 + z^2$, $0 \le y \le 1$ and $S_2$ the disk $x^2 + z^2 \le 1$, $y = 1$. Since $S$ is a closed surface, we use the outward orientation.

On $S_1$: $\mathbf{F}(\mathbf{r}(x, z)) = (x^2 + z^2)\mathbf{j} - z\mathbf{k}$ and $\mathbf{r}_x \times \mathbf{r}_z = 2x\mathbf{i} - \mathbf{j} + 2z\mathbf{k}$ (since the $\mathbf{j}$-component must be negative on $S_1$). Then

$$\iint_{S_1} \mathbf{F} \cdot d\mathbf{S} = \iint_{x^2 + z^2 \le 1} \left[ -(x^2 + z^2) - 2z^2 \right] dA = -\int_0^{2\pi} \int_0^1 (r^2 + 2r^2 \cos^2\theta) \, r \, dr \, d\theta$$

$$= -\int_0^{2\pi} \tfrac{1}{4}(1 + 2\cos^2\theta) \, d\theta = -\left( \tfrac{\pi}{2} + \tfrac{\pi}{2} \right) = -\pi$$

On $S_2$: $\mathbf{F}(\mathbf{r}(x, z)) = \mathbf{j} - z\mathbf{k}$ and $\mathbf{r}_z \times \mathbf{r}_x = \mathbf{j}$. Then $\iint_{S_2} \mathbf{F} \cdot d\mathbf{S} = \iint_{x^2 + z^2 \le 1} (1) \, dA = \pi$.

Hence $\iint_S \mathbf{F} \cdot d\mathbf{S} = -\pi + \pi = 0$.

**29.** Here $S$ consists of the six faces of the cube as labeled in the figure. On $S_1$:

$\mathbf{F} = \mathbf{i} + 2y\mathbf{j} + 3z\mathbf{k}$, $\mathbf{r}_y \times \mathbf{r}_z = \mathbf{i}$ and $\iint_{S_1} \mathbf{F} \cdot d\mathbf{S} = \int_{-1}^1 \int_{-1}^1 dy \, dz = 4$;

$S_2$: $\mathbf{F} = x\mathbf{i} + 2\mathbf{j} + 3z\mathbf{k}$, $\mathbf{r}_z \times \mathbf{r}_x = \mathbf{j}$ and $\iint_{S_2} \mathbf{F} \cdot d\mathbf{S} = \int_{-1}^1 \int_{-1}^1 2 \, dx \, dz = 8$;

$S_3$: $\mathbf{F} = x\mathbf{i} + 2y\mathbf{j} + 3\mathbf{k}$, $\mathbf{r}_x \times \mathbf{r}_y = \mathbf{k}$ and $\iint_{S_3} \mathbf{F} \cdot d\mathbf{S} = \int_{-1}^1 \int_{-1}^1 3 \, dx \, dy = 12$;

$S_4$: $\mathbf{F} = -\mathbf{i} + 2y\mathbf{j} + 3z\mathbf{k}$, $\mathbf{r}_z \times \mathbf{r}_y = -\mathbf{i}$ and $\iint_{S_4} \mathbf{F} \cdot d\mathbf{S} = 4$;

$S_5$: $\mathbf{F} = x\mathbf{i} - 2\mathbf{j} + 3z\mathbf{k}$, $\mathbf{r}_x \times \mathbf{r}_z = -\mathbf{j}$ and $\iint_{S_5} \mathbf{F} \cdot d\mathbf{S} = 8$;

$S_6$: $\mathbf{F} = x\mathbf{i} + 2y\mathbf{j} - 3\mathbf{k}$, $\mathbf{r}_y \times \mathbf{r}_x = -\mathbf{k}$ and $\iint_{S_6} \mathbf{F} \cdot d\mathbf{S} = \int_{-1}^1 \int_{-1}^1 3 \, dx \, dy = 12$.

Hence $\iint_S \mathbf{F} \cdot d\mathbf{S} = \sum_{i=1}^6 \iint_{S_i} \mathbf{F} \cdot d\mathbf{S} = 48$.

**31.** Here $S$ consists of four surfaces: $S_1$, the top surface (a portion of the circular cylinder $y^2 + z^2 = 1$); $S_2$, the bottom surface (a portion of the $xy$-plane); $S_3$, the front half-disk in the plane $x = 2$, and $S_4$, the back half-disk in the plane $x = 0$.

On $S_1$: The surface is $z = \sqrt{1 - y^2}$ for $0 \le x \le 2$, $-1 \le y \le 1$ with upward orientation, so

$$\iint_{S_1} \mathbf{F} \cdot d\mathbf{S} = \int_0^2 \int_{-1}^1 \left[ -x^2(0) - y^2 \left( -\frac{y}{\sqrt{1 - y^2}} \right) + z^2 \right] dy \, dx = \int_0^2 \int_{-1}^1 \left( \frac{y^3}{\sqrt{1 - y^2}} + 1 - y^2 \right) dy \, dx$$

$$= \int_0^2 \left[ -\sqrt{1 - y^2} + \tfrac{1}{3}(1 - y^2)^{3/2} + y - \tfrac{1}{3}y^3 \right]_{y=-1}^{y=1} dx = \int_0^2 \tfrac{4}{3} \, dx = \tfrac{8}{3}$$

On $S_2$: The surface is $z = 0$ with downward orientation, so

$$\iint_{S_2} \mathbf{F} \cdot d\mathbf{S} = \int_0^2 \int_{-1}^1 (-z^2)\, dy\, dx = \int_0^2 \int_{-1}^1 (0)\, dy\, dx = 0$$

On $S_3$: The surface is $x = 2$ for $-1 \le y \le 1$, $0 \le z \le \sqrt{1 - y^2}$, oriented in the positive $x$-direction. Regarding $y$ and $z$ as parameters, we have $\mathbf{r}_y \times \mathbf{r}_z = \mathbf{i}$ and

$$\iint_{S_3} \mathbf{F} \cdot d\mathbf{S} = \int_{-1}^1 \int_0^{\sqrt{1-y^2}} x^2\, dz\, dy = \int_{-1}^1 \int_0^{\sqrt{1-y^2}} 4\, dz\, dy = 4A\,(S_3) = 2\pi$$

On $S_4$: The surface is $x = 0$ for $-1 \le y \le 1$, $0 \le z \le \sqrt{1 - y^2}$, oriented in the negative $x$-direction. Regarding $y$ and $z$ as parameters, we use $-\,(\mathbf{r}_y \times \mathbf{r}_z) = -\mathbf{i}$ and

$$\iint_{S_4} \mathbf{F} \cdot d\mathbf{S} = \int_{-1}^1 \int_0^{\sqrt{1-y^2}} x^2\, dz\, dy = \int_{-1}^1 \int_0^{\sqrt{1-y^2}} (0)\, dz\, dy = 0$$

Thus $\iint_S \mathbf{F} \cdot d\mathbf{S} = \frac{8}{3} + 0 + 2\pi + 0 = 2\pi + \frac{8}{3}$.

**33.** We use Formula 4 with $z = 3 - 2x^2 - y^2$ $\Rightarrow$ $\partial z / \partial x = -4x$, $\partial z / \partial y = -2y$. The boundaries of the region

$3 - 2x^2 - y^2 \ge 0$ are $-\sqrt{\frac{3}{2}} \le x \le \sqrt{\frac{3}{2}}$ and $-\sqrt{3 - 2x^2} \le y \le \sqrt{3 - 2x^2}$, so we use a CAS (with precision reduced to

seven or fewer digits; otherwise the calculation may take a long time) to calculate

$$\iint_S x^2 y^2 z^2\, dS = \int_{-\sqrt{3/2}}^{\sqrt{3/2}} \int_{-\sqrt{3-2x^2}}^{\sqrt{3-2x^2}} x^2 y^2 (3 - 2x^2 - y^2)^2 \sqrt{16x^2 + 4y^2 + 1}\, dy\, dx \approx 3.4895$$

**35.** If $S$ is given by $y = h(x, z)$, then $S$ is also the level surface $f(x, y, z) = y - h(x, z) = 0$.

$\mathbf{n} = \dfrac{\nabla f(x, y, z)}{|\nabla f(x, y, z)|} = \dfrac{-h_x\,\mathbf{i} + \mathbf{j} - h_z\,\mathbf{k}}{\sqrt{h_x^2 + 1 + h_z^2}}$, and $-\mathbf{n}$ is the unit normal that points to the left. Now we proceed as in the

derivation of (10), using Formula 4 to evaluate

$$\iint_S \mathbf{F} \cdot d\mathbf{S} = \iint_S \mathbf{F} \cdot \mathbf{n}\, dS = \iint_D (P\mathbf{i} + Q\mathbf{j} + R\mathbf{k}) \frac{\dfrac{\partial h}{\partial x}\mathbf{i} - \mathbf{j} + \dfrac{\partial h}{\partial z}\mathbf{k}}{\sqrt{\left(\dfrac{\partial h}{\partial x}\right)^2 + 1 + \left(\dfrac{\partial h}{\partial z}\right)^2}} \sqrt{\left(\frac{\partial h}{\partial x}\right)^2 + 1 + \left(\frac{\partial h}{\partial z}\right)^2}\, dA$$

where $D$ is the projection of $S$ onto the $xz$-plane. Therefore $\displaystyle\iint_S \mathbf{F} \cdot d\mathbf{S} = \iint_D \left(P\frac{\partial h}{\partial x} - Q + R\frac{\partial h}{\partial z}\right) dA$.

**37.** $m = \iint_S K\, dS = K \cdot 4\pi \left(\frac{1}{2}a^2\right) = 2\pi a^2 K$; by symmetry $M_{xz} = M_{yz} = 0$, and

$M_{xy} = \iint_S zK\, dS = K \int_0^{2\pi} \int_0^{\pi/2} (a\cos\phi)(a^2 \sin\phi)\, d\phi\, d\theta = 2\pi K a^3 \left[-\frac{1}{4}\cos 2\phi\right]_0^{\pi/2} = \pi K a^3$.

Hence $(\overline{x}, \overline{y}, \overline{z}) = \left(0, 0, \frac{1}{2}a\right)$.

**39.** (a) $I_z = \iint_S (x^2 + y^2)\rho(x, y, z)\, dS$

(b) $I_z = \iint_S (x^2 + y^2)\left(10 - \sqrt{x^2 + y^2}\right) dS = \iint_{1 \le x^2 + y^2 \le 16} (x^2 + y^2)\left(10 - \sqrt{x^2 + y^2}\right) \sqrt{2}\, dA$

$\qquad = \int_0^{2\pi} \int_1^4 \sqrt{2}\,(10r^3 - r^4)\, dr\, d\theta = 2\sqrt{2}\,\pi\left(\frac{4329}{10}\right) = \frac{4329}{5}\sqrt{2}\,\pi$

**41.** The rate of flow through the cylinder is the flux $\iint_S \rho \mathbf{v} \cdot \mathbf{n} \, dS = \iint_S \rho \mathbf{v} \cdot d\mathbf{S}$. We use the parametric representation

$\mathbf{r}(u, v) = 2\cos u \, \mathbf{i} + 2\sin u \, \mathbf{j} + v \, \mathbf{k}$ for $S$, where $0 \le u \le 2\pi$, $0 \le v \le 1$, so $\mathbf{r}_u = -2\sin u \, \mathbf{i} + 2\cos u \, \mathbf{j}$, $\mathbf{r}_v = \mathbf{k}$, and the

outward orientation is given by $\mathbf{r}_u \times \mathbf{r}_v = 2\cos u \, \mathbf{i} + 2\sin u \, \mathbf{j}$. Then

$$\iint_S \rho \mathbf{v} \cdot d\mathbf{S} = \rho \int_0^{2\pi} \int_0^1 \left(v\,\mathbf{i} + 4\sin^2 u\,\mathbf{j} + 4\cos^2 u\,\mathbf{k}\right) \cdot (2\cos u\,\mathbf{i} + 2\sin u\,\mathbf{j}) \, dv \, du$$

$$= \rho \int_0^{2\pi} \int_0^1 \left(2v\cos u + 8\sin^3 u\right) dv \, du = \rho \int_0^{2\pi} \left(\cos u + 8\sin^3 u\right) du$$

$$= \rho\left[\sin u + 8\left(-\tfrac{1}{3}\right)\left(2 + \sin^2 u\right)\cos u\right]_0^{2\pi} = 0 \text{ kg/s}$$

**43.** $S$ consists of the hemisphere $S_1$ given by $z = \sqrt{a^2 - x^2 - y^2}$ and the disk $S_2$ given by $0 \le x^2 + y^2 \le a^2$, $z = 0$.

On $S_1$: $\mathbf{E} = a\sin\phi\cos\theta\,\mathbf{i} + a\sin\phi\sin\theta\,\mathbf{j} + 2a\cos\phi\,\mathbf{k}$,

$\mathbf{T}_\phi \times \mathbf{T}_\theta = a^2\sin^2\phi\cos\theta\,\mathbf{i} + a^2\sin^2\phi\sin\theta\,\mathbf{j} + a^2\sin\phi\cos\phi\,\mathbf{k}$. Thus

$$\iint_{S_1} \mathbf{E} \cdot d\mathbf{S} = \int_0^{2\pi}\int_0^{\pi/2}\left(a^3\sin^3\phi + 2a^3\sin\phi\cos^2\phi\right) d\phi \, d\theta$$

$$= \int_0^{2\pi}\int_0^{\pi/2}\left(a^3\sin\phi + a^3\sin\phi\cos^2\phi\right) d\phi \, d\theta = (2\pi)a^3\left(1 + \tfrac{1}{3}\right) = \tfrac{8}{3}\pi a^3$$

On $S_2$: $\mathbf{E} = x\,\mathbf{i} + y\,\mathbf{j}$, and $\mathbf{r}_y \times \mathbf{r}_x = -\mathbf{k}$ so $\iint_{S_2} \mathbf{E} \cdot d\mathbf{S} = 0$. Hence the total charge is $q = \varepsilon_0 \iint_S \mathbf{E} \cdot d\mathbf{S} = \tfrac{8}{3}\pi a^3 \varepsilon_0$.

**45.** $K\nabla u = 6.5(4y\,\mathbf{j} + 4z\,\mathbf{k})$. $S$ is given by $\mathbf{r}(x, \theta) = x\,\mathbf{i} + \sqrt{6}\cos\theta\,\mathbf{j} + \sqrt{6}\sin\theta\,\mathbf{k}$ and since we want the inward heat flow, we

use $\mathbf{r}_x \times \mathbf{r}_\theta = -\sqrt{6}\cos\theta\,\mathbf{j} - \sqrt{6}\sin\theta\,\mathbf{k}$. Then the rate of heat flow inward is given by

$$\iint_S (-K\nabla u) \cdot d\mathbf{S} = \int_0^{2\pi}\int_0^4 -(6.5)(-24) \, dx \, d\theta = (2\pi)(156)(4) = 1248\pi.$$

**47.** Let $S$ be a sphere of radius $a$ centered at the origin. Then $|\mathbf{r}| = a$ and $\mathbf{F}(\mathbf{r}) = c\mathbf{r}/|\mathbf{r}|^3 = (c/a^3)(x\,\mathbf{i} + y\,\mathbf{j} + z\,\mathbf{k})$. A

parametric representation for $S$ is $\mathbf{r}(\phi, \theta) = a\sin\phi\cos\theta\,\mathbf{i} + a\sin\phi\sin\theta\,\mathbf{j} + a\cos\phi\,\mathbf{k}$, $0 \le \phi \le \pi$, $0 \le \theta \le 2\pi$. Then

$\mathbf{r}_\phi = a\cos\phi\cos\theta\,\mathbf{i} + a\cos\phi\sin\theta\,\mathbf{j} - a\sin\phi\,\mathbf{k}$, $\mathbf{r}_\theta = -a\sin\phi\sin\theta\,\mathbf{i} + a\sin\phi\cos\theta\,\mathbf{j}$, and the outward orientation is given

by $\mathbf{r}_\phi \times \mathbf{r}_\theta = a^2\sin^2\phi\cos\theta\,\mathbf{i} + a^2\sin^2\phi\sin\theta\,\mathbf{j} + a^2\sin\phi\cos\phi\,\mathbf{k}$. The flux of $\mathbf{F}$ across $S$ is

$$\iint_S \mathbf{F} \cdot d\mathbf{S} = \int_0^\pi \int_0^{2\pi} \frac{c}{a^3}\left(a\sin\phi\cos\theta\,\mathbf{i} + a\sin\phi\sin\theta\,\mathbf{j} + a\cos\phi\,\mathbf{k}\right)$$

$$\cdot \left(a^2\sin^2\phi\cos\theta\,\mathbf{i} + a^2\sin^2\phi\sin\theta\,\mathbf{j} + a^2\sin\phi\cos\phi\,\mathbf{k}\right) d\theta \, d\phi$$

$$= \frac{c}{a^3}\int_0^\pi\int_0^{2\pi} a^3\left(\sin^3\phi + \sin\phi\cos^2\phi\right) d\theta \, d\phi = c\int_0^\pi\int_0^{2\pi}\sin\phi \, d\theta \, d\phi = 4\pi c$$

Thus the flux does not depend on the radius $a$.

## 13.7 Stokes' Theorem

1. Both $H$ and $P$ are oriented piecewise-smooth surfaces that are bounded by the simple, closed, smooth curve $x^2 + y^2 = 4$,

    $z = 0$ (which we can take to be oriented positively for both surfaces). Then $H$ and $P$ satisfy the hypotheses of Stokes'

    Theorem, so by (3) we know $\iint_H \operatorname{curl} \mathbf{F} \cdot d\mathbf{S} = \int_C \mathbf{F} \cdot d\mathbf{r} = \iint_P \operatorname{curl} \mathbf{F} \cdot d\mathbf{S}$ (where $C$ is the boundary curve).

3. The paraboloid $z = x^2 + y^2$ intersects the cylinder $x^2 + y^2 = 4$ in the circle $x^2 + y^2 = 4$, $z = 4$. This boundary curve $C$

    should be oriented in the counterclockwise direction when viewed from above, so a vector equation of $C$ is

    $\mathbf{r}(t) = 2\cos t\,\mathbf{i} + 2\sin t\,\mathbf{j} + 4\,\mathbf{k}$, $0 \leq t \leq 2\pi$. Then $\mathbf{r}'(t) = -2\sin t\,\mathbf{i} + 2\cos t\,\mathbf{j}$,

    $\mathbf{F}(\mathbf{r}(t)) = (4\cos^2 t)(16)\,\mathbf{i} + (4\sin^2 t)(16)\,\mathbf{j} + (2\cos t)(2\sin t)(4)\,\mathbf{k} = 64\cos^2 t\,\mathbf{i} + 64\sin^2 t\,\mathbf{j} + 16\sin t\,\cos t\,\mathbf{k}$,

    and by Stokes' Theorem,

    $$\iint_S \operatorname{curl} \mathbf{F} \cdot d\mathbf{S} = \int_C \mathbf{F} \cdot d\mathbf{r} = \int_0^{2\pi} \mathbf{F}(\mathbf{r}(t)) \cdot \mathbf{r}'(t)\, dt = \int_0^{2\pi} (-128\cos^2 t\,\sin t + 128\sin^2 t\,\cos t + 0)\, dt$$

    $$= 128\left[\tfrac{1}{3}\cos^3 t + \tfrac{1}{3}\sin^3 t\right]_0^{2\pi} = 0$$

5. $C$ is the square in the plane $z = -1$. By (3), $\iint_{S_1} \operatorname{curl} \mathbf{F} \cdot d\mathbf{S} = \oint_C \mathbf{F} \cdot d\mathbf{r} = \iint_{S_2} \operatorname{curl} \mathbf{F} \cdot d\mathbf{S}$ where $S_1$ is the original cube

    without the bottom and $S_2$ is the bottom face of the cube. $\operatorname{curl} \mathbf{F} = x^2 z\,\mathbf{i} + (xy - 2xyz)\,\mathbf{j} + (y - xz)\,\mathbf{k}$. For $S_2$, we choose

    $\mathbf{n} = \mathbf{k}$ so that $C$ has the same orientation for both surfaces. Then $\operatorname{curl} \mathbf{F} \cdot \mathbf{n} = y - xz = x + y$ on $S_2$, where $z = -1$. Thus

    $\iint_{S_2} \operatorname{curl} \mathbf{F} \cdot d\mathbf{S} = \int_{-1}^1 \int_{-1}^1 (x + y)\, dx\, dy = 0$ so $\iint_{S_1} \operatorname{curl} \mathbf{F} \cdot d\mathbf{S} = 0$.

7. $\operatorname{curl} \mathbf{F} = -2z\,\mathbf{i} - 2x\,\mathbf{j} - 2y\,\mathbf{k}$ and we take the surface $S$ to be the planar region enclosed by $C$, so $S$ is the portion of the plane

    $x + y + z = 1$ over $D = \{(x, y) \mid 0 \leq x \leq 1, 0 \leq y \leq 1 - x\}$. Since $C$ is oriented counterclockwise, we orient $S$ upward.

    Using Equation 13.6.10, we have $z = g(x, y) = 1 - x - y$, $P = -2z$, $Q = -2x$, $R = -2y$, and

    $$\int_C \mathbf{F} \cdot d\mathbf{r} = \iint_S \operatorname{curl} \mathbf{F} \cdot d\mathbf{S} = \iint_D [-(-2z)(-1) - (-2x)(-1) + (-2y)]\, dA$$

    $$= \int_0^1 \int_0^{1-x} (-2)\, dy\, dx = -2\int_0^1 (1 - x)\, dx = -1$$

9. $\operatorname{curl} \mathbf{F} = (xe^{xy} - 2x)\,\mathbf{i} - (ye^{xy} - y)\,\mathbf{j} + (2z - z)\,\mathbf{k}$ and we take $S$ to be the disk $x^2 + y^2 \leq 16$, $z = 5$. Since $C$ is oriented

    counterclockwise (from above), we orient $S$ upward. Then $\mathbf{n} = \mathbf{k}$ and $\operatorname{curl} \mathbf{F} \cdot \mathbf{n} = 2z - z$ on $S$, where $z = 5$. Thus

    $$\oint_C \mathbf{F} \cdot d\mathbf{r} = \iint_S \operatorname{curl} \mathbf{F} \cdot \mathbf{n}\, dS = \iint_S (2z - z)\, dS = \iint_S (10 - 5)\, dS = 5(\text{area of } S) = 5(\pi \cdot 4^2) = 80\pi$$

11. (a) The curve of intersection is an ellipse in the plane $x + y + z = 1$ with unit normal $\mathbf{n} = \frac{1}{\sqrt{3}}(\mathbf{i} + \mathbf{j} + \mathbf{k})$,

    $\operatorname{curl} \mathbf{F} = x^2\,\mathbf{j} + y^2\,\mathbf{k}$, and $\operatorname{curl} \mathbf{F} \cdot \mathbf{n} = \frac{1}{\sqrt{3}}(x^2 + y^2)$. Then

    $$\oint_C \mathbf{F} \cdot d\mathbf{r} = \iint_S \frac{1}{\sqrt{3}}(x^2 + y^2)\, dS = \iint_{x^2 + y^2 \leq 9} (x^2 + y^2)\, dx\, dy = \int_0^{2\pi} \int_0^3 r^3\, dr\, d\theta = 2\pi\left(\tfrac{81}{4}\right) = \tfrac{81\pi}{2}$$

(b)

(c) One possible parametrization is $x = 3 \cos t$, $y = 3 \sin t$, $z = 1 - 3 \cos t - 3 \sin t$, $0 \le t \le 2\pi$.

**13.** The boundary curve $C$ is the circle $x^2 + y^2 = 16$, $z = 4$ oriented in the clockwise direction as viewed from above (since $S$ is oriented downward). We can parametrize $C$ by $\mathbf{r}(t) = 4 \cos t \, \mathbf{i} - 4 \sin t \, \mathbf{j} + 4 \, \mathbf{k}$, $0 \le t \le 2\pi$, and then

$\mathbf{r}'(t) = -4 \sin t \, \mathbf{i} - 4 \cos t \, \mathbf{j}$. Thus $\mathbf{F}(\mathbf{r}(t)) = 4 \sin t \, \mathbf{i} + 4 \cos t \, \mathbf{j} - 2 \, \mathbf{k}$, $\mathbf{F}(\mathbf{r}(t)) \cdot \mathbf{r}'(t) = -16 \sin^2 t - 16 \cos^2 t = -16$, and

$$\oint_C \mathbf{F} \cdot d\mathbf{r} = \int_0^{2\pi} \mathbf{F}(\mathbf{r}(t)) \cdot \mathbf{r}'(t) \, dt = \int_0^{2\pi} (-16) \, dt = -16 \, (2\pi) = -32\pi$$

Now curl $\mathbf{F} = 2 \, \mathbf{k}$, and the projection $D$ of $S$ on the $xy$-plane is the disk $x^2 + y^2 \le 16$, so by Equation 13.6.10 with

$z = g(x, y) = \sqrt{x^2 + y^2}$ [and multiplying by $-1$ for the downward orientation] we have

$$\iint_S \operatorname{curl} \mathbf{F} \cdot d\mathbf{S} = - \iint_D (-0 - 0 + 2) \, dA = -2 \cdot A(D) = -2 \cdot \pi(4^2) = -32\pi$$

**15.** The boundary curve $C$ is the circle $x^2 + z^2 = 1$, $y = 0$ oriented in the counterclockwise direction as viewed from the positive $y$-axis. Then $C$ can be described by $\mathbf{r}(t) = \cos t \, \mathbf{i} - \sin t \, \mathbf{k}$, $0 \le t \le 2\pi$, and $\mathbf{r}'(t) = -\sin t \, \mathbf{i} - \cos t \, \mathbf{k}$. Thus

$\mathbf{F}(\mathbf{r}(t)) = - \sin t \, \mathbf{j} + \cos t \, \mathbf{k}$, $\mathbf{F}(\mathbf{r}(t)) \cdot \mathbf{r}'(t) = -\cos^2 t$, and $\oint_C \mathbf{F} \cdot d\mathbf{r} = \int_0^{2\pi} (-\cos^2 t) \, dt = -\frac{1}{2} t - \frac{1}{4} \sin 2t \big]_0^{2\pi} = -\pi$.

Now curl $\mathbf{F} = -\mathbf{i} - \mathbf{j} - \mathbf{k}$, and $S$ can be parametrized (see Example 12.6.1) by

$\mathbf{r}(\phi, \theta) = \sin \phi \, \cos \theta \, \mathbf{i} + \sin \phi \, \sin \theta \, \mathbf{j} + \cos \phi \, \mathbf{k}$, $0 \le \theta \le \pi$, $0 \le \phi \le \pi$. Then

$\mathbf{r}_\phi \times \mathbf{r}_\theta = \sin^2 \phi \, \cos \theta \, \mathbf{i} + \sin^2 \phi \, \sin \theta \, \mathbf{j} + \sin \phi \, \cos \phi \, \mathbf{k}$ and

$$\iint_S \operatorname{curl} \mathbf{F} \cdot d\mathbf{S} = \iint_{x^2 + z^2 \le 1} \operatorname{curl} \mathbf{F} \cdot (\mathbf{r}_\phi \times \mathbf{r}_\theta) \, dA = \int_0^\pi \int_0^\pi (-\sin^2 \phi \, \cos \theta - \sin^2 \phi \, \sin \theta - \sin \phi \, \cos \phi) \, d\theta \, d\phi$$

$$= \int_0^\pi (-2 \sin^2 \phi - \pi \sin \phi \, \cos \phi) \, d\phi = \left[ \tfrac{1}{2} \sin 2\phi - \phi - \tfrac{\pi}{2} \sin^2 \phi \right]_0^\pi = -\pi$$

**17.** It is easier to use Stokes' Theorem than to compute the work directly. Let $S$ be the planar region enclosed by the path of the particle, so $S$ is the portion of the plane $z = \frac{1}{2} y$ for $0 \le x \le 1$, $0 \le y \le 2$, with upward orientation.

curl $\mathbf{F} = 8y \, \mathbf{i} + 2z \, \mathbf{j} + 2y \, \mathbf{k}$ and

$$\oint_C \mathbf{F} \cdot d\mathbf{r} = \iint_S \operatorname{curl} \mathbf{F} \cdot d\mathbf{S} = \iint_D \left[ -8y \, (0) - 2z \left( \tfrac{1}{2} \right) + 2y \right] dA = \int_0^1 \int_0^2 \left( 2y - \tfrac{1}{2} y \right) dy \, dx$$

$$= \int_0^1 \int_0^2 \tfrac{3}{2} y \, dy \, dx = \int_0^1 \left[ \tfrac{3}{4} y^2 \right]_{y=0}^{y=2} dx = \int_0^1 3 \, dx = 3$$

**19.** Assume $S$ is centered at the origin with radius $a$ and let $H_1$ and $H_2$ be the upper and lower hemispheres, respectively, of $S$.

Then $\iint_S \operatorname{curl} \mathbf{F} \cdot d\mathbf{S} = \iint_{H_1} \operatorname{curl} \mathbf{F} \cdot d\mathbf{S} + \iint_{H_2} \operatorname{curl} \mathbf{F} \cdot d\mathbf{S} = \oint_{C_1} \mathbf{F} \cdot d\mathbf{r} + \oint_{C_2} \mathbf{F} \cdot d\mathbf{r}$ by Stokes' Theorem. But $C_1$ is the

circle $x^2 + y^2 = a^2$ oriented in the counterclockwise direction while $C_2$ is the same circle oriented in the clockwise direction.

Hence $\oint_{C_2} \mathbf{F} \cdot d\mathbf{r} = - \oint_{C_1} \mathbf{F} \cdot d\mathbf{r}$ so $\iint_S \operatorname{curl} \mathbf{F} \cdot d\mathbf{S} = 0$ as desired.

## 13.8 The Divergence Theorem

**1.** $\operatorname{div} \mathbf{F} = 3 + x + 2x = 3 + 3x$, so

$\iiint_E \operatorname{div} \mathbf{F} \, dV = \int_0^1 \int_0^1 \int_0^1 (3x + 3) \, dx \, dy \, dz = \frac{9}{2}$ (notice the triple integral is

three times the volume of the cube plus three times $\overline{x}$).

To compute $\iint_S \mathbf{F} \cdot d\mathbf{S}$, on

$S_1$: $\mathbf{n} = \mathbf{i}$, $\mathbf{F} = 3\mathbf{i} + y\mathbf{j} + 2z\mathbf{k}$, and $\iint_{S_1} \mathbf{F} \cdot d\mathbf{S} = \iint_{S_1} 3 \, dS = 3$;

$S_2$: $\mathbf{F} = 3x\mathbf{i} + x\mathbf{j} + 2xz\mathbf{k}$, $\mathbf{n} = \mathbf{j}$ and $\iint_{S_2} \mathbf{F} \cdot d\mathbf{S} = \iint_{S_2} x \, dS = \frac{1}{2}$;

$S_3$: $\mathbf{F} = 3x\mathbf{i} + xy\mathbf{j} + 2x\mathbf{k}$, $\mathbf{n} = \mathbf{k}$ and $\iint_{S_3} \mathbf{F} \cdot d\mathbf{S} = \iint_{S_3} 2x \, dS = 1$;

$S_4$: $\mathbf{F} = \mathbf{0}$, $\iint_{S_4} \mathbf{F} \cdot d\mathbf{S} = 0$; $S_5$: $\mathbf{F} = 3x\mathbf{i} + 2x\mathbf{k}$, $\mathbf{n} = -\mathbf{j}$ and $\iint_{S_5} \mathbf{F} \cdot d\mathbf{S} = \iint_{S_5} 0 \, dS = 0$;

$S_6$: $\mathbf{F} = 3x\mathbf{i} + xy\mathbf{j}$, $\mathbf{n} = -\mathbf{k}$ and $\iint_{S_6} \mathbf{F} \cdot d\mathbf{S} = \iint_{S_6} 0 \, dS = 0$. Thus $\iint_S \mathbf{F} \cdot d\mathbf{S} = \frac{9}{2}$.

**3.** $\operatorname{div} \mathbf{F} = 0 + 1 + 0 = 1$, so $\iiint_E \operatorname{div} \mathbf{F} \, dV = \iiint_E 1 \, dV = V(E) = \frac{4}{3}\pi \cdot 4^3 = \frac{256}{3}\pi$. $S$ is a sphere of radius 4 centered at

the origin which can be parametrized by $\mathbf{r}(\phi, \theta) = \langle 4\sin\phi\cos\theta, 4\sin\phi\sin\theta, 4\cos\phi \rangle$, $0 \le \phi \le \pi$, $0 \le \theta \le 2\pi$ (similar to

Example 12.6.1). Then

$$\mathbf{r}_\phi \times \mathbf{r}_\theta = \langle 4\cos\phi\cos\theta, 4\cos\phi\sin\theta, -4\sin\phi \rangle \times \langle -4\sin\phi\sin\theta, 4\sin\phi\cos\theta, 0 \rangle$$
$$= \langle 16\sin^2\phi\cos\theta, 16\sin^2\phi\sin\theta, 16\cos\phi\sin\phi \rangle$$

and $\mathbf{F}(\mathbf{r}(\phi, \theta)) = \langle 4\cos\phi, 4\sin\phi\sin\theta, 4\sin\phi\cos\theta \rangle$. Thus

$\mathbf{F} \cdot (\mathbf{r}_\phi \times \mathbf{r}_\theta) = 64\cos\phi\sin^2\phi\cos\theta + 64\sin^3\phi\sin^2\theta + 64\cos\phi\sin^2\phi\cos\theta = 128\cos\phi\sin^2\phi\cos\theta + 64\sin^3\phi\sin^2\theta$

and

$$\iint_S \mathbf{F} \cdot d\mathbf{S} = \iint_D \mathbf{F} \cdot (\mathbf{r}_\phi \times \mathbf{r}_\theta) \, dA = \int_0^{2\pi} \int_0^\pi (128\cos\phi\sin^2\phi\cos\theta + 64\sin^3\phi\sin^2\theta) \, d\phi \, d\theta$$
$$= \int_0^{2\pi} \left[ \frac{128}{3}\sin^3\phi\cos\theta + 64\left(-\frac{1}{3}(2 + \sin^2\phi)\cos\phi\right)\sin^2\theta \right]_{\phi=0}^{\phi=\pi} d\theta$$
$$= \int_0^{2\pi} \frac{256}{3}\sin^2\theta \, d\theta = \frac{256}{3}\left[\frac{1}{2}\theta - \frac{1}{4}\sin 2\theta\right]_0^{2\pi} = \frac{256}{3}\pi$$

**5.** $\operatorname{div} \mathbf{F} = \frac{\partial}{\partial x}(xye^z) + \frac{\partial}{\partial y}(xy^2z^3) + \frac{\partial}{\partial z}(-ye^z) = ye^z + 2xyz^3 - ye^z = 2xyz^3$, so by the Divergence Theorem,

$$\iint_S \mathbf{F} \cdot d\mathbf{S} = \iiint_E \operatorname{div} \mathbf{F} \, dV = \int_0^3 \int_0^2 \int_0^1 2xyz^3 \, dz \, dy \, dx = 2\int_0^3 x \, dx \int_0^2 y \, dy \int_0^1 z^3 \, dz$$
$$= 2\left[\frac{1}{2}x^2\right]_0^3 \left[\frac{1}{2}y^2\right]_0^2 \left[\frac{1}{4}z^4\right]_0^1 = 2\left(\frac{9}{2}\right)(2)\left(\frac{1}{4}\right) = \frac{9}{2}$$

**7.** $\operatorname{div} \mathbf{F} = 3y^2 + 0 + 3z^2$, so using cylindrical coordinates with $y = r\cos\theta$, $z = r\sin\theta$, $x = x$ we have

$$\iint_S \mathbf{F} \cdot d\mathbf{S} = \iiint_E (3y^2 + 3z^2) \, dV = \int_0^{2\pi} \int_0^1 \int_{-1}^2 (3r^2\cos^2\theta + 3r^2\sin^2\theta) \, r \, dx \, dr \, d\theta$$
$$= 3\int_0^{2\pi} d\theta \int_0^1 r^3 \, dr \int_{-1}^2 dx = 3(2\pi)\left(\frac{1}{4}\right)(3) = \frac{9\pi}{2}$$

**9.** $\operatorname{div} \mathbf{F} = 2x\sin y - x\sin y - x\sin y = 0$, so by the Divergence Theorem, $\iint_S \mathbf{F} \cdot d\mathbf{S} = \iiint_E 0 \, dV = 0$.

**11.** div $\mathbf{F} = y^2 + 0 + x^2 = x^2 + y^2$ so

$$\iint_S \mathbf{F} \cdot d\mathbf{S} = \iiint_E (x^2 + y^2)\, dV = \int_0^{2\pi} \int_0^2 \int_{r^2}^4 r^2 \cdot r\, dz\, dr\, d\theta = \int_0^{2\pi} \int_0^2 r^3 (4 - r^2)\, dr\, d\theta$$

$$= \int_0^{2\pi} d\theta \int_0^2 (4r^3 - r^5)\, dr = 2\pi \left[ r^4 - \tfrac{1}{6}r^6 \right]_0^2 = \tfrac{32}{3}\pi$$

**13.** div $\mathbf{F} = 12x^2 z + 12y^2 z + 12z^3$ so

$$\iint_S \mathbf{F} \cdot d\mathbf{S} = \iiint_E 12z(x^2 + y^2 + z^2)\, dV = \int_0^{2\pi} \int_0^{\pi} \int_0^R 12(\rho \cos \phi)(\rho^2)\rho^2 \sin \phi\, d\rho\, d\phi\, d\theta$$

$$= 12\int_0^{2\pi} d\theta \int_0^{\pi} \sin \phi\, \cos \phi\, d\phi \int_0^R \rho^5\, d\rho = 12(2\pi)\left[\tfrac{1}{2}\sin^2 \phi\right]_0^{\pi} \left[\tfrac{1}{6}\rho^6\right]_0^R = 0$$

**15.** $\iint_S \mathbf{F} \cdot d\mathbf{S} = \iiint_E \sqrt{3 - x^2}\, dV = \int_{-1}^1 \int_{-1}^1 \int_0^{2 - x^4 - y^4} \sqrt{3 - x^2}\, dz\, dy\, dx = \frac{341}{60}\sqrt{2} + \frac{81}{20}\sin^{-1}\left(\frac{\sqrt{3}}{3}\right)$

**17.** For $S_1$ we have $\mathbf{n} = -\mathbf{k}$, so $\mathbf{F} \cdot \mathbf{n} = \mathbf{F} \cdot (-\mathbf{k}) = -x^2 z - y^2 = -y^2$ (since $z = 0$ on $S_1$). So if $D$ is the unit disk, we get

$\iint_{S_1} \mathbf{F} \cdot d\mathbf{S} = \iint_{S_1} \mathbf{F} \cdot \mathbf{n}\, dS = \iint_D (-y^2)\, dA = -\int_0^{2\pi} \int_0^1 r^2 (\sin^2 \theta)\, r\, dr\, d\theta = -\frac{1}{4}\pi$. Now since $S_2$ is closed, we can use

the Divergence Theorem. Since div $\mathbf{F} = \frac{\partial}{\partial x}\left(z^2 x\right) + \frac{\partial}{\partial y}\left(\frac{1}{3}y^3 + \tan z\right) + \frac{\partial}{\partial z}\left(x^2 z + y^2\right) = z^2 + y^2 + x^2$, we use spherical

coordinates to get $\iint_{S_2} \mathbf{F} \cdot d\mathbf{S} = \iiint_E \text{div}\, \mathbf{F}\, dV = \int_0^{2\pi} \int_0^{\pi/2} \int_0^1 \rho^2 \cdot \rho^2 \sin \phi\, d\rho\, d\phi\, d\theta = \frac{2}{5}\pi$. Finally

$\iint_S \mathbf{F} \cdot d\mathbf{S} = \iint_{S_2} \mathbf{F} \cdot d\mathbf{S} - \iint_{S_1} \mathbf{F} \cdot d\mathbf{S} = \frac{2}{5}\pi - \left(-\frac{1}{4}\pi\right) = \frac{13}{20}\pi$.

**19.** The vectors that end near $P_1$ are longer than the vectors that start near $P_1$, so the net flow is inward near $P_1$ and div $\mathbf{F}(P_1)$ is negative. The vectors that end near $P_2$ are shorter than the vectors that start near $P_2$, so the net flow is outward near $P_2$ and div $\mathbf{F}(P_2)$ is positive.

**21.**

From the graph it appears that for points above the $x$-axis, vectors starting near a particular point are longer than vectors ending there, so divergence is positive.

The opposite is true at points below the $x$-axis, where divergence is negative.

$\mathbf{F}(x, y) = \langle xy, x + y^2 \rangle \quad \Rightarrow \quad \text{div}\, \mathbf{F} = \frac{\partial}{\partial x}(xy) + \frac{\partial}{\partial y}(x + y^2) = y + 2y = 3y$.

Thus div $\mathbf{F} > 0$ for $y > 0$, and div $\mathbf{F} < 0$ for $y < 0$.

**23.** Since $\dfrac{\mathbf{x}}{|\mathbf{x}|^3} = \dfrac{x\mathbf{i} + y\mathbf{j} + z\mathbf{k}}{(x^2 + y^2 + z^2)^{3/2}}$ and $\dfrac{\partial}{\partial x}\left(\dfrac{x}{(x^2 + y^2 + z^2)^{3/2}}\right) = \dfrac{(x^2 + y^2 + z^2) - 3x^2}{(x^2 + y^2 + z^2)^{5/2}}$ with similar expressions

for $\dfrac{\partial}{\partial y}\left(\dfrac{y}{(x^2 + y^2 + z^2)^{3/2}}\right)$ and $\dfrac{\partial}{\partial z}\left(\dfrac{z}{(x^2 + y^2 + z^2)^{3/2}}\right)$, we have

$$\text{div}\left(\frac{\mathbf{x}}{|\mathbf{x}|^3}\right) = \frac{3(x^2 + y^2 + z^2) - 3(x^2 + y^2 + z^2)}{(x^2 + y^2 + z^2)^{5/2}} = 0, \text{ except at } (0,0,0) \text{ where it is undefined.}$$

**25.** $\iint_S \mathbf{a} \cdot \mathbf{n}\, dS = \iiint_E \text{div}\, \mathbf{a}\, dV = 0$ since div $\mathbf{a} = 0$.

**27.** $\iint_S \text{curl}\, \mathbf{F} \cdot d\mathbf{S} = \iiint_E \text{div}(\text{curl}\, \mathbf{F})\, dV = 0$ by Theorem 13.5.11.

**29.** $\iint_S (f \nabla g) \cdot \mathbf{n}\, dS = \iiint_E \text{div}(f \nabla g)\, dV = \iiint_E (f \nabla^2 g + \nabla g \cdot \nabla f)\, dV$ by Exercise 13.5.25.

**31.** If $\mathbf{c} = c_1\,\mathbf{i} + c_2\,\mathbf{j} + c_3\,\mathbf{k}$ is an arbitrary constant vector, we define $\mathbf{F} = f\mathbf{c} = fc_1\,\mathbf{i} + fc_2\,\mathbf{j} + fc_3\,\mathbf{k}$. Then

$$\operatorname{div} \mathbf{F} = \operatorname{div} f\mathbf{c} = \frac{\partial f}{\partial x}\,c_1 + \frac{\partial f}{\partial y}\,c_2 + \frac{\partial f}{\partial z}\,c_3 = \nabla f \cdot \mathbf{c} \text{ and the Divergence Theorem says } \iint_S \mathbf{F} \cdot d\mathbf{S} = \iiint_E \operatorname{div} \mathbf{F}\, dV \quad \Rightarrow$$

$\iint_S \mathbf{F} \cdot \mathbf{n}\, dS = \iiint_E \nabla f \cdot \mathbf{c}\, dV$. In particular, if $\mathbf{c} = \mathbf{i}$ then $\iint_S f\mathbf{i} \cdot \mathbf{n}\, dS = \iiint_E \nabla f \cdot \mathbf{i}\, dV \quad \Rightarrow$

$$\iint_S fn_1\, dS = \iiint_E \frac{\partial f}{\partial x}\, dV \text{ (where } \mathbf{n} = n_1\,\mathbf{i} + n_2\,\mathbf{j} + n_3\,\mathbf{k}\text{). Similarly, if } \mathbf{c} = \mathbf{j} \text{ we have } \iint_S fn_2\, dS = \iiint_E \frac{\partial f}{\partial y}\, dV,$$

and $\mathbf{c} = \mathbf{k}$ gives $\displaystyle\iint_S fn_3\, dS = \iiint_E \frac{\partial f}{\partial z}\, dV$. Then

$$\iint_S f\mathbf{n}\, dS = \left(\iint_S fn_1\, dS\right)\mathbf{i} + \left(\iint_S fn_2\, dS\right)\mathbf{j} + \left(\iint_S fn_3\, dS\right)\mathbf{k}$$

$$= \left(\iiint_E \frac{\partial f}{\partial x}\, dV\right)\mathbf{i} + \left(\iiint_E \frac{\partial f}{\partial y}\, dV\right)\mathbf{j} + \left(\iiint_E \frac{\partial f}{\partial z}\, dV\right)\mathbf{k} = \iiint_E \left(\frac{\partial f}{\partial x}\,\mathbf{i} + \frac{\partial f}{\partial y}\,\mathbf{j} + \frac{\partial f}{\partial z}\,\mathbf{k}\right) dV$$

$$= \iiint_E \nabla f\, dV \quad \text{as desired.}$$

# 13 Review

## CONCEPT CHECK

**1.** See Definitions 1 and 2 in Section 13.1. A vector field can represent, for example, the wind velocity at any location in space, the speed and direction of the ocean current at any location, or the force vectors of Earth's gravitational field at a location in space.

**2.** (a) A conservative vector field $\mathbf{F}$ is a vector field which is the gradient of some scalar function $f$.

(b) The function $f$ in part (a) is called a potential function for $\mathbf{F}$, that is, $\mathbf{F} = \nabla f$.

**3.** (a) See Definition 13.2.2.

(b) We normally evaluate the line integral using Formula 13.2.3.

(c) The mass is $m = \int_C \rho\,(x, y)\, ds$, and the center of mass is $(\overline{x}, \overline{y})$ where $\overline{x} = \frac{1}{m}\int_C x\rho\,(x, y)\, ds$, $\overline{y} = \frac{1}{m}\int_C y\rho\,(x, y)\, ds$.

(d) See (5) and (6) in Section 13.2 for plane curves; we have similar definitions when $C$ is a space curve
   [see the equation preceding (10) in Section 13.2].

(e) For plane curves, see Equations 13.2.7. We have similar results for space curves
   [see the equation preceding (10) in Section 13.2].

**4.** (a) See Definition 13.2.13.

(b) If $\mathbf{F}$ is a force field, $\int_C \mathbf{F} \cdot d\mathbf{r}$ represents the work done by $\mathbf{F}$ in moving a particle along the curve $C$.

(c) $\int_C \mathbf{F} \cdot d\mathbf{r} = \int_C P\, dx + Q\, dy + R\, dz$

**5.** See Theorem 13.3.2.

**6.** (a) $\int_C \mathbf{F} \cdot d\mathbf{r}$ is independent of path if the line integral has the same value for any two curves that have the same initial and terminal points.

(b) See Theorem 13.3.4.

**7.** See the statement of Green's Theorem on page 934.

**8.** See Equations 13.4.5.

**9.** (a) $\operatorname{curl} \mathbf{F} = \left( \dfrac{\partial R}{\partial y} - \dfrac{\partial Q}{\partial z} \right) \mathbf{i} + \left( \dfrac{\partial P}{\partial z} - \dfrac{\partial R}{\partial x} \right) \mathbf{j} + \left( \dfrac{\partial Q}{\partial x} - \dfrac{\partial P}{\partial y} \right) \mathbf{k} = \nabla \times \mathbf{F}$

   (b) $\operatorname{div} \mathbf{F} = \dfrac{\partial P}{\partial x} + \dfrac{\partial Q}{\partial y} + \dfrac{\partial R}{\partial z} = \nabla \cdot \mathbf{F}$

   (c) For curl $\mathbf{F}$, see the discussion accompanying Figure 1 on page 944 as well as Figure 6 and the accompanying discussion on page 964. For div $\mathbf{F}$, see the discussion following Example 5 on page 945 as well as the discussion preceding (8) on page 971.

**10.** See Theorem 13.3.6; see Theorem 13.5.4.

**11.** (a) See (1) in Section 13.6.

   (b) We normally evaluate the surface integral using Formula 13.6.2.

   (c) See Formula 13.6.4.

   (d) The mass is $m = \iint_S \rho(x, y, z) \, dS$ and the center of mass is $(\bar{x}, \bar{y}, \bar{z})$ where $\bar{x} = \frac{1}{m} \iint_S x\rho(x, y, z) \, dS$, $\bar{y} = \frac{1}{m} \iint_S y\rho(x, y, z) \, dS$, $\bar{z} = \frac{1}{m} \iint_S z\rho(x, y, z) \, dS$.

**12.** (a) See Figures 6 and 7 and the accompanying discussion in Section 13.6. A Möbius strip is a nonorientable surface; see Figures 4 and 5 and the accompanying discussion on page 953.

   (b) See Definition 13.6.8.

   (c) See Formula 13.6.9.

   (d) See Formula 13.6.10.

**13.** See the statement of Stokes' Theorem on page 961.

**14.** See the statement of the Divergence Theorem on page 967.

**15.** In each theorem, we have an integral of a "derivative" over a region on the left side, while the right side involves the values of the original function only on the boundary of the region.

## TRUE-FALSE QUIZ

**1.** False; div $\mathbf{F}$ is a scalar field.

**3.** True, by Theorem 13.5.3 and the fact that div $\mathbf{0} = 0$.

**5.** False. See Exercise 13.3.35. (But the assertion is true if $D$ is simply-connected; see Theorem 13.3.6.)

**7.** True. Apply the Divergence Theorem and use the fact that div $\mathbf{F} = 0$.

EXERCISES

---

**1.** (a) Vectors starting on $C$ point in roughly the direction opposite to $C$, so the tangential component $\mathbf{F} \cdot \mathbf{T}$ is negative.
Thus $\int_C \mathbf{F} \cdot d\mathbf{r} = \int_C \mathbf{F} \cdot \mathbf{T}\, ds$ is negative.

(b) The vectors that end near $P$ are shorter than the vectors that start near $P$, so the net flow is outward near $P$ and $\operatorname{div} \mathbf{F}(P)$ is positive.

**3.** $\int_C yz \cos x\, ds = \int_0^\pi (3\cos t)(3\sin t)\cos t \sqrt{(1)^2 + (-3\sin t)^2 + (3\cos t)^2}\, dt = \int_0^\pi (9\cos^2 t \sin t)\sqrt{10}\, dt$
$$= 9\sqrt{10}\left(-\tfrac{1}{3}\cos^3 t\right)\Big]_0^\pi = -3\sqrt{10}\,(-2) = 6\sqrt{10}$$

**5.** $\int_C y^3\, dx + x^2\, dy = \int_{-1}^1 \left[y^3(-2y) + (1 - y^2)^2\right] dy = \int_{-1}^1 (-y^4 - 2y^2 + 1)\, dy$
$$= \left[-\tfrac{1}{5}y^5 - \tfrac{2}{3}y^3 + y\right]_{-1}^1 = -\tfrac{1}{5} - \tfrac{2}{3} + 1 - \tfrac{1}{5} - \tfrac{2}{3} + 1 = \tfrac{4}{15}$$

**7.** $C$: $x = 1 + 2t \;\Rightarrow\; dx = 2\,dt$, $y = 4t \;\Rightarrow\; dy = 4\,dt$, $z = -1 + 3t \;\Rightarrow\; dz = 3\,dt$, $0 \le t \le 1$.
$$\int_C xy\, dx + y^2\, dy + yz\, dz = \int_0^1 [(1 + 2t)(4t)(2) + (4t)^2(4) + (4t)(-1 + 3t)(3)]\, dt$$
$$= \int_0^1 (116t^2 - 4t)\, dt = \left[\tfrac{116}{3}t^3 - 2t^2\right]_0^1 = \tfrac{116}{3} - 2 = \tfrac{110}{3}$$

**9.** $\mathbf{F}(\mathbf{r}(t)) = e^{-t}\mathbf{i} + t^2(-t)\mathbf{j} + (t^2 + t^3)\mathbf{k}$, $\mathbf{r}'(t) = 2t\,\mathbf{i} + 3t^2\,\mathbf{j} - \mathbf{k}$ and
$$\int_C \mathbf{F} \cdot d\mathbf{r} = \int_0^1 (2te^{-t} - 3t^5 - (t^2 + t^3))\, dt = \left[-2te^{-t} - 2e^{-t} - \tfrac{1}{2}t^6 - \tfrac{1}{3}t^3 - \tfrac{1}{4}t^4\right]_0^1 = \tfrac{11}{12} - \tfrac{4}{e}.$$

**11.** $\frac{\partial}{\partial y}\left[(1 + xy)e^{xy}\right] = 2xe^{xy} + x^2 y e^{xy} = \frac{\partial}{\partial x}\left[e^y + x^2 e^{xy}\right]$ and the domain of $\mathbf{F}$ is $\mathbb{R}^2$, so $\mathbf{F}$ is conservative. Thus there exists a function $f$ such that $\mathbf{F} = \nabla f$. Then $f_y(x, y) = e^y + x^2 e^{xy}$ implies $f(x, y) = e^y + xe^{xy} + g(x)$ and then $f_x(x, y) = xye^{xy} + e^{xy} + g'(x) = (1 + xy)e^{xy} + g'(x)$. But $f_x(x, y) = (1 + xy)e^{xy}$, so $g'(x) = 0 \;\Rightarrow\; g(x) = K$. Thus $f(x, y) = e^y + xe^{xy} + K$ is a potential function for $\mathbf{F}$.

**13.** Since $\frac{\partial}{\partial y}(4x^3 y^2 - 2xy^3) = 8x^3 y - 6xy^2 = \frac{\partial}{\partial x}(2x^4 y - 3x^2 y^2 + 4y^3)$ and the domain of $\mathbf{F}$ is $\mathbb{R}^2$, $\mathbf{F}$ is conservative. Furthermore $f(x, y) = x^4 y^2 - x^2 y^3 + y^4$ is a potential function for $\mathbf{F}$.  $t = 0$ corresponds to the point $(0, 1)$ and $t = 1$ corresponds to $(1, 1)$, so $\int_C \mathbf{F} \cdot d\mathbf{r} = f(1, 1) - f(0, 1) = 1 - 1 = 0$.

**15.** $C_1$: $\mathbf{r}(t) = t\mathbf{i} + t^2\mathbf{j}$, $-1 \le t \le 1$;
$C_2$: $\mathbf{r}(t) = -t\mathbf{i} + \mathbf{j}$, $-1 \le t \le 1$.

Then

$$\int_C xy^2\, dx - x^2 y\, dy = \int_{-1}^1 (t^5 - 2t^5)\, dt + \int_{-1}^1 t\, dt$$
$$= \left[-\tfrac{1}{6}t^6\right]_{-1}^1 + \left[\tfrac{1}{2}t^2\right]_{-1}^1 = 0$$

Using Green's Theorem, we have

$$\int_C xy^2\, dx - x^2 y\, dy = \iint_D \left[\frac{\partial}{\partial x}(-x^2 y) - \frac{\partial}{\partial y}(xy^2)\right] dA = \iint_D (-2xy - 2xy)\, dA = \int_{-1}^1 \int_{x^2}^1 -4xy\, dy\, dx$$
$$= \int_{-1}^1 \left[-2xy^2\right]_{y=x^2}^{y=1}\, dx = \int_{-1}^1 (2x^5 - 2x)\, dx = \left[\tfrac{1}{3}x^6 - x^2\right]_{-1}^1 = 0$$

**17.** $\int_C x^2 y\, dx - xy^2\, dy = \iint\limits_{x^2+y^2\le 4} \left[ \frac{\partial}{\partial x}(-xy^2) - \frac{\partial}{\partial y}(x^2y) \right] dA = \iint\limits_{x^2+y^2\le 4} (-y^2 - x^2)\, dA = -\int_0^{2\pi}\int_0^2 r^3\, dr\, d\theta = -8\pi$

**19.** If we assume there is such a vector field $\mathbf{G}$, then $\operatorname{div}(\operatorname{curl}\mathbf{G}) = 2 + 3z - 2xz$. But $\operatorname{div}(\operatorname{curl}\mathbf{F}) = 0$ for all vector fields $\mathbf{F}$.

Thus such a $\mathbf{G}$ cannot exist.

**21.** For any piecewise-smooth simple closed plane curve $C$ bounding a region $D$, we can apply Green's Theorem to

$\mathbf{F}(x, y) = f(x)\,\mathbf{i} + g(y)\,\mathbf{j}$ to get $\int_C f(x)\, dx + g(y)\, dy = \iint_D \left[ \frac{\partial}{\partial x} g(y) - \frac{\partial}{\partial y} f(x) \right] dA = \iint_D 0\, dA = 0$.

**23.** $\nabla^2 f = 0$ means that $\dfrac{\partial^2 f}{\partial x^2} + \dfrac{\partial^2 f}{\partial y^2} = 0$. Now if $\mathbf{F} = f_y\,\mathbf{i} - f_x\,\mathbf{j}$ and $C$ is any closed path in $D$, then applying Green's

Theorem, we get

$$\int_C \mathbf{F}\cdot d\mathbf{r} = \int_C f_y\, dx - f_x\, dy = \iint_D \left[ \frac{\partial}{\partial x}(-f_x) - \frac{\partial}{\partial y}(f_y) \right] dA$$

$$= -\iint_D (f_{xx} + f_{yy})\, dA = -\iint_D 0\, dA = 0$$

Therefore the line integral is independent of path, by Theorem 13.3.3.

**25.** $z = f(x, y) = x^2 + y^2$ with $0 \le x^2 + y^2 \le 4$ so $\mathbf{r}_x \times \mathbf{r}_y = -2x\,\mathbf{i} - 2y\,\mathbf{j} + \mathbf{k}$ (using upward orientation). Then

$$\iint_S z\, dS = \iint\limits_{x^2+y^2\le 4} (x^2 + y^2)\sqrt{4x^2 + 4y^2 + 1}\, dA$$

$$= \int_0^{2\pi}\int_0^2 r^3\sqrt{1 + 4r^2}\, dr\, d\theta = \tfrac{1}{60}\pi\left(391\sqrt{17} + 1\right)$$

(Substitute $u = 1 + 4r^2$ and use tables.)

**27.** Since the sphere bounds a simple solid region, the Divergence Theorem applies and

$$\iint_S \mathbf{F}\cdot d\mathbf{S} = \iiint_E \operatorname{div}\mathbf{F}\, dV = \iiint_E (z - 2)\, dV = \iiint_E z\, dV - 2\iiint_E dV$$

$$= 0 \begin{bmatrix} \text{odd function in } z \\ \text{and } E \text{ is symmetric} \end{bmatrix} - 2\cdot V(E) = -2\cdot\tfrac{4}{3}\pi(2)^3 = -\tfrac{64}{3}\pi$$

*Alternate solution:* $\mathbf{F}(\mathbf{r}(\phi, \theta)) = 4\sin\phi\,\cos\theta\,\cos\phi\,\mathbf{i} - 4\sin\phi\,\sin\theta\,\mathbf{j} + 6\sin\phi\,\cos\theta\,\mathbf{k}$,

$\mathbf{r}_\phi \times \mathbf{r}_\theta = 4\sin^2\phi\,\cos\theta\,\mathbf{i} + 4\sin^2\phi\,\sin\theta\,\mathbf{j} + 4\sin\phi\,\cos\phi\,\mathbf{k}$, and

$\mathbf{F}\cdot(\mathbf{r}_\phi \times \mathbf{r}_\theta) = 16\sin^3\phi\,\cos^2\theta\,\cos\phi - 16\sin^3\phi\,\sin^2\theta + 24\sin^2\phi\,\cos\phi\,\cos\theta$. Then

$$\iint_S \mathbf{F}\cdot d\mathbf{S} = \int_0^{2\pi}\int_0^\pi (16\sin^3\phi\,\cos\phi\,\cos^2\theta - 16\sin^3\phi\,\sin^2\theta + 24\sin^2\phi\,\cos\phi\,\cos\theta)\, d\phi\, d\theta$$

$$= \int_0^{2\pi} \tfrac{4}{3}(-16\sin^2\theta)\, d\theta = -\tfrac{64}{3}\pi$$

**29.** Since $\operatorname{curl}\mathbf{F} = \mathbf{0}$, $\iint_S (\operatorname{curl}\mathbf{F})\cdot d\mathbf{S} = 0$. We parametrize $C$: $\mathbf{r}(t) = \cos t\,\mathbf{i} + \sin t\,\mathbf{j}$, $0 \le t \le 2\pi$ and

$\oint_C \mathbf{F}\cdot d\mathbf{r} = \int_0^{2\pi}(-\cos^2 t\,\sin t + \sin^2 t\,\cos t)\, dt = \left. \tfrac{1}{3}\cos^3 t + \tfrac{1}{3}\sin^3 t \right]_0^{2\pi} = 0$.

**31.** The surface is given by $x + y + z = 1$ or $z = 1 - x - y$, $0 \le x \le 1$, $0 \le y \le 1 - x$ and $\mathbf{r}_x \times \mathbf{r}_y = \mathbf{i} + \mathbf{j} + \mathbf{k}$. Then

$\oint_C \mathbf{F}\cdot d\mathbf{r} = \iint_S \operatorname{curl}\mathbf{F}\cdot d\mathbf{S} = \iint_D (-y\,\mathbf{i} - z\,\mathbf{j} - x\,\mathbf{k})\cdot(\mathbf{i} + \mathbf{j} + \mathbf{k})\, dA = \iint_D (-1)\, dA = -(\text{area of } D) = -\tfrac{1}{2}$.

**33.** $\iiint_E \text{div}\,\mathbf{F}\,dV = \iiint\limits_{x^2+y^2+z^2 \le 1} 3\,dV = 3(\text{volume of sphere}) = 4\pi$. Then

$\mathbf{F}(\mathbf{r}(\phi,\theta)) \cdot (\mathbf{r}_\phi \times \mathbf{r}_\theta) = \sin^3\phi\cos^2\theta + \sin^3\phi\sin^2\theta + \sin\phi\cos^2\phi = \sin\phi$ and

$\iint_S \mathbf{F}\cdot d\mathbf{S} = \int_0^{2\pi}\int_0^\pi \sin\phi\,d\phi\,d\theta = (2\pi)(2) = 4\pi$.

**35.** Because $\text{curl}\,\mathbf{F} = \mathbf{0}$, $\mathbf{F}$ is conservative, so there exists a function $f$ such that $\nabla f = \mathbf{F}$. Then $f_x(x,y,z) = 3x^2yz - 3y$

implies $f(x,y,z) = x^3yz - 3xy + g(y,z) \quad \Rightarrow \quad f_y(x,y,z) = x^3z - 3x + g_y(y,z)$. But $f_y(x,y,z) = x^3z - 3x$, so

$g(y,z) = h(z)$ and $f(x,y,z) = x^3yz - 3xy + h(z)$. Then $f_z(x,y,z) = x^3y + h'(z)$ but $f_z(x,y,z) = x^3y + 2z$,

so $h(z) = z^2 + K$ and a potential function for $\mathbf{F}$ is $f(x,y,z) = x^3yz - 3xy + z^2$. Hence

$\int_C \mathbf{F}\cdot d\mathbf{r} = \int_C \nabla f \cdot d\mathbf{r} = f(0,3,0) - f(0,0,2) = 0 - 4 = -4$.

**37.** By the Divergence Theorem, $\iint_S \mathbf{F}\cdot\mathbf{n}\,dS = \iiint_E \text{div}\,\mathbf{F}\,dV = 3(\text{volume of } E) = 3(8-1) = 21$.

# ☐ FOCUS ON PROBLEM SOLVING

**1.** Let $S_1$ be the portion of $\Omega(S)$ between $S(a)$ and $S$, and let $\partial S_1$ be its boundary. Also let $S_L$ be the lateral surface of $S_1$ [that is, the surface of $S_1$ except $S$ and $S(a)$]. Applying the Divergence Theorem we have $\displaystyle\iint_{\partial S_1} \frac{\mathbf{r}\cdot\mathbf{n}}{r^3}\,dS = \iiint_{S_1} \nabla\cdot\frac{\mathbf{r}}{r^3}\,dV$.

But
$$\nabla\cdot\frac{\mathbf{r}}{r^3} = \left\langle \frac{\partial}{\partial x}, \frac{\partial}{\partial y}, \frac{\partial}{\partial z}\right\rangle \cdot \left\langle \frac{x}{(x^2+y^2+z^2)^{3/2}}, \frac{y}{(x^2+y^2+z^2)^{3/2}}, \frac{z}{(x^2+y^2+z^2)^{3/2}}\right\rangle$$
$$= \frac{(x^2+y^2+z^2-3x^2) + (x^2+y^2+z^2-3y^2) + (x^2+y^2+z^2-3z^2)}{(x^2+y^2+z^2)^{5/2}} = 0$$

$\Rightarrow \displaystyle\iint_{\partial S_1}\frac{\mathbf{r}\cdot\mathbf{n}}{r^3}\,dS = \iiint_{S_1} 0\,dV = 0$. On the other hand, notice that for the surfaces of $\partial S_1$ other than $S(a)$ and $S$,

$\mathbf{r}\cdot\mathbf{n} = 0 \;\Rightarrow$

$$0 = \iint_{\partial S_1}\frac{\mathbf{r}\cdot\mathbf{n}}{r^3}\,dS = \iint_{S}\frac{\mathbf{r}\cdot\mathbf{n}}{r^3}\,dS + \iint_{S(a)}\frac{\mathbf{r}\cdot\mathbf{n}}{r^3}\,dS + \iint_{S_L}\frac{\mathbf{r}\cdot\mathbf{n}}{r^3}\,dS = \iint_{S}\frac{\mathbf{r}\cdot\mathbf{n}}{r^3}\,dS + \iint_{S(a)}\frac{\mathbf{r}\cdot\mathbf{n}}{r^3}\,dS \;\Rightarrow$$

$\displaystyle\iint_{S}\frac{\mathbf{r}\cdot\mathbf{n}}{r^3}\,dS = -\iint_{S(a)}\frac{\mathbf{r}\cdot\mathbf{n}}{r^3}\,dS$. Notice that on $S(a)$, $r = a \;\Rightarrow\; \mathbf{n} = -\dfrac{\mathbf{r}}{r} = -\dfrac{\mathbf{r}}{a}$ and $\mathbf{r}\cdot\mathbf{r} = r^2 = a^2$, so

that $-\displaystyle\iint_{S(a)}\frac{\mathbf{r}\cdot\mathbf{n}}{r^3}\,dS = \iint_{S(a)}\frac{\mathbf{r}\cdot\mathbf{r}}{a^4}\,dS = \iint_{S(a)}\frac{a^2}{a^4}\,dS = \frac{1}{a^2}\iint_{S(a)}\,dS = \frac{\text{area of }S(a)}{a^2} = |\Omega(S)|$.

Therefore $|\Omega(S)| = \displaystyle\iint_{S}\frac{\mathbf{r}\cdot\mathbf{n}}{r^3}\,dS$.

**3.** Let $\mathbf{F} = \mathbf{a}\times\mathbf{r} = \langle a_1, a_2, a_3\rangle \times \langle x, y, z\rangle = \langle a_2 z - a_3 y, a_3 x - a_1 z, a_1 y - a_2 x\rangle$. Then $\operatorname{curl}\mathbf{F} = \langle 2a_1, 2a_2, 2a_3\rangle = 2\mathbf{a}$, and $\iint_S 2\mathbf{a}\cdot d\mathbf{S} = \iint_S \operatorname{curl}\mathbf{F}\cdot d\mathbf{S} = \int_C \mathbf{F}\cdot d\mathbf{r} = \int_C (\mathbf{a}\times\mathbf{r})\cdot d\mathbf{r}$ by Stokes' Theorem.

**5.** The given line integral $\frac{1}{2}\int_C (bz - cy)\,dx + (cx - az)\,dy + (ay - bx)\,dz$ can be expressed as $\int_C \mathbf{F}\cdot d\mathbf{r}$ if we define the vector field $\mathbf{F}$ by $\mathbf{F}(x, y, z) = P\mathbf{i} + Q\mathbf{j} + R\mathbf{k} = \frac{1}{2}(bz - cy)\mathbf{i} + \frac{1}{2}(cx - az)\mathbf{j} + \frac{1}{2}(ay - bx)\mathbf{k}$. Then define $S$ to be the planar interior of $C$, so $S$ is an oriented, smooth surface. Stokes' Theorem says $\int_C \mathbf{F}\cdot d\mathbf{r} = \iint_S \operatorname{curl}\mathbf{F}\cdot d\mathbf{S} = \iint_S \operatorname{curl}\mathbf{F}\cdot\mathbf{n}\,dS$. Now

$$\operatorname{curl}\mathbf{F} = \left(\frac{\partial R}{\partial y} - \frac{\partial Q}{\partial z}\right)\mathbf{i} + \left(\frac{\partial P}{\partial z} - \frac{\partial R}{\partial x}\right)\mathbf{j} + \left(\frac{\partial Q}{\partial x} - \frac{\partial P}{\partial y}\right)\mathbf{k}$$
$$= \left(\tfrac{1}{2}a + \tfrac{1}{2}a\right)\mathbf{i} + \left(\tfrac{1}{2}b + \tfrac{1}{2}b\right)\mathbf{j} + \left(\tfrac{1}{2}c + \tfrac{1}{2}c\right)\mathbf{k} = a\mathbf{i} + b\mathbf{j} + c\mathbf{k} = \mathbf{n}$$

so $\operatorname{curl}\mathbf{F}\cdot\mathbf{n} = \mathbf{n}\cdot\mathbf{n} = |\mathbf{n}|^2 = 1$, hence $\iint_S \operatorname{curl}\mathbf{F}\cdot\mathbf{n}\,dS = \iint_S dS$ which is simply the surface area of $S$. Thus, $\int_C \mathbf{F}\cdot d\mathbf{r} = \frac{1}{2}\int_C (bz - cy)\,dx + (cx - az)\,dy + (ay - bx)\,dz$ is the plane area enclosed by $C$.

# APPENDIXES

## D  PRECISE DEFINITIONS OF LIMITS

**1.** (*Note:* This is Exercise 25 in the full version of the text.)

Let $\varepsilon > 0$. We want to find $\delta > 0$ such that

$$\left| \frac{xy}{\sqrt{x^2 + y^2}} - 0 \right| < \varepsilon \qquad \text{whenever} \qquad 0 < \sqrt{x^2 + y^2} < \delta$$

that is,

$$\frac{|xy|}{\sqrt{x^2 + y^2}} < \varepsilon \qquad \text{whenever} \qquad 0 < \sqrt{x^2 + y^2} < \delta$$

But $|x| = \sqrt{x^2} \le \sqrt{x^2 + y^2}$ and $|y| = \sqrt{y^2} \le \sqrt{x^2 + y^2}$, so

$$\frac{|xy|}{\sqrt{x^2 + y^2}} \le \frac{\left(\sqrt{x^2 + y^2}\right)^2}{\sqrt{x^2 + y^2}} = \sqrt{x^2 + y^2}$$

Thus, if we choose $\delta = \varepsilon$ and let $0 < \sqrt{x^2 + y^2} < \delta$, then

$$\left| \frac{xy}{\sqrt{x^2 + y^2}} - 0 \right| \le \sqrt{x^2 + y^2} < \delta = \varepsilon$$

Hence, by Definition 1 (Definition 5 in the full version of the text),

$$\lim_{(x,y) \to (0,0)} \frac{xy}{\sqrt{x^2 + y^2}} = 0$$

## H  POLAR COORDINATES

### H.1  Curves in Polar Coordinates

**1.** (a) $\left(2, \frac{\pi}{3}\right)$

By adding $2\pi$ to $\frac{\pi}{3}$, we obtain the point $\left(2, \frac{7\pi}{3}\right)$. The direction opposite $\frac{\pi}{3}$ is $\frac{4\pi}{3}$, so $\left(-2, \frac{4\pi}{3}\right)$ is a point that satisfies the $r < 0$ requirement.

(b) $\left(1, -\frac{3\pi}{4}\right)$

$r > 0$: $\left(1, -\frac{3\pi}{4} + 2\pi\right) = \left(1, \frac{5\pi}{4}\right)$

$r < 0$: $\left(-1, -\frac{3\pi}{4} + \pi\right) = \left(-1, \frac{\pi}{4}\right)$

(c) $\left(-1, \frac{\pi}{2}\right)$

$\boldsymbol{r} > \boldsymbol{0}$: $\left(-(-1), \frac{\pi}{2} + \pi\right) = \left(1, \frac{3\pi}{2}\right)$

$\boldsymbol{r} < \boldsymbol{0}$: $\left(-1, \frac{\pi}{2} + 2\pi\right) = \left(-1, \frac{5\pi}{2}\right)$

**3.** (a)

$x = 1\cos\pi = 1(-1) = -1$ and

$y = 1\sin\pi = 1(0) = 0$ give us

the Cartesian coordinates $(-1, 0)$.

(b)

$x = 2\cos\left(-\frac{2\pi}{3}\right) = 2\left(-\frac{1}{2}\right) = -1$ and

$y = 2\sin\left(-\frac{2\pi}{3}\right) = 2\left(-\frac{\sqrt{3}}{2}\right) = -\sqrt{3}$

give us $\left(-1, -\sqrt{3}\right)$.

(c)

$x = -2\cos\frac{3\pi}{4} = -2\left(-\frac{\sqrt{2}}{2}\right) = \sqrt{2}$ and

$y = -2\sin\frac{3\pi}{4} = -2\left(\frac{\sqrt{2}}{2}\right) = -\sqrt{2}$

gives us $\left(\sqrt{2}, -\sqrt{2}\right)$.

**5.** (a) $x = 2$ and $y = -2$ $\Rightarrow$ $r = \sqrt{2^2 + (-2)^2} = 2\sqrt{2}$ and $\theta = \tan^{-1}\left(\frac{-2}{2}\right) = -\frac{\pi}{4}$. Since $(2, -2)$ is in the fourth

quadrant, the polar coordinates are (i) $\left(2\sqrt{2}, \frac{7\pi}{4}\right)$ and (ii) $\left(-2\sqrt{2}, \frac{3\pi}{4}\right)$.

(b) $x = -1$ and $y = \sqrt{3}$ $\Rightarrow$ $r = \sqrt{(-1)^2 + \left(\sqrt{3}\right)^2} = 2$ and $\theta = \tan^{-1}\left(\frac{\sqrt{3}}{-1}\right) = \frac{2\pi}{3}$. Since $\left(-1, \sqrt{3}\right)$ is in the second

quadrant, the polar coordinates are (i) $\left(2, \frac{2\pi}{3}\right)$ and (ii) $\left(-2, \frac{5\pi}{3}\right)$.

**7.** The curves $r = 1$ and $r = 2$ represent circles with center $O$ and radii 1 and 2. The region in the plane

satisfying $1 \leq r \leq 2$ consists of both circles and the shaded region between them in the figure.

**9.** The region satisfying $0 \le r < 4$ and $-\pi/2 \le \theta < \pi/6$ does

not include the circle $r = 4$ nor the line $\theta = \frac{\pi}{6}$.

**11.** $2 < r < 3$,   $\frac{5\pi}{3} \le \theta \le \frac{7\pi}{3}$

**13.** $r = 3\sin\theta \;\Rightarrow\; r^2 = 3r\sin\theta \;\Leftrightarrow\; x^2 + y^2 = 3y \;\Leftrightarrow\; x^2 + \left(y - \frac{3}{2}\right)^2 = \left(\frac{3}{2}\right)^2$, a circle of radius $\frac{3}{2}$ centered at $\left(0, \frac{3}{2}\right)$.

The first two equations are actually equivalent since $r^2 = 3r\sin\theta \;\Rightarrow\; r(r - 3\sin\theta) = 0 \;\Rightarrow\; r = 0$ or $r = 3\sin\theta$. But

$r = 3\sin\theta$ gives the point $r = 0$ (the pole) when $\theta = 0$. Thus, the single equation $r = 3\sin\theta$ is equivalent to the compound

condition ($r = 0$ or $r = 3\sin\theta$).

**15.** $r = \csc\theta \;\Leftrightarrow\; r = \dfrac{1}{\sin\theta} \;\Leftrightarrow\; r\sin\theta = 1 \;\Leftrightarrow\; y = 1$, a horizontal line 1 unit above the $x$-axis.

**17.** $x = -y^2 \;\Leftrightarrow\; r\cos\theta = -r^2\sin^2\theta \;\Leftrightarrow\; \cos\theta = -r\sin^2\theta \;\Leftrightarrow\; r = -\dfrac{\cos\theta}{\sin^2\theta} = -\cot\theta\csc\theta$.

**19.** $x^2 + y^2 = 2cx \;\Leftrightarrow\; r^2 = 2cr\cos\theta \;\Leftrightarrow\; r^2 - 2cr\cos\theta = 0 \;\Leftrightarrow\; r(r - 2c\cos\theta) = 0 \;\Leftrightarrow\; r = 0$ or $r = 2c\cos\theta$.

$r = 0$ is included in $r = 2c\cos\theta$ when $\theta = \frac{\pi}{2} + n\pi$, so the curve is represented by the single equation $r = 2c\cos\theta$.

**21.** (a) The description leads immediately to the polar equation $\theta = \frac{\pi}{6}$, and the Cartesian equation $y = \tan\left(\frac{\pi}{6}\right)x = \frac{1}{\sqrt{3}}x$ is

slightly more difficult to derive.

(b) The easier description here is the Cartesian equation $x = 3$.

**23.** $\theta = -\pi/6$

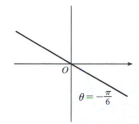

**25.** $r = \sin\theta \;\Leftrightarrow\; r^2 = r\sin\theta \;\Leftrightarrow\; x^2 + y^2 = y \;\Leftrightarrow\;$

$x^2 + \left(y - \frac{1}{2}\right)^2 = \left(\frac{1}{2}\right)^2$. The reasoning here is the same

as in Exercise 13. This is a circle of radius $\frac{1}{2}$ centered at $\left(0, \frac{1}{2}\right)$.

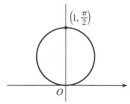

**27.** $r = 2(1 - \sin\theta)$. This curve is a cardioid.

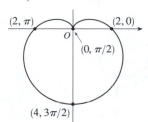

**29.** $r = \theta$,   $\theta \ge 0$

**31.** $r = 4\sin 3\theta$

**33.** $r = 2\cos 4\theta$

**35.** $r = 1 - 2\sin\theta$

**37.** $r^2 = 9\sin 2\theta$

**39.** $r = 2\cos\left(\frac{3}{2}\theta\right)$

**41.** $r = 1 + 2\cos 2\theta$

**43.** For $\theta = 0$, $\pi$, and $2\pi$, $r$ has its minimum value of about 0.5. For $\theta = \frac{\pi}{2}$ and $\frac{3\pi}{2}$, $r$ attains its maximum value of 2.

We see that the graph has a similar shape for $0 \le \theta \le \pi$ and $\pi \le \theta \le 2\pi$.

 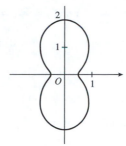

**45.** $x = (r) \cos \theta = (4 + 2 \sec \theta) \cos \theta = 4 \cos \theta + 2$. Now, $r \to \infty \Rightarrow$

$(4 + 2 \sec \theta) \to \infty \Rightarrow \theta \to \left(\frac{\pi}{2}\right)^{-}$ or $\theta \to \left(\frac{3\pi}{2}\right)^{+}$ [since we need only

consider $0 \le \theta < 2\pi$], so $\displaystyle\lim_{r \to \infty} x = \lim_{\theta \to \pi/2^{-}} (4 \cos \theta + 2) = 2$. Also,

$r \to -\infty \Rightarrow (4 + 2 \sec \theta) \to -\infty \Rightarrow \theta \to \left(\frac{\pi}{2}\right)^{+}$ or $\theta \to \left(\frac{3\pi}{2}\right)^{-}$, so

$\displaystyle\lim_{r \to -\infty} x = \lim_{\theta \to \pi/2^{+}} (4 \cos \theta + 2) = 2$. Therefore, $\displaystyle\lim_{r \to \pm\infty} x = 2 \Rightarrow x = 2$ is a vertical asymptote.

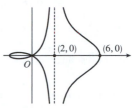

**47.** (a) We see that the curve $r = 1 + c \sin \theta$ crosses itself at the origin, where $r = 0$ (in fact the inner loop corresponds to

negative $r$-values,) so we solve the equation of the limaçon for $r = 0 \Leftrightarrow c \sin \theta = -1 \Leftrightarrow \sin \theta = -1/c$. Now if

$|c| < 1$, then this equation has no solution and hence there is no inner loop. But if $c < -1$, then on the interval $(0, 2\pi)$

the equation has the two solutions $\theta = \sin^{-1}(-1/c)$ and $\theta = \pi - \sin^{-1}(-1/c)$, and if $c > 1$, the solutions are

$\theta = \pi + \sin^{-1}(1/c)$ and $\theta = 2\pi - \sin^{-1}(1/c)$. In each case, $r < 0$ for $\theta$ between the two solutions, indicating a loop.

(b) For $0 < c < 1$, the dimple (if it exists) is characterized by the fact that $y$ has a local maximum at $\theta = \frac{3\pi}{2}$. So we

determine for what $c$-values $\dfrac{d^2 y}{d\theta^2}$ is negative at $\theta = \frac{3\pi}{2}$, since by the Second Derivative Test this indicates a maximum:

$y = r \sin \theta = \sin \theta + c \sin^2 \theta \Rightarrow \dfrac{dy}{d\theta} = \cos \theta + 2c \sin \theta \cos \theta = \cos \theta + c \sin 2\theta \Rightarrow \dfrac{d^2 y}{d\theta^2} = -\sin \theta + 2c \cos 2\theta$.

At $\theta = \frac{3\pi}{2}$, this is equal to $-(-1) + 2c(-1) = 1 - 2c$, which is negative only for $c > \frac{1}{2}$. A similar argument shows that

for $-1 < c < 0$, $y$ only has a local minimum at $\theta = \frac{\pi}{2}$ (indicating a dimple) for $c < -\frac{1}{2}$.

**49.** $r = 1/\theta \Rightarrow x = r \cos \theta = (\cos \theta)/\theta$, $y = r \sin \theta = (\sin \theta)/\theta \Rightarrow$

$$\frac{dy}{dx} = \frac{dy/d\theta}{dx/d\theta} = \frac{\sin \theta (-1/\theta^2) + (1/\theta) \cos \theta}{\cos \theta (-1/\theta^2) - (1/\theta) \sin \theta} \cdot \frac{\theta^2}{\theta^2} = \frac{-\sin \theta + \theta \cos \theta}{-\cos \theta - \theta \sin \theta}$$

When $\theta = \pi$, $\dfrac{dy}{dx} = \dfrac{-0 + \pi(-1)}{-(-1) - \pi(0)} = \dfrac{-\pi}{1} = -\pi$.

**51.** $r = \cos 2\theta \Rightarrow x = r \cos \theta = \cos 2\theta \cos \theta$, $y = r \sin \theta = \cos 2\theta \sin \theta \Rightarrow$

$$\frac{dy}{dx} = \frac{dy/d\theta}{dx/d\theta} = \frac{\cos 2\theta \cos \theta + \sin \theta (-2 \sin 2\theta)}{\cos 2\theta (-\sin \theta) + \cos \theta (-2 \sin 2\theta)}$$

When $\theta = \dfrac{\pi}{4}$, $\dfrac{dy}{dx} = \dfrac{0(\sqrt{2}/2) + (\sqrt{2}/2)(-2)}{0(-\sqrt{2}/2) + (\sqrt{2}/2)(-2)} = \dfrac{-\sqrt{2}}{-\sqrt{2}} = 1$.

**53.** $r = 3\cos\theta \implies x = r\cos\theta = 3\cos\theta\cos\theta, \; y = r\sin\theta = 3\cos\theta\sin\theta \implies$

$\frac{dy}{d\theta} = -3\sin^2\theta + 3\cos^2\theta = 3\cos2\theta = 0 \implies 2\theta = \frac{\pi}{2}$ or $\frac{3\pi}{2} \iff \theta = \frac{\pi}{4}$ or $\frac{3\pi}{4}$.

So the tangent is horizontal at $\left(\frac{3}{\sqrt{2}}, \frac{\pi}{4}\right)$ and $\left(-\frac{3}{\sqrt{2}}, \frac{3\pi}{4}\right)$ $\left[\text{same as } \left(\frac{3}{\sqrt{2}}, -\frac{\pi}{4}\right)\right]$.

$\frac{dx}{d\theta} = -6\sin\theta\cos\theta = -3\sin2\theta = 0 \implies 2\theta = 0$ or $\pi \iff \theta = 0$ or $\frac{\pi}{2}$. So the tangent is vertical at $(3, 0)$ and $\left(0, \frac{\pi}{2}\right)$.

**55.** $r = 1 + \cos\theta \implies x = r\cos\theta = \cos\theta\,(1 + \cos\theta), \; y = r\sin\theta = \sin\theta\,(1 + \cos\theta) \implies$

$\frac{dy}{d\theta} = (1 + \cos\theta)\cos\theta - \sin^2\theta = 2\cos^2\theta + \cos\theta - 1 = (2\cos\theta - 1)(\cos\theta + 1) = 0 \implies \cos\theta = \frac{1}{2}$ or $-1 \implies$

$\theta = \frac{\pi}{3}, \pi,$ or $\frac{5\pi}{3} \implies$ horizontal tangent at $\left(\frac{3}{2}, \frac{\pi}{3}\right), (0, \pi),$ and $\left(\frac{3}{2}, \frac{5\pi}{3}\right)$.

$\frac{dx}{d\theta} = -(1 + \cos\theta)\sin\theta - \cos\theta\sin\theta = -\sin\theta\,(1 + 2\cos\theta) = 0 \implies \sin\theta = 0$ or $\cos\theta = -\frac{1}{2} \implies$

$\theta = 0, \pi, \frac{2\pi}{3},$ or $\frac{4\pi}{3} \implies$ vertical tangent at $(2, 0), \left(\frac{1}{2}, \frac{2\pi}{3}\right),$ and $\left(\frac{1}{2}, \frac{4\pi}{3}\right)$.

Note that the tangent is horizontal, not vertical when $\theta = \pi$, since $\lim\limits_{\theta \to \pi} \dfrac{dy/d\theta}{dx/d\theta} = 0$.

**57.** $r = a\sin\theta + b\cos\theta \implies r^2 = ar\sin\theta + br\cos\theta \implies x^2 + y^2 = ay + bx \implies$

$x^2 - bx + \left(\frac{1}{2}b\right)^2 + y^2 - ay + \left(\frac{1}{2}a\right)^2 = \left(\frac{1}{2}b\right)^2 + \left(\frac{1}{2}a\right)^2 \implies \left(x - \frac{1}{2}b\right)^2 + \left(y - \frac{1}{2}a\right)^2 = \frac{1}{4}(a^2 + b^2)$, and this is a circle

with center $\left(\frac{1}{2}b, \frac{1}{2}a\right)$ and radius $\frac{1}{2}\sqrt{a^2 + b^2}$.

Note for Exercises 59–62: Maple is able to plot polar curves using the `polarplot` command, or using the `coords=polar` option in a regular `plot` command. In Mathematica, use `PolarPlot`. In Derive, change to `Polar` under `Options  State`. If your graphing device cannot plot polar equations, you must convert to parametric equations. For example, in Exercise 59, $x = r\cos\theta = [1 + 2\sin(\theta/2)]\cos\theta$, $y = r\sin\theta = [1 + 2\sin(\theta/2)]\sin\theta$.

**59.** $r = e^{\sin\theta} - 2\cos(4\theta)$.

The parameter interval is $[0, 2\pi]$.

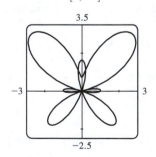

**61.** $r = 2 - 5\sin(\theta/6)$.

The parameter interval is $[-6\pi, 6\pi]$.

**63.** It appears that the graph of $r = 1 + \sin\left(\theta - \frac{\pi}{6}\right)$ is the same shape as the graph of $r = 1 + \sin\theta$, but rotated counterclockwise about the origin by $\frac{\pi}{6}$. Similarly, the graph of $r = 1 + \sin\left(\theta - \frac{\pi}{3}\right)$ is rotated by $\frac{\pi}{3}$. In general, the graph of $r = f(\theta - \alpha)$ is the same shape as that of $r = f(\theta)$, but rotated counterclockwise through $\alpha$ about the origin. That is, for any point $(r_0, \theta_0)$ on the curve $r = f(\theta)$, the point $(r_0, \theta_0 + \alpha)$ is on the curve $r = f(\theta - \alpha)$, since $r_0 = f(\theta_0) = f((\theta_0 + \alpha) - \alpha)$.

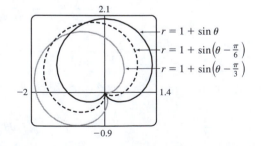

$r = 1 + \sin\theta$
$r = 1 + \sin\left(\theta - \frac{\pi}{6}\right)$
$r = 1 + \sin\left(\theta - \frac{\pi}{3}\right)$

**65.** (a) $r = \sin n\theta$.

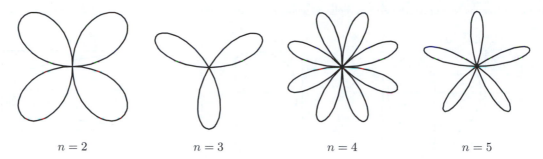

$n = 2$        $n = 3$        $n = 4$        $n = 5$

From the graphs, it seems that when $n$ is even, the number of loops in the curve (called a rose) is $2n$, and when $n$ is odd, the number of loops is simply $n$. This is because in the case of $n$ odd, every point on the graph is traversed twice, due to the fact that

$$r(\theta + \pi) = \sin[n(\theta + \pi)] = \sin n\theta \cos n\pi + \cos n\theta \sin n\pi = \begin{cases} \sin n\theta & \text{if } n \text{ is even} \\ -\sin n\theta & \text{if } n \text{ is odd} \end{cases}$$

(b) The graph of $r = |\sin n\theta|$ has $2n$ loops whether $n$ is odd or even, since $r(\theta + \pi) = r(\theta)$.

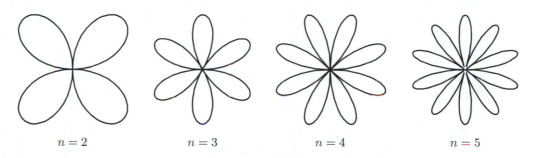

$n = 2$        $n = 3$        $n = 4$        $n = 5$

**67.** $r = \dfrac{1 - a\cos\theta}{1 + a\cos\theta}$. We start with $a = 0$, since in this case the curve is simply the circle $r = 1$.

As $a$ increases, the graph moves to the left, and its right side becomes flattened. As $a$ increases through about $0.4$, the right side seems to grow a dimple, which upon closer investigation (with narrower $\theta$-ranges) seems to appear at $a \approx 0.42$ [the actual value is $\sqrt{2} - 1$]. As $a \to 1$, this dimple becomes more pronounced, and the curve begins to stretch out horizontally, until at $a = 1$ the denominator vanishes at $\theta = \pi$, and the dimple becomes an actual cusp. For $a > 1$ we must choose our parameter interval carefully, since $r \to \infty$ as $1 + a\cos\theta \to 0 \iff \theta \to \pm\cos^{-1}(-1/a)$. As $a$ increases from 1, the curve splits into two parts. The left part has a loop, which grows larger as $a$ increases, and the right part grows broader vertically, and its left tip develops a dimple when $a \approx 2.42$ [actually, $\sqrt{2} + 1$]. As $a$ increases, the dimple grows more and more pronounced. If $a < 0$, we get the same graph as we do for the corresponding positive $a$-value, but with a rotation through $\pi$ about the pole, as happened when $c$ was replaced with $-c$ in Exercise 66.

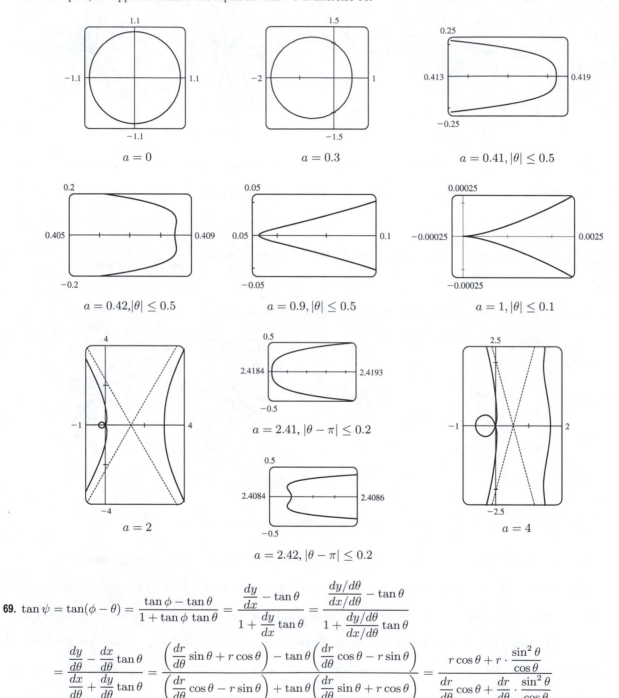

$a = 0$

$a = 0.3$

$a = 0.41, |\theta| \leq 0.5$

$a = 0.42, |\theta| \leq 0.5$

$a = 0.9, |\theta| \leq 0.5$

$a = 1, |\theta| \leq 0.1$

$a = 2$

$a = 2.41, |\theta - \pi| \leq 0.2$

$a = 2.42, |\theta - \pi| \leq 0.2$

$a = 4$

**69.** $\tan\psi = \tan(\phi - \theta) = \dfrac{\tan\phi - \tan\theta}{1 + \tan\phi\tan\theta} = \dfrac{\dfrac{dy}{dx} - \tan\theta}{1 + \dfrac{dy}{dx}\tan\theta} = \dfrac{\dfrac{dy/d\theta}{dx/d\theta} - \tan\theta}{1 + \dfrac{dy/d\theta}{dx/d\theta}\tan\theta}$

$= \dfrac{\dfrac{dy}{d\theta} - \dfrac{dx}{d\theta}\tan\theta}{\dfrac{dx}{d\theta} + \dfrac{dy}{d\theta}\tan\theta} = \dfrac{\left(\dfrac{dr}{d\theta}\sin\theta + r\cos\theta\right) - \tan\theta\left(\dfrac{dr}{d\theta}\cos\theta - r\sin\theta\right)}{\left(\dfrac{dr}{d\theta}\cos\theta - r\sin\theta\right) + \tan\theta\left(\dfrac{dr}{d\theta}\sin\theta + r\cos\theta\right)} = \dfrac{r\cos\theta + r\cdot\dfrac{\sin^2\theta}{\cos\theta}}{\dfrac{dr}{d\theta}\cos\theta + \dfrac{dr}{d\theta}\cdot\dfrac{\sin^2\theta}{\cos\theta}}$

$= \dfrac{r\cos^2\theta + r\sin^2\theta}{\dfrac{dr}{d\theta}\cos^2\theta + \dfrac{dr}{d\theta}\sin^2\theta} = \dfrac{r}{dr/d\theta}$

## H.2  Areas and Lengths in Polar Coordinates

**1.** $r = \theta^2$, $0 \le \theta \le \frac{\pi}{4}$.  $A = \int_0^{\pi/4} \frac{1}{2} r^2 \, d\theta = \int_0^{\pi/4} \frac{1}{2} (\theta^2)^2 \, d\theta = \int_0^{\pi/4} \frac{1}{2} \theta^4 \, d\theta = \left[\frac{1}{10} \theta^5\right]_0^{\pi/4} = \frac{1}{10} \left(\frac{\pi}{4}\right)^5 = \frac{1}{10{,}240} \pi^5$

**3.** $r = \sin\theta$, $\frac{\pi}{3} \le \theta \le \frac{2\pi}{3}$.

$$A = \int_{\pi/3}^{2\pi/3} \frac{1}{2} \sin^2\theta \, d\theta = \frac{1}{4} \int_{\pi/3}^{2\pi/3} (1 - \cos 2\theta) \, d\theta = \frac{1}{4}\left[\theta - \frac{1}{2}\sin 2\theta\right]_{\pi/3}^{2\pi/3} = \frac{1}{4}\left[\frac{2\pi}{3} - \frac{1}{2}\sin\frac{4\pi}{3} - \frac{\pi}{3} + \frac{1}{2}\sin\frac{2\pi}{3}\right]$$

$$= \frac{1}{4}\left[\frac{2\pi}{3} - \frac{1}{2}\left(-\frac{\sqrt{3}}{2}\right) - \frac{\pi}{3} + \frac{1}{2}\left(\frac{\sqrt{3}}{2}\right)\right] = \frac{1}{4}\left(\frac{\pi}{3} + \frac{\sqrt{3}}{2}\right) = \frac{\pi}{12} + \frac{\sqrt{3}}{8}$$

**5.** $r = \sqrt{\theta}$, $0 \le \theta \le 2\pi$.  $A = \int_0^{2\pi} \frac{1}{2} r^2 \, d\theta = \int_0^{2\pi} \frac{1}{2}\left(\sqrt{\theta}\right)^2 d\theta = \int_0^{2\pi} \frac{1}{2}\theta \, d\theta = \left[\frac{1}{4}\theta^2\right]_0^{2\pi} = \pi^2$

**7.** $r = 4 + 3\sin\theta$, $-\frac{\pi}{2} \le \theta \le \frac{\pi}{2}$.

$$A = \int_{-\pi/2}^{\pi/2} \frac{1}{2}\left((4 + 3\sin\theta)^2\right) d\theta = \frac{1}{2}\int_{-\pi/2}^{\pi/2} (16 + 24\sin\theta + 9\sin^2\theta)\, d\theta$$

$$= \frac{1}{2}\int_{-\pi/2}^{\pi/2} (16 + 9\sin^2\theta)\, d\theta \qquad \text{[by Theorem 5.5.6(b)]}$$

$$= \frac{1}{2}\cdot 2\int_0^{\pi/2}\left[16 + 9\cdot\frac{1}{2}(1 - \cos 2\theta)\right] d\theta \qquad \text{[by Theorem 5.5.6(a)]}$$

$$= \int_0^{\pi/2}\left(\frac{41}{2} - \frac{9}{2}\cos 2\theta\right) d\theta = \left[\frac{41}{2}\theta - \frac{9}{4}\sin 2\theta\right]_0^{\pi/2} = \left(\frac{41\pi}{4} - 0\right) - (0 - 0) = \frac{41\pi}{4}$$

**9.** The curve goes through the pole when $\theta = \pi/4$, so we'll find the area for

$0 \le \theta \le \pi/4$ and multiply it by 4.

$A = 4\int_0^{\pi/4} \frac{1}{2} r^2 \, d\theta = 2\int_0^{\pi/4} (4\cos 2\theta)\, d\theta$

$= 8\int_0^{\pi/4} \cos 2\theta \, d\theta = 4\left[\sin 2\theta\right]_0^{\pi/4} = 4$

**11.** One-sixth of the area lies above the polar axis and is bounded by the curve

$r = 2\cos 3\theta$ for $\theta = 0$ to $\theta = \pi/6$.

$$A = 6\int_0^{\pi/6} \frac{1}{2}(2\cos 3\theta)^2 \, d\theta = 12\int_0^{\pi/6} \cos^2 3\theta \, d\theta$$

$$= \frac{12}{2}\int_0^{\pi/6}(1 + \cos 6\theta)\, d\theta$$

$$= 6\left[\theta + \frac{1}{6}\sin 6\theta\right]_0^{\pi/6} = 6\left(\frac{\pi}{6}\right) = \pi$$

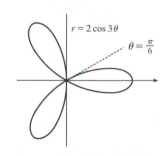

**13.** $A = \int_0^{2\pi} \frac{1}{2}(1 + 2\sin 6\theta)^2 \, d\theta = \frac{1}{2}\int_0^{2\pi}(1 + 4\sin 6\theta + 4\sin^2 6\theta)\, d\theta$

$= \frac{1}{2}\int_0^{2\pi}\left[1 + 4\sin 6\theta + 4\cdot\frac{1}{2}(1 - \cos 12\theta)\right] d\theta$

$= \frac{1}{2}\int_0^{2\pi}(3 + 4\sin 6\theta - 2\cos 12\theta)\, d\theta$

$= \frac{1}{2}\left[3\theta - \frac{2}{3}\cos 6\theta - \frac{1}{6}\sin 12\theta\right]_0^{2\pi}$

$= \frac{1}{2}\left[\left(6\pi - \frac{2}{3} - 0\right) - \left(0 - \frac{2}{3} - 0\right)\right] = 3\pi$

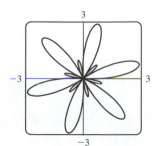

**15.** The shaded loop is traced out from $\theta = 0$ to $\theta = \pi/2$.

$$A = \int_0^{\pi/2} \tfrac{1}{2} r^2 \, d\theta = \tfrac{1}{2} \int_0^{\pi/2} \sin^2 2\theta \, d\theta$$

$$= \tfrac{1}{2} \int_0^{\pi/2} \tfrac{1}{2}(1 - \cos 4\theta) \, d\theta = \tfrac{1}{4} \left[\theta - \tfrac{1}{4}\sin 4\theta\right]_0^{\pi/2}$$

$$= \tfrac{1}{4}\left(\tfrac{\pi}{2}\right) = \tfrac{\pi}{8}$$

$r = \sin 2\theta$

**17.**

$r = 1 + 2\sin\theta$ (rect.)

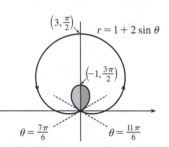
$r = 1 + 2\sin\theta$

This is a limaçon, with inner loop traced out between $\theta = \frac{7\pi}{6}$ and $\frac{11\pi}{6}$ [found by solving $r = 0$].

$$A = 2 \int_{7\pi/6}^{3\pi/2} \tfrac{1}{2}(1 + 2\sin\theta)^2 \, d\theta = \int_{7\pi/6}^{3\pi/2} \left(1 + 4\sin\theta + 4\sin^2\theta\right) d\theta = \int_{7\pi/6}^{3\pi/2} \left[1 + 4\sin\theta + 4 \cdot \tfrac{1}{2}(1 - \cos 2\theta)\right] d\theta$$

$$= \left[\theta - 4\cos\theta + 2\theta - \sin 2\theta\right]_{7\pi/6}^{3\pi/2} = \left(\tfrac{9\pi}{2}\right) - \left(\tfrac{7\pi}{2} + 2\sqrt{3} - \tfrac{\sqrt{3}}{2}\right) = \pi - \tfrac{3\sqrt{3}}{2}$$

**19.** $2\cos\theta = 1 \;\Rightarrow\; \cos\theta = \tfrac{1}{2} \;\Rightarrow\; \theta = \tfrac{\pi}{3}$ or $\tfrac{5\pi}{3}$.

$$A = 2 \int_0^{\pi/3} \tfrac{1}{2}\left[(2\cos\theta)^2 - 1^2\right] d\theta = \int_0^{\pi/3} (4\cos^2\theta - 1) \, d\theta$$

$$= \int_0^{\pi/3} \left\{4\left[\tfrac{1}{2}(1 + \cos 2\theta)\right] - 1\right\} d\theta = \int_0^{\pi/3} (1 + 2\cos 2\theta) \, d\theta$$

$$= \left[\theta + \sin 2\theta\right]_0^{\pi/3} = \tfrac{\pi}{3} + \tfrac{\sqrt{3}}{2}$$

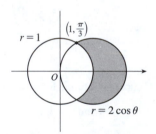
$r = 1$  $\left(1, \tfrac{\pi}{3}\right)$  $O$  $r = 2\cos\theta$

**21.** $3\cos\theta = 1 + \cos\theta \;\Leftrightarrow\; \cos\theta = \tfrac{1}{2} \;\Rightarrow\; \theta = \tfrac{\pi}{3}$ or $-\tfrac{\pi}{3}$.

$$A = 2 \int_0^{\pi/3} \tfrac{1}{2}\left[(3\cos\theta)^2 - (1 + \cos\theta)^2\right] d\theta$$

$$= \int_0^{\pi/3} (8\cos^2\theta - 2\cos\theta - 1) \, d\theta = \int_0^{\pi/3} \left[4(1 + \cos 2\theta) - 2\cos\theta - 1\right] d\theta$$

$$= \int_0^{\pi/3} (3 + 4\cos 2\theta - 2\cos\theta) \, d\theta = \left[3\theta + 2\sin 2\theta - 2\sin\theta\right]_0^{\pi/3}$$

$$= \pi + \sqrt{3} - \sqrt{3} = \pi$$

$\theta = \tfrac{\pi}{3}$  $r = 1 + \cos\theta$  $r = 3\cos\theta$

**23.** $\sqrt{3}\cos\theta = \sin\theta \;\Rightarrow\; \sqrt{3} = \dfrac{\sin\theta}{\cos\theta} \;\Rightarrow\; \tan\theta = \sqrt{3} \;\Rightarrow\; \theta = \tfrac{\pi}{3}$.

$$A = \int_0^{\pi/3} \tfrac{1}{2}(\sin\theta)^2 \, d\theta + \int_{\pi/3}^{\pi/2} \tfrac{1}{2}\left(\sqrt{3}\cos\theta\right)^2 d\theta$$

$$= \int_0^{\pi/3} \tfrac{1}{2} \cdot \tfrac{1}{2}(1 - \cos 2\theta) \, d\theta + \int_{\pi/3}^{\pi/2} \tfrac{1}{2} \cdot 3 \cdot \tfrac{1}{2}(1 + \cos 2\theta) \, d\theta$$

$$= \tfrac{1}{4}\left[\theta - \tfrac{1}{2}\sin 2\theta\right]_0^{\pi/3} + \tfrac{3}{4}\left[\theta + \tfrac{1}{2}\sin 2\theta\right]_{\pi/3}^{\pi/2}$$

$$= \tfrac{1}{4}\left[\left(\tfrac{\pi}{3} - \tfrac{\sqrt{3}}{4}\right) - 0\right] + \tfrac{3}{4}\left[\left(\tfrac{\pi}{2} + 0\right) - \left(\tfrac{\pi}{3} + \tfrac{\sqrt{3}}{4}\right)\right]$$

$$= \tfrac{\pi}{12} - \tfrac{\sqrt{3}}{16} + \tfrac{\pi}{8} - \tfrac{3\sqrt{3}}{16} = \tfrac{5\pi}{24} - \tfrac{\sqrt{3}}{4}$$

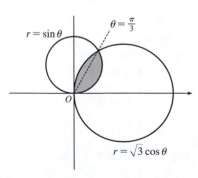
$r = \sin\theta$  $\theta = \tfrac{\pi}{3}$  $O$  $r = \sqrt{3}\cos\theta$

**25.** $\sin 2\theta = \cos 2\theta \quad \Rightarrow \quad \dfrac{\sin 2\theta}{\cos 2\theta} = 1 \quad \Rightarrow \quad \tan 2\theta = 1 \quad \Rightarrow \quad 2\theta = \frac{\pi}{4} \quad \Rightarrow$

$\theta = \frac{\pi}{8} \quad \Rightarrow$

$$A = 8 \cdot 2 \int_0^{\pi/8} \tfrac{1}{2} \sin^2 2\theta \, d\theta = 8 \int_0^{\pi/8} \tfrac{1}{2}(1 - \cos 4\theta) \, d\theta$$

$$= 4\big[\theta - \tfrac{1}{4}\sin 4\theta\big]_0^{\pi/8} = 4\big(\tfrac{\pi}{8} - \tfrac{1}{4}\cdot 1\big) = \tfrac{\pi}{2} - 1$$

**27.** The darker shaded region (from $\theta = 0$ to $\theta = 2\pi/3$) represents $\frac{1}{2}$ of the desired area plus $\frac{1}{2}$ of the area of the inner loop.

From this area, we'll subtract $\frac{1}{2}$ of the area of the inner loop (the lighter shaded region from $\theta = 2\pi/3$ to $\theta = \pi$), and then

double that difference to obtain the desired area.

$$A = 2\Big[\int_0^{2\pi/3} \tfrac{1}{2}\big(\tfrac{1}{2} + \cos\theta\big)^2 \, d\theta - \int_{2\pi/3}^{\pi} \tfrac{1}{2}\big(\tfrac{1}{2} + \cos\theta\big)^2 \, d\theta\Big]$$

$$= \int_0^{2\pi/3}\big(\tfrac{1}{4} + \cos\theta + \cos^2\theta\big)\, d\theta - \int_{2\pi/3}^{\pi}\big(\tfrac{1}{4} + \cos\theta + \cos^2\theta\big)\, d\theta$$

$$= \int_0^{2\pi/3}\big[\tfrac{1}{4} + \cos\theta + \tfrac{1}{2}(1 + \cos 2\theta)\big]\, d\theta$$

$$\qquad\qquad - \int_{2\pi/3}^{\pi}\big[\tfrac{1}{4} + \cos\theta + \tfrac{1}{2}(1 + \cos 2\theta)\big]\, d\theta$$

$$= \Big[\tfrac{\theta}{4} + \sin\theta + \tfrac{\theta}{2} + \tfrac{\sin 2\theta}{4}\Big]_0^{2\pi/3} - \Big[\tfrac{\theta}{4} + \sin\theta + \tfrac{\theta}{2} + \tfrac{\sin 2\theta}{4}\Big]_{2\pi/3}^{\pi}$$

$$= \Big(\tfrac{\pi}{6} + \tfrac{\sqrt{3}}{2} + \tfrac{\pi}{3} - \tfrac{\sqrt{3}}{8}\Big) - \Big(\tfrac{\pi}{4} + \tfrac{\pi}{2}\Big) + \Big(\tfrac{\pi}{6} + \tfrac{\sqrt{3}}{2} + \tfrac{\pi}{3} - \tfrac{\sqrt{3}}{8}\Big)$$

$$= \tfrac{\pi}{4} + \tfrac{3}{4}\sqrt{3} = \tfrac{1}{4}\big(\pi + 3\sqrt{3}\big)$$

**29.** $2\sin 2\theta = 1 \quad \Rightarrow \quad \sin 2\theta = \tfrac{1}{2} \quad \Rightarrow \quad 2\theta = \tfrac{\pi}{6}, \tfrac{5\pi}{6}, \tfrac{13\pi}{6}, \text{ or } \tfrac{17\pi}{6}.$

By symmetry, the eight points of intersection are given by

$(1, \theta)$, where $\theta = \tfrac{\pi}{12}, \tfrac{5\pi}{12}, \tfrac{13\pi}{12}$, and $\tfrac{17\pi}{12}$, and

$(-1, \theta)$, where $\theta = \tfrac{7\pi}{12}, \tfrac{11\pi}{12}, \tfrac{19\pi}{12}$, and $\tfrac{23\pi}{12}$.

[There are many ways to describe these points.]

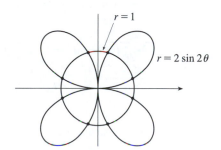

**31.** The pole is a point of intersection. $\sin\theta = \sin 2\theta = 2\sin\theta\cos\theta \quad \Leftrightarrow$

$\sin\theta\,(1 - 2\cos\theta) = 0 \quad \Leftrightarrow \quad \sin\theta = 0 \text{ or } \cos\theta = \tfrac{1}{2} \quad \Rightarrow$

$\theta = 0, \pi, \tfrac{\pi}{3}, \text{ or } -\tfrac{\pi}{3} \quad \Rightarrow \quad$ the other intersection points are $\Big(\tfrac{\sqrt{3}}{2}, \tfrac{\pi}{3}\Big)$

and $\Big(\tfrac{\sqrt{3}}{2}, \tfrac{2\pi}{3}\Big)$  [by symmetry].

**33.**

From the first graph, we see that the pole is one point of intersection. By zooming in or using the cursor, we find the $\theta$-values

of the intersection points to be $\alpha \approx 0.88786 \approx 0.89$ and $\pi - \alpha \approx 2.25$. (The first of these values may be more easily

estimated by plotting $y = 1 + \sin x$ and $y = 2x$ in rectangular coordinates; see the second graph.) By symmetry, the total

area contained is twice the area contained in the first quadrant, that is,

$$A = 2 \int_0^\alpha \tfrac{1}{2}(2\theta)^2 \, d\theta + 2 \int_\alpha^{\pi/2} \tfrac{1}{2}(1 + \sin\theta)^2 \, d\theta = \int_0^\alpha 4\theta^2 \, d\theta + \int_\alpha^{\pi/2} \left[ 1 + 2\sin\theta + \tfrac{1}{2}(1 - \cos 2\theta) \right] d\theta$$

$$= \left[ \tfrac{4}{3}\theta^3 \right]_0^\alpha + \left[ \theta - 2\cos\theta + \left( \tfrac{1}{2}\theta - \tfrac{1}{4}\sin 2\theta \right) \right]_\alpha^{\pi/2} = \tfrac{4}{3}\alpha^3 + \left[ \left( \tfrac{\pi}{2} + \tfrac{\pi}{4} \right) - \left( \alpha - 2\cos\alpha + \tfrac{1}{2}\alpha - \tfrac{1}{4}\sin 2\alpha \right) \right] \approx 3.4645$$

**35.** $L = \int_a^b \sqrt{r^2 + (dr/d\theta)^2} \, d\theta = \int_0^{\pi/3} \sqrt{(3\sin\theta)^2 + (3\cos\theta)^2} \, d\theta = \int_0^{\pi/3} \sqrt{9(\sin^2\theta + \cos^2\theta)} \, d\theta$

$= 3 \int_0^{\pi/3} d\theta = 3 \left[ \theta \right]_0^{\pi/3} = 3\left( \tfrac{\pi}{3} \right) = \pi.$

As a check, note that the circumference of a circle with radius $\tfrac{3}{2}$ is $2\pi\left( \tfrac{3}{2} \right) = 3\pi$, and since $\theta = 0$ to $\pi = \tfrac{\pi}{3}$ traces out $\tfrac{1}{3}$ of the

circle (from $\theta = 0$ to $\theta = \pi$), $\tfrac{1}{3}(3\pi) = \pi.$

**37.** $L = \int_a^b \sqrt{r^2 + (dr/d\theta)^2} \, d\theta = \int_0^{2\pi} \sqrt{(\theta^2)^2 + (2\theta)^2} \, d\theta = \int_0^{2\pi} \sqrt{\theta^4 + 4\theta^2} \, d\theta$

$= \int_0^{2\pi} \sqrt{\theta^2(\theta^2 + 4)} \, d\theta = \int_0^{2\pi} \theta \sqrt{\theta^2 + 4} \, d\theta$

Now let $u = \theta^2 + 4$, so that $du = 2\theta \, d\theta \quad \left[ \theta \, d\theta = \tfrac{1}{2} \, du \right]$ and

$\int_0^{2\pi} \theta \sqrt{\theta^2 + 4} \, d\theta = \int_4^{4\pi^2 + 4} \tfrac{1}{2}\sqrt{u} \, du = \tfrac{1}{2} \cdot \tfrac{2}{3} \left[ u^{3/2} \right]_4^{4(\pi^2 + 1)} = \tfrac{1}{3}[ 4^{3/2}(\pi^2 + 1)^{3/2} - 4^{3/2} ] = \tfrac{8}{3}[ (\pi^2 + 1)^{3/2} - 1 ]$

**39.** The curve $r = 3\sin 2\theta$ is completely traced with $0 \leq \theta \leq 2\pi$.

$r^2 + \left( \tfrac{dr}{d\theta} \right)^2 = (3\sin 2\theta)^2 + (6\cos 2\theta)^2 \quad \Rightarrow$

$L = \int_0^{2\pi} \sqrt{9\sin^2 2\theta + 36\cos^2 2\theta} \, d\theta \approx 29.0653$

$r = 3 \sin 2\theta$

$(0, 0) = (0, 2\pi)$

## I   Complex Numbers

**1.** $(5 - 6i) + (3 + 2i) = (5 + 3) + (-6 + 2)i = 8 + (-4)i = 8 - 4i$

**3.** $(2 + 5i)(4 - i) = 2(4) + 2(-i) + (5i)(4) + (5i)(-i) = 8 - 2i + 20i - 5i^2 = 8 + 18i - 5(-1)$
$$= 8 + 18i + 5 = 13 + 18i$$

**5.** $\overline{12 + 7i} = 12 - 7i$

**7.** $\dfrac{1 + 4i}{3 + 2i} = \dfrac{1 + 4i}{3 + 2i} \cdot \dfrac{3 - 2i}{3 - 2i} = \dfrac{3 - 2i + 12i - 8(-1)}{3^2 + 2^2} = \dfrac{11 + 10i}{13} = \dfrac{11}{13} + \dfrac{10}{13}i$

**9.** $\dfrac{1}{1 + i} = \dfrac{1}{1 + i} \cdot \dfrac{1 - i}{1 - i} = \dfrac{1 - i}{1 - (-1)} = \dfrac{1 - i}{2} = \dfrac{1}{2} - \dfrac{1}{2}i$

**11.** $i^3 = i^2 \cdot i = (-1)i = -i$

**13.** $\sqrt{-25} = \sqrt{25}\, i = 5i$

**15.** $\overline{12 - 5i} = 12 + 15i$ and $|12 - 15i| = \sqrt{12^2 + (-5)^2} = \sqrt{144 + 25} = \sqrt{169} = 13$

**17.** $\overline{-4i} = \overline{0 - 4i} = 0 + 4i = 4i$ and $|-4i| = \sqrt{0^2 + (-4)^2} = \sqrt{16} = 4$

**19.** $4x^2 + 9 = 0 \ \Leftrightarrow \ 4x^2 = -9 \ \Leftrightarrow \ x^2 = -\frac{9}{4} \ \Leftrightarrow \ x = \pm\sqrt{-\frac{9}{4}} = \pm\sqrt{\frac{9}{4}}\, i = \pm\frac{3}{2}i.$

**21.** By the quadratic formula, $x^2 + 2x + 5 = 0 \ \Leftrightarrow \ x = \dfrac{-2 \pm \sqrt{2^2 - 4(1)(5)}}{2(1)} = \dfrac{-2 \pm \sqrt{-16}}{2} = \dfrac{-2 \pm 4i}{2} = -1 \pm 2i.$

**23.** By the quadratic formula, $z^2 + z + 2 = 0 \ \Leftrightarrow \ z = \dfrac{-1 \pm \sqrt{1^2 - 4(1)(2)}}{2(1)} = \dfrac{-1 \pm \sqrt{-7}}{2} = -\dfrac{1}{2} \pm \dfrac{\sqrt{7}}{2}i.$

**25.** For $z = -3 + 3i$, $r = \sqrt{(-3)^2 + 3^2} = 3\sqrt{2}$ and $\tan\theta = \frac{3}{-3} = -1 \ \Rightarrow \ \theta = \frac{3\pi}{4}$ (since $z$ lies in the second quadrant).

Therefore, $-3 + 3i = 3\sqrt{2}\left(\cos\frac{3\pi}{4} + i\sin\frac{3\pi}{4}\right)$.

**27.** For $z = 3 + 4i$, $r = \sqrt{3^2 + 4^2} = 5$ and $\tan\theta = \frac{4}{3} \ \Rightarrow \ \theta = \tan^{-1}\left(\frac{4}{3}\right)$  (since $z$ lies in the first quadrant). Therefore,

$3 + 4i = 5\{\cos[\tan^{-1}(\frac{4}{3})] + i\sin[\tan^{-1}(\frac{4}{3})]\}$.

**29.** For $z = \sqrt{3} + i$, $r = \sqrt{(\sqrt{3})^2 + 1^2} = 2$ and $\tan\theta = \frac{1}{\sqrt{3}} \ \Rightarrow \ \theta = \frac{\pi}{6} \ \Rightarrow \ z = 2\left(\cos\frac{\pi}{6} + i\sin\frac{\pi}{6}\right)$.

For $w = 1 + \sqrt{3}\, i$, $r = 2$ and $\tan\theta = \sqrt{3} \ \Rightarrow \ \theta = \frac{\pi}{3} \ \Rightarrow \ w = 2\left(\cos\frac{\pi}{3} + i\sin\frac{\pi}{3}\right)$.

Therefore, $zw = 2 \cdot 2\left[\cos\left(\frac{\pi}{6} + \frac{\pi}{3}\right) + i\sin\left(\frac{\pi}{6} + \frac{\pi}{3}\right)\right] = 4\left(\cos\frac{\pi}{2} + i\sin\frac{\pi}{2}\right)$,

$z/w = \frac{2}{2}\left[\cos\left(\frac{\pi}{6} - \frac{\pi}{3}\right) + i\sin\left(\frac{\pi}{6} - \frac{\pi}{3}\right)\right] = \cos\left(-\frac{\pi}{6}\right) + i\sin\left(-\frac{\pi}{6}\right)$, and $1 = 1 + 0i = 1(\cos 0 + i\sin 0) \ \Rightarrow$

$1/z = \frac{1}{2}\left[\cos\left(0 - \frac{\pi}{6}\right) + i\sin\left(0 - \frac{\pi}{6}\right)\right] = \frac{1}{2}\left[\cos\left(-\frac{\pi}{6}\right) + i\sin\left(-\frac{\pi}{6}\right)\right]$. For $1/z$, we could also use the formula that precedes

Example 5 to obtain $1/z = \frac{1}{2}\left(\cos\frac{\pi}{6} - i\sin\frac{\pi}{6}\right)$.

**31.** For $z = 2\sqrt{3} - 2i$, $r = \sqrt{\left(2\sqrt{3}\right)^2 + (-2)^2} = 4$ and $\tan\theta = \frac{-2}{2\sqrt{3}} = -\frac{1}{\sqrt{3}} \Rightarrow \theta = -\frac{\pi}{6} \Rightarrow$

$z = 4\left[\cos\left(-\frac{\pi}{6}\right) + i\sin\left(-\frac{\pi}{6}\right)\right]$. For $w = -1 + i$, $r = \sqrt{2}$, $\tan\theta = \frac{1}{-1} = -1 \Rightarrow \theta = \frac{3\pi}{4} \Rightarrow$

$w = \sqrt{2}\left(\cos\frac{3\pi}{4} + i\sin\frac{3\pi}{4}\right)$. Therefore, $zw = 4\sqrt{2}\left[\cos\left(-\frac{\pi}{6} + \frac{3\pi}{4}\right) + i\sin\left(-\frac{\pi}{6} + \frac{3\pi}{4}\right)\right] = 4\sqrt{2}\left(\cos\frac{7\pi}{12} + i\sin\frac{7\pi}{12}\right)$,

$z/w = \frac{4}{\sqrt{2}}\left[\cos\left(-\frac{\pi}{6} - \frac{3\pi}{4}\right) + i\sin\left(-\frac{\pi}{6} - \frac{3\pi}{4}\right)\right] = \frac{4}{\sqrt{2}}\left[\cos\left(-\frac{11\pi}{12}\right) + i\sin\left(-\frac{11\pi}{12}\right)\right] = 2\sqrt{2}\left(\cos\frac{13\pi}{12} + i\sin\frac{13\pi}{12}\right)$, and

$1/z = \frac{1}{4}\left[\cos\left(-\frac{\pi}{6}\right) - i\sin\left(-\frac{\pi}{6}\right)\right] = \frac{1}{4}\left(\cos\frac{\pi}{6} + i\sin\frac{\pi}{6}\right)$.

**33.** For $z = 1 + i$, $r = \sqrt{2}$ and $\tan\theta = \frac{1}{1} = 1 \Rightarrow \theta = \frac{\pi}{4} \Rightarrow z = \sqrt{2}\left(\cos\frac{\pi}{4} + i\sin\frac{\pi}{4}\right)$. So by De Moivre's Theorem,

$$(1 + i)^{20} = \left[\sqrt{2}\left(\cos\frac{\pi}{4} + i\sin\frac{\pi}{4}\right)\right]^{20} = (2^{1/2})^{20}\left(\cos\frac{20 \cdot \pi}{4} + i\sin\frac{20 \cdot \pi}{4}\right) = 2^{10}(\cos 5\pi + i\sin 5\pi)$$
$$= 2^{10}[-1 + i(0)] = -2^{10} = -1024$$

**35.** For $z = 2\sqrt{3} + 2i$, $r = \sqrt{\left(2\sqrt{3}\right)^2 + 2^2} = \sqrt{16} = 4$ and $\tan\theta = \frac{2}{2\sqrt{3}} = \frac{1}{\sqrt{3}} \Rightarrow \theta = \frac{\pi}{6} \Rightarrow z = 4\left(\cos\frac{\pi}{6} + i\sin\frac{\pi}{6}\right)$.

So by De Moivre's Theorem,

$$\left(2\sqrt{3} + 2i\right)^5 = \left[4\left(\cos\frac{\pi}{6} + i\sin\frac{\pi}{6}\right)\right]^5 = 4^5\left(\cos\frac{5\pi}{6} + i\sin\frac{5\pi}{6}\right) = 1024\left[-\frac{\sqrt{3}}{2} + \frac{1}{2}i\right] = -512\sqrt{3} + 512i.$$

**37.** $1 = 1 + 0i = 1(\cos 0 + i\sin 0)$. Using Equation 3 with $r = 1$, $n = 8$, and $\theta = 0$, we have

$w_k = 1^{1/8}\left[\cos\left(\frac{0 + 2k\pi}{8}\right) + i\sin\left(\frac{0 + 2k\pi}{8}\right)\right] = \cos\frac{k\pi}{4} + i\sin\frac{k\pi}{4}$, where $k = 0, 1, 2, \ldots, 7$.

$w_0 = 1(\cos 0 + i\sin 0) = 1$, $w_1 = 1\left(\cos\frac{\pi}{4} + i\sin\frac{\pi}{4}\right) = \frac{1}{\sqrt{2}} + \frac{1}{\sqrt{2}}i$,

$w_2 = 1\left(\cos\frac{\pi}{2} + i\sin\frac{\pi}{2}\right) = i$, $w_3 = 1\left(\cos\frac{3\pi}{4} + i\sin\frac{3\pi}{4}\right) = -\frac{1}{\sqrt{2}} + \frac{1}{\sqrt{2}}i$,

$w_4 = 1(\cos\pi + i\sin\pi) = -1$, $w_5 = 1\left(\cos\frac{5\pi}{4} + i\sin\frac{5\pi}{4}\right) = -\frac{1}{\sqrt{2}} - \frac{1}{\sqrt{2}}i$,

$w_6 = 1\left(\cos\frac{3\pi}{2} + i\sin\frac{3\pi}{2}\right) = -i$, $w_7 = 1\left(\cos\frac{7\pi}{4} + i\sin\frac{7\pi}{4}\right) = \frac{1}{\sqrt{2}} - \frac{1}{\sqrt{2}}i$

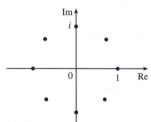

**39.** $i = 0 + i = 1\left(\cos\frac{\pi}{2} + i\sin\frac{\pi}{2}\right)$. Using Equation 3 with $r = 1$, $n = 3$, and $\theta = \frac{\pi}{2}$, we have

$w_k = 1^{1/3}\left[\cos\left(\frac{\frac{\pi}{2} + 2k\pi}{3}\right) + i\sin\left(\frac{\frac{\pi}{2} + 2k\pi}{3}\right)\right]$, where $k = 0, 1, 2$.

$w_0 = \left(\cos\frac{\pi}{6} + i\sin\frac{\pi}{6}\right) = \frac{\sqrt{3}}{2} + \frac{1}{2}i$

$w_1 = \left(\cos\frac{5\pi}{6} + i\sin\frac{5\pi}{6}\right) = -\frac{\sqrt{3}}{2} + \frac{1}{2}i$

$w_2 = \left(\cos\frac{9\pi}{6} + i\sin\frac{9\pi}{6}\right) = -i$

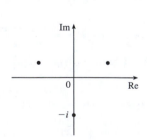

**41.** Using Euler's formula (6) with $y = \frac{\pi}{2}$, we have $e^{i\pi/2} = \cos\frac{\pi}{2} + i\sin\frac{\pi}{2} = 0 + 1i = i$.

**43.** Using Euler's formula (6) with $y = \frac{\pi}{3}$, we have $e^{i\pi/3} = \cos\frac{\pi}{3} + i\sin\frac{\pi}{3} = \frac{1}{2} + \frac{\sqrt{3}}{2}i$.

**45.** Using Equation 7 with $x = 2$ and $y = \pi$, we have $e^{2+i\pi} = e^2 e^{i\pi} = e^2(\cos\pi + i\sin\pi) = e^2(-1 + 0) = -e^2$.

**47.** Take $r = 1$ and $n = 3$ in De Moivre's Theorem to get

$$[1(\cos\theta + i\sin\theta)]^3 = 1^3(\cos 3\theta + i\sin 3\theta)$$

$$(\cos\theta + i\sin\theta)^3 = \cos 3\theta + i\sin 3\theta$$

$$\cos^3\theta + 3(\cos^2\theta)(i\sin\theta) + 3(\cos\theta)(i\sin\theta)^2 + (i\sin\theta)^3 = \cos 3\theta + i\sin 3\theta$$

$$\cos^3\theta + (3\cos^2\theta \sin\theta)i - 3\cos\theta \sin^2\theta - (\sin^3\theta)i = \cos 3\theta + i\sin 3\theta$$

$$(\cos^3\theta - 3\sin^2\theta \cos\theta) + (3\sin\theta \cos^2\theta - \sin^3\theta)i = \cos 3\theta + i\sin 3\theta$$

Equating real and imaginary parts gives $\cos 3\theta = \cos^3\theta - 3\sin^2\theta \cos\theta$ and $\sin 3\theta = 3\sin\theta \cos^2\theta - \sin^3\theta$.

**49.** $F(x) = e^{rx} = e^{(a+bi)x} = e^{ax+bxi} = e^{ax}(\cos bx + i\sin bx) = e^{ax}\cos bx + i(e^{ax}\sin bx) \quad \Rightarrow$

$$F'(x) = (e^{ax}\cos bx)' + i(e^{ax}\sin bx)'$$

$$= (ae^{ax}\cos bx - be^{ax}\sin bx) + i(ae^{ax}\sin bx + be^{ax}\cos bx)$$

$$= a[e^{ax}(\cos bx + i\sin bx)] + b[e^{ax}(-\sin bx + i\cos bx)]$$

$$= ae^{rx} + b[e^{ax}(i^2\sin bx + i\cos bx)]$$

$$= ae^{rx} + bi[e^{ax}(\cos bx + i\sin bx)] = ae^{rx} + bie^{rx} = (a + bi)e^{rx} = re^{rx}$$